GeV-TeV GAMMA RAY ASTROPHYSICS WORKSHOP

Related Titles from the AIP Conference Proceedings Subseries on Astronomy and Astrophysics

516 26th International Cosmic Ray Conference, ICRC XXVI, Invited, Rapporteur, and Highlight Papers
Edited by Brenda L. Dingus, David B. Kieda, and Michael H. Salamon
May 2000, 1-56396-939-4

510 The Fifth Compton Symposium
Edited by Mark L. McConnell and James M. Ryan, March 2000, 1-56396-932-7

499 Small Missions for Energetic Astrophysics: Ultraviolet to Gamma-Ray
Edited by Steven P. Brumby, December 1999, 1-56396-912-2

471 Solar Wind Nine: Proceedings of the Ninth International Solar Wind Conference
Edited by Shadia Rifai Habbal, Ruth Esser, Joseph V. Hollweg, Philip A. Isenberg, May 1999, 1-56396-865-7

433 Workshop on Observing Giant Cosmic Ray Air Showers from $>10^{20}$ eV Particles from Space
Edited by John F. Krizmanic, Jonathan F. Ormes, and Robert E. Streitmatter, June 1998, 1-56396-788-X

428 Gamma-Ray Bursts: 4th Huntsville Symposium
Edited by Charles A. Meegan, Robert D. Preece, and Thomas M. Koshut, May 1998, 1-56396-766-9

410 The Fourth Compton Symposium
Edited by Charles D. Dermer, Mark S. Strickman, and James D. Kurfess, December 1997, 2 vol. set, 1-56396-659-X

To learn more about these titles, or the AIP Conference Proceedings Series, please visit the webpage **http://www.aip.org/catalog/aboutconf.html**

GeV-TeV GAMMA RAY ASTROPHYSICS WORKSHOP

Towards a Major Atmospheric
Cherenkov Detector VI

Snowbird, Utah   13–16 August 1999

EDITORS
Brenda L. Dingus
Michael H. Salamon
David B. Kieda
University of Utah

Melville, New York
AIP CONFERENCE PROCEEDINGS ■ 515

Editors:

Brenda L. Dingus
University of Wisconsin-Madison
Department of Physics
1150 University Avenue
Madison, WI 53706-1390
USA

E-mail: dingus@alizarin.physics.wisc.edu

Michael H. Salamon
David B. Kieda

University of Utah
Physics Department
115 South 1400 East Room 201
Salt Lake City, UT 84112-0830
USA

E-mail: salamon@physics.utah.edu
       kieda@physics.utah.edu

The articles on pp. 3–15, 105–110, and 492–499 were authored by U. S. Government employees and are not covered by the below mentioned copyright.

Authorization to photocopy items for internal or personal use, beyond the free copying permitted under the 1978 U.S. Copyright Law (see statement below), is granted by the American Institute of Physics for users registered with the Copyright Clearance Center (CCC) Transactional Reporting Service, provided that the base fee of $17.00 per copy is paid directly to CCC, 222 Rosewood Drive, Danvers, MA 01923. For those organizations that have been granted a photocopy license by CCC, a separate system of payment has been arranged. The fee code for users of the Transactional Reporting Service is: 1-56396-938-6/00/$17.00.

© 2000 American Institute of Physics

Individual readers of this volume and nonprofit libraries, acting for them, are permitted to make fair use of the material in it, such as copying an article for use in teaching or research. Permission is granted to quote from this volume in scientific work with the customary acknowledgment of the source. To reprint a figure, table, or other excerpt requires the consent of one of the original authors and notification to AIP. Republication or systematic or multiple reproduction of any material in this volume is permitted only under license from AIP. Address inquiries to Office of Rights and Permissions, Suite 1NO1, 2 Huntington Quadrangle, Melville, N.Y. 11747-4502; phone: 516-576-2268; fax: 516-576-2450; e-mail: rights@aip.org.

L.C. Catalog Card No. 00-102242
ISBN 1-56396-938-6
ISSN 0094-243X
Printed in the United States of America

CONTENTS

Preface xiii

INTRODUCTORY REVIEW

VHE Astronomy Before the New Millennium 3
    T. C. Weekes

BLAZARS AND BL LACS

Understanding Blazar Jets Through Their Multifrequency Emission 19
    R. M. Sambruna
Leptonic Jet Models of Blazars: Broadband Spectra and
Spectral Variability 31
    M. Böttcher
Hadronic Blazar Models and Correlated X-ray/TeV Flares 41
    J. P. Rachen
X-Ray Selected BL Lacs and Blazars 53
    E. S. Perlman
Spectral Variability in the Blazar PKS 0528+134 66
    R. Mukherjee and M. Böttcher
Observation of BL Lac Type AGNs with the Equatorial
Mounted HEGRA Cherenkov Telescope................................. 71
    M. Schilling, O. Mang, and G. Rauterberg for the HEGRA Collaboration
Search for TeV Gamma-Emissions from BL Lacertae with the HEGRA
Equatorial Mount Cherenkov Telescope................................. 76
    O. Mang, G. Rauterberg, and M. Schilling for the HEGRA Collaboration
Upper Limits on Low Redshift AGN..................................... 81
    H. Bojahr for the HEGRA Collaboration
TeV Emission from PKS 2155-304.. 86
    P. M. Chadwick, K. Lyons, T. J. L. McComb, K. J. Orford,
    J. L. Osborne, S. M. Rayner, S. E. Shaw, and K. E. Turver
Flux Limits for TeV Emission from AGNs............................... 91
    P. M. Chadwick, K. Lyons, T. J. L. McComb, K. J. Orford,
    J. L. Osborne, S. M. Rayner, S. E. Shaw, and K. E. Turver
Observations of TeV Emission from PKS 2005-489 and PKS 0548-322 96
    P. M. Chadwick, K. Lyons, T. J. L. McComb, K. J. Orford,
    J. L. Osborne, S. M. Rayner, S. E. Shaw, and K. E. Turver
The Implications of Galaxy Formation Models for the TeV
Observations of Current Detectors 100
    L. M. Boone, J. S. Bullock, J. R. Primack, and D. A. Williams
Extragalactic Background Light Absorption Signal in the
0.26–10 TeV Spectra of Blazars.. 105
    V. V. Vassiliev

ON MARKARIAN 421 AND 501

VHE γ-ray Spectral Properties of the Blazars Mrk 501 and Mrk 421 from the CAT Observations in 1997 and 1998 113
    F. Piron for the CAT Collaboration

Multiwavelength Observations of Mrk 501 in 1997 119
    M. Böttcher, D. Petry, and V. Connaughton

Comparison Between HEGRA and RXTE ASM Data from Mkn 501 124
    J. C. González and D. Kranich

Periodicity Analysis of Markarian 501 Flaring Activity in 1997 129
    S. J. Fegan, I. H. Bond, S. M. Bradbury, A. C. Breslin,
    J. H. Buckley, A. M. Burdett, M. Carson, D. A. Carter-Lewis,
    M. Catanese, M. F. Cawley, S. Dunlea, M. D'Vali, D. J. Fegan,
    J. P. Finley, J. A. Gaidos, T. A. Hall, A. M. Hillas, D. Horan,
    J. Knapp, F. Krennrich, S. Le Bohec, R. W. Lessard, C. Masterson,
    B. McKernan, J. Quinn, H. J. Rose, F. W. Samuelson, G. H. Sembroski,
    V. V. Vassiliev, and T. C. Weekes

Periodicity of TeV Gamma-Ray Emissions from Mrk501 134
    S. Hayashi, D. Nishikawa, N. Chamoto, M. Chikawa, Y. Hayashi,
    N. Hayashida, K. Hibino, H. Hirasawa, K. Honda, N. Hotta, N. Inoue,
    F. Ishikawa, N. Ito, S. Kabe, F. Kajino, T. Kashiwagi, S. Kakizawa,
    S. Kawakami, Y. Kawasaki, N. Kawasumi, H. Kitamura, K. Kuramochi,
    E. Kusano, H. Lafoux, E. C. Loh, K. Mase, T. Matsuyama, K. Mizutani,
    Y. Morizane, M. Nagano, J. Nishimura, T. Nishiyama, M. Nishizawa,
    T. Ouchi, H. Ohoka, M. Ohnishi, S. Osone, T. Saito, N. Sakaki,
    M. Sakata, M. Sasano, H. Shimodaira, A. Shiomi, P. Sokolsky,
    T. Takahashi, S. F. Taylor, M. Takeda, M. Teshima, R. Torii,
    M. Tsukiji, Y. Uchihori, T. Yamamoto, Y. Yamamoto, K. Yasui,
    S. Yoshida, H. Yoshii, and T. Yuda

Detection of TeV Gamma Rays from Mrk 501 at High Flaring State of Activity in 1997 with the Tibet Air Shower Array 139
    M. Amenomori, S. Ayabe, P. Y. Cao, Danzengluobu, L. K. Ding,
    Z. Y. Fen, Y. Fu, H. W. Guo, M. He, K. Hibino, N. Hotta, Q. Huang,
    A. X. Huo, K. Izu, H. Y. Jia, F. Kajino, K. Kasahara, Y. Katayose,
    Labaciren, J. Y. Li, H. Lu, S. L. Lu, G. X. Luo, X. R. Meng,
    K. Mizutani, J. Mu, H. Nanjo, M. Nishizawa, M. Ohnishi, I. Ohta,
    T. Ouchi, J. R. Ren, T. Saito, M. Sakata, T. Sasaki, Z. Z. Shi,
    M. Shibata, A. Shiomi, T. Shirai, H. Sugimoto, K. Taira, Y. H. Tan,
    N. Tateyama, S. Torii, T. Utsugi, C. R. Wang, H. Wang, X. W. Xu,
    Y. Yamamoto, G. C. Yu, A. F. Yuan, T. Yuda, C. S. Zhang,
    H. M. Zhang, J. L. Zhang, N. J. Zhang, X. Y. Zhang, Zhaxisangzhu,
    Zhaxiciren, and W. D. Zhou

Results from Milagrito on TeV Emission by Active Galactic Nuclei 144
    D. A. Williams for the Milagro Collaboration

Modeling the April 1997 Flare of Mkn 501 149
    A. Mücke and R. J. Protheroe

Spectral Measurements of TeV γ-Ray Emission from Mkn501
and Mkn421 Using the HEGRA Stereoscopic System of IACTS 154
  M. Panter and H. Krawczynski for the HEGRA Collaboration
Effect of Intergalactic Absorption in the TeV γ-Ray Spectrum
of Mkn 501 159
  A. K. Konopelko
The Probable Binary Galaxy System MKN 421: Kinematics & Structure
from Optical Observations 165
  P. W. Gorham, L. van Zee, S. C. Unwin, and C. S. Jacobs

SUPERNOVA REMNANTS

Modelling Hard Gamma-Ray Emission from Supernova Remnants 173
  M. G. Baring
Environmental and Age Limits on Particle Acceleration
in Supernova Remnants... 183
  L. O'C. Drury, J. Kirk, and P. Duffy
The SNR W28 at TeV Energies 187
  G. P. Rowell, T. Naito, S. A. Dazeley, P. G. Edwards, S. Gunji,
  T. Hara, J. Holder, A. Kawachi, T. Kifune, Y. Matsubara,
  Y. Mizumoto, M. Mori, H. Muraishi, Y. Muraki, K. Nishijimi,
  S. Ogio, J. R. Patterson, M. D. Roberts, T. Sako, K. Sakurazawa,
  R. Susukita, T. Tamura, T. Tanimori, G. J. Thornton, S. Yanagita,
  T. Yoshida, and T. Yoshikoshi
Recent Results from the CANGAROO 3.8 m Telescope 192
  T. Yoshikoshi for CANGAROO
HEGRA Observations of Galactic Sources............................... 197
  H. Völk for the HEGEA Collaboration
The Supernova Remnants as the Main Sources of Gamma-Quanta
with Energies more than 1 TeV in our Galactic 205
  V. G. Sinitsyna
TeV Emission from Supernovae... 210
  P. M. Chadwick, K. Lyons, T. J. L. McComb, K. J. Orford,
  J. L. Osborne, S. M. Rayner, S. E. Shaw, and K. E. Turver
Stereoscopic Observations of the Crab Nebula with the HEGRA
System of Imaging Air Čerenkov Telescopes........................... 215
  A. K. Konopelko and G. Pühlhofer for the HEGRA Collaboration

GAMMA RAY BURSTS

High Energy Radiation from Gamma Ray Bursts 225
  C. D. Dermer and J. Chiang
High-Energy Spectral Signatures in Gamma-Ray Bursts................. 238
  M. G. Baring

First Results of a Study of TeV Emission from GRBs in Milagrito............ 243
    J. E. McEnery, R. Atkins, W. Benbow, D. Berley, M. L. Chen,
    D. G. Coyne, B. L. Dingus, D. E. Dorfan, R. W. Ellsworth,
    D. Evans, A. Falcone, L. Fleysher, R. Fleysher, G. Gisler,
    J. A. Goodman, T. J. Haines, C. M. Hoffman, S. Hugenberger,
    L. A. Kelley, I. Leonor, M. McConnell, J. F. McCullough,
    R. S. Miller, A. I. Mincer, M. F. Morales, P. Nemethy,
    J. M. Ryan, B. Shen, A. Shoup, C. Sinnis, A. J. Smith,
    G. W. Sullivan, T. Tumer, K. Wang, M. O. Wascko,
    S. Westerhoff, D. A. Williams, T. Yang, and G. B. Yodh

On the Conversion of Blast Wave Energy into Radiation in AGN and GRBs............ 249
    M. Pohl and R. Schlickeiser

Detection Techniques of μs Gamma-Ray Bursts Using Ground Based Telescopes............ 253
    F. Krennrich, S. Le Bohec, and T. Weekes

OTHER SOURCES

EGRET Unidentified Sources............ 261
    I. A. Grenier

TeV Observations of X-Ray Binaries............ 271
    P. M. Chadwick, K. Lyons, T. J. L. McComb, K. J. Orford,
    J. L. Osborne, S. M. Rayner, S. E. Shaw, and K. E. Turver

Flux Limits for TeV Emission from Pulsars............ 276
    P. M. Chadwick, K. Lyons, T. J. L. McComb, K. J. Orford,
    J. L. Osborne, S. M. Rayner, S. E. Shaw, and K. E. Turver

The Diffusive Galactic GeV/TeV Gamma-Ray Background: Sources vs. Transport............ 281
    H. J. Völk

GeV Gamma-Ray Sources............ 288
    R. C. Lamb and D. J. Macomb

The New Metagalactic Source of Gamma-Quanta with Energy $>10^{12}$ eV............ 293
    V. G. Sinitsyna

IMAGING ATMOSPHERE CHERENKOV OBSERVATORIES

The Status of the Whipple 10m GRANITE III Upgrade Program............ 301
    J. P. Finley, I. H. Bond, S. M. Bradbury, A. C. Breslin,
    J. H. Buckley, A. M. Burdett, M. Carson, D. A. Carter-Lewis,
    M. Catanese, M. F. Cawley, S. Dunlea, M. D'Vali, D. J. Fegan,
    S. J. Fegan, J. A. Gaidos, T. A. Hall, A. M. Hillas, D. Horan,
    J. Knapp, F. Krennrich, S. LeBohec, R. W. Lessard, C. Masterson,
    B. McKernan, J. Quinn, H. J. Rose, F. W. Samuelson,
    G. H. Sembroski, V. V. Vassiliev, and T. C. Weekes

Calibration of the CAT Telescope............ 308
    F. Piron for the CAT Collaboration

Initial Performance of CANGAROO-II 7m Telescope 313
    H. Kubo, S. A. Dazeley, P. G. Edwards, S. Gunji, S. Hara,
    T. Hara, J. Jinbo, A. Kawachi, T. Kifune, J. Kushida,
    Y. Matsubara, Y. Mizumoto, M. Mori, M. Moriya,
    H. Muraishi, Y. Muraki, T. Naito, K. Nishijima,
    J. R. Patterson, M. D. Roberts, G. P. Rowell,
    T. Sako, K. Sakurazawa, Y. Sato, R. Susukita,
    T. Tamura, T. Tanimori, S. Yanagita, T. Yoshida,
    T. Yoshikoshi, and A. Yuki

Data Analysis Techniques for Stereo IACT Systems 318
    W. Hofmann

**Optimum Spacing Between Imaging Atmospheric Čerenkov
Telescopes in the Future 50 GeV Multi-Telescope Arrays** 323
    A. K. Konopelko

"Convergent Observations" with Stereoscopic HEGRA CT System 328
    H. Lampeitl and W. Hofmann for the HEGRA Collaboration

**Experimental Results on the Optimum Spacing of Stereoscopic
Imaging Atmospheric Cherenkov Telescopes** 333
    W. Hofmann for the HEGRA Collaboration

Kernel Analysis in TeV Gamma-Ray Selection 338
    P. Moriarty and F. W. Samuelson

Observations at Large Zenith Angles 343
    F. Schröder for the HEGRA Collaboration and D. Heck

Monte Carlo Simulations for High Zenith Angle 348
    A. Ibarra, J. C. González, J. Cortina, J. A. Barrio, and V. Fonseca

**Geomagnetic Effects on the Performance of Atmospheric
Čerenkov Telescopes** 353
    P. M. Chadwick, K. Lyons, T. J. L. McComb, K. J. Orford,
    J. L. Osborne, S. M. Rayner, S. E. Shaw, and K. E. Turver

**A Concept for the Readout of Multichannel Detectors
by Using Analog Signal Transmission Via Optical Fibres
Coupled to a Fast CCD** 358
    R. Mirzoyan, E. Lorenz, and J. Rose

The FADC Readout of the MAGIC Telescope 363
    J. Cortina, E. Lorenz, and R. Mirzoyan for the MAGICT Collaboration

The New Data Acquisition System of the First Telescope in HEGRA ... 368
    J. Cortina, J. A. Barrio, and G. Rauterberg for the HEGRA Collaboration

**GigaHertz Analogue Memories in Ground-based
Gamma-Ray Astronomy** 373
    M. Punch, J.-P. Denance, P. Nayman, F. Toussenel, M. Rivoal,
    J.-P. Tavernet, P. Goret, L.-M. Chounet, B. Degrange,
    P. Espigat, P. Fleury, C. Renault, and P. Vincent

**A Cherenkov Camera with Integrated Electronics
Based on the "Smart Pixel" Concept** 378
    N. Bulian, T. Hirsch, W. Hofmann, T. Kihm, A. Kohnle,
    M. Panter, and M. Stein

An Optical Reflector System for the CANGAROO-II Telescope 383
    A. Kawachi for the CANGAROO Collaboration

A Study of the Effect of Polarizing Filters in Imaging
Cherenkov Telescopes 388
    I. de la Calle, J. L. Contreras, and V. Fonseca
Sensing Atmospheric Conditions Using MIR Radiometers 393
    P. M. Chadwick, K. Lyons, T. J. L. McComb, K. J. Orford,
    J. L. Osborne, S. M. Rayner, S. E. Shaw, and K. E. Turver

OTHER TeV OBSERVATORIES

STACEE-32: Design, Performance, and Preliminary Results 401
    R. A. Ong
STACEE: Instrument Performance and Future Plans 411
    C. E. Covault, D. Bhattacharya, L. Boone, M. C. Chantell,
    Z. Conner, M. Dragovan, D. Gingrich, D. Gregorich,
    D. S. Hanna, R. Mukherjee, R. A. Ong, S. Oser, K. Ragan,
    R. A. Scalzo, C. G. Théoret, T. O. Tumer, D. A. Williams,
    and J. A. Zweerink
First Observations with CELESTE 416
    D. A. Smith and M. de Naurois
The Solar Two Gamma-Ray Observatory: Astronomy between
20–300 GeV 426
    J. A. Zweerink, D. Bhattacharya, G. Mohanty, U. Mohideen,
    A. Radu, R. Rieben, V. Souchkov, H. Tom, and T. O. Tumer
The Physics Potential of Ground-Based Gamma-Ray Astronomy
below 50 GeV 431
    N. Magnussen
The CLUE Experiment Running with 8 Telescopes;
Observations of Gamma Sources and Runs on Moon 436
    D. Bastieri, B. Bartoli, C. Bigongiari, R. Biral, M. A. Ciocci, D. Cosulich,
    M. Cresti, D. Kartashov, F. Liello, N. Malakov, M. Mariotti, G. Marsella,
    A. Menzione, R. Paoletti, G. Parlavecchio, L. Peruzzo, A. Piccioli,
    F. Rosso, R. Sacco, A. Saggion, G. Sartori, P. Sartori,
    C. Sbarra, A. Scribano, A. Stamerra, N. Turini, and F. Zetti
Results from the Milagrito Experiment 441
    A. J. Smith, R. Atkins, W. Benbow, D. Berley, M. L. Chen,
    D. G. Coyne, B. L. Dingus, D. E. Dorfan, R. W. Ellsworth,
    D. Evans, A. Falcone, L. Fleysher, R. Fleysher, G. Gisler,
    J. A. Goodman, T. J. Haines, C. M. Hoffman, S. Hugenberger,
    L. A. Kelley, I. Leonor, M. McConnell, J. F. McCullough,
    J. E. McEnery, R. S. Miller, A. I. Mincer, M. F. Morales,
    P. Nemethy, J. M. Ryan, B. Shen, A. Shoup, G. Sinnis,
    G. W. Sullivan, T. Tumer, K. Wang, M. O. Wascko,
    S. Westerhoff, D. A. Williams, T. Yang, and G. B. Yodh
    (The Milagro Collaboration)
Computer Animation of Extensive Air Showers Interacting
with the Milagro Water Cherenkov Detector............................ 448
    M. F. Morales

Gamma Hadron Separation in Milagro 453
  G. B. Yodh for the Milagro Collaboration
Search for Multi-TeV Gamma-Ray Emission from Nearby SNRs
with the Tibet Air Shower Array 459
  M. Amenomori, S. Ayabe, P. Y. Cao, Danzengluobu, L. K. Ding,
  Z. Y. Feng, Y. Fu, H. W. Guo, M. He, K. Hibino, N. Hotta,
  Q. Huang, A. X. Huo, K. Izu, H. Y. Jia, F. Kajino, K. Kasahara,
  Y. Katayose, Labaciren, J. Y. Li, H. Lu, S. L. Lu, G. X. Luo,
  X. R. Meng, K. Mizutani, J. Mu, H. Nanjo, M. Nishizawa,
  M. Ohnishi, I. Ohta, T. Ouchi, J. R. Ren, T. Saito, M. Sakata,
  T. Sasaki, Z. Z. Shi, M. Shibata, A. Shiomi, T. Shirai, H. Sugimoto,
  K. Taira, Y. H. Tan, N. Tateyama, S. Torii, T. Utsugi, C. R. Wang,
  H. Wang, H. Y. Wang, P. X. Wang, X. W. Xu, Y. Yamamoto,
  G. C. Yu, A. F. Yuan, T. Yuda, C. S. Zhang, H. M. Zhang,
  J. L. Zhang, N. J. Zhang, X. Y. Zhang, Zhaxisangzhu,
  Zhaxiciren, and W. D. Zhou

FUTURE OBSERVATORIES

The AGILE Gamma-Ray Astronomy Mission 467
  S. Mereghetti, G. Barbiellini, G. Budini, P. Caraveo, E. Costa,
  V. Cocco, G. Di Cocco, M. Feroci, C. Labanti, F. Longo,
  E. Morelli, A. Morselli, A. Pellizzoni, F. Perotti, P. Picozza,
  M. Prest, P. Soffitta, L. Soli, M. Tavani, E. Vallazza,
  and S. Vercellone
The Capabilities of the AMS as GeV γ-Ray Detector 474
  R. Battiston
The CANGAROO-III Project 485
  M. Mori, S. A. Dazeley, P. G. Edwards, S. Gunji, S. Hara, T. Hara,
  J. Jinbo, A. Kawachi, T. Kifune, H. Kubo, J. Kushida, Y. Matsubara,
  Y. Mizumoto, M. Moriya, H. Muraishi, Y. Muraki, T. Naito, K. Nishijima,
  J. R. Patterson, M. D. Roberts, G. P. Rowell, T. Sako, K. Sakurazawa,
  Y. Sato, R. Susukita, T. Tamura, T. Tanimori, S. Yanagita, T. Yoshida,
  T. Yoshikoshi, and A. Yuki
The Gamma-Ray Large Area Space Telescope (GLAST) 492
  D. A. Kniffen, D. L. Bertsch, and N. Gehrels
The High Energy Stereoscopic System (HESS) Project 500
  W. Hofmann for the HESS Collaboration
The MAGIC Telescope Project 510
  E. Lorenz for the Magic Collaboration
VERITAS: Very Energetic Radiation Imaging Telescope Array System 515
  F. Krennrich, S. M. Bradbury, I. H. Bond, A. C. Breslin,
  J. H. Buckley, D. A. Carter-Lewis, M. Catanese, B. L. Dingus,
  D. J. Fegan, J. P. Finley, J. Gaidos, J. Grindlay, A. M. Hillas,
  G. Hermann, P. Kaaret, D. Kieda, J. Knapp, S. LeBohec,
  R. W. Lessard, J. Lloyd-Evans, D. Müller, R. Ong, J. Quinn,
  H. J. Rose, M. Salamon, G. H. Sembroski, S. Swordy,
  V. V. Vassiliev, and T. C. Weekes

List of Attendees..529

Author Index...533

PREFACE

This workshop on GeV-TeV γ-ray astrophysics was held at Snowbird, Utah USA from the 13th to 16th of August 1999, just preceding the 26th International Cosmic Ray Conference in Salt Lake City, Utah. This meeting was the 6th in a series of workshops that emphasize the development of future instrumentation. The previous workshops were held in Palaiseau, France (1992), Calgary, Canada (1993), Tokyo, Japan (1994), Padova, Italy (1995), and Kruger National Park, South Africa (1997). One hundred and fifty scientists attended this workshop, and 99 papers were presented as talks or posters.

As evidenced in this workshop, the field of GeV-TeV gamma-ray astrophysics has an exciting future. Existing observatories are being improved, energy thresholds are being lowered, and new detection and analysis techniques are being developed. Within the next decade several new observatories, both on the Earth and in space, will become operational. These will provide orders of magnitude more sources, and will allow, for the first time, complete coverage of the GeV-TeV energy spectrum with significant sensitivity. The opportunities for multiwavelength campaigns will also increase. The knowledge gained from these observations greatly enhances our understanding of some of the highest energy particle accelerators in the Universe.

We particularly wish to thank the members of the international scientific organizing committee—Felix Aharonian, Peter Biermann, Giovanni Bignami, Okkie DeJager, Patrick Fleury, Alice Harding, Tadashi Kifune, Rene Ong, and Trevor Weekes – for their active participation in the planning of this meeting. We are also grateful to Hamamatsu Corporation and NASA for their financial assistance. We thank Nancy Waterman of the University of Utah Institute for High Energy Astrophysics for help with the compilation of these proceedings, and Linda Williams of the University of Utah's Conferences Department for the smooth running of the conference. Finally, we thank all the participants for making this a productive and successful workshop.

                                          The Local Organizing Committee
                                          Brenda L. Dingus
                                          Michael H. Salamon
                                          David B. Kieda

INTRODUCTORY REVIEW

VHE Astronomy before the New Millennium

Trevor C. Weekes

*Whipple Observatory, Harvard-Smithsonian Center for Astrophysics,
P.O. Box 97, Amado, AZ 85645-0097 U.S.A.
e-mail: tweekes@cfa.harvard.edu*

I INTRODUCTION

In planning this workshop the organizers suggested that a review of new TeV γ-ray observations at the beginning might obviate a need for separate submissions from individual groups (which would be presented anyway at the ICRC) and thus permit the program to be devoted to technical developments, interpretations and new programs. In practice, this did not work too well since many groups wished to personally present their own results at the workshop also. Therefore I tried to present some synthesis of the results as seen in August, 1999. In principle, these are pre-workshop but in practice since many of the ICRC papers were available (courtesy of astro-ph) and others were sent directly to me, I have been able to include many of the most recent results. As usual, the printed ICRC papers do not contain the whole story since many authors still regard the pre-conference papers as place holders with the real results to be presented orally at the conference. Hence the proceedings are archaic as soon as they are printed (or recorded on CD)!

I will concentrate on the TeV observations reported since the last workshop (Kruger Park, August, 1997). I will not discuss the impressive technical improvements reported from many Very High Energy (VHE) observatories, nor describe the new VHE observatories that have come on-line (or are due shortly to come on-line) nor the many interpretations now reported for the VHE observations. Suffice to say that VHE astronomy is still observation-driven and that theoretical VHE astrophysics still lags as a predictive discipline.

More general reviews of VHE γ-ray astronomy can be found elsewhere [1,2].

II STANDARDS OF CREDIBILITY

It is well known that VHE γ-ray astronomy has had a rather murky past in that many sources were claimed in the eighties which were never verified. In fairness it should be noted that in the balloon era of 100 MeV γ-ray astronomy (in the sixties) there were also many unverified claims which were only resolved with the advent of satellite astronomy and the use of reliable statistical methods. Nevertheless VHE γ-ray astronomy suffers from the problems that the detector medium is the atmosphere over which the experimenter has no control and that non-statistical effects due to atmospheric/detector conditions are a perpetual problem. This is primarily so for optical techniques but it has yet to be demonstrated that air shower arrays are free of such effects. In such circumstances, some subjective bias in the data analysis, i.e., in data selection, is almost inevitable. It should also be remembered that this is a relatively new discipline (it is only ten years since the publication of the first verifiable observation) and hence its exponents are still feeling their way. In these circumstances it behooves all involved to adopt a conservative attitude to claims for the detection of new sources.

It was at the ICRC in Hobart, Australia in 1971 that Prof. A.E. Chudakov suggested that the acceptable standard of credibility for a new VHE γ-ray source should be 5 σ. I have not been able to find a reference for this statement and I was not at the conference but I remember well the dismay with which we greeted this news on the return of our senior colleagues who attended. At that stage we had managed to accumulate a 3 σ result on the Crab Nebula [3] which we thought bordered on the edge of credibility. A 5 σ result seemed an impossible standard.

Years later I met Prof. Chudakov at the 14th Texas Conference on Relativistic Astrophysics; I reminded him of his 5 σ credibility criterion (which had not been met on another source in the interval) and proudly announced that with the atmospheric Cherenkov imaging technique we now had a 9 σ signal from the Crab. He paused for a moment and then said drolly: "I think I should have said 10 σ". He was right of course; a 5 σ detection is not always entirely credible. But he was wrong also, for a 10 σ detection by a single experiment using a new technique is almost as liable to be a systematic effect as a 5 σ detection.

To be convincing, I suggest that we adopt a standard where a *really* credible source requires a 5 σ detection coupled with an equally significant verification by another experiment. Ideally the "σ" should include the best estimate of potential systematic effects. This would qualify as a Grade A detection. A Grade B detection would then be the same but without the independent verification. Grade C would be a strong detection but with some qualifications: time variability or some other factors which introduce extra degrees of freedom.

III THE 1997 TEV SOURCE CATALOG

In his comprehensive rapporteur talk at the 25th ICRC, Chaman Bhat listed eleven sources of TeV gamma-rays based on observations reported at that meeting [4]. These were the SNR remnants, the Crab Nebula, PRS 1706-44, Vela, and SN1006; the AGNs, Markarian 421, Markarian 501 and 1ES2344+514; the binary, Centaurus X-3; and the pulsars, PSR B 1259-63, PSR B 1509-58, and PSR 1105-61. As we shall see below, all except the three pulsars (about which there have been no further reports of confirmed detections) appear in the 1999 catalog.

TABLE 1. Flux from the Crab Nebula

Group	VHE Spectrum (10^{-11} photons cm^{-2} s^{-1} TeV^{-1})	E_{th} (TeV)
Whipple (1991) [5]	$(25(E/0.4\text{TeV}))^{-2.4\pm0.3}$	0.4
Whipple (1998) [7]	$(3.2\pm0.7)(E/\text{TeV})^{(-2.49\pm0.06_{stat}\pm0.04_{syst})}$	0.3
HEGRA (1999) [8]	$(2.7\pm0.2\pm0.8)(E/\text{TeV})^{-2.60\pm0.05_{stat}\pm0.05_{syst}}$	0.5
CAT (1999) [9]	$(2.7\pm0.17\pm0.40)(E/\text{TeV})^{-2.57\pm0.14_{stat}\pm0.08_{syst}}$	0.25
CANGAROO (1998) [10]	$(2.01\pm0.36)\text{x}10^{-2})(E/7\text{TeV})^{-2.53\pm0.18}$	7
Tibet HD (1999) [11]	$(4.61\pm0.90)\text{x}10^{-1}(E/3\text{TeV})^{-2.62\pm0.17}$	3

IV SOURCE SUMMARY

A Galactic Sources

The Crab Nebula

There is remarkable agreement now between the absolute fluxes and spectral shapes reported from observations of the Crab Nebula by imaging ACTs; the results from the Whipple, HEGRA, CAT and CANGAROO experiments are shown in Table 1. These are also in agreement with the flux reported in the first detections of the Crab [6,5] but this must be considered fortuitous in view of the large error bars in these early measurements. At least in the 300 GeV to 3 TeV range it is clear that the Crab can now be considered a standard candle and a grade A source.

New observations of the Crab Nebula have been reported at both high and low energies. CELESTE, with a threshold energy of 50 GeV, observed it for just three hours [13] whereas STACEE, with an interim threshold of 75 GeV, had a 7 σ detection in 50 hours of observation [12]. Neither experiment could quote a flux value and neither experiment saw any evidence for a pulsed component from the Crab pulsar.

At higher energies the Crab was seen for the first time by a conventional air shower array (the Tibet High Density Array at 4.5 km) [11]. The energy threshold was 3 TeV and the flux deduced (see Table 1) was a factor of 2-3 higher than that seen in ACT experiments.

PSR 1706-44

Following the TeV detection of this source by the CANGAROO group [14] and its confirmation by the Durham group [15], there have been no new reports of observations. No periodic emission is seen and it is believed that the VHE emission comes from a weak plerion. Although weaker than the Crab this may be the standard candle for the southern hemisphere and merits a grade A ranking.

SN1006

In 1997 the CANGAROO Collaboration reported the observation of TeV γ-ray emission from the shell-type SNR, SN 1006 [16]. Observations taken in 1996 and

1997 indicated a statistically significant excess from the northeast rim of the SNR shell. The flux at $> 1.7 \pm 0.5$ TeV was $(4.6 \pm 0.6(sys) \pm 1.4(stat) \times 10^{-12}$ photons cm^{-2} s^{-1}. The observations were motivated by the observation of non-thermal X- rays by the *ASCA* experiment [17]. It represented the first direct evidence of acceleration of particles to TeV energies in the shocks of SNRs.

At this workshop there was the disturbing report from the Durham group of the failure to detect this source in 40 hours of observation. Their upper limit at 300 GeV was 1.7×10^{-12} photons cm^{-2}s^{-1} and at 1.5 TeV was 1.3×10^{-12} photons cm^{-2}s^{-1}, barely compatible with the CANGAROO observation. They point out that the presence of a bright star near the SNR complicates the measurement. Because of this report I assign this source a B− grade.

Vela

The CANGAROO group reported the detection of a 6σ signal from the vicinity of the Vela pulsar [18]. The integral γ-ray flux above 2.5 TeV is 2.5×10^{-12} photons cm^{-2} s^{-1}. There is no evidence for periodicity and the flux limit is about a factor of ten less than the steady flux. The signal is offset (by 0.14°) from the pulsar position which makes it more likely that the source is a synchrotron nebula. Since this offset position is coincident with the birthplace of the pulsar it is suggested that the progenitor electrons are relics of the initial supernova explosion and they have survived because the magnetic field was weak.

Again the source was not confirmed by observations by the Durham group (J.Osborne, this workshop). The upper limit to the γ-ray flux above 300 GeV is 5×10^{-11} photons cm^{-2} s^{-1}. Given the differences in energy and the uncertainties in flux estimates in the two experiments, the Durham group felt the two results were compatible. However it would have been reassuring to see the independent confirmation. I give this one a B grade.

RJX1713.7-3946

The detection of TeV gamma-rays from this shell-type SNR was reported by the CANGAROO group for the first time this year [19]. The observations were motivated by the observation of a hard X-ray power-law spectrum by *ASCA* [20]. In this respect, it is very similar to SN1006 but is three times brighter in X-rays. It has a characteristic dimension of 70 arc-min, lies at a distance of 1.1 kpc and has an estimated age of 2,100 years. The γ-ray flux above 2 TeV is 3×10^{-12} photons cm^{-2} s^{-1} with a 5 σ significance. There is evidence that the source is extended in the same direction as the X-ray source. This is clearly a grade B source.

Cassiopeia A

It is natural that the strongest source in the radio sky should have been one of the first targets of TeV observations [21]. It is appropriate that it should have been eventually detected as a TeV source; however this only came after a very long exposure by the HEGRA group [22]. As with SN1006 and RXJ1713.7-3946, these observations were motivated by observations of a hard X-ray power-law spectrum

TABLE 2. Observations of Cassiopeia A

Group	E_{th} (TeV)	Exposure (hours)	Integral Flux (10^{-11} photons cm^{-2} s^{-1})
Whipple	500	7.5	< 0.66
CAT	400	24.4	< 0.74
HEGRA	1000	127.9	≈0.3

[23]. The source is a classical shell-SNR of diameter 2.2 arc-min which is effectively point-like to a γ-ray telescope. It is believed to be 300 years old and there is no active pulsar at its center; however there may be a neutron star. The HEGRA observations were made in 1997 and 1998 and comprised some 130 hours on the source. The flux above 1 TeV has not yet been determined but must be $\approx 3 \times 10^{-12}$ photons cm^{-2} s^{-1}. The total detection was just less than 5 σ and it is probably the weakest TeV source detected to date.

Upper limits to the TeV emission have been reported by the CAT [24] and Whipple [25] groups. These were at lower energies but, because the exposures were much shorter, the upper limits are compatible with the HEGRA detection. The three results are summarized in Table 2.

Because the detection is below the magic 5 σ level and Cassiopeia A is the weakest source yet detected (and hence more susceptible to systematic effects), I give this a C grade for now.

Centaurus X-3

New observations were reported on the high mass X-ray binary, Cen X-3 [26]. The system contains a 4.8 s pulsar in orbit with a period of 2.1 days. Originally reported as a source of sporadic outbursts of pulsed emission [27,28], it was later found to be a source of steady (unpulsed) weak emission [29]. At this time it was also seen as an unpulsed GeV EGRET source [30]. The new observations, taken in 1998 and 1999 by the Durham group, do not add to the overall statistical significance of the detections which remain somewhat marginal; hence I give it a C grade.

B Extragalactic Sources

Markarian 421

Markarian 421 was one of the weakest AGNs detected by EGRET; it was also the closest BL Lac at z = 0.031. It was the first TeV source detected [31]. It is also the AGN in which the clearest correlations have been found over multiple wavelengths (see [2]) and in which the shortest time variations have been seen [32]. At discovery, its intensity was approximately 30% of the Crab; however it has flared to levels more than ten times greater than the Crab.

In 1998 there were extensive multiwavelength campaigns on this source between various ground-based γ-ray observatories and the *ASCA* and *BeppoSAX* X-ray satellites [33,34]. The most interesting event was the flare seen on April 21, 1998 by the Whipple Observatory [35] and the *BeppoSAX* telescopes. Although the flare was observed to rise and peak at the same time in both telescopes, the TeV signal decayed within a few hours whereas the X-ray signal persisted for half a day. It is difficult to model this behavior.

The energy spectrum of Markarian 421 has been reported by several groups. There is general agreement that it can be fit by a simple power law. While the absolute flux has little meaning since it varies with time, the differential power-law spectral index should be comparable in different experiments unless it is also variable with time. There is good agreement on the indices obtained thus far by CAT ($-2.96 \pm 0.13 \pm 0.05$) [36]; HEGRA ($-3.09 \pm 0.07 \pm 0.10$) [37]; 7TA (-2.81) [38]. However the Whipple group gets consistently harder spectra [39] particularly during flaring e.g (-2.54 ± 0.04) on May 7, 1996. Preliminary analysis of non-flaring data gives a similar result. Obviously further work is required here to ensure that the analysis is free of large systematic errors.

Despite its variability Markarian 421 is well-established and merits an A grade.

Markarian 501

Markarian 501, a BL Lac at z=0.034, was the first extragalactic γ-ray source detected first at TeV energies. Originally detected as a weak source (8% Crab) [40] it has been intensively monitored by ground-based telescopes since then. The TeV outburst from Markarian 501 in 1997 merited a Highlight session at the 25th ICRC [41]. Sadly while the conference was taking place the source was already in decline and it has been relatively quiescent ever since. Most of the interest in the source since that time has been in a detailed analysis of the high intensity signal, in particular in the derivation of an accurate energy spectrum.

The 1997 outburst data has been summarized in a number of publications [42–45]. Variations with doubling times as short as two hours have been reported [42] but in general the variations are not as short as those seen in Markarian 421. There were no significant new results from multiwavelength campaigns.

Spectral measurements were in agreement in that the energy spectrum could not be satisfactorily represented by a simple power law. The Whipple and HEGRA groups [46,39]) reported that they observed no change in spectral shape with source intensity; in contrast, the CAT group [47] using a simple Hardness ratio found that the spectrum hardened as the intensity increased. It is not clear whether this is a real change or the result of the systematics in a new analysis method (which has not yet been tested on any other source). In addition the CAT group reported that the weak intensity emission observed in 1998 could be best fit by a simple power law with differential spectral index -2.97 ± 0.20.

Markarian 501 is the archetypical extreme BL Lac, characterized by its low luminosity, its high synchrotron peak (up to 100 keV), and its high Compton peak (up to TeV) [51,49]. As the strongest source (for a few months in 1997) thus far observed, it clearly merits an A grade.

1ES2344+514

Although less well-studied, this X-ray-selected BL Lac at z=0.044 is superficially very similar to the above two sources. Recent X- ray observations by *Beppo-SAX* [50] emphasize this similarity: time variability on times scales of hours has been seen and the putative synchrotron spectrum peaks at energies greater than 10 keV. It was reported as a TeV source [51] primarily on the basis of a flare seen in one night at the 6 σ level; the average flux over that night was F_γ (>350 GeV) = $(6.6 \pm 1.9) \times 10^{-11}$ photons cm^{-2} s^{-1} which was 60% of the Crab. The averaged flux (including the flare) was at the 5.8 σ level. The source was not detected in the 1996/7 observing season.

Based on the observed behavior of Markarian 421 and Markarian 501 it might have been expected that continued monitoring of 1ES2344+514 would have confirmed this detection and given more information about its properties at high energies. In practice, continued monitoring by the Whipple group and HEGRA [52] have not confirmed either the flaring or steady emission; hence this source which would have been ranked B in 1997 must now have a C grade.

PKS2155-304

The above three sources are in the northern hemisphere; it had been predicted that PKS2155-304 would be the best candidate for TeV emission in the southern hemisphere. An X-ray-selected BL Lac, it has been detected by EGRET and has been the object of numerous multiwavelength observing campaigns. The Durham group detected it in 1996 and 1997 [53]; the November 1997 observations were particularly interesting as they coincided with observations by EGRET and RXTE which indicated that the source was active at this time.

More recent observations by the Durham group (J.Osborne, this workshop) have not detected the source. Because of its relatively large redshift (z=0.116), the energy spectrum of this source is of particular interest; however none is yet available. This appears to be a highly probable source and thus merits a B grade.

1ES1959+650

The Utah Seven Telescope Array group have reported the detection of the BL Lac, 1ES1959+650 based on 57 hours of observation in 1998 [38]. As with the four AGNs listed above, this is an X- ray-selected BL Lac; its redshift is 0.048. The energy threshold for these observations was 600 GeV. The flux level was not reported but the total signal was at the 3.9 σ level. This is not normally considered high enough to claim the detection of a new source; however, within this database there were two epochs which were selected a posteriori which gave signals above the canonical 5 σ level. This source has not yet been confirmed by any other group; it was observed by the Whipple group but no flux was detected [55]. It is therefore awarded a B− grade.

3C66A

This is potentially the most exciting TeV detection of an AGN as it is quite different from the other AGNs. The source is a radio- selected, EGRET-detected, BL Lac and the redshift is 0.44, i.e., much more distant than the other objects. The Crimean Astrophysical Observatory group using the GT-48 telescope detected this source at the 5 σ level in 1996 [56]. The flux above 900 GeV was $(3\pm1) \times 10^{-11}$ photons cm^{-2} s^{-1}. There were previous and later upper limits to the TeV emission from the source, e.g. F_γ (> 350 GeV) < 1.9×10^{-11} photons- cm^{-2}-s^{-1} from Whipple in 1993 [57]. Confirmation of this detection is urgently required; until then it must be considered a grade C source.

V PERIODICITY IN 1997 SIGNAL FROM MRK 501

Several groups have reported on the apparent periodicity in the TeV γ-ray signal from Markarian 501 (TA, HEGRA, Whipple). The best data base is that of the HEGRA group since they observed during part of the bright period of the moon with one of their telescopes and hence have a database that is less prone to aliases. The reported periodicities occured at 12.7 day [45] and 23-24 day [58,59] and S.Fegan, this workshop, and were arrived at using the Lomb method which is recommended for observations made at irregular intervals. The epoch chosen by the HEGRA group for periodicity analysis is *a posteriori* but coincides with the bulk of the TeV observations and the peak in the γ-ray signal intensity. There is no evidence for periodicity outside this interval, either in 1997 or in other years. A visual inspection shows that the γ-ray signal has a few clearly defined flares with several time constants and the most obvious is at 23 days.

Since all the γ-ray experiments were observing at approximately the same time, they must see the same time variations; hence reports from the separate experiments do not constitute independent confirmations. The real question is whether the observed "periodicity" is really statistically significant given the large number of time variations. It is difficult to arrive at the real statistical significance of the observed effect.

Similar periodicity is seen in the X-ray signal by ASM/ RXTE and it has been suggested that this constitutes independent evidence for the periodicity. However, correlation between the X-ray and TeV γ-ray signals from Markarian 421 and 501 on a variety of time-scales now seems to be well-established so that the independent analysis of the RXTE database only confirms this correlation, not the statistical significance of the periodicity.

The conclusion is that while there is apparent periodicity in the TeV/X-ray signals from Markarian 501 for a five month epoch in 1997, it is almost impossible to arrive at a definitive conclusion about its statistical significance.

Those who survived the many pseudo-periodicities seen in the "a posteriori"

TABLE 3. The TeV Source Catalog c.1999

Source	Type	Discovery	EGRET	Grade
Galactic				
Crab Nebula	Plerion	1989	yes	A
PSR 1706-44	Plerion?	1995	no	A
Vela	Plerion?	1997	no	B
SN1006	Shell	1997	no	B−
RXJ1713.7-3946	Shell	1999	no	B
Casssiopea A	Shell	1999	no	C
Centuarus X-3	Binary	1999	yes	C
Extragalactic				
Markarian 421	XBL z=0.031	1992	yes	A
Markarian 501	XBL z=0.034	1995	yes	A
1ES2344+514	XBL z=0.044	1997	no	C
PKS2155-304	XBL z=0.116	1999	yes	B
1ES1959+650	XBL z=0.048	1999	no	B−
3C66A	RBL z=0.44	1998	yes	C

analysis of the TeV observation of binaries in the previous decade will perhaps be forgiven a little skepticism in the discussion of this new and potentially important result. It is unlikely that it will become credible until the phenomenon is observed again, either in Markarian 501 or another BL Lac.

VI GAMMA RAY BURST GRB970417A

At this workshop there was the first report of the detection of a γ-ray burst at TeV energies by Milagrito, the first stage of the Milagro experiment (J.McEnery, this workshop). In some 15 months of operation 54 possible GRBs were within the FOV of the detector and from one of these , GRB970417a, 18 events were detected during the duration of the BATSE burst (9 s) where the background was 3.46 events. Allowing for trials, the probability of the observation being a statistical fluctuation was calculated to be 1.5×10^{-3}. The energy threshold was about 1 TeV and the flux a few times 10^{-12} photons- cm^{-2}-s^{-1}. Given the importance of a detection of a TeV burst it would seem wise to await a confirmation from the full, more sensitive, Milagro array before too many conclusions are drawn. The BATSE fluence of GRB970417a was 1.5×10^{-7} erg cm^{-2}, a rather weak burst, and there was nothing otherwise unusual about it. There was no precise position and hence no information on X-ray or optical counterparts nor any indication as to distance.

TABLE 4. Status of HE/VHE Sources

Energy Range	10 MeV - 10 GeV	300 GeV - 30 TeV
Platform	Space	Ground
Discrete Sources		
Type	No. of Sources	No. of Sources
AGNs	75	6
Normal Galaxies	1	0
Pulsars	5?	0
SNR Shell	4?	3
SNR Plerion	1	3
Binaries	1	1
Total identified	87	13
Unidentified	165	0
Total	250	13
Other Sources		
Galactic Plane	Yes	No
Extragalactic Diffuse	Yes	No
All Sky Survey	Yes	No
Gamma Ray Bursts	5	1?

VII HE/VHE STATUS AND OUTLOOK

Based on the above discussion the 1999 TeV Source catalog is derived (Table 3); it is disappointing that it is not much larger than the 1997 Catalog.

As we conclude this century it is worthwhile to summarize the progress in HE and VHE astronomy and compare the achievements in each band. The recent publication of the 3rd EGRET Catalog summarizes the field at energies from 30 MeV to 10 GeV.

In Table 4 the number of sources reported in various categories is compared. The 100 MeV sources are from the Third EGRET Catalog [60] and from [61]. Don Kniffen has pointed out (this workshop) that only a handful of the EGRET sources could be classified with an A grade. The TeV sources are from Table 3; note in this context C must be considered a passing grade.

Although EGRET has still some sensitivity the mission is essentially over and not much change can be expected in the observational picture until the launch of GLAST in 2005 (hopefully). The intermediate missions, AMS and AGILE, which are described elsewhere in these proceedings, will not significantly improve on the EGRET flux sensitivity and can be considered place-holders for GLAST. The latter will offer an improvement of a factor of 10-20 in most parameters compared to EGRET.

In contrast to the drought expected in MeV-GeV γ-ray observations in the immediate future, ground-based γ-ray astronomy has never been more active. There are already nine atmospheric Cherenkov imaging telescopes in operation and two low threshold air shower arrays; one can expect to see steady improvements in

TABLE 5. Future Roadmap for HE/VHE Gamma Ray Astronomy

Energy	MeV 10-100 Space	GeV 0.1-1 Space	GeV 1-10 Space	GeV 10-100 Space/Ground	TeV 0.1-1 Ground	TeV 1-10+ Ground
Year						
1999	*Comptel*	(EGRET)		**********	**9ACITs*	***+2ASA
2000	****				CEL/STAC	**********	**********
2001	****			**********	**********	**********
2002	Integral			**********	**********	**********
2003	**	AMS/AGILE	**********	*MAGIC**	**********	**********
2003	**	**********	**********	**********	HESS/CAN	**********
2004	**	**********	**********	**********	VERITAS*	**********
2005	*GLAST**	**GLAST**	**GLAST*	*GLAST**	**********	**********
2006	*********	**********	**********	**********	**********	**********
2007	*********	**********	**********	**********	**********	**********
2008	*********	**********	**********	**********	**********	**********

sensitivity in these telescopes over the next decade. In addition, there are several other Cherenkov experiments coming on-line e.g., STACEE, CELESTE, Solar-Two, Pachmari, etc. Four major Cherenkov imaging telescopes/arrays are scheduled for completion by 2003, well in advance of GLAST and with significant overlap in the 30-300 GeV range. One can expect to see a steady increase in the GeV-TeV source catalog from ground-based observations so that even if the GLAST launch were to be delayed there would be a healthy increase in activity in studies of γ-ray astrophysics at these high energies.

VIII WHERE HAVE ALL THE HADRONS GONE?

It is a matter of some disappointment for the many cosmic-ray physicists who entered the field of high energy γ-ray astronomy that none of the sources thus far detected, either at HE or VHE energies, can be positively identified with hadron progenitors. In the early days it was widely believed that γ-ray astronomy would finally solve the mystery of the origin of the cosmic radiation. However with the exception of the galactic plane (and the Large Magellanic Cloud) where we observe not the source of cosmic radiation but its interaction during propagation, every one of the sources detected so far can be attributed to a source in which electrons are the progenitor particles. In no source is the much heralded "bump" in the energy spectrum near 70 MeV seen. In some cases there are proponents of plausible models in which hadrons are the progenitors but there are equally vociferous proponents who would advocate electron models and in many cases these seem the more plausible. Thus in the 40 plus years since the publication of Morrison's seminal paper [62] while we have learnt some interesting astrophysics we have come no closer to a definitive model of cosmic-ray origins.

IX   ACKNOWLEDGEMENTS

Research in VHE γ-ray astronomy at the Smithsonian Astrophysical Observatory is supported by a grant from the U.S. Department of Energy. Mike Catanese and Vladimir Vassiliev read the manuscript and supplied many critical comments.

REFERENCES

1. Ong, R. A. 1998, Physics Reports, 305, 93
2. Catanese, M., Weekes, T.C. 1998, Publ. Ast. Soc. Pac. 111, 1193.
3. Fazio, G.G. et al. 1972, Ap.J.Lett., 175, L117
4. Bhat, C.L. 1997, Proc. 25th ICRC (Durban), 8, 211
5. Vacanti, G., et al. 1991, Ap.J., 377, 467
6. Weekes, T.C., et al. 1989, Ap.J., 342, 379
7. Hillas, A. M., et al. 1998, Ap.J., 503, 744
8. Konopelko, A. et al. 1999, 26th ICRC (Salt Lake City), OG 2.2.01, 3, 444
9. Musquere, A. et al. 1999, 26th ICRC (Salt Lake City), OG 2.2.05, 3, 460
10. Tanimori, T., et al. 1998b, Ap.J.Lett., 492, L33
11. Amenomori, M. et al. 1999, 26th ICRC (Salt Lake City), OG 2.2.04, 3, 456
12. Oser, S., 1999, 26th ICRC (Salt Lake City), 3, 464
13. Musquere, A. et al. 1999, 26th ICRC (Salt Lake City), 3, 527
14. Kifune, T., et al. 1995, Ap.J.Lett., 438, L91
15. Chadwick, P. M., et al. 1997, Proc. 25th ICRC (Durban), 3, 189
16. Tanimori, T., et al. 1998a, Ap.J.Lett., 497, L25
17. Koyama, M., et al. 1995, Nature, 378, 255
18. Yoshikoshi, T., et al. 1997, Ap.J.Lett., 487, L65
19. Muraisi, H. et al. 1999, 26th ICRC (Salt Lake City), OG 2.2.20, 3, 500
20. Koyama, K. et al. 1997, PASJ, 49, L7.
21. Chudakov, A.E. et al. 1965, Transl. Cons.Bur., Lebedev Phys. Inst. 26, 99
22. Puelhlhofer, G. et al. 1999, 26th ICRC (Salt Lake City), OG 2.2.17, 3, 492
23. Allen, G. E., et al. 1995, Ap.J.Lett., 487, L97
24. Goret, P. et al. 1999, 26th ICRC (Salt Lake City), OG 2.2.18, 3, 49
25. Lessard, R.W. et al. 1999, 26th ICRC (Salt Lake City), OG 2.2.16, 3, 488
26. Chadwick, P.M. et al. 1999, 26th ICRC (Salt Lake City), OG 2.4.9, 4, 72
27. Carraminana, A., et al. 1989, "Timing Neutron Stars", Kluwer Acad. Press, 389
28. Raubenheimer, B.C. et al. 1989, Ap.J., 336, 349
29. Chadwick, P.M. et al. 1998, Ap.J., 503, 391
30. Vestrand, et al. 1997, Ap.J.Lett., 483, L49
31. Punch, M., et al. 1992, Nature, 358, 477
32. Gaidos, J. A., et al. 1996, Nature, 383, 319

33. Takahashi, T., et al., 1999, Astropart. Phys., 11, 177.
34. Maraschi, L., et al., 1999, Astropart. Phys., 11, 189
35. Catanese, M. et al. 1999, 26th ICRC (Salt Lake City), OG 2.1.03, 3, 305
36. Piron, F. et al. 1999, 26th ICRC (Salt Lake City), OG 2.1.09, 3, 326
37. Aharonian, F. A., et al. 1999c, Astron.Astropys., submitted (astro-ph/9905032)
38. Kajino, F. et al. 1999, 26th ICRC (Salt Lake City), OG 2.1.21, 3, 370
39. Krennrich, F. et al. 1999, 26th ICRC (Salt Lake City), OG 2.1.02, 3, 301
40. Quinn, J., et al. 1996, Ap.J.Lett., 456, L83
41. Protheroe, R. J., et al., 1997, Proc. 25th ICRC (Durban), 8, 317
42. Quinn, J., et al. 1999, Ap.J., in press
43. Aharonian, F.A. et al. 1999, Astron.Astrophys., 342, 69
44. Punch, M. et al. 1997, 25th ICRC (Durban), 3, 253
45. Hayashida, N. et al. 1998, Ap.J., 504, L71
46. Aharonian, F. A. et al. 1999, 26th ICRC (Salt Lake City), OG 2.1.16, 3, 350.
47. Tavernet, J.P. et al. 1997, 26th ICRC (Salt Lake City), OG 2.1.08, 3, 322
48. Catanese, M., et al. 1998, Ap.J., 501, 616
49. Ghisellini, G., et al. 1998, Mon.Not.Roy.Ast.Soc., 301, 451
50. Giommi, M. 1999, preprint.
51. Catanese, M., et al. 1998, Ap.J., 501, 616
52. Konopelko, A. et al. 1999, 26th ICRC (Salt Lake City), OG 2.1.38, 3, 426
53. Chadwick, P.M. et al. 1999, 26th ICRC (Salt Lake City), OG 2.1.13, 3, 338
54. Chadwick, P.M. et al. 1999, Ap.J., 513, 161
55. Catanese, M. et al. 1997, 25th ICRC (Durban), 3, 277.
56. Neshpor, Y. I., et al. 1998, Astron. Letts., 24, 134
57. Kerrick, A. D., et al. 1995b, Ap.J., 452, 588
58. Nishikawa, D. et al. 1999, 26th ICRC (Salt Lake City), OG 2.1.17, 3, 354
59. Kranich, D. et al. 1999, 26th ICRC (Salt Lake City), OG 2.1.18, 3, 358
60. Hartman, R.C et al. 1999, Ap.J.Suppl., 123, 79
61. Esposito, J.A. et al. 1996, Ap.J., 461, 820
62. Morrison, P. 1958, Il Nuovo Cimento, 7, 858.

BLAZARS AND BL LACS

Understanding Blazar Jets Through Their Multifrequency Emission

Rita M. Sambruna*

*Pennsylvania State University
Dept. of Astronomy & Astrophys
525 Davey Lab, University Park, PA 16802
(rms@astro.psu.edu)

Abstract.
Being dominated by non-thermal (synchrotron and inverse Compton) emission from a relativistic jet, blazars offer important clues to the structure and radiative processes in extragalactic jets. Crucial information is provided by blazars' spectral energy distributions from radio to gamma-rays (GeV and TeV energies), their trends with bolometric luminosity, and their correlated variability properties. This review is focussed on recent multiwavelength monitorings of confirmed and candidate TeV blazars and the constraints they provide for the radiative properties of the emitting particles. I also present recent observations of the newly discovered class of "blue quasars" and the implications for current blazars' unification schemes.

THE BLAZAR FAMILY

Blazars are radio-loud Active Galactic Nuclei (AGNs) with polarized, luminous, and rapidly variable non-thermal continuum emission, extending from radio to gamma-rays (GeV and TeV energies), from a relativistic jet oriented close to the line of sight. As such, they are rare laboratories to study the physics and structure of extragalactic jets, present in all radio-loud AGNs [1].

Strong clues are provided by blazars' spectral energy distributions (SEDs). These are typically double-humped (Fig. 1; from [2,3]), with the first component peaking at IR/optical wavelengths in "red blazars" (also called Low-energy peaked blazars, LBLs) and at UV/X-rays in "blue blazars" (or HBLs, High-energy peaked blazars) [1]. Its rapid variability and high polarization leave little doubt that it is due to synchrotron emission from relativistic electrons in the jet. The second spectral

---

[1]) A practical way to discriminate between LBLs and HBLs is though their radio-to-X-ray spectral indices, α_{rx}, with $\alpha_{rx} > 0.8$ in LBLs and $\alpha_{rx} < 0.8$ in HBLs [4]. Flat Spectrum Radio Quasars (FSRQs) have SEDs similar to LBLs, but are more luminous and have stronger optical emission lines [5].

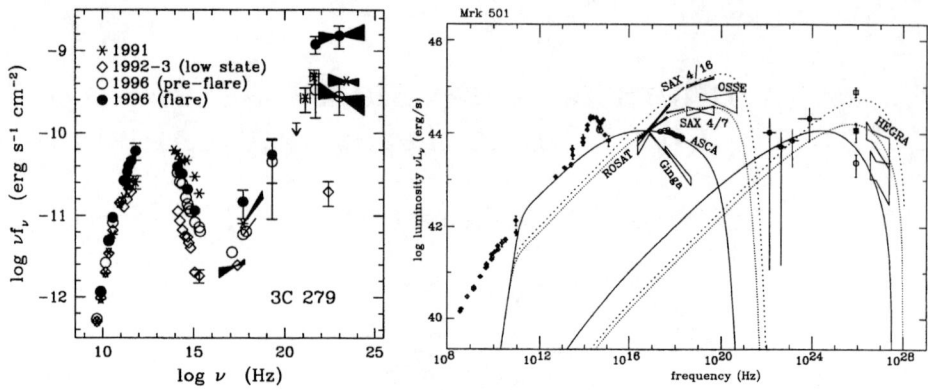

FIGURE 1. Spectral energy distributions (SEDs) of the red [Left, (a)] and blue [Right, (b):] blazars 3C 279 and Mrk 501, respectively (from [2,3]). Blazars' SEDs typically have two broad humps, the first peaking anywhere from IR/optical (in red blazars like 3C279) to hard X-rays (in blue blazars like Mrk 501) and due to synchrotron emission from a relativistic jet. The second component, extending to gamma-rays, is less well understood. A popular explanation is inverse Compton scattering of ambient seed photons off the jet's electrons. The largest variability amplitudes are observed above the peaks in both sources.

component extends from X-rays to gamma-rays, and its origin is less well understood. In the leptonic models, it could be due to inverse Compton (IC) scattering off the electrons of photons either internal or external to the jet (synchrotron-self Compton, SSC and external Compton, EC, respectively; see, e.g., [6]). Here I will assume the leptonic models, but acknowledge that an alternative is provided by the hadronic scenario (e.g., [7]).

Red and blue blazars are just the extrema of a continuous distribution of SEDs [8,9]. Indeed, deep multicolor surveys [10,11] are finding an increasing number of sources with intermediate SED shapes, and new trends with jet bolometric luminosity are discovered [8,12]. Specifically, the lower-luminosity blue blazars have higher synchrotron and IC peak frequencies, lower ratios of the IC to synchrotron peak fluxes, and weaker or absent optical emission lines than their more luminous red counterparts.

A possible interpretation is that the different SEDs are due to different predominant electrons' cooling mechanisms [13]. In a homogeneous scenario, the synchrotron peak frequency $\nu_{peak} \propto \gamma_{peak}^2$, where γ_{peak} is the electron energy determined by the competition between acceleration and cooling. Because of the lower energy densities, in line-less blue blazars the balance between heating and cooling is achieved at larger γ_{peak}. In contrast, in red blazars the electrons are more efficiently cooled due to the additional EC component and reach a lower final γ_{peak}. The emerging scenario is that blue blazars are SSC-dominated, while the EC mechanism dominates the production of gamma-rays in red blazars. While there are a few caveats to this picture [14], the clear message is that the spectral diversity of blazars' jets cannot be explained by beaming/orientation effects *only*, but require instead a change of physical parameters and/or a different jet environment [8,15].

PROBING BLAZARS' PARADIGM: MULTIWAVELENGTH VARIABILITY OF TEV BLAZARS

Correlated multiwavelength variability provides a way to test the cooling paradigm since the various synchrotron and IC models make different predictions for the relative flare amplitudes and shape, and the time lags. Since the same population of electrons is responsible for emitting both spectral components (in a homogeneous scenario), correlated variability of the fluxes at the low- and high-energy peaks with no lags is expected (see [16] and references therein). In blue blazars, the TeV emission is largely due to lower-energy electrons scattering low-energy (IR) photons, with the higher-energy electrons ($\gamma > \gamma_{peak}$) producing the harder TeV photons [17]. The X-ray light curve should track the TeV one, with relative amplitude for the flares at the two energies in a linear relationship [18]. Thus, *TeV blazars probe the spectrum of the emitting particles*. This is amply demonstrated by the case of Mrk 421 and Mrk 501, the two brightest and best-studied TeV blazars.

Mrk 421

Mrk 421 was extensively studied during multifrequency campaigns conducted in 1994, 1995, and 1998. The early monitorings with ASCA, Whipple, EUVE, and ground-based telescopes established that the X-ray and TeV emission is well correlated on longer ($\sim$ days) timescales, with amplitude generally decreasing with increasing wavelength [19], although within a rather sparse sampling. An intensive campaign, continuous over a period of seven days, was performed in 1998 April, involving ASCA, SAX, EUVE, and various TeV telescopes [20]. The ASCA and TeV light curves are shown in Fig. 2a. Complex X-ray flux and spectral variability is detected by ASCA, with short ($\sim$ 0.5 day) flares superposed on a longer trend and intra-day variations. The TeV light curve, disrupted by unfortunate episodes of bad weather, tracks the general trend observed at X-rays.

A few days before the start of the ASCA observations and partly overlapping with it, Mrk 421 was observed with Whipple and SAX, with a strong flare observed at both wavelengths (Fig. 2b). The new and exciting result is the first detection of X-ray/TeV correlated variability on timescales of *hours* [21], strongly supporting the idea that the same electron population is responsible for emitting the X-rays

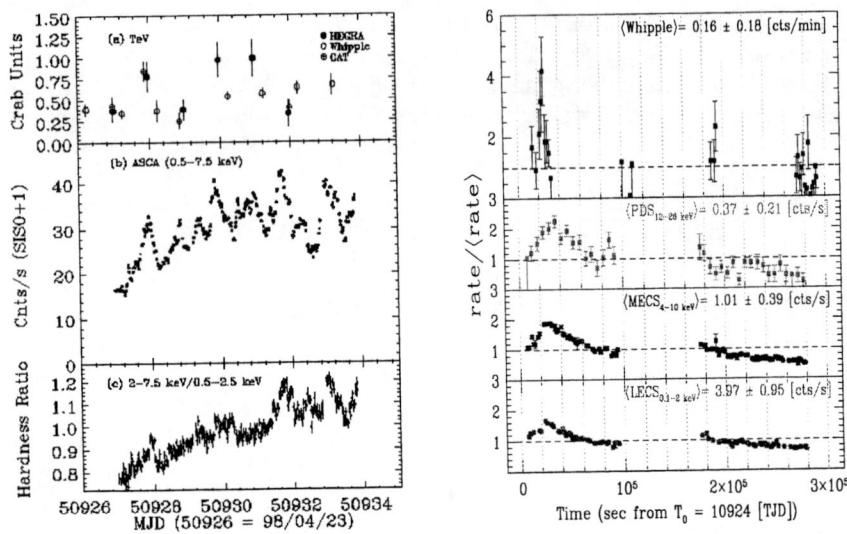

FIGURE 2. Multiwavelength observations of Mrk 421 in 1998 April. [Left, (a):] Simultaneous TeV and X-ray continuous monitoring over a period of seven days . Reprinted from [20], © 1999, with permission from Elsevier Science. Despite the gaps in the sampling, the TeV light curve tracks the X-ray variations over longer (∼ days) timescales.

Complex flux and spectral variations are observed at X-rays, with several short flares of ∼ one day superposed over a longer trend and intra-day variations. [Right, (b):] Whipple and SAX observations of a TeV/X-ray flare at the beginning of the 1998 April monitoring [21]. The curves are binned at 28 minutes. Correlated X-ray and TeV flux changes on timescales of hours are detected, strengthening the idea that the same electron population is responsible for emitting the X-rays via synchrotron and the TeV photons via inverse Compton.

via synchrotron and the TeV photons via IC. Note the different decay times of the flare, much faster at TeV than at X-rays (Fig. 2b), difficult to explain in the context of a simple homogeneous model [21].

Mrk 501

Mrk 501 attracted much attention in 1997 April when it underwent a spectacular flare at TeV energies [22–24], well correlated with a similarly-structured X-ray flare detected by RXTE. No delays longer than one day were detected between the X-ray and TeV emission [25]. A remarkable spectral behavior was observed in the X-rays (Fig. 1b), where an unusually flat (photon index $\Gamma_X \sim 1.8$) X-ray continuum was measured by SAX and RXTE during the TeV flare [26,25]. This implies a shift of the synchrotron peak toward higher energies by more than two orders of magnitude, almost certainly reflecting a large increase of the electron energy [26], or the injection of a new electron population on top a quiescent one [3]. Later RXTE

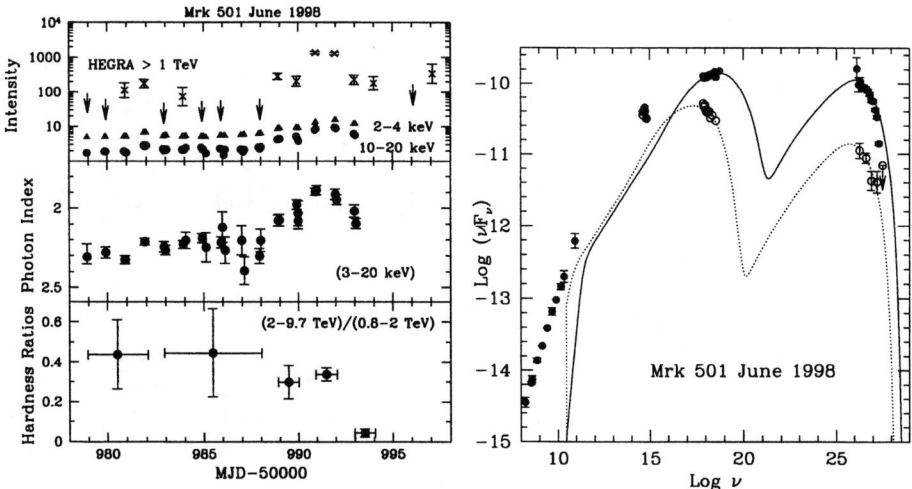

FIGURE 3. Multiwavelength campaign of Mrk 501 in 1998 June with RXTE and HEGRA [29]. [Left, (a):] TeV and X-ray light curves in different energy ranges (top panel). A large-amplitude flare is detected at TeV energies, well-correlated to a shorter-amplitude flare in the X-rays. Large X-ray spectral variations are observed (middle panel), while the TeV hardness ratios stay constant during the flare and soften one day after (bottom panel). [Right, (b):] Spectral energy distribution during the flare (filled circles) and during quiescence (open circles), fitted with the SSC model (solid and dotted lines; see [29] for details). The large spectral variability at X-rays implies a shift of the synchrotron peak at ≥ 50 keV in ~ 2 days, faster than in 1997. This indicates a similar energizing mechanism at both epochs operating on different timescales.

observations in 1997 July found the source still in a high and hard X-ray state [27], indicating a persistent energizing mechanism. A similarly flat X-ray continuum was observed in another weaker TeV blazar, 1ES2344+514 [28].

An interesting new behavior was observed during our latest 2-week RXTE and HEGRA monitoring of Mrk 501 in 1998 June [29], when 100% overlap between the X-rays and TeV light curves was achieved (Fig. 3a). A strong (factor 20 or more), short-lived ($\sim$ two days) TeV flare is detected, well correlated to a lower-amplitude, broader flare in the X-rays. As in 1997 April, large X-ray spectral variations are observed in 1998 June, with the X-ray continuum flattening to $\Gamma_X = 1.9$ at the peak of the TeV flare. This implies a similar shift to ≥ 50 keV of the synchrotron peak, but on much faster timescales (Fig. 3b). However, while in 1997 April the TeV spectrum hardened during the flare [24], as it did in the X-rays, in 1998 June the TeV hardness ratios stayed relatively constant during the flare and softened one day later (Fig. 3a). The correspondence between the X-ray and TeV spectra is no longer present during the 1998 June flare.

Intra-hour TeV variability is also detected, with a doubling timescale for the

TeV flux of ∼ 30 min [29]. No correlation with the X-ray light curve on such short timescales was possible, due to unfortunate gaps in the RXTE sampling. The short TeV variability timescale implies a size of the emitting region $R \leq 5 \times 10^{14}$ cm and a Doppler factor of the emitting plasma $\delta \geq 10$, similar to Mrk 421 [30].

ACCELERATION AND COOLING IN BLAZARS JETS

Multiwavelength correlated variability of the synchrotron emission provides tight clues to the structure of the jet through the study of the energy-dependence of the flare and the accompanying spectral variations. The rise and decay times of the synchrotron flux depend on a few source typical timescales [31], while the accompany spectral variability is a strong diagnostic of the electron acceleration versus cooling processes [32], with characteristic patterns predicted for the hardness ratios as a function of total intensity depending on the timescales of the two processes. When cooling dominates, the radiative time can be approximated by the lag between the shorter and longer synchrotron wavelengths, providing an estimate of the magnetic field B of the source via $t_{lag} \sim t_{cool} \propto E^{-0.5} \delta^{-0.5} B^{-1.5}$, where E is in keV, B in Gauss, and δ is the Doppler factor of the emitting plasma [19,33].

Two excellent laboratories to study the energy propagation of the synchrotron flare are PKS 2155−304 and PKS 2005−489, since they are bright and rapidly variable at all observed wavelengths from optical to X-rays. They were the targets of recent monitorings, as described below.

Multiwavelength observations of PKS 2155−304

PKS 2155−304 was detected at TeV energies by the Durham group [34] in 1997 November, during a period of intense X-ray activity [35,36], although the source was not bright enough to allow a detailed TeV light curve. The synchrotron peak in the SED, usually in the EUV/soft X-ray energy range, shifted forward one order of magnitude [37], indicating a more modest acceleration event than in Mrk 501. Because of its intermediate SED, the correlated TeV and X-ray variability properties of PKS 2155−304 could be different than the Markarian objects, and this source qualifies as a high-priority candidate for future X-ray/TeV monitoring campaigns.

In 1991 November, PKS 2155−304 was observed with 4.5-day continuous monitoring from optical to UV and X-rays. It exhibited small-amplitude (10%), energy-*independent* variability, with well-correlated flares from optical to X-rays on timescales of a few hours and the shorter wavelengths leading the longer ones [38]. However, in a subsequent campaign in 1994 May, PKS 2155−304 showed a substantially different behavior. The source was observed continuously for ∼ 10 days in UV and EUV, and for 2 days in the X-rays and optical [33], exhibiting energy-*dependent* variations. A well-defined X-ray flare was observed, followed by broader, lower-amplitude flares at EUV and UV by ∼ 1 and 2 days, respectively. In X-rays, the harder energies lead the softer ones by one hour [39,40], implying a

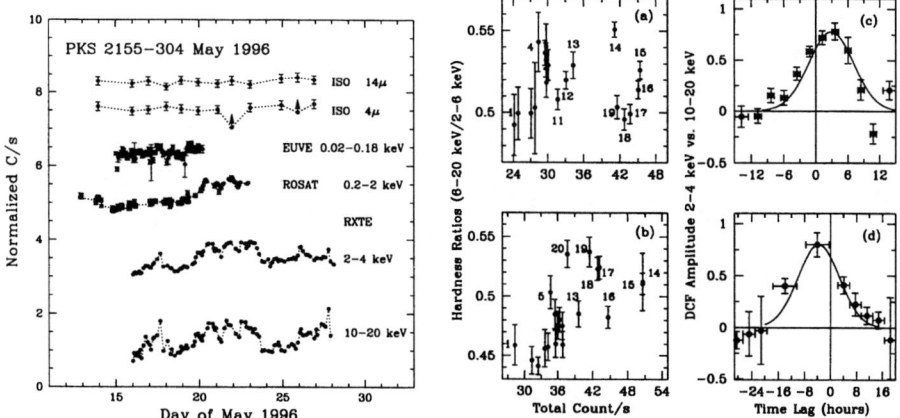

FIGURE 4. Multiwavelength monitoring of PKS 2155−304 in 1996 May [41]. [Left, (a):] Multiwavelength light curves, showing the excellent sampling at X-rays and the poorer coverage at the longer wavelengths. Complex flux and spectral variability was detected in the X-rays, with energy-dependent amplitude and different flares exhibiting different spectral behaviors in the hardness ratio versus flux diagram. Hysteresis loops of both "clockwise" and "anti-clockwise" signs are observed. [Middle, (b):] Two examples of clockwise (upper panel) and anti-clockwise (lower panel) loops for the May 18.5–20.2 and 24.2–26.9 flares, respectively. [Right, (c):] Discrete Correlation Function applied to the same flares. Soft and hard lags are detected for the clockwise and anti-clockwise loops, respectively, of ∼ a few hours. This complex spectral behavior is a powerful diagnostic of the acceleration and cooling processes in the jet.

magnetic field of $B \sim 0.1$ Gauss for $\delta = 10$, similar to Mrk 421 [19,33]. This apparent progression of the X-ray flare to longer wavelengths in 1994 May was explained well by an acceleration (or equivalent) event in the context of synchrotron radiation, with the time delay reflecting either synchrotron loss timescales or physical inhomogeneities of the emission region [33]. However, the quantitatively different behaviors between the campaigns in 1991 and 1994 show that the variability properties of blazars are complex, involving different modes and likely reflecting an underlying complexity in the jet structure and/or in the emission mechanisms.

PKS 2155−304 was monitored again in 1996 May from IR to X-rays (Fig. 4a; from [41]). In the X-rays excellent sampling was achieved with RXTE, and a complex flux and spectral behavior was observed, with short (∼ 1-2 days), energy-dependent flares superposed to a longer-timescale trend. Inspection of the X-ray hardness ratios versus flux shows that the individual flares exhibit different spectral variability patterns. Hysteresis loops of opposite signs, both in a "clockwise" (C) and "anti-clockwise" (A) sense, are observed (Fig. 4b,c). Analysis of individual flares with various correlation methods shows that C loops correspond to soft lags (hard energies varying first) while A loops correspond to hard lags (soft energies

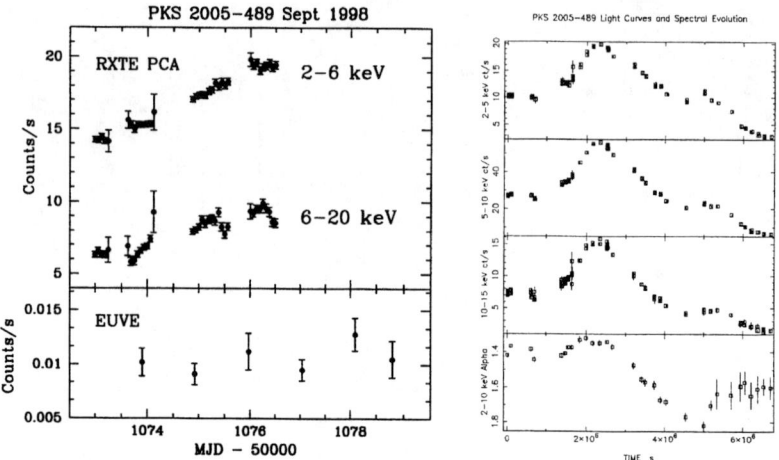

FIGURE 5. Multiwavelength observations of the TeV candidate PKS 2005–489. [Left, (a):] Our RXTE and EUVE monitoring in 1998 September. At EUV energies, the source was too faint and no variability is observed in the daily-binned light curve. Despite the gaps in the RXTE sampling, energy-dependent variability is apparent at X-rays, with a general flux increase of 30% or more in amplitude. The harder energies vary before the softer ones, consistent with cooling dominating the flux variability. [Right, (b):] One month later, PKS 2005–489 underwent a strong, long-lasting X-ray flare which was well sampled by RXTE, exhibiting spectral variations on timescales of hours [45], with similar variability patterns as in 1998 September.

varying first).

This behavior is consistent with a model where energetic electrons are injected in the source via a shock and escape into the emission region, where they cool [32]. When acceleration is faster than cooling, the latter dominates variability and, because of its energy dependence, the harder energies are emitted first, with C loops/soft lags observed. If instead the acceleration is slower ($t_{acc} \sim t_{cool}$), the electrons need to work their way up in energy and the softer energies are emitted first, yielding A loops and hard lags. For the first time, we are observing electron acceleration, which, together with cooling, is responsible for the observed X-ray variability properties of PKS 2155–304 in 1996 May.

The TeV candidate PKS 2005–489

PKS 2005–489 qualifies as a TeV candidate because of its SED, similar to Mrk 421, and its proximity (z=0.071) [42,43]. Observations by the CANGAROO group yielded only an upper limit to the TeV flux [44]. We monitored this source in 1998 September with RXTE and EUVE, to study the energy-dependence of the

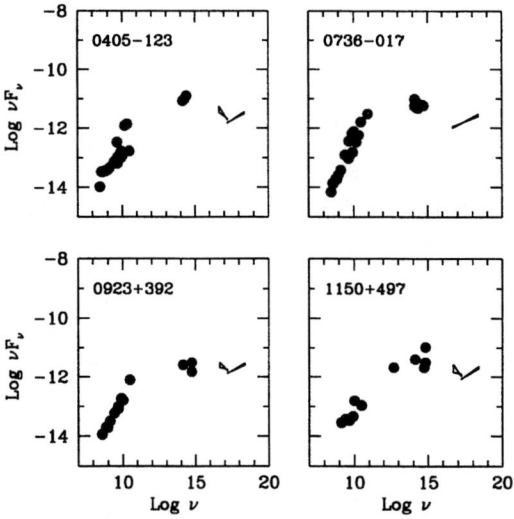

FIGURE 6. Spectral energy distributions of four blue quasars, from our recent ASCA observations and archival ROSAT and longer-wavelength data [48]. The ASCA continua are flat, implying an upturn at energies > 2 keV. The nature of the optical-to-soft X-ray emission is not well constrained, and both a thermal (from the accretion disk) and non-thermal (synchrotron emission from the jet) origin is possible. It is important to discriminate between these two scenarios with future observations, since a non-thermal origin would represent a challenge for current blazar unification schemes.

synchrotron flares. Unfortunately, the source was faint at EUV wavelengths, with a count rate of ~ 0.01 c/s; no variability is observed in the daily-binned light curve (Fig. 5a). The X-ray light curves in a soft (2–6 keV) and hard (6–20 keV) energy band are shown in Fig. 5a. Despite the gaps, due to the RXTE observational constraints, it is apparent that the variability at hard X-rays is faster than at soft X-rays, consistent with cooling dominating the synchrotron flux changes. This is confirmed by the analysis of the hardness ratios versus flux, where only clockwise loops are observed. A similar behavior was also observed one month later during a larger-amplitude, longer-lasting flare, when spectral variations occurred on timescales of a few hours [45] (Fig. 5b).

BLUE QUASARS: UNDERMINING THE BLAZAR PARADIGM?

A simple and yet powerful probe of the blazar paradigm described above is provided by the X-ray spectra of the various blazar classes. In blue blazars, where the synchrotron peak falls at high frequencies, the X-rays are dominated by the high-energy tail of the synchrotron component and their X-ray continua should thus be steep and convex, as a result of radiative losses. On the other end of the spectral sequence, FSRQs are dominated in the X-rays by the emerging Compton component and their X-ray continua should be flatter. This is confirmed by observations of large samples of sources with ROSAT and ASCA [9,46], which yield flat (photon index $\Gamma_X \sim 1.5$) X-ray continua for FSRQs, steep ($\Gamma_X \sim 2.5$) and downward-curved continua for HBLs, and intermediate slopes for LBLs ($\Gamma_X \sim 2.0$).

It was thus surprising when a sub-group of FSRQs, with otherwise "canonical" properties, was observed to have unusually steep ROSAT spectra, $\Gamma_{0.1-2.4~keV} > 2$, similar to HBLs [9,47]. These objects were dubbed "blue quasars" to indicate that they could be the quasar counterparts of HBLs, contrary to the predictions of the blazar paradigm which purports that blue blazars are essentially line-less. Recent deep multicolor surveys, including our own [10], are finding an increasing number of these sources, which now amount to a non-negligible fraction of the total blazar population.

What is the true nature of blue quasars? A first simple test is to measure their hard X-ray continua. If synchrotron dominates the optical through X-ray emission, as in HBLs, the X-ray emission above 2 keV should be as steep or even steeper than at softer energies. We performed exploratory ASCA observations of four blue quasars from our ROSAT sample [9], selected because relatively nearby ($z < 1$) and *not* yet detected at gamma-rays, thus avoiding an *a priori* bias toward flat X-ray slopes. We find that their ASCA spectra are consistent with flat X-ray continua above 2 keV, $\Gamma_{2-10~keV} \sim 1.5$, similar to the "canonical" FSRQs of the red type and implying an upturn in the SEDs (Fig. 6), most likely the onset of the Compton tail [48].

In Fig. 6, the nature of the emission below 1 keV is not well constrained, and both a thermal and non-thermal origin is possible. To this regard, blue quasars could be similar to 3C 273 and 3C 345, two quasar-like blazars where a thermal "blue bump" from the accretion disk is notoriously present, intruding into the soft X-rays (see [9] and references therein). It is thus entirely plausible that the large α_{rx} observed in blue blazars (or at least some of them) is due to a strong thermal contribution from the accretion disk, as in 3C 273 and 3C 345. Future simultaneous optical-UV-X-ray spectra will have the potential to better constrain the nature of blue quasars and their role in the blazar family.

CONCLUSIONS AND FUTURE PROSPECTS

Recent multiwavelength campaigns of blazars expanded the currently available database, from which we are learning important new lessons. Detailed modeling of the SEDs of bright gamma-ray blazars of the red and blue types tend to support the current cooling paradigm, where the different blazars flavors are related to the predominant cooling mechanisms of the electrons at the higher energies (EC in more luminous red sources, SSC in lower-luminosity blue ones). Future larger statistical samples are needed, to fully address the observational biases, especially in gamma-rays. In particular, it will be important to expand the sample of TeV blazars, which currently includes only a handful of sources.

Correlated multiwavelength variability is the key to understanding the structure of blazars' jets. In TeV blazars, the X-rays are well correlated to the TeV emission down to timescales of days and hours (in Mrk 421), supporting a model where the same electrons are responsible for the emission at both wavelengths. It will be important to determine the shortest timescales on which this correlation holds, to pin down the electron energy distribution and the location of the emitting region(s) in the jet. This awaits well-sampled gamma-ray light curves, which will be afforded by the next higher-sensitivity missions (HESS, VERITAS, MAGIC, CANGAROO II at TeV and GLAST at GeV). Broader-band, higher quality gamma-ray spectra will also be available, allowing a more precise location of the Compton peak.

An outstanding still unanswered question is the jet composition (electrons/positrons versus protons). Single-epoch SEDs of blazars are adequately modeled by both the leptonic and hadronic models, with different tuning of the parameters. However, while the leptonic models make specific predictions for the correlated variability properties, more extensive modeling is currently needed in the context of the proton jet models. A first effort was presented at this meeting [7].

Finally, coordinated X-ray and longer-wavelength observations of the synchrotron component in blue blazars strongly suggests that acceleration, cooling, and escape are the dominant mechanisms responsible for the observed variability properties. The case of PKS 2155-304 shows that different flaring modes could be present in a single source, stressing the importance of multi-epoch monitorings to obtain a complete picture of the physical processes in blazar jets.

This work was funded through NASA contract NAS-38252 and NASA grant NAG5-7276. I thank Felix Aharonian and the HEGRA team for allowing me to show the 1998 TeV data of Mrk 501, Eric Feigelson for comments, and Lester Chou for help with the RXTE data reduction.

REFERENCES

1. Urry, C.M. and Padovani, P., *PASP* **10**, 803 (1995).
2. Wehrle, A. et al. *ApJ* **497**, 178 (1998).
3. Kataoka, J. et al., *ApJ* **514**, 138 (1999).
4. Padovani, P. and Giommi, P., *ApJ* **444**, 567 (1995).
5. Scarpa, R. and Falomo, R., *A&A* **325**, 109 (1997).
6. Böttcher, M. (1999), this volume.
7. Rachen, J. (1999), this volume.
8. Sambruna, R.M., Maraschi, L., and Urry, C.M., *ApJ* **463**, 444 (1996).

9. Sambruna, R.M. *ApJ* **487**, 536 (1997).
10. Perlman, E.S. (1999), this volume.
11. Laurent-Muehleisen, S. et al., *ApJS* **118**, 127 (1998).
12. Fossati, G. et al., *MNRAS* **299**, 433 (1998).
13. Ghisellini, G. et al., *MNRAS* **301**, 451 (1998).
14. Urry, C.M., *Astroparticle Physics* **11**, 159 (1999).
15. Georganopoulos, M. and Marscher, A., *ApJ* **506**, 621 (1998).
16. Sambruna, R.M., "Coordinated RXTE and multiwavelength observations of blazars", in *Proceedings of the Frascati '99 Workshop on Multifrequency behavior of high-energy cosmic sources*, May 24–29, Vulcano, Italy, MEMSAIt (1999) in press.
17. Tavecchio, F., Maraschi, L., & Ghisellini, G., *ApJ* **509**, 608 (1998).
18. Coppi, P. and Aharonian, F., *ApJ* **521**, L33 (1999).
19. Takahashi, T. et al., *ApJ* **470**, L89 (1996).
20. Takahashi, T. et al., *Astroparticle Physics* **11**, 177 (1999).
21. Maraschi, L. et al., *ApJ Letters* (1999), in press.
22. Catanese, M. et al., *ApJ* **487**, 143 (1997).
23. Aharonian, F. et al., *A&A* **342**, 69 (1999).
24. Djannati-Atai, A. et al., *A&A* (1999) in press (astro-ph/9906060).
25. Krawczynski, H. et al., *A&A* (1999), in press.
26. Pian, E. et al. *ApJ* **486**, 770 (1998).
27. Lamer, G. and Wagner, S. *A&AL* **331**, 13 (1998).
28. Giommi, P., Padovani, P. and Perlman, E.S., *MNRAS*, in press (1999) (astro-ph/9907377).
29. Sambruna, R.M. et al., *ApJ Letters* (1999), subm.
30. Gaidos, J.A. et al., *Nature* **383**, 319 (1996).
31. Chiaberge, M. and Ghisellini, G., *MNRAS* **306**, 551 (1999).
32. Kirk, J.G., Riegler, F.M., and Mastichiadis, A., *A&A* **333**, 452 (1998).
33. Urry, C.M. et al. *ApJ* **486**, 799 (1997).
34. Chadwick, P. et al., *ApJ* **513**, 161 (1999).
35. Chiappetti, L. et al., *ApJ* **521**, 552 (1999).
36. Chadwick, P. et al. (1999), this volume.
37. Bertone, E. et al. (1999), in prep.
38. Edelson, R.A. et al., *ApJ* **438**, 120 (1995).
39. Kataoka, J. et al., *ApJ* (1999), in press.
40. Zhang, Y.H. et al. *ApJ* (1999), in press (astro-ph/9907325).
41. Sambruna, R.M. et al., (1999) in prep.
42. Sambruna, R.M. et al., *ApJ* **449**, 567 (1995).
43. Stecker, F.A., De Jager, O.C., and Salamon, M.H., *ApJ* **473**, L75 (1996).
44. Roberts, M.D. et al., *A&A* **343**, 691 (1999).
45. Perlman, E.S. et al., *ApJ* **523**, L11 (1999).
46. Kubo, H. et al., *ApJ* **504**, 693 (1998).
47. Padovani, P., Giommi, P., and Fiore, F., *MNRAS* **284**, 569 (1997).
48. Sambruna, R.M., Chou, L., and Urry, C.M., *ApJ* (1999), subm.

Leptonic Jet Models of Blazars: Broadband Spectra and Spectral Variability

Markus Böttcher*

*Space Physics and Astronomy Department
Rice University, MS 108
6100 S. Main Street
Houston, TX 77005 - 1892
USA

Abstract. The current status of leptonic jet models for gamma-ray blazars is reviewed. Differences between the quasar and BL-Lac subclasses of blazars may be understood in terms of the dominance of different radiation mechanisms in the gamma-ray regime. Spectral variability patterns of different blazar subclasses appear to be significantly different and require different intrinsic mechanisms causing gamma-ray flares. As examples, recent results of long-term multiwavelength monitoring of PKS 0528+134 and Mrk 501 are interpreted in the framework of leptonic jet models. A simple quasi-analytic toy model for broadband spectral variability of blazars is presented.

INTRODUCTION

Recent high-energy detections and simultaneous broadband observations of blazars, determining their spectra and spectral variability, are posing strong constraints on currently popular jet models of blazars. 66 blazars have been detected by EGRET at energies above 100 MeV [1], the two nearby high-frequency peaked BL Lac objects (HBLs) Mrk 421 and Mrk 501 are now multiply confirmed sources of multi-GeV – TeV radiation [2–5], and the TeV detections of PKS 2155-314 [6] and 1ES 2344+514 [7] are awaiting confirmation. Most EGRET-detected blazars exhibit rapid variability [8], in some cases on intraday and even sub-hour (e. g., [9]) timescales, where generally the most rapid variations are observed at the highest photon frequencies.

The broadband spectra of blazars consist of at least two clearly distinct spectral components. The first one extends in the case of flat-spectrum radio quasars (FSRQs) from radio to optical/UV frequencies, in the case of HBLs up to soft and even hard X-rays, and is consistent with non-thermal synchrotron radiation from ultrarelativistic electrons. The second spectral component emerges at γ-ray energies and peaks at several MeV – a few GeV in most quasars, while in the case of some HBLs the peak of this component appears to be located at TeV energies.

The bolometric luminosity of EGRET-detected quasars and some low-frequency peaked BL Lac objects (LBLs) during flares is dominated by the γ-ray emission. If this emission were isotropic, it would correspond to enormous luminosities (up to $\sim 10^{49}$ erg s^{-1}) which, in combination with the short observed timescales (implying a small size of the emission region) would lead to a strong modification of the emissivity spectra by $\gamma\gamma$ absorption, in contradiction to the observed smooth power-laws at EGRET energies. This has motivated the concept of relativistic beaming of radiation emitted by ultrarelativistic particles moving at relativistic bulk speed along a jet (for a review of these arguments, see [10]). While it is generally accepted that blazar emission originates in relativistic jets, the radiation mechanisms responsible for the observed γ radiation are still under debate. It is not clear yet whether in these jets protons are the primarily accelerated particles, which then produce the γ radiation via photo-pair and photo-pion production, followed by π^0 decay and synchrotron emission by secondary particles (e. g., [11]), or electrons (and positrons) are accelerated directly and produce γ-rays in Compton scattering interactions with the various target photon fields in the jet [12–15].

In this review, I will describe the current status of blazar models based on leptons (electrons and/or pairs; in the following, the term "electrons" refers to both electrons and positrons) as the primary constituents of the jet which are responsible for the γ-ray emission. Hadronic jet models are discussed in a separate paper by J. Rachen [16]. In Section 2, I will give a description of the model and discuss the different γ-ray production mechanisms and their relevance for different blazar classes. In Section 3, I will review recent progress in understanding intrinsic differences between different blazar classes and present state-of-the-art model calculations, using a leptonic jet model, to undermine the general theoretical concept. In Section 4, I will discuss broadband spectral variability of individual blazars and their interpretation in the framework of leptonic jet models. A simple quasi-analytical toy model for blazar broadband spectral variability will be presented in section 5.

MODEL DESCRIPTION AND RADIATION MECHANISMS

The basic geometry of leptonic blazar jet models is illustrated in Fig. 1. At the center of the AGN, an accretion disk around a supermassive, probably rotating, black hole is powering a relativistic jet. Along this pre-existing jet structure, occasionally blobs of ultrarelativistic electrons are ejected at relativistic bulk velocity.

The electrons are emitting synchrotron radiation, which will be observable at IR – UV or even X-ray frequencies, and hard X-rays and γ-rays via Compton scattering processes. Possible target photon fields for Compton scattering are the synchrotron photons produced within the jet (the SSC process, [12,17,18]), the UV – soft X-ray emission from the disk — either entering the jet directly (the

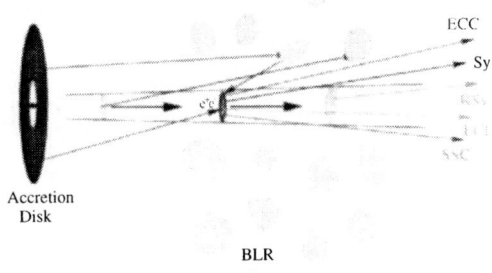

FIGURE 1. Illustration of the model geometry and the relevant γ radiation mechanisms for leptonic jet models.

ECD [External Comptonization of Direct disk radiation] process; [13,19]) or after reprocessing at the broad line regions or other circumnuclear material (the ECC [External Comptonization of radiation from Clouds] process; [14,20,21]), or jet synchrotron radiation reflected at the broad line regions (the RSy [Reflected Synchrotron] mechanism; [22–24]).

The relative importance of these components may be estimated by comparing the energy densities of the respective target photon fields. Denoting by u'_B the co-moving energy density of the magnetic field, the energy density of the synchrotron radiation field, governing the luminosity of the SSC component, may be estimated by $u'_{sy} \approx u'_B \tau_T \gamma_e^2$, where $\tau_T = n'_{e,B} R'_B \sigma_T$ is the Thomson depth of the relativistic plasma blob and γ_e is the average Lorentz factor of electrons in the blob. The SSC spectrum exhibits a broad hump without strong spectral break, peaking around $\langle \epsilon \rangle_{SSC} \approx (B'/B_{cr}) D \gamma_e^4 \approx \langle \epsilon \rangle_{sy} \gamma_e^2$, where B' is the co-moving magnetic field, $B_{cr} = 4.414 \cdot 10^{13}$ G, and $D = (\Gamma [1 - \beta_\Gamma \cos \theta_{obs}])^{-1}$ is the Doppler factor associated with the bulk motion of the blob. Throughout this paper, all photon energies are described by the dimensionless quantity $\epsilon = h\nu/(m_e c^2)$.

If the blob is sufficiently far from the central engine of the AGN so that the accretion disk can be approximated as a point source of photons, its photon energy density (in the co-moving frame) is $u'_D \approx L_D/(4\pi z^2 c \Gamma^2)$, where L_D is the accretion disk luminosity, and z is the height of the blob above the accretion disk. The ECD spectrum can exhibit a strong spectral break, depending on the existence of a low-energy cutoff in the electron distribution function, and peaks at $\langle \epsilon \rangle_{ECD} \approx \langle \epsilon \rangle_D (D/\Gamma) \gamma_e^2$, where $\langle \epsilon \rangle_D$ is the average photon energy of the accretion disk radiation (typically of order 10^{-5} for Shakura-Sunyaev type accretion disks [25] around black holes of $\sim 10^8 - 10^{10} M_\odot$).

Part of the accretion disk and the synchrotron radiation will be reprocessed by circumnuclear material in the broad line region and can re-enter the jet. Since

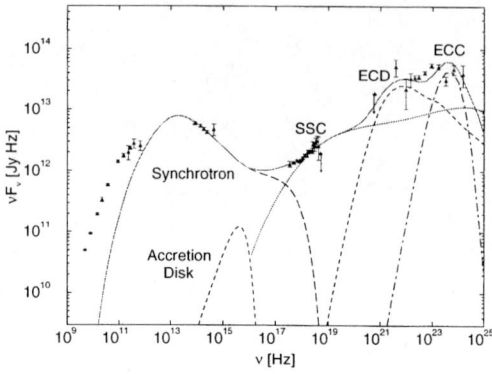

FIGURE 2. Fit to the simultaneous broadband spectrum of the FSRQ 3C279 during its very bright γ-ray flare in September 1991. See [29] for model parameters.

this reprocessed radiation is nearly isotropic in the rest-frame of the AGN, it will be strongly blue-shifted into the rest-frame of the relativistically moving plasma blob. Thus, assuming that a fraction a_{BLR} of the radiation is rescattered into the jet trajectory, we find for the energy density of rescattered accretion disk photons: $u'_{ECC} \approx L_D \, a_{BLR} \, \Gamma^2 / (4\pi \, \langle r \rangle^2_{BLR} \, c)$, where $\langle r \rangle_{BLR}$ is the average distance of the BLR material from the central black hole. The ECC photon spectrum peaks around $\langle \epsilon \rangle_{ECC} \approx \langle \epsilon \rangle_D \, D \, \Gamma \, \gamma_e^2 \approx \langle \epsilon \rangle_{ECD} \, \Gamma^2$.

For the synchrotron mirror mechanism, additional constraints due to light travel time effects need to be taken into account in order to estimate the reflected synchrotron photon energy density (for a detailed discussion see [24]), which is well approximated by $u'_{RSy} \approx u'_{sy} \, 4\Gamma^3 \, a_{BLR} (R'_B / \Delta r_{BLR}) (1 - 2\Gamma R'_B / z)$, where Δr_{BLR} is a measure of the geometrical thickness of the broad line region. Similar to the SSC spectrum, the RSy spectrum does not show a strong spectral break. It peaks around $\langle \epsilon \rangle_{RSy} \approx (B'/B_{cr}) \, D \, \Gamma^2 \, \gamma_e^4 \approx \langle \epsilon \rangle_{SSC} \, \Gamma^2$.

TRENDS BETWEEN DIFFERENT BLAZAR CLASSES

There appears to be a more or less continuous sequence in the broadband spectral properties of blazars, ranging from FSRQs over LBLs to HBLs, which was first presented in a systematic way in [26]. While in FSRQs the synchrotron and γ-ray peaks are typically located at infrared and MeV – GeV energies, respectively, they are shifted towards higher frequencies in BL Lacs, occurring at medium to even hard X-rays and at multi-GeV – TeV energies in some HBLs. The bolometric luminosity of FSRQs is — at least during γ-ray high states — strongly dominated by the γ-ray emission, while in HBLs the relative power outputs in synchrotron and γ-ray emission are comparable.

Detailed modeling of several blazars has indicated that this sequence appears to

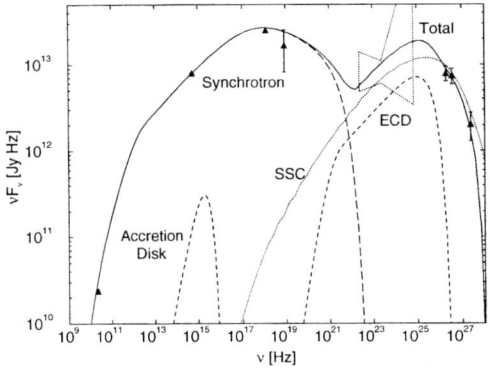

FIGURE 3. Fit to a weekly-averaged broadband spectrum of the extreme HBL Mrk 501 during a high γ-ray state centered on MJD 50564 in 1997. The dotted "bow-tie" curve indicates the highest flux ever measured by EGRET, which provides an upper limit for the possible contribution of external Comptonization. See [32] for model parameters.

be related to the relative contribution of the external Comptonization mechanisms ECD and ECC to the γ-ray spectrum. While most FSRQs are successfully modelled with external Comptonization models (e. g., [21,27–29]), the broadband spectra of HBLs are consistent with pure SSC models (e. g., [30–32]). BL Lacertae, a LBL, appears to be intermediate between these two extremes, requiring an external Comptonization component to explain the EGRET spectrum [33,34]. Figs. 2 and 3 illustrate detailed modeling results of two objects located at opposite ends of this sequence of blazars, using the jet radiation transfer code described in [15,34].

A physical interpretation of this sequence in the framework of a unified jet model for blazars was given in [35]. Assume that the average energy of electrons, γ_e, is determined by the balance of an energy-independent acceleration rate $\dot\gamma_{acc}$ and radiative losses, $\dot\gamma_{rad} \approx -(4/3)\, c\, \sigma_T \, (u'/m_e c^2)\, \gamma^2$, where the target photon density u' is the sum of the sources intrinsic to the jet, $u'_B + u'_{sy}$ plus external photon sources, $u'_{ECD} + u'_{ECC} + u'_{RSy}$. The average electron energy will then be $\gamma_e \propto (\dot\gamma_{acc}/u')^{1/2}$. If one assumes that the properties determining the acceleration rate of relativistic electrons do not vary significantly between different blazar subclasses, then an increasing energy density of the external radiation field will obviously lead to a stronger radiation component due to external Comptonization, but also to a decreasing average electron energy γ_e, implying that the peak frequencies of both spectral components are displaced towards lower frequencies.

SPECTRAL VARIABILITY OF BLAZARS

Between flaring and non-flaring states, blazars show very distinct spectral variability. Not only does the emission at the highest frequencies generally vary on

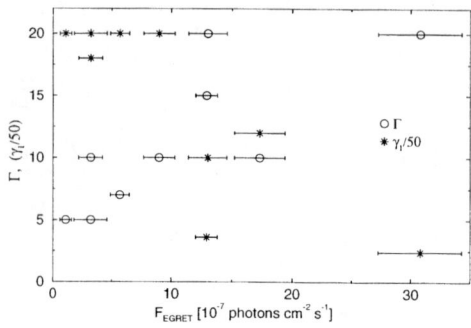

FIGURE 4. The dependence of the fit parameters Γ and γ_1 (low-energy cut-off of the electron distribution) on the EGRET flux for fits to simultaneous broadband spectra of PKS 0528+134 (see [28]).

the shortest time scales, but also the flaring amplitudes are significantly different among different wavelength bands. FSRQs often show spectral hardening of their γ-ray spectra during γ-ray flares (e. g., [36,39,40]), and the flaring amplitude in γ-rays is generally larger than in all other wavelength bands. The concept of multi-component γ-ray spectra of quasars, as first suggested for PKS 0528+134 in [36], offers a plausible explanation for this spectral variability due to the different beaming patterns of different radiation mechanisms, as pointed out in [37]. This has been applied to PKS 0528+134 in [38] and [28].

The results of [28] indicate that γ-ray flaring states of PKS 0528+134 are consistent with an increasing bulk Lorentz factor Γ of ejected jet material, while at the same time the low-energy cutoff γ_1 of the electron distribution injected into the jet is lowered. This is in agreement with the physical picture that due to an increasing Γ, the quasi-isotropic external photon field is more strongly Lorentz boosted into the blob rest frame, leading to stronger external Compton losses, implying a lower value of γ_1. The external Compton γ-ray components depend much more strongly on the bulk Lorentz factor than the synchrotron and SSC components do. This leads naturally to a hardening of the γ-ray spectrum, if the SSC mechanisms plays an important or even dominant role in the X-ray — soft γ-ray regime, while external Comptonization is the dominant radiation mechanism at higher γ-ray energies. The results of this study on PKS 0528+134 are discussed in more detail in [41].

While this flaring mechanism is plausible for FSRQs, short-timescale, correlated X-ray and γ-ray flares of the HBLs Mrk 421 and Mrk 501 [30,31] and synchrotron flares of other HBLs (e. g., PKS 2155-304, [42,43]) have been explained successfully in the context of SSC models where flares are related to an increase of the maximum electron energy, γ_2, and a hardening of the electron spectrum. How the spectral evolution in synchrotron flares can be used to constrain the magnetic field and the physics of injection and acceleration of relativistic pairs, has been described in

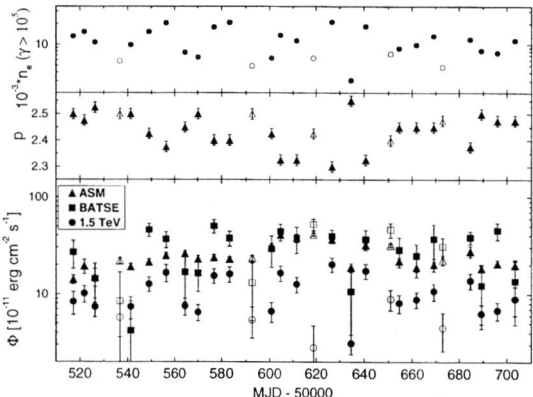

FIGURE 5. Temporal variation of the fit parameters $n_e(\gamma > 10^5)$ (density of high-energy electrons) and p (spectral index of injected electron distribution) compared to the weekly averaged light curves from RXTE ASM, BATSE and HEGRA (see [32]).

detail in the previous talk by R. Sambruna [44].

Comparing detailed spectral fits to weekly averaged broadband spectra of Mrk 501 [32] over a period of 6 months, we have found that TeV and hard X-ray high states on intermediate timescales are consistent with a hardening of the electron spectrum (decreasing spectral index) and an increasing number density of high-energy electrons, while the value of γ_2 has only minor influence on the weekly averaged spectra. Fig. 5 shows how the spectral index of the injected electron distribution and the density of high-energy electrons resulting from our fits are varying in comparison to the RXTE ASM, BATSE, and HEGRA 1.5 TeV light curves. For a more detailed discussion of this analysis see [45].

These variability studies seem to indicate that due to the different dominant γ radiation mechanisms in quasars and HBLs also the physics of γ-ray flares and extended high states is considerably different. While in FSRQs the γ-ray emission and its flaring behavior appears to be dominated by conditions of the external radiation field, this influence is unimportant in the case of HBLs where emission lines are very weak or absent, implying that the BLR might be very dilute, leading to a very weak external radiation field, which becomes negligible compared to the synchrotron radiation field intrinsic to the jet.

A TOY MODEL FOR SPECTRAL VARIABILITY

On the basis of the estimates of the photon energy densities of the various radiation fields and the peak energies of the diverse radiation components given in Section 2, one can develop a very simple, quasi-analytic toy model which allows us to study the influence of parameter variations on the predicted broadband spectrum of a blazar. For construction of this toy model, I assume that the magnetic field

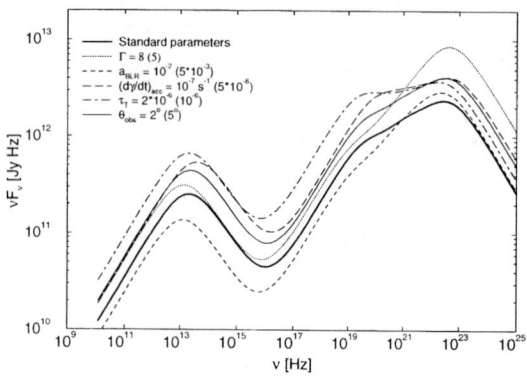

FIGURE 6. Thick solid curve: Toy model calculation representative of the broadband spectrum of PKS 0528+134. Parameters: $\Gamma = 5$, $\theta_{obs} = 5°$, $\dot\gamma_{acc} = 5 \cdot 10^{-8}$ s^{-1}, $L_D = 6 \cdot 10^{46}$ erg s^{-1}, $\langle\epsilon\rangle_D = 10^{-5}$, $a_{BLR} = 5 \cdot 10^{-3}$, $z = 2 \cdot 10^{17}$ cm, $R'_B = 3 \cdot 10^{16}$ cm, $\tau_T = 10^{-6}$. For the other curves, a single parameter, as indicated by the label, has been changed (value of the original calculation in parantheses).

within the jet is in equipartition with the ultrarelativistic electrons, and that the light-travel time effects affecting the efficiency of the synchrotron mirror mechanism can be parametrized by a correction factor $f_{ltt} \lesssim 0.1$ so that $u'_{RSy} \approx u'_{sy} a_{BLR} \Gamma^3 f_{ltt}$. Then, the entire broadband spectrum, accounting for all synchrotron and inverse-Compton components, is determined by 9 parameters: Γ, θ_{obs}, $\dot\gamma_{acc}$, L_D, $\langle\epsilon\rangle_D$, a_{BLR}, the scale height z of energy dissipation in the jet, R'_B, and τ_T. In most cases, several of these parameters can be constrained by independent observations. The radiation spectra of each individual component are approximated by double power-laws with a smooth transition.

The thick solid curve in Fig. 6 shows a toy model calculation with the location of peak frequencies and the relative contributions of the γ radiation components as found in our fits to PKS 0528+134 based on detailed simulations [28]. The hard X-ray to soft γ-ray spectrum below ~ 1 MeV is dominated by the SSC mechanism, while at higher energies, the ECC mechanism is dominant. The other curves in Fig. 6 indicate the effect of single parameter changes on the broadband spectrum which could be thought of as the cause of flares at γ-ray energies.

From Fig. 6, one can see that an increasing bulk Lorentz factor leads to a strong flare at γ-ray energies, while only moderate flaring at infrared and X-ray frequencies results (dotted curve). A shift of the synchrotron peak towards lower frequencies is predicted [46]. If the BLR albedo a_{BLR} increases (short-dashed curve), the result is a slight increase in the power output at high-energy γ-rays, while due to enhanced external-Compton cooling the flux in the synchrotron and SSC components even decreases. An increased acceleration rate $\dot\gamma_{acc}$ (long-dashed curve), leads to a strong synchrotron and SSC flare, where the largest variability amplitude is expected at MeV energies, while only moderate variability at higher frequencies is predicted.

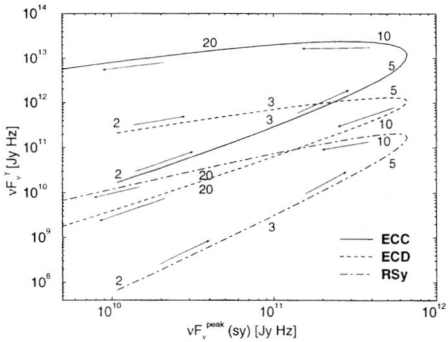

FIGURE 7. Variation of the power output in synchrotron and the external Comptonization components if the bulk Lorentz factor Γ is the only parameter changing during a flare. The numbers along the curves are the respective values of Γ at that point, and the arrows indicate the sense of evolution if Γ is increasing.

If the density of relativistic electrons in the blob increases during a flare (dot-dashed curve), the variability amplitude is again predicted to be largest at MeV energies and the peak frequencies of all components are expected to be shifted slightly towards lower frequencies. A decreasing observing angle — which could be a consequence of a bending jet — leads to a shift of the entire broadband spectrum towards higher fluxes and slightly higher peak frequencies.

Obviously, definite conclusions can not be drawn from this simplistic toy-model analysis. However, it may support previous results that an increasing bulk Lorentz factor is a viable and plausible explanation for the spectral variability observed in PKS 0528+134 and possibly also in other FSRQs.

Fig. 7 illustrates the variation of the power output in the different external Compton γ-ray components with respect to the synchrotron component, if Γ is the only parameter changing during a γ-ray flare. Most notably, a very steep relation between $\nu F_\nu^{peak}(sy)$ and $\nu F_\nu^{peak}(ECC)$ is predicted for values of the Lorentz factor close to the critical Γ, at which the observer is looking at the superluminal angle. This dependence can be significantly steeper than quadratic, $\Delta \nu F_\nu^{peak}(sy) \propto \left[\Delta \nu F_\nu^{peak}(ECC) \right]^\alpha$ with $\alpha > 2$, which has recently been observed in the prominent 1996 flare of 3C279 [40].

REFERENCES

1. Hartman, R. C., et al., *ApJ*, in press (1999a).
2. Punch, M., et al., *Nature*, **358**, 477 (1992).
3. Petry, D., et al., *A&A*, **311**, L13 (1996).
4. Quinn, J., et al., 1996, *ApJ*, **456**, L83 (1996).
5. Bradbury, S. M., et al., 1997, *A&A*, **320**, L5 (1997).

6. Chadwick, P. M., et al., *ApJ*, in press (1999).
7. Catanese, M., et al., *ApJ*, **501**, 616 (1998).
8. Mukherjee, R., et al., *ApJ*, **490**, 116 (1997).
9. Gaidos, J. A., et al., *Nature*, **383**, 319 (1996).
10. Schlickeiser, R., *Space Sci. Rev.* **75**, 299 (1996).
11. Mannheim, K., *A&A*, **269**, 67 (1993).
12. Marscher, A. P., & Gear, W. K., *ApJ*, **298**, 114 (1985).
13. Dermer, C. D., Schlickeiser, R., & Mastichiadis, A., *A&A*, **256**, L27 (1992).
14. Sikora, M., Begelman, MM. C., & Rees, M. J, *ApJ*, **421**, 153 (1994).
15. Böttcher, M., Mause, H., & Schlickeiser, R., *A&A*, **324**, 395 (1997).
16. Rachen, J., these proceedings (1999).
17. Maraschi, L., Ghisellini, G., & Celotti, A., *ApJ*, **397**, L5 (1992).
18. Bloom, S. D., & Marscher, A. P., *ApJ*, **461**, 657 (1996).
19. Dermer, C. D., & Schlickeiser, R., *ApJ*, **416**, 458 (1993).
20. Blandford, R. D., & Levinson, A., *ApJ*, **441**, 79 (1995).
21. Dermer, C. D., Sturner, S. J., & Schlickeiser, R., *ApJS*, **109**, 103 (1997).
22. Ghisellini, G., & Madau, P., *MNRAS*, **280**, 67 (1996).
23. Bednarek, W., *A&A*, **342**, 69 (1998).
24. Böttcher, M., & Dermer, C. D., *ApJ*, **501**, L51 (1998).
25. Shakura, N. I., & Sunyaev, R. A., *A&A*, **24**, 337 (1973).
26. Fossati, G., et al., *MNRAS*, **289**, 136 (1997).
27. Sambruna, R., et al., *ApJ*, **474**, 639 (1997).
28. Mukherjee, R., et al., *ApJ*, **527**, in press (1999).
29. Hartman, R. C., et al., in preparation (1999b).
30. Mastichiadis, A., & Kirk, J. G., *A&A*, **320**, 19 (1997).
31. Pian, E., et al., *ApJ*, **492**, L17 (1998).
32. Petry, D., et al., *ApJ*, submitted (1999).
33. Madejski, G., et al., *ApJ*, **521**, 145 (1999).
34. Böttcher, M., & Bloom, S. D., *AJ*, submitted (1999).
35. Ghisellini, G., et al., *MNRAS*, **301**, 451 (1998).
36. Collmar, W., et al., *A&A*, **328**, 33 (1997).
37. Dermer, C. D., *ApJ*, **446**, L63 (1995).
38. Böttcher, M., & Collmar, W., *A&A*, **329**, L57 (1998).
39. Hartman, R. C., et al., *ApJ*, **461**, 698 (1996).
40. Wehrle, A. E., et al., *ApJ*, **497**, 178 (1998).
41. Mukherjee, R., & Böttcher, M., these proceedings (1999).
42. Georganopoulos, M., & Marscher, A. P., *ApJ*, **506**, L11 (1998).
43. Kataoka, J., et al. *ApJ*, in press (1999).
44. Sambruna, R., these proceedings (1999).
45. Böttcher, M., Petry, D., & Connaughton, V., these proceedings (1999).
46. Böttcher, M., *ApJ*, **515**, L21 (1999).

Hadronic blazar models and correlated X-ray/TeV flares

Jörg P. Rachen

Sterrenkundig Instituut, Universiteit Utrecht, 3508 TA Utrecht, The Netherlands

Abstract. The hypothesis that AGN jets might be the sources of the ultra-high energy cosmic rays has originally motivated the venture of TeV gamma ray astronomy. Surprisingly, after the discovery of TeV emission from blazars the attention has shifted to more traditional explanations which do not involve energetic hadrons, and there is even common believe that a hadronic interpretation is disfavored by observations. It is shown here that this is not the case, and that the currently observed spectra and variability features of blazars can be perfectly understood within hadronic blazar models. I also discuss how hadronic models might be observationally distinguished from common leptonic models, and point out some interesting aspects which could be relevant for the understanding of the differences between blazar classes.

WHY HADRONIC MODELS?

AGN jets and the origin of cosmic rays

Cosmic rays are observed up to the enormous energy of 3×10^{20} eV, and both theoretical and observational arguments suggest an extragalactic origin of these most energetic particles [1]. Since the cosmic ray arrival directions are largely randomized by Galactic and extragalactic magnetic fields, their sources cannot be easily identified from direct observations. However, if we assume that they gain their energy by acceleration (rather than by quantum processes, i.e., the decay of superheavy particles) their extreme maximum energy allows to derive quite restrictive selection criteria for possible acceleration sites. Most acceleration scenarios in astrophysics, for example Fermi acceleration [2], assume that particles are magnetically confined for some time $t_{\rm acc}$ in the accelerating region. This implies three fundamental constraints on the maximum proton energy,

$$E_{\rm cr} < eBR\Gamma, \tag{1a}$$

$$E_{\rm cr} < \tfrac{3}{2} m_p c^2 \Gamma (m_p/m_e) \sqrt{\eta B_{\rm c}/\alpha_{\rm f} B} \tag{1b}$$

$$E_{\rm cr} < \eta eB\lambda_{\rm GZK}\Gamma, \tag{1c}$$

which we call (a) the *confinement limit*, (b) the *synchrotron limit*, and (c) the *GZK limit*. The confinement limit simply expresses that the particle gyro-radius $r_{\rm L}$ in the magnetic field B is smaller than the system size R. The synchrotron limit expresses that $t_{\rm acc}$ is smaller than the synchrotron loss time, where $B_c \approx 4 \times 10^{13}$ G is the critical magnetic field, and $\alpha_f \approx 1/137$ the fine structure constant. The parameter $\eta = \Delta E/E$ is the fractional energy gain in a Larmor time, $t_L = E/eBc$, and one can show that for Fermi acceleration generally $\eta < 1$ (see [3, App. D] for some discussion and references). The GZK limit expresses that proton acceleration must be faster than photohadronic losses at the microwave background, called the Greisen-Zatsepin-Kuzmin or GZK effect [4], and $\lambda_{GZK} \sim 10$ Mpc is the appropriate attenuation length of protons with $E > 10^{20}$ eV [5]. The GZK limit dominates over the confinement limit for sources on supercluster scales ($\gtrsim 10$ Mpc), but it applies also to somewhat smaller scales (e.g. clusters of galaxies) since in such objects usually $\eta \ll 1$ [6].

Only very few cosmic sources have been found which may satisfy all three conditions for $E_{cr} \gtrsim 10^{20}$ eV, and all of them are connected to strong shocks in relativistic flows ($\Gamma \sim 10-1000$): (a) the termination shocks of extended jets in radio galaxies [7], (b) compact jets in blazars [8,9], and (c) internal or external shocks in fireballs proposed to induce Gamma-Ray Bursts [10]. All of them are known as powerful emitters of non-thermal photons, in particular gamma rays. If we adopt the assumption that there is a universal ratio between cosmic ray and non-thermal photon emission (which is rather simplistic, but not unreasonable as an estimate), we could use their total contribution to the observed extragalactic gamma rays as a scale of their total power as cosmic ray sources. This clearly favors AGN over GRB, because AGN are known to produce 10-30% of all extragalactic gamma-rays (counting both resolved sources and diffuse background), while resolved GRB contribute less than 1%. Due to the GZK effect, another important selection criterion for UHECR sources is proximity. This disfavors powerful quasars, and also GRB if their cosmological distribution follows the star forming activity, since in both cases most power is emitted at large redshifts. In contrast, BL Lac objects and their proposed unbeamed counterpart, FR-I radio galaxies [11], seem to be more frequent in the current epoch than at large redshifts [12]. Since their jets may also satisfy the condition for the acceleration of UHECR, and they contribute significantly to the non-thermal radiation in the universe, blazars and FR-I radio galaxies may be regarded the primary candidates for the origin of the highest energy cosmic rays.

Hadronic vs. leptonic gamma-ray emission in blazars

Energetic hadrons can lead to the emission of gamma rays via pp interactions with surrounding gas, or $p\gamma$ interactions with ambient photons. This leads to the production of secondary $e^\pm$ pairs, or mesons like $\pi^\pm$ and π^0, which eventually decay into $e^\pm$ pairs, photons and neutrinos. Electrons or pairs can produce high energy photons by synchrotron (or Compton) processes. The photons can either escape from the jet or produce new pairs in $\gamma\gamma \to e^+e^-$ processes, which subsequently radiate a new generation of photons. In particular *synchrotron-pair cascades* of this kind are important if interactions of UHECR protons in AGN jets are considered, where they shift the energy in secondary radiation down from the extreme proton energies to the observable gamma-ray regime. This mechanism has been coined PIC for "proton induced cascades" [13].

The term "hadronic blazar models" is used for a large variety of models for the gamma-ray production in blazars, not all of which involve UHE cosmic rays. All in common is just that they propose energetic protons as the main carrier of dissipated energy in the jet, rather than so-called "leptonic models", which assume the bulk of the energy available for radiation in electrons or $e^\pm$ pairs. Some hadronic models assume gamma-ray production in pp interactions invoked in the collision of jets with surrounding gas clouds, or in a very massive jet itself [14]. In a different class of models, protons have been suggested as being responsible just for the injection of energetic electrons, which then produce the observed photon emission by a synchrotron-self Compton (SSC) mechanism [15]. These models offer an interesting explanation for the observed strong variability in gamma-ray blazars due to intrinsic instabilities. They usually require only moderate proton energies, but very large densities of relativistic protons in the jet.

Following the motivation given above, I want to focus here on a different kind of hadronic models in which UHE cosmic rays in the jet interact with low energy target photons. They are split into two classes: (a) target photons are produced by synchrotron-radiating electrons co-accelerated with the protons [8], and (b) external target photons are present in the vicinity of the jet, as for example thermal photons emitted from an accretion disk or a warm dust torus [16]. In a realistic scenario, photons from both sources would be present. The local ratio of the external and internal photon energy density in a reference frame comoving with the jet differs hereby from the the ratio of the observed external and internal luminosities by a factor $\Gamma_{\rm jet}^6$ (the corresponding number ratio of external to internal photons scales with $\Gamma_{\rm jet}^4$). It is therefore hard to constrain the local comoving density of external photons in the jet by observations. However, an indirect constraint on the external photon density has been pointed out by Protheroe and Biermann [17]: using a common accretion disk/torus model to estimate the radiation fields in AGN, they show that the emission of TeV photons would be strongly suppressed by $\gamma\gamma$ absorption, if the emission region is close enough to the AGN that external photons could be relevant for $p\gamma$ interactions. This is inconsistent with the observed spectra in TeV blazars like Mrk 501 or Mrk 421 (see below). Nevertheless, external photons can still be important in high luminosity EGRET blazars (like 3C 279) [9].

For the discussion of TeV blazars, we can therefore concentrate on so-called *synchrotron-self proton induced cascade* (SS-PIC) models, in which UHE protons interact with synchrotron photons emitted by electrons accelerated *in the same process* as the protons. One clear consequence of such models is that the gamma-ray emission due to the PIC process is in competition with SSC emission of the electrons. The relation of the luminosities of both processes is expressed by

$$\frac{L_{\rm PIC}}{L_{\rm SSC}} \lesssim 0.1 \frac{u_p(\hat{E}_p)}{u_{\rm ph}} \left[\frac{B}{1\,{\rm G}}\right]^3 \left[\frac{R}{10^{16}\,{\rm cm}}\right]^2, \qquad (2)$$

where $u_p = N_p^2 dN_p/dE_p$ is the energy density of protons at the maximum energy, $\hat{E}_p$, and $u_{\rm ph}$ is the bolometric energy density of the target photons (i.e., the *synchrotron* photons emitted by primary electrons), and B and R are magnetic field and size of the emission region, respectively. The relation gives in fact only an upper limit on the PIC contribution, since approximate equality assumes $\hat{E}_p = eBR$. We see that, if the UHE proton content of the jet is significant ($u_p(\hat{E}_p) \geq u_{\rm ph}$), and the maximum energy sufficient to explain the observed UHECR ($eBR\Gamma_{\text{jet}} \gtrsim 10^{20}$ eV, assuming a jet Lorentz factor $\Gamma_{\text{jet}} \sim 10$), PIC emission will dominate over SSC at compact jet scales ($R \lesssim 10^{16}$ cm) — and vice versa. For the likely case of adiabatic scaling of a dominantly transversal magnetic field with the jet radius, $B \propto R^{-1}$, the relative contribution of SSC emission would increase on larger scales. Therefore both processes may contribute to the observed radiation, although emission which is variable on short time scales (requiring small R) would tend to be dominated by PIC. Defining clear signatures to distinguish PIC from SSC emission can therefore make the hypothesis that blazars and FR-I radio galaxies are the sources of UHE cosmic rays testable by GeV-TeV gamma-ray observations.

SPECTRAL PROPERTIES OF SS-PIC MODELS

Synchrotron-pair cascades in power law target spectra

At frequencies below the X-ray band, blazar spectra are well represented by broken power laws, in some cases even by single power laws down to the sub-mm regime [18]. These photons play a triple role in SS-PIC models: (i) they are targets for the initial $p\gamma$ interactions, (ii) they are targets for the propagation of synchrotron pair cascades, and (iii) they allow clues on the particle spectrum of the primary accelerator. Let us, for simplicity, assume that the soft target photon number spectrum is described by a single power law, $dN_{\text{ph}}/dE \propto E^{-\alpha_t - 1}$, $\alpha_t \sim 1$ is called the energy index of the target spectrum. The stationary electron spectrum producing these photons by synchrotron radiation must then have the form $dN_e/dE \propto Q_{e,\text{acc}} t_{\text{syn}} \propto E^{-\alpha_e - 1}$ with an energy index $\alpha_e = 2\alpha_t$, where $Q_{\text{acc}} \propto E^{-\alpha_{\text{acc}} - 1}$ is the number of electrons injected by the accelerator per unit time, and $t_{\text{syn}} \propto E^{-1}$ is the synchrotron cooling time of the electrons. This means that the spectrum injected by the accelerator is a power law with energy index $\alpha_{\text{acc}} = \alpha_e - 1$. In contrast to electrons, cooling of protons is usually dominated by adiabatic [3] or advection losses [8]. Hence, the proton cooling time $\bar{t}_p$ can be considered as energy independent, and the stationary proton spectrum is $dN_p/dE \propto Q_{p,\text{acc}}(E)$. If the accelerated protons have the same spectrum as the electrons, the proton power law energy index is $\alpha_p \propto \alpha_{\text{acc}} = 2\alpha_t - 1$, but we discuss also different choices of α_p.

The opacity for the photohadronic production is $\tau_{p\gamma} = \bar{t}_p/t_{p\gamma}$, where $t_{p\gamma} \propto E^{\alpha_t}$ is the according loss time scale. The major channel for hadronic gamma production is $p\gamma \to \pi^0 + \ldots$ with the subsequent decay $\pi^0 \to \gamma\gamma$. The resulting photons are dominantly absorbed by the soft target photons through $\gamma\gamma$ pair production. The $\gamma\gamma$ opacity has the same energy dependence as $\tau_{p\gamma}$ for constant $\bar{t}_p$, that is $\tau_{\gamma\gamma} \propto E^{-\alpha_t}$, hence the stationary energy distribution of the primary photons from pion decay is $dN_\gamma^{[0]}/dE = dN_p/dE(\tau_{\gamma\gamma}/\tau_{p\gamma}) \propto dN_p/dE$. Saturated production of pairs in $\gamma\gamma$ collisions (i.e. $\tau_{\gamma\gamma} \gg 1$) then leads to [19]

$$\frac{dN_\pm^{[1]}}{dx} \propto \frac{1}{x^2} \int_{2x}^{\hat{x}_0} dx' \tau_{\gamma\gamma}(x') \frac{dN_\gamma^{[0]}}{dx'} \propto x^{-\alpha_\pm - 1} = \begin{cases} x^{\alpha_t - \alpha_p - 2} & \text{for } \alpha_p > \alpha_t \\ x^{-2} & \text{for } \alpha_p \leq \alpha_t \end{cases} \quad (3)$$

where x is the energy of photons or pairs in units of $m_e c^2$, and $\hat{x}_0 = \hat{\gamma}_p m_\pi/2m_e \sim 100\hat{\gamma}_p$ is the maximum x of the primary injected photons. The pairs produce a new generation of photons by synchrotron radiation, which are distributed in the stationary, saturated case as

$$\frac{dN_\gamma^{[1]}}{dx_1} \propto \frac{Q_{\text{syn}}^{[1]}(x_1)}{\tau_{\gamma\gamma}(x_1)} \propto x_1^{-\alpha_t - 3/2} \left[\frac{x^2 dN_\pm(x)}{dx} \right]_{x = x_1^{1/2}}, \quad (4)$$

i.e., as a power law with energy index $\alpha_1 = \frac{1}{2}\alpha_\pm + \alpha_t$. $Q_{\text{syn}}^{[1]}(x) \propto x^{-\alpha_\pm/2}$ is the injected number of synchrotron photons per unit time at energy $m_e c^2 x$ by the first generation of pairs. The stationary photons can inject a second generation of pairs, and so on. This *synchrotron-pair cascade* continues, and with each step the characteristic photon energy is reduced by

$$x_n \approx \frac{B}{4B_c} x_{n-1}^2 \quad \text{for} \quad Bx_{n-1} < 4B_c. \quad (5)$$

The latter condition expresses the classical limit of synchrotron radiation and is mostly fulfilled in hadronic blazar models; in the non-classical case the cascade propagates approximately as $x_n = \frac{1}{2}x_{n-1}$. Thus, the energy is rapidly reduced in each step once $x \ll B_c/B \sim 10^{12}$, and since $\tau_{\gamma\gamma} \propto x^{\alpha_t}$ the opacity for the production of subsequent cascade generations quickly decreases. The cascade becomes unsaturated at a photon energy $\sim m_e c^2 x_{\gamma\gamma}$, defined by $\tau_{\gamma\gamma}(x_{\gamma\gamma}) = 1$, and dies out rapidly for $x < x_{\gamma\gamma}$. The photon spectrum emerging from the emitter is [19,13]

$$\frac{dN_{\gamma,\text{em}}}{dx} \propto \sum_{n=1}^{n^*+1} Q_{\text{syn}}^{[n]}(x) \left[\frac{1 - e^{-\tau_{\gamma\gamma}(x)}}{\tau_{\gamma\gamma}(x)} \right] \propto \begin{cases} x^{\alpha_t - \alpha_1 - 1} & \text{for} \quad x \ll x_{\gamma\gamma} \\ x^{-\alpha_1 - 1} & \text{for} \quad x \gg x_{\gamma\gamma} \end{cases}, \quad (6)$$

where the sum extends over all cascade generations and n^* is the last generation with a highest photon energy $\hat{x}_{n^*} > x_{\gamma\gamma}$ ($\hat{x}_n$ is determined by repeated application of Eq. (5) on $\hat{x}_0$). It can be shown that for typical blazar conditions $n^* = 3$–4, and that $Q_{\text{syn}}^{[n^*]}$ dominates the emission around $x_{\gamma\gamma}$ [20].

The term in brackets in Eq. (6) has the meaning of a mean escape probability of the photons from the emission region, and is strictly correct only for a plane-parallel geometry. However, the asymptotic behavior of the spectrum would be the same in any geometry, that is, $x_{\gamma\gamma}$ marks a break in the spectrum by $\Delta\alpha = \alpha_t$, which we call the *opacity break*. This broken power law shape is typical for photospheric emission, which is in the nature of the SS-PIC models where the soft photons are responsible both for the production and absorption of gamma rays. It is directly observable unless there is significant absorption by external photons surrounding the jet, which would lead to an exponential cutoff $\propto \exp(-\tau_{\gamma\gamma,\text{ext}})$. The observed GeV– TeV spectra of TeV-blazars support in most cases a broken power law shape, as expected from the SS-PIC model for $m_e c^2 x_{\gamma\gamma} \Gamma_{\text{jet}} \sim 1\,\text{TeV}$, but are inconsistent with a rapid cutoff in this regime [18]. This has been used as an argument against external PIC models for TeV blazars [17]. An observed opacity break in the TeV regime is also expected from a direct determination of $\tau_{\gamma\gamma}$ for typical blazar parameters [8], and from general considerations concerning the efficiency of hadronic blazar models [9].

Robust features of hadronic blazar spectra

If we ignore the details of the cascade spectra in the vicinity of cutoffs or breaks, the spectral propagation can be described in terms of a simple algebra. If we call $\alpha_n(x_n)$ the local spectral index in the vicinity $x \sim x_n$, we can introduce the relations

$$\alpha_n(x_n) = f_+[\alpha_{n-1}(x_{n-1})] \quad \text{for} \quad x_n > x_{\gamma\gamma} \tag{7a}$$
$$\alpha_n(x_n) = f_-[\alpha_{n-1}(x_{n-1})] \quad \text{for} \quad x_n < x_{\gamma\gamma} \tag{7b}$$

with

$$f_\pm[\alpha_n] = \max\left\{\tfrac{1}{2}(\alpha_n \pm \alpha_t + 1), \tfrac{1}{2}\right\}, \tag{7c}$$

where the dimensionless photon energy x_n propagates following Eq. (5). The spectrum of the k-th cascade is then given by $f_+^k[\alpha_p]$ above the opacity break at $x_{\gamma\gamma}$, and by $f_-[f_+^{k-1}[\alpha_p]]$ below (but above $x'_{\gamma\gamma} = Bx_{\gamma\gamma}^2/4B_c \ll x_{\gamma\gamma}$), where f_+^k denotes the k-fold iterative application of f_+.

If we ask which cascade generation is dominant at a given energy we encounter a problem related to the energy propagation equation (5): for any given primary spectral feature at x_0, the position of its "image" in the n-th cascade is dependent on the magnetic field in the emitter by $x_n \propto B^{2^n-1}$. This means, for example, that the cutoff of the 4-th cascade generation is $\hat{x}_4 \propto B^{15}$, so that a variation of B by a factor of 2 over the emitting region would "smear out" the value of $\hat{x}_4$ by more than 4 orders of magnitude. Fortunately, it is still possible to make some quite robust predictions on hadronic blazar spectra, since the cascade generation spectra converge quickly with k to $f_+^\infty[\alpha_p] = 1+\alpha_t$, and $f_-[f_+^\infty[\alpha_p]] = 1$, independent of α_p. Moreover, the power contained in each cascade generation is approximately equal for $n < n^* + 1$ and rapidly decreases for larger n, so that the spectrum below $x_{\gamma\gamma}$ is not strongly changed by adding generations with $n > n^*$. This allows the statement that *hadronic gamma-ray blazar spectra are described by a broken power law with energy indices $\alpha \approx 1$ below, and $\alpha \approx 1 + \alpha_t$ above a break observed at ~ 1 TeV*, where α_t is the IR-X energy index of the source.

So far we did not consider the cutoff in the target photon spectrum, $\hat{x}_t$, which is generally observed at an energy $m_e c^2 \hat{x}_t \Gamma_{\rm jet}$ between in the optical and X-ray regimes. We consider the target photon spectrum for $x_t > \hat{x}_t$ as a power law with a steep energy index, $\alpha'_t > \alpha_t + 1$. Gamma rays with $x_n < \hat{x}_t^{-1}$ then induce cascade photons below $\tilde{x} = B/(4B_c \hat{x}_t^2)$ with an index $\alpha'_{n+1} \approx f'_-[f_+^\infty[\alpha_p]] = \tfrac{1}{2}$, where f'_- is the cascade operator defined in Eq. (7c) replacing α_t with α'_t (for $\tilde{x} < x_{\gamma\gamma}$). Since this result in independent of the detailed value of α'_t as long as $\alpha'_t > \alpha_t + 1$, we can consider it as valid for any steep cutoff, e.g., an exponential one. The energy $m_e c^2 \tilde{x} \Gamma_{\rm jet}$ marks a break in the spectrum, which is observable above the cutoff of the primary electron emission ($\tilde{x} > \hat{x}_t$) if

$$\hat{\epsilon}_t = m_e c^2 \Gamma_{\rm jet} \hat{x}_t \lesssim 100\,{\rm eV} \left[\frac{\Gamma_{\rm jet}}{10}\right] \left[\frac{B}{\rm G}\right]^{1/3}. \tag{8}$$

For $B < 100\,{\rm G}$ this transition corresponds to the observationally motivated distinction between so-called low-energy-cutoff blazars (LBLs) and high-energy-cutoff blazars

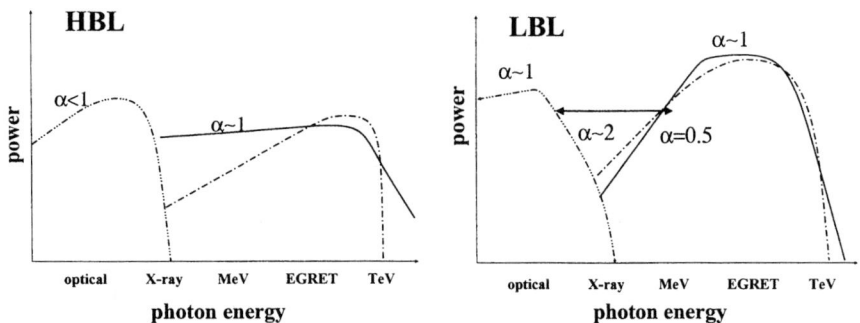

FIGURE 1. Generic SS-PIC spectra for HBL (left) and LBL (right) type blazars. Dashed high-energy lines show a typical spectral shape of an SSC model for comparison.

(HBLs) [11]. In the SS-PIC models we therefore expect fundamentally different shapes for the high energy spectra of these classes, as illustrated in Figure 1. Obviously, the spectrum for LBLs can easily be confused with an SSC spectrum, while a clear difference is predicted for HBLs in the MeV gamma-ray regime.

Narrow cascades: proton and muon synchrotron radiation

Apart from π^0 decay, energetic photons can also be produced by synchrotron radiation of the protons. Moreover, it has been shown that also the most energetic muons produced by the decay of photohadronically produced charged mesons can lose a significant fraction of their energy in synchrotron radiation in magnetic fields typical for SS-PIC blazar models, prior to their decay [3]. The total power ratio of synchrotron radiation compared to the π^0 cascade injected by protons of the energy Ecr in the observer's frame is

$$\frac{L_{p,\text{syn}}}{L_{\pi^0}} \sim 4 \left[\frac{E_{\text{cr}}}{10^{20}\,\text{eV}}\right]^{1-\alpha_t} \left[\frac{B}{10\,\text{G}}\right]^2 \left[\frac{R}{10^{16}\,\text{cm}}\right]^2 \left[\frac{\Gamma_{\text{jet}}}{10}\right]^3 \left[\frac{L_{\text{IR}}}{10^{44}\,\text{erg/s}}\right]^{-1}, \quad (9)$$

where L_{IR} is the observed infrared luminosity of the blazar. For muon synchrotron radiation the ratio is $L_{\mu,\text{syn}}/L_{\pi^0} \approx 2$ because of the photohadronic branching ratios [21].

If the jet contains cosmic ray protons up to an energy $\hat{E}_{\text{cr}} \sim 3\times 10^{20}$ eV in the observer's frame, $B \gtrsim 10$ G and the typical parameters assumed in Eq. (9) otherwise, proton synchrotron radiation extends up to observable energies of $B\hat{x}_p^2 m_e^3 \Gamma_{\text{jet}}/(4B_c m_p^3) \gtrsim 300$ GeV. Muon synchrotron radiation requires a minimal dimensionless energy $x^* = E_\mu/m_e c^2 \Gamma_{\text{jet}} \sim 5\times 10^{11}$ for $B \sim 10$ G [3], and therefore extends from 30 GeV to 3 TeV using the same parameters. Since we have argued above that the opacity break is in the same regime, a significant fraction of this radiation can be reprocessed in a further cascade generation and would appear in the hard X-ray regime up to energies of ~ 10 keV and ~ 100 keV for reprocessed proton and muon synchrotron radiation, respectively.

The interesting aspect about these cascades becomes obvious when we consider their spectrum below the opacity break. Here, we assume for simplicity that proton syn-

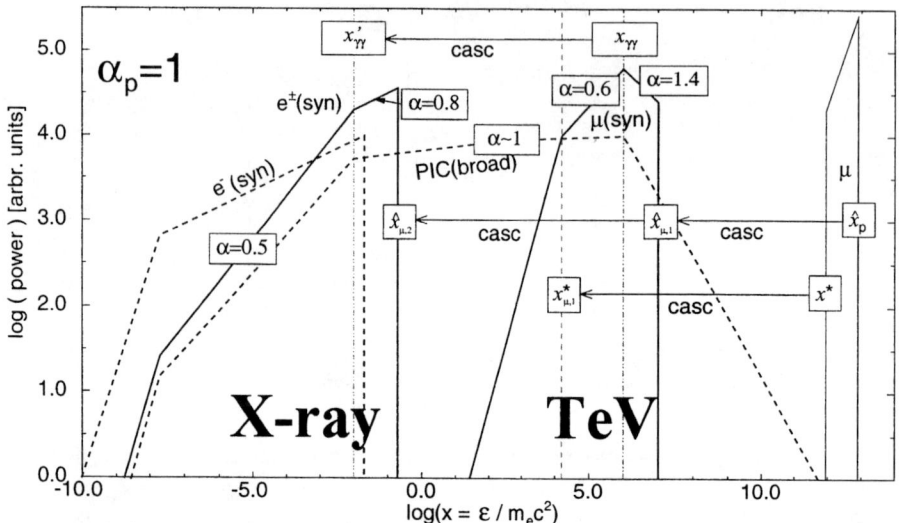

FIGURE 2. Schematic spectra of narrow cascades superposing the broad band emission of ordinary π^0 cascades and primary electron synchrotron radiation — Spectral indices correspond to μ-induced cascades for $\alpha_p = 1$, arrows indicate the cascading of spectral features. The proton-synchrotron cascade is omitted for clarity, but has a similar structure at photon energies about a factor of 10 lower (see also [23]).

chrotron radiation is significant, but not dominant over adiabatic or advection losses. In this case, the proton synchrotron spectrum has an energy index $\alpha_{p,1} = \frac{1}{2}\alpha_p \sim 0.5$. For muons synchrotron cooling is always dominant (otherwise it is suppressed due to muon decay), and we obtain $\alpha_{\mu,1} = f^-[\alpha_p] \sim 0.5$ for $x_{\gamma\gamma} > x > x_1^* = Bx^{*2}m_e^3/(4B_c m_\mu^3)$, and $\alpha_{\mu,1} = -\frac{1}{3}$ below from synchrotron emission below the characteristic frequency. Hence, both processes produce spectra very much flatter than those arising from ordinary, π^0 induced cascades. Above the opacity break, the spectra steepen by $\alpha_t \sim 1$, which means that the peak of the luminosity is reached at the opacity break. Here, the dominance of this component over the broad π^0 cascades is even stronger than expected from the power ratio factors derived above, since it is first generation synchrotron emission, while the π^0 emission from the same protons is broadened through the cascading process. The spectral indices expected from the narrow cascades below and above the opacity break (Fig. 2) are consistent with the typical EGRET and TeV indices, respectively, observed in Mrk 421 and Mrk 501 [18]. The spectral indices of the second cascade are obtained from a second application of the operator f^-, yielding again values around 0.5 steepening by $\frac{1}{2}\alpha_t$ above an X-ray break observable at $m_e c^2 \Gamma_{\rm jet} x'_{\gamma\gamma}$ with $x'_{\gamma\gamma} \approx Bx_{\gamma\gamma}^2/4B_c$ (see Fig. 2). These compare well to the indices observed by Beppo-SAX in three prominent flares of Mrk 501 in April 1997 [22]. The luminosity in the X-ray peak depends sensitively on the energy of the opacity break, which determines how much energy of the TeV peak is reprocessed. For example, in the SS-PIC model by Mücke and Protheroe [23] the opacity break is at $m_e c^2 x_{\gamma\gamma} \Gamma_{\rm jet} \sim 25\,{\rm TeV}$, which explains why their X-ray peak from reprocessed proton synchrotron radiation is strongly suppressed. For suitable parameters, however, *narrow PIC emission can produce a two-bump spectrum with comparable peaks the TeV and the X-ray regime*, as illustrated in Figure 2.

VARIABILITY AND CORRELATED FLARES

Flares and the quiescent background

A major result of the simultaneous multiwaveband observations of Mrk 421 and Mrk 501 was that their variability in the X-ray and TeV-regime is largely correlated, and stronger than in other wavebands. This has been used as an argument against a hadronic interpretation of their gamma-ray emission, since ordinary cascade models expect a quite model independent spectral index (as explained above), which implies that the gamma-ray variability of such sources should be comparable at all gamma-ray energies.

Obviously, this picture changes if we consider narrow cascades. In order to play a dominant role in the emission, these require conditions which allow to produce cosmic rays of extremely high energies. Let us assume now that such conditions are not always present in the jet, but only in some confined regions for a limited time. Of course, we may still assume that also in other regions the jet accelerates protons and electrons, albeit not to such high energies. This would lead to a permanent "glow" of the jet, which is dominated by the emission of primary electrons and π^0 induced cascades with a spectrum comparable to the case illustrated in Fig. 1 for HBLs, which is consistent with observations of the quiescent emission of Mrk 421. A short flare which contains UHE protons able to produce narrow cascades, and which is energetic enough to compete with the total emission of the rest of the jet, would then cause the strongest variability in the energy regimes where the flare spectrum peaks, and these are the X-ray and TeV regimes. At other wavebands, the variability may be low, or not present at all since the emission there may continue to be dominated by the quiescent background (see Fig. 3).

Opacity and variability

The assumption of the existence of an opacity break in or below the TeV regime is vital for a successful explanation of blazar spectra in an SS-PIC model. This has some very interesting implications if the emission is variable. Let us assume a nearly plane-parallel geometry of length R for a region emitting photons simultaneously over its entire volume, over a time scale $t_{\rm rad} \ll R/c$. In the optically thin case, this would induce a flare of duration R/c owing to the run-time differences of photons. If the emitter is opaque at some some energy x, i.e., $\tau_{\gamma\gamma}(x) \gg 1$, only photons emitted within a distance $R/\tau_{\gamma\gamma}$ can reach the observer, causing a flare of duration $R/c\tau_{\gamma\gamma}$.

Applying this to SS-PIC models with an opacity break below TeV, we would therefore expect that the observed multi-TeV variability becomes systematically faster with increasing photon energy, while the variability time scales are energy independent at X-ray or EGRET energies that are below the opacity break. Of course, the simple relation $T_{\rm var} = T_{\rm var,0}/\tau_{\gamma\gamma}$ for $\tau_{\gamma\gamma} > 1$, and $T_{\rm var} = T_{\rm var,0}$ for $\tau_{\gamma\gamma} < 1$ is unlikely to apply

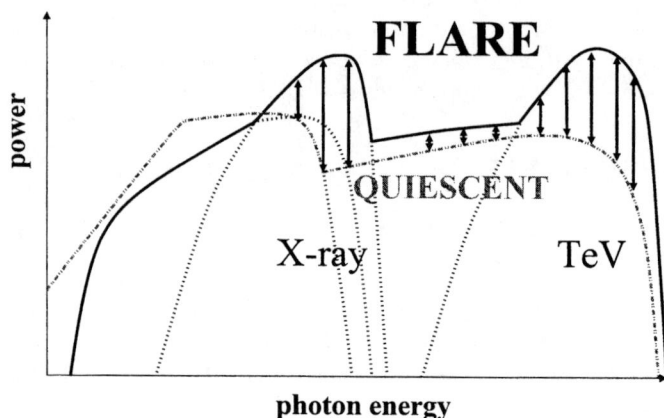

FIGURE 3. Correlated variability on blazars as a result of flaring, narrow PIC emission superposing a non-peaked background from the surrounding jet.

to SS-PIC, since both cascade injection and propagation involves time scales $\sim R/c$. Nonetheless, the effect remains qualitatively in the sense that it reduces the variability time from $\sim t_{\rm rad} + R/c$ to $\sim t_{\rm rad}$ for $\tau_{\gamma\gamma} \gg 1$. This may explain some recent simultaneous X-ray/TeV observations of Mrk 501 which indicate a TeV variability about a factor of 3 shorter than at three different X-ray energies, which all show the same time curve [18]. In contrast, optically thin SSC models would expect that there is some X-ray energy which shows the same variability as observed in the TeV.

SUPPLEMENTARY ASPECTS

Particle acceleration in the SS-PIC model

Some of the properties of the SS-PIC model discussed above on a purely heuristic base (i.e., the assumption that UHECR are produced) can be put on a more rigid foundation if we include the acceleration process. For jets the most reasonable assumption is Fermi acceleration at shocks or plasma turbulence [2]. If a particle with mass m is accelerated on a time scale $t_{\rm acc} = \eta^{-1} t_{\rm L}$, where η has the same meaning as in Eqs. (1), then the cutoff of the synchrotron radiation emitted by the particle satisfies the condition

$$\hat{\epsilon} \lesssim \pi \eta\, mc^2/\alpha_{\rm f} \sim 400\eta\, mc^2 \,, \tag{10}$$

which is reached if the particle energy limited by synchrotron cooling ($t_{\rm acc} = t_{\rm syn}$). Applying Eq. (10) to synchrotron radiating protons we find a maximum observed photon energy of $\approx [4\,{\rm TeV}]\eta_p$ (assuming $\Gamma_{\rm jet} \sim 10$). For Fermi acceleration we can write $\eta(E) \sim [\beta\, \delta B(r_{\rm L})/B]^2$, where β is the velocity of the shock or plasma wave, and $\delta B(r_{\rm L})/B$ is the fractional magnetic field turbulence on the scale of the particle gyro-radius, $r_{\rm L} = E/eB$ (see [3, App. D], and references therein). Obviously, in order to explain photons observed above 300 GeV by proton synchrotron radiation we need $\eta(\hat{E}_{\rm cr}) > 0.1$, which requires relativistic shocks and strong turbulence on the largest scales in the system. In this picture, correlated flares in blazars are due to the appearance of transient relativistic shocks in a turbulent flow, while the background emission is due to continuous acceleration at omnipresent weak shocks or plasma turbulence.

Applying Eq. (10) to accelerated electrons, we find a maximum synchrotron photon energy of $\approx [2\,{\rm GeV}]\eta_e$ for $\Gamma_{\rm jet} \sim 10$, which implies $\eta_e \sim 10^{-9}-10^{-5}$ to explain the observed IR-X cutoffs in blazars. The large difference between η_p and η_e can be understood from the theory of plasma turbulence, noting that the electrons probe very much smaller scales of the turbulence than the protons. If the turbulence is described by $\eta(E) \propto [\delta B(r_{\rm L})/B]^2 \propto E^y$, we find

$$\hat{\epsilon}_{e,\rm syn} \sim \hat{\epsilon}_{p,\rm syn}\left[m_e/m_p\right]^{(3y+2)/(2-y)}, \tag{11}$$

which is obviously independent of Doppler boosting. Biermann and Strittmatter [24] pointed out that the near infrared cutoffs observed in many quasars (e.g., LBLs) can be understood if they accelerate protons to the GZK limit, and the plasma turbulence is described by a Kolmogorov spectrum ($y = \frac{2}{3}$). In terms of Eq. (11), we would obtain $\hat{\epsilon}_{e,\rm syn} \sim 1\,{\rm eV}$ for $y = \frac{2}{3}$ and $\hat{\epsilon}_{p,\rm syn} = 10\,{\rm GeV}$, which is consistent with the observations of LBLs and would imply $\eta_p \sim 10^{-3}$. The Kolmogorov spectrum applies to fully developed hydrodynamical turbulence if the magnetic field does not significantly contribute to the total energy density of the fluid. In a magnetically dominated plasma a Kraichnan turbulence spectrum would be expected [24], that is $y = \frac{1}{2}$. Combining this with $\hat{\epsilon}_{p,\rm syn} \sim 300\,{\rm GeV}$ which we required for blazars with correlated X-ray/TeV variability, we find $\hat{\epsilon}_{e,\rm syn} \sim 10\,{\rm keV}$, which corresponds to X-ray cutoffs typically observed in HBLs (note that the flare emission up to >100 keV in Mrk 501 is explained by hadronic cascades in this model). This allows an interesting explanation of the physical difference between these blazar classes: while LBL jets are energetically dominated by the hydrodynamic motion of the plasma, and involve only non-relativistic shocks and/or weak to moderate turbulence, HBL jets are strongly turbulent, magnetically dominated flows, in which also relativistic shocks occur. More aspects of particle acceleration in the SS-PIC model are discussed in [23].

Gamma rays, cosmic rays and neutrinos

One prediction which is unique to hadronic blazar models is the production of energetic neutrinos with a luminosity comparable to gamma-rays. The detection of such neutrinos, in particular if correlated with blazar flares, would thus be a "smoking gun" for the hadronic scenario; unfortunately, the neutrino fluxes from single blazar flares are that low that this can hardly be expected within the next decades [3]. However, we would expect to find *diffuse* VHE neutrino fluxes comparable to the the diffuse extragalactic gamma ray background (DEGRB), if a considerable fraction of the extragalactic gamma rays are of hadronic origin. This would also imply that a significant fraction of the UHECR flux is produced by the same process [9]. To utilize these relations to decide the nature of gamma-ray emission in blazars, we need to determine which fraction to the DEGRB they contribute — which is again a task for gamma-ray astronomy.

In conclusion, although VHE neutrino observations will be very important to clarify the origin of cosmic rays, I still see the currently better technical possibilities to decide this question on the side of gamma-ray astronomy. To do this by revealing the nature of gamma-ray emission from blazars will require (i) a complete coverage of the gamma-ray wavebands from MeV to multi-TeV energies with sufficient spectral resolution, together with more campaigns allowing truly simultaneous multifrequency monitoring, and (ii) a comprehensive discussion of the data in the view of possible leptonic *and* hadronic explanations.

Acknowledgments. I wish to thank A. Achterberg, A. Atoyan, R. Bingham, J. Kirk, K. Mannheim, A. Mastichiadis, A. Mücke, and R. Sambruna for interesting and helpful discussions, and the LOC for support allowing me to visit the meeting. This work was supported in part by the EU-TMR network Astro-Plasma Physics, under contract number ERBFMRX-CT98-0168.

REFERENCES

1. P. L. Biermann, J. Phys. G **23**, 1 (1997), and references therein.
2. L. O'C. Drury, Rep. Prog. Phys. **46**, 973 (1983), and references therein.
3. J. P. Rachen and P. Mészáros, Phys. Rev. D **58**, 123005 (1998).
4. K. Greisen, Phys. Rev. Lett. **16**, 748; G. Zatsepin and V. Kuzmin, JETP Lett. **4**, 78 (1966).
5. F. W. Stecker, Phys. Rev. Lett. **21**, 1016 (1968).
6. C. A. Norman *et al.*, ApJ **454**, 60 (1995); H. Kang *et al.*, MNRAS **286**, 257 (1997).
7. J. P. Rachen and P. L. Biermann, A&A **272**, 161 (1993).
8. K. Mannheim, A&A **269**, 67 (1993); Space Sci. Rev. **75**, 331 (1996).
9. K. Mannheim *et al.*, Phys. Rev. D submitted, astro-ph/9812398.
10. M. Vietri, ApJ **453**, 883; E. Waxman, Phys. Rev. Lett. **75**, 386 (1995).
11. C. M. Urry and P. Padovani, PASP **107**, 803 (1995).
12. N. Bade *et al.*, A&A **334**, 459 (1998).
13. K. Mannheim *et al.*, A&A **251**, 723 (1991).
14. A. Dar and A. Laor, ApJ **478**, L5 (1997); M. Pohl, these proceedings.
15. J. G. Kirk and A. Mastichiadis, Nature **360**, 135 (1992); D. Kazanas and A. Mastichiadis, ApJ **518**, L17 (1999).
16. R. Protheroe, in *Accretion Phenomena and Related Outflows*, ed. D. Wickramasinghe *et al.*, IAU Colloquium 163, p. 585 (1997).
17. R. Protheroe and P. L. Biermann, Astropart. Phys. **6**, 293 (1997).
18. R. Sambruna, these proceedings, and references therein.
19. R. Svensson, MNRAS **227**, 403 (1987).
20. K. Mannheim, Phys. Rev. D **48**, 2408 (1993).
21. A. Mücke *et al.*, Proc. 19th Texas Symposium, Paris, 1998, astro-ph/9905153.
22. E. Pian *et al.*, ApJ **497**, L17 (1998).
23. A. Mücke and R. Protheroe, these proceedings, and private communication.
24. P. L. Biermann and P. A. Strittmatter, ApJ **322**, 643 (1987), and references therein.

X-ray Selected BL Lacs and Blazars

Eric S. Perlman

*Department of Physics and Astronomy
The Johns Hopkins University
3400 North Charles Street
Baltimore, MD 21218, USA
email: perlman@pha.jhu.edu*

Abstract.
With their rapid, violent variability and broad featureless continuum emission, blazars have puzzled astronomers for over two decades. Today blazars represent the only extragalactic objects detected in high-energy gamma-rays. Their spectral energy distributions (SEDs) are characteristically double-humped, with lower-energy emission originating as synchrotron radiation in a relativistically beamed jet, and higher-energy emission due to inverse-Compton processes. This has accentuated the biases inherent in any survey to favor objects which are bright in the survey band, and should serve as a cautionary note both to those designing new surveys as well as theorists attempting to model blazar properties. The location of the synchrotron peak determines which blazar population is dominant at GeV and TeV energies. At GeV energies, low-energy peaked, high luminosity objects, which have high L_C/L_S ratios, dominate, while at TeV energies, high-energy peaked objects are all that is seen. I review the differences between low-energy peaked and high-energy peaked blazars, and models to explain those differences. I also look at efforts to bridge the gap between these classes with new surveys. Two new surveys have detected a large population of high-energy peaked emission line blazars (FSRQ), with properties somewhat different from previously known objects. This discovery has the potential to revolutionize blazar physics in a way comparable to the discovery of X-ray selected BL Lacs ten years ago by *Einstein*. I cull from the new and existing surveys a list of $z < 0.1$ high-energy peaked blazars which should be targets for new TeV telescopes. Among these are several high-energy peaked FSRQ.

I INTRODUCTION

Blazars have the most extreme properties of any class of active galactic nuclei. In every wavelength range, their properties are dominated by a broad, highly variable continuum. This continuum has a characteristic, double-humped shape (Figure 1), indicative of two emission processes. At lower energies, synchrotron radiation dominates the energy budget, but at X-ray through gamma-ray energies, inverse-Compton processes increasingly dominate the properties we observe. The rapid, violent variability that is the hallmark of these objects (blazars can vary in bright-

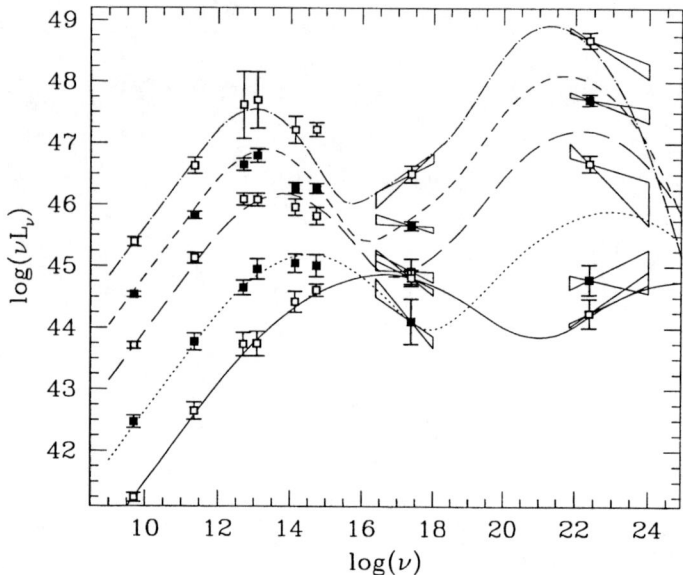

FIGURE 1. The variation of the average spectral energy distribution of blazars with radio luminosity (reprinted with permission from Blackwell 2000 [9]). The low-energy component, due to synchrotron radiation, peaks in the infrared for "red", low-energy peaked blazars, and at UV/X-ray energies for "blue", high-energy peaked blazars. Note how the location of the synchrotron peak varies with luminosity. In both "red" and "blue" blazars, the emissions at higher energies are dominated by Comptonization (see the reviews and articles herein, e.g. [2,29,41]).

ness by factors of ten or more, and doubling on timescales of hours is seen in their lightcurves; see the review herein by Rita Sambruna [44]), forces us to explain their properties as a consequence of viewing a relativistic jet moving very close to our line of sight (see [56,20] for reviews).

The blazar class covers a very wide range in luminosity as well as peak frequency. More luminous objects tend to peak at lower frequencies, but there is a wide scatter in this relation [9]. Historically, optical spectroscopic properties have been used to separate blazars into two divisions: flat-spectrum radio quasars (FSRQ) have strong, broad emission lines, while BL Lacs have very faint or no emission lines. However, this distinction now appears arbitrary, as recent work has shown a continuous distribution of emission line luminosities and equivalent widths [47].

I will review the surveys which have been used to find blazars as well as their biases, and show how this has produced two populations with somewhat different properties. I will for pedagogical reasons adopt the traditional division between BL Lacs and FSRQ, but I believe that one of the most important tasks the new surveys must undertake is to define new, physically based classes for blazars. I will describe the latest crop of surveys and their findings, and "round up" a herd of objects which should be targets for the new generation of VHE gamma-ray observatories.

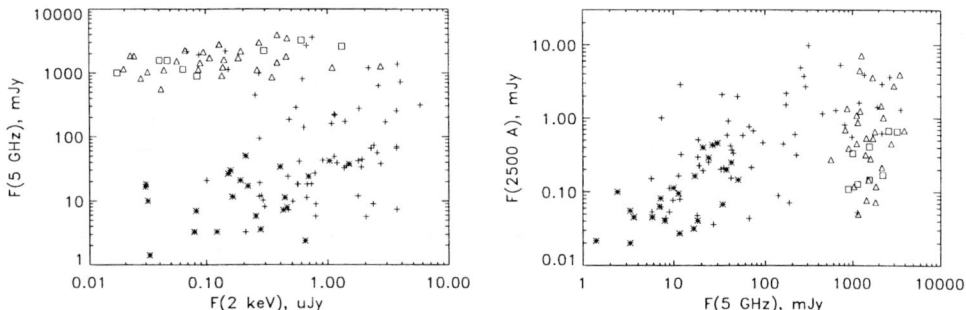

FIGURE 2. Two views of the radio-optical-X-ray flux parameter space for previous samples of BL Lacs. 1 Jy BL Lacs are shown plotted as triangles, S4 BL Lacs are shown plotted as squares, Slew BL Lacs are shown as pluses, and EMSS BL Lacs are shown as asterisks. Note the large gaps which existed prior to 1995 in our coverage of this parameter space.

II SURVEY METHODS AND BIASES

Because of their rareness, blazars have an unfortunate history of divisions "invented" because of observed properties or selection methods which may not have any physical basis. The result has been confusion not only over how to define subclasses and their properties, but indeed over the definition of the blazar class itself! The BL Lac/FSRQ division is an example of this phenomenon; another is the definition of "radio-selected" and "X-ray selected" BL Lac classes, based on the survey in which an object was found. Yet several well known objects turn up in both radio and X-ray surveys, for example Mkn 421, Mkn 501 and BL Lac.

Our understanding is helped considerably if we take a step back and try to understand the biases inherent in single-band surveys. The key point (which seems obvious but is in fact surprisingly subtle) is that any survey selects preferentially objects that are bright in the survey band. Thus the overwhelming majority of blazars selected in X-rays peak in the UV or X-rays, while nearly all blazars selected in the radio peak at much lower (IR) energies (Figure 1). The subtlety lies in the fact that these two methods attack opposite ends of parameter space (Figure 2), and *do* seem to find objects with somewhat different properties (more about this in §III). Thus while inquiries into blazar properties have achieved much by using X-ray and radio selected samples, they have hardly delved into what connects them.

Today we speak of "high-energy peaked" and "low-energy peaked" blazars, referring to objects which peak at (respectively) UV/X-ray or infrared energies. The disparity in the peak frequencies indicates significant differences in jet physics. This is expected on theoretical grounds, since the characteristic electron energy for synchrotron emission is directly related to the magnetic field ($\gamma_{peak} \propto \sqrt{\nu_{peak}/B}$). Moreover, the trends we find with luminosity (decreasing ν_{peak}, increasing emission line luminosity) indicate substantially more cooling in more luminous objects.

The location of the synchrotron peak is intimately connected to which kinds of objects are observed to dominate at GeV and TeV gamma-ray energies. At GeV energies, lower-energy peaked, high-luminosity objects dominate, as they have much higher ratios L_C/L_S. These objects, however, do not make electrons with $\gamma \sim 10^{6-7}$, which are required for X-ray synchrotron emission, probably because of increased cooling. Thus at TeV energies, objects which peak in the UV/X-rays (i.e., high-energy peaked or X-ray selected blazars) are all that is seen.

Unfortunately, current surveys do not contain enough information to tell us which kind of blazar (high or low-energy peaked) is more common. This is because current complete samples cover very shallow dynamic ranges (Figure 2). There is a general indication that high-luminosity objects are less common [56]. However, deeper surveys are needed, because finding the absolute number of either kind of object requires correcting current surveys for the objects it does *not* find – and to do this, we must go deep enough in both radio and X-ray so that radio surveys start detecting significant numbers of high-energy peaked blazars, and vice versa.

III THE PROPERTIES OF RED AND BLUE BLAZARS

Over the last decade, many workers have delved into the properties of high-energy peaked and low-energy peaked blazars, by using samples of BL Lacs selected in the radio and X-rays. BL Lacs were used in this work because until very recently there were no FSRQ known to peak at UV/X-ray energies([35,32,34]; §IV). These works found significant differences between the properties of blue, high-energy peaked BL Lacs (HBLs) and red, low-energy peaked BL Lacs (LBLs):

- HBLs are less luminous in radio and bolometrically [46,8,9].

- HBLs are less core-dominated in the radio than LBLs [37,38,26,21,43].

- HBLs are less polarized than LBLs, with a smaller duty cycle, and tend to have a preferred position angle of polarization, while LBLs do not [18].

- Occupy a different region of X-ray-optical-radio parameter space (Figure 2), and in fact a unique region of parameter space in X-ray-optical and radio-optical spectral index space (Figure 3, [54]).

- HBLs tend to have steeper X-ray and optical-X-ray continua than LBLs [40,46,57,22,33].

- HBLs are distributed differently in space, with more objects or more luminous objects (current samples cannot discriminate between these possibilities) at low redshifts [30,59,40,1,14,43]; while LBLs are consistent with either a uniform distribution with redshift or more objects at high redshift [50].

As with their properties, the relationship between HBLs and LBLs has been a subject of active debate in the literature. At first it was thought that they were related through viewing angle (e.g., [11] and refs. therein). This explained many properties in a natural way, for example the differences in polarization behavior and radio core dominance (though see [43] for new counter-evidence), as well as the observed difference in space density (which could have been a selection effect, however; see §II). A second model was proposed by Padovani & Giommi [15,31], under which HBLs and LBLs represent two ends of a continuous distribution of synchrotron peak frequencies. It turns out that both descriptions have problems. Rita Sambruna showed in her thesis [46] that differences in in viewing angle cannot produce a variation of 10^4 in peak frequency. And while the "different peak frequencies" description is accurate phenomenologically, it cannot by itself explain the differences in radio core-dominance [21,55] and polarization PA [55] behavior.

This question is still open, but a modern view is evolving which basically says that both the viewing angle and different spectral energy distributions pictures have a piece of the puzzle. Two competing models now ascribe the HBL-LBL relationship to combinations of luminosity and viewing angle [10], or luminosity and peak frequency [12]. Current data cannot distinguish between these models, although further investigation of the polarization differences with larger samples and in multi-wavelength campaigns, offer in my view the best hope for doing so.

IV THE NEW SURVEYS: BRIDGING THE GAPS

As I discussed in §II, the existing complete samples of blazars suffer from several problems. First of all they are small: typically a few dozen objects at most. There are also various concerns about completeness, particularly at the lowest luminosities (e.g., [4,27,28,40,42]). But the most difficult problems have to do with the small dynamic ranges covered in flux and luminosity (cf. Figure 2).

In this section we review the latest information on existing samples. However, we will concentrate largely on four new surveys which are bridging the gaps in our coverage of parameter space. These surveys are allowing us to for the first time actively pursue the connections between blazar classes and get at the real physics. The most exciting discovery of these surveys is the existence of a large population of high-energy peaked FSRQ. We will reanalyze two existing samples in the light of these findings, and show that their makeup is consistent with the new surveys.

The existing samples of blazars are listed below:

- *Einstein* Slew Survey [39,36]: $F(X) \gtrsim 10^{-11}$ erg cm^{-2} s^{-1}, $F(R) > 1$ mJy, 50% of the sky. 66 BL Lacs, 19 FSRQ. Includes all known TeV emitters.

- *Einstein* EMSS [54,30,43,36]: $F(X) > 2 \times 10^{-13}$ erg cm^{-2} s^{-1}, $F(R) > 1$ mJy, 2% of the sky. 43 BL Lacs, 16 FSRQ.

- 1 Jy [50,49,51]: $F(R) > 1$ Jy, 60% of sky. 37 BL Lacs, 222 FSRQ, but not completely identified.

- S5 [53]: $F(R) > 250$ mJy, $\delta > 70°$. 11 BL Lacs, 20 FSRQ, but not completely identified.

- S4 [52]: $F(R) > 500$ mJy, $35° < \delta < 70°$. 7 BL Lac, 56 FSRQ, but not completely identified.

Four new surveys are in progress, filling vast new regions of parameter space (Figures 3,4). These surveys are listed in Table 1.

TABLE 1. New Surveys for Blazars

Survey	BL Lacs	$z < 0.1$	FSRQ	$z < 0.1$	Flux Limit F(0.1-2 keV) erg cm^{-2} s^{-1}	F(R) mJy	Types of Objects
DXRBS [35,23,34]	40	5[a]	218[b]	0	10^{-14}	25-50	most LBL/intermediate
RGB [3,24,25]	127	14[c]	252[d]	9	10^{-12}	25	most HBL/intermediate
NVSS-RASS [14]	58[e]	1[f]	?	?	10^{-12}	3.5	all HBL
REX [5]	?[g]	?	?	?	10^{-14}	3	mostly HBL

[a] 36 BL Lacs with redshifts.
[b] IDs nearly complete; 49 HBL-like FSRQs with $\alpha_{rx} < 0.78$ and $\alpha_{ro} < 0.6$.
[c] 49 BL Lacs with redshifts.
[d] IDs nearly complete; 96 HBL-like FSRQs with $\alpha_{rx} < 0.78$ and $\alpha_{ro} < 0.6$.
[e] Spectroscopy ongoing: 155 candidates, 85% efficiency expected.
[f] 36 with redshifts
[g] Candidate list not released; surveys ROSAT pointed database

The depth and size of these new surveys has allowed them to probe much deeper into the luminosity function of blazars than ever before. Close examination of Figures 3 and 4, and comparison with Figure 2 reveals two key discoveries. First of all, the number of FSRQ with radio luminosity $L_R < 10^{26.5}$ W Hz^{-1} has increased nearly ten-fold, and for the first time luminosities below $10^{25.5}$ W Hz^{-1} are being reached (an equally large expansion of the number of low luminosity BL Lacs is also taking place). This is important because the knee in the radio luminosity function of FSRQ is located at or near $10^{26.5}$ W Hz^{-1}, and the location of the knee and the shape of the luminosity function below the knee are very poorly constrained due to the paucity of low luminosity objects in current samples. The second, equally important discovery, is that because these surveys have plugged the holes in our coverage of X-ray-optical-radio parameter space, they have revealed large numbers of blazars in regions of parameter space where previously very few were known [35,24,34,25]. These objects fall into two categories: intermediate BL Lacs, which have peak energies $\sim 1-10$ eV, and X-ray bright FSRQ, which some authors had predicted did not exist due to the observed continuity in broadband spectral properties between the blazars known in complete samples in 1996 [46]. These two discoveries are in fact not completely independent of one another, because the lowest luminosities are dominated by high-energy peaked objects (Figure 4).

The X-ray bright FSRQ are in fact particularly important for our knowledge of the class, because they overlap significantly in radio and bolometric luminosities

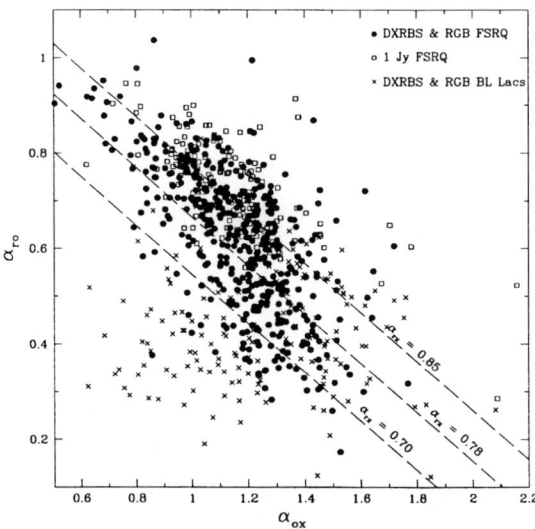

FIGURE 3. X-ray-optical and Optical-Radio Spectral indices of FSRQ and BL Lacs discovered in radio and X-ray surveys[34,35]. Note the vastly different ranges of parameter spaces covered by historical radio surveys (1 Jy: open squares) compared to X-ray surveys (DXRBS and RGB: see §4). Prior to 1996, in fact, the parameter space covered by X-ray and radio techniques was almost *completely* disjoint, as shown in Figure 2.

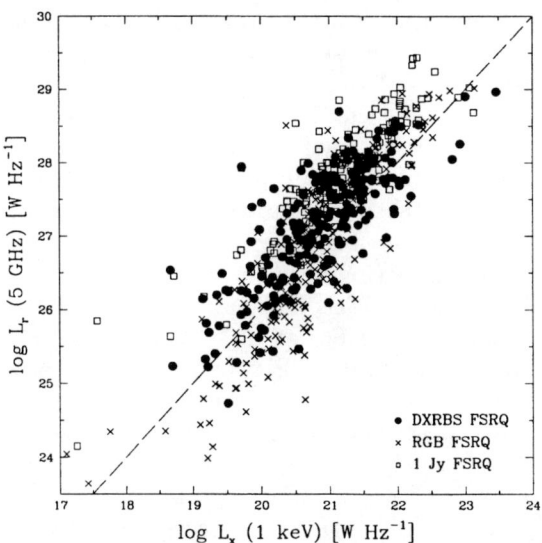

FIGURE 4. X-ray and Radio luminosities for FSRQs[34,35]. Note that high-energy peaked FSRQs (right of the dashed line) are less luminous in the radio (for the same X-ray luminosity). DXRBS and RGB sample unexplored regions of parameter space, containing objects $\sim 20\times$ less luminous than previous surveys.

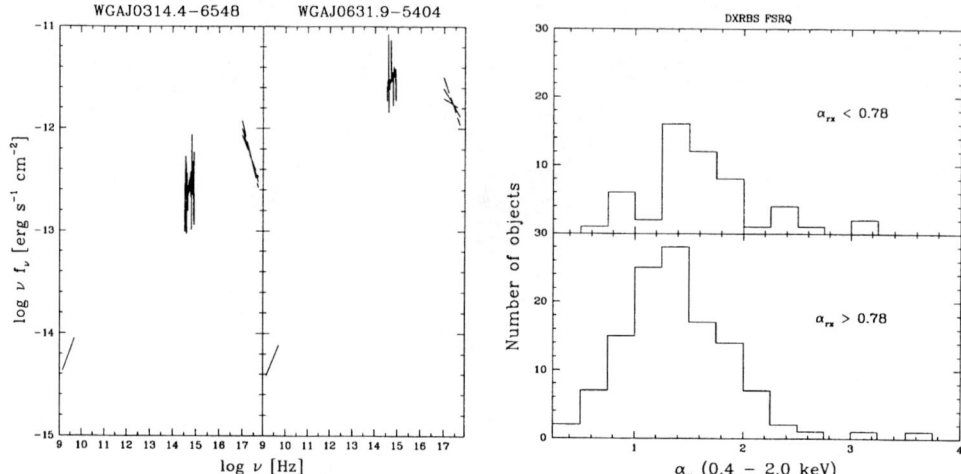

FIGURE 5. At left, the radio to X-ray SED of two high-energy peaked FSRQ [35]. The data clearly point to a spectral peak in the UV/X-ray, similar to HBL BL Lacs. At right, ROSAT spectral indices of DXRBS FSRQ with radio-to-X-ray spectral indices above and below 0.78 [35], which is equivalent to the diagonal line at $\log L_x/L_R = 10^{-6}$ on Figure 4. The spectra of the X-ray brighter objects are steeper than those of more radio-loud "traditional" FSRQ.

with the low-energy peaked BL Lacs. Investigations into their properties and comparisons with previously known FSRQ hold the promise of truly understanding the connections between different classes of blazars. A similar thing can be said for the intermediate BL Lacs; but since the range of parameter space opened up is larger for the emission line objects and the potential impact on VHE gamma-ray astronomy from them is greater, I will concentrate on them in this paper.

The DXRBS collaboration has begun an investigation into the properties of X-ray bright FSRQ. We find that these objects differ in important ways from their lower-energy peaked cousins (in most cases paralleling radio/X-ray selected BL Lac differences). These differences (Figure 5) include ROSAT spectra that are steeper than known FSRQ [34,45], and SEDs that appear to peak in the UV/X-rays, based on ROSAT, optical and radio catalog data [34]. They also overlap significantly with HBL BL Lacs on the (α_{ro}, α_{ox}) plane (Figure 3). However, observations of several of these objects at harder X-ray energies with ASCA and SAX have revealed that most (but not all) have rather flat spectra, more similar to other FSRQ than HBL BL Lacs ([44,6], although note that the selection criteria were different).

This last observation is a fly in the ointment, because the most natural explanation of the radio to X-ray SEDs derived from survey data is synchrotron radiation from a single population of electrons, as with HBL BL Lacs. If indeed this paradigm holds, one would expect steeper hard X-ray spectra, dominated by the steep tail of the particle distribution. The observation that many of these objects appear to have flat hard X-ray spectra is difficult to explain in this context (note that the curvature is in the opposite sense for "blue bump" emission, however). There are several possibilities. For example, if Compton cooling is stronger in these objects (due to higher electron density, for example) than HBL BL Lacs, Comptonization could begin to dominate the energy budget at energies of a few keV instead of tens to hundreds of keV as it does in the HBL BL Lacs. It is also possible that there might be some Compton reflected emission from the accretion disk in the hard X-rays, as there is for Seyfert galaxies (e.g., [17]). However, no spectral curvature is seen at a few keV [6,44]; nor do we see a strong Fe Kα line from the inner reaches of the accretion disk, as is seen in Seyferts. There is also a question of selection: Padovani & Giommi [15,31] showed that (assuming a roughly parabolic SED), a hypothetical object moves diagonally down on the $(\alpha_{ox}, \alpha_{ro})$ plane (roughly along the locus of the 1 Jy sample) until its peak reaches the optical, after which it begins to move horizontally to the left (roughly along the locus of the X-ray selected surveys). Thus one may need to make two cuts (in α_{ro} and α_{rx}, as below) rather than one in order to select high-energy peaked objects.

If indeed the radio to X-ray continuum in high-energy peaked FSRQ is produced by synchrotron radiation, one might expect that Comptonized emission would peak at around 1 TeV (similar to HBL BL Lacs), so that these objects should be targets for new TeV observatories. However, this assumes that the balance of mechanisms in Compton cooling is not too different in these objects than it is in HBL BL Lacs.

To produce a list of TeV candidates, I analyzed the DXRBS and RGB samples, as well as existing surveys, which previously had unknown numbers of FSRQ. I found significant numbers of FSRQ in both the EMSS (16) and Slew (19); About 30% of these appear to be high-energy peaking objects ([36]; 7/16 in the EMSS and 4/19 in the Slew), consistent with the DXRBS and RGB results [34]. I have selected from these surveys $z < 0.1$ objects with $\alpha_{ro} < 0.6$ and $\alpha_{rx} < 0.78$ and $F(X) > 10^{-12}$ erg cm^{-2} s^{-1}. The result is shown in Table 2[1].

Importantly, lists such as Table 2 should *not* serve as a be-all and end-all for future TeV surveys. This is particularly true for southern hemisphere observers, since most blazar surveys unfortunately cover very little of the southern sky! It is very important that large angle TeV surveys be carried out in the near future, and I am encouraged to see that one is now being planned [7]. It has become a truism that every time a large-area survey is done in a new waveband, some completely unexpected discovery is made. It is also equally true that without large-area surveys, it is not possible to correctly derive physics for an entire class. An example of this comes from comparing the ratio of GeV to radio emission typical for EGRET detected blazars (~ 700) with that required of all blazars in order to produce the diffuse GeV background (~ 70; [19,48]). If we based all our modeling of how the GeV continuum of blazars is produced upon the assumption that nature only makes blazars with GeV/radio ratios in the hundreds, our model would not be accurate for the vast majority of sources. The same holds in the TeV [58].

---

[1] The NVSS-RASS survey will add a few candidates when their survey is complete; a first list of TeV candidates is expected to be released soon in [13]

TABLE 2. Suggested Candidates for TeV Emission

Name	RA	Dec	$F(0.3\text{--}3.5\text{ keV})^a$ [32]	$F(R)$	z	ID
	(J2000)		10^{-12} erg cm^{-2} s^{-1}	mJy		
1ES0033+595	00 35 52.7	+59 50 04	75.1	66	0.086	BL Lac
RGB0110+418	01 10 04.8	+41 49 50	2.5^b	40	0.096	BL Lac
RGB0152+017	01 52 39.7	+01 47 18	5.0^b	51	0.080	BL Lac
RGB0153+712	01 53 25.9	+71 15 07	3.1^b	643	0.022	BL Lac
RGB0214+517	02 14 17.9	+51 44 52	13.0^b	291	0.049	BL Lac
RGB0314+247	03 14 02.7	+24 44 31	2.1^b	6	0.054	BL Lac
1ES0548−322	05 50 41.9	−32 16 11	44.1	170	0.069	BL Lac
RGB0656+426	06 56 10.7	+42 37 02	3.9^b	480	0.059	BL Lac
Mkn 180	11 36 26.4	+70 09 28	7.1	94	0.046	BL Lac
RGB1532+302	15 32 02.2	+30 16 28	5.9^b	42	0.064	BL Lac
RGBJ1610+671	16 10 02.6	+67 10 29	4.8^b	36	0.067	BL Lac
1ES1727+502	17 28 18.5	+50 13 11	13.7	159	0.055	BL Lac
1ES1741+196	17 43 57.5	+19 35 10	24.6	223	0.083	BL Lac
1ES1959+650	19 59 59.9	+65 08 55	83.4	252	0.048	BL Lac
PKS2005−489	20 09 25.3	−48 49 53	16.1	1192	0.071	BL Lac
1ES2321+419	23 23 52.0	+42 11 00	2.2	19	0.059	BL Lac
RGB 2322+346	23 22 44.0	+34 36 14	2.2^b	78	0.098	BL Lac
III Zw 2	00 10 31.0	+10 58 28	4.8	420	0.090	FSRQ
B2 0138+398	01 41 57.8	+39 23 30	1.1^b	115	0.080	FSRQ
B2 0321+33	03 24 41.2	+34 10 45	6.6	364	0.062	FSRQ
RGB1413+436	14 13 43.7	+43 39 45	4.5^b	39	0.090	FSRQ
PG 2209+184	22 11 53.7	+ 18 41 51	8.4	134	0.070	FSRQ

^{a} corrected for galactic absorption
^{b} uncorrected 0.1-2.4 keV Flux taken from [25]; no 0.3-3.5 keV flux measurement available

V CONCLUSIONS

It is safe to say that there are many key open questions in blazar research. Future surveys will be crucial in addressing many of these. Here are a few examples:

- What is the nature of the HBL-LBL relationship (in both BL Lacs *and* FSRQ), and the BL Lac-FSRQ relationship? These questions can only be addressed properly when radio selected samples have large numbers of HBL type objects, and X-ray selected samples have large numbers of LBL type objects. Gamma-ray surveys can also help here: each model for the production of gamma-rays has a different dependence on viewing angle and Lorentz Γ [2,41,29].

- What constrains emission line luminosity and what role does the emission line luminosity play in gamma-ray production in various types of blazars? One prerequisite for answering this question now exists: large surveys which contain blazars of all X-ray/radio continuum and emission line luminosity classes. But the other does not: a gamma-ray database that includes variability information on statistically significant numbers of objects at all luminosities.

- What are the gamma-ray spectral energy distributions of all blazars? The prerequisite for answering this survey is gamma-ray surveys in the MeV, GeV and TeV ranges, which are deep enough to contain statistically significant numbers of blazars of all continuum and emission line luminosity classes.

- Are blazars (whether aligned or misaligned) really the only extragalactic objects which produce detectable quantities of gamma-rays with energies greater than ~ 10 MeV? This question is intimately related to the issue of what produces the gamma-ray background [19].

These questions, and the new issues they raise, are intimately connected with, and equally as important, as those that will be addressed (and raised) in multi-wavelength campaigns and modelling efforts.

This paper summarizes work I have done in collaboration with several other scientists. I would particularly like to acknowledge my DXRBS collaborators, Paolo Padovani, Paolo Giommi, Hermine Landt and Rita Sambruna.

REFERENCES

1. Bade, N., Beckmann, V., Douglas, N. G., Barthel, P. D., Engels, D., Cordis, L., Nass, P., & Voges, W., 1998, A & A 334, 459.
2. Böttcher, M., 1999, these proceedings.
3. Brinkmann, W., Siebert, J., Feigelson, E. D., Kollgaard, R. I., Laurent-Muehleisen, S. A., Reich, W., Fuerst, E., Reich, P., Voges, W., Trümper, J., McMahon, R., 1997, A & A 323, 739.
4. Browne, I. W. A., & Marchã, M. J. M., 1993, MNRAS 261, 795.
5. Caccianiga, A., Maccacaro, T., Wolter, A., Della Ceca, R., Gioia, I. M., 1999, ApJ 513, 51.
6. Costamante,L., Padovani, P., Perlman, E., Giommi, P., 1999, in preparation.
7. Dingus, B., these proceedings.
8. Fossati, G., Celotti, A., Ghisellini, G., & Maraschi, L., 1997, MNRAS 289, 136.
9. Fossati, G., Maraschi, L., Celotti, A., Comastri, A., & Ghisellini, G., 1998, MNRAS 299, 433.
10. Georganopoulos, M., & Marscher, A. P., 1998, ApJ 506, 621.
11. Ghisellini, G., Padovani, P., Celotti, A., & Maraschi, L., 1993, ApJ 407, 65.
12. Ghisellini, G., Celotti, A., Fossati, G., Maraschi, L., & Comastri, A., 1998, MNRAS 301, 451.
13. Giommi, P., Menna, M. T., & Padovani, P., in preparation.
14. Giommi, P., Menna, M. T., & Padovani, P., 1999, MNRAS in press (astro-ph/9907014).
15. Giommi, P., & Padovani, P., 1994 MNRAS 268, L51.
16. Giommi, P., Tagliaferri, G., Bueurmann, K., Branduardi-Raymont, G., Brissenden, R., Graser, U., Mason, K. O., Mittaz, J. D. P., Murdin, P., Pooley, G., Thomas, H.-C., & Tuohy, I., 1991, ApJ 378, 77.

17. Grandi, P., Haardt, F., Ghisellini, G., Grove, E. J., Maraschi, L., & Urry, C. M., 1998, ApJ 498, 220.
18. Jannuzi, B. T., Smith, P. S., & Elston, R., 1993, ApJ 428, 130.
19. Kazanas, D., & Perlman, E., 1997, ApJ 476, 7.
20. Kollgaard, R. I., 1994, Vist. Astron 38, 29.
21. Kollgaard, R. I., Palma, C., Laurent-Muehleisen, S. A., & Feigelson, E. D., 1996, ApJ 465, 115.
22. Lamer, G., Brunner, H., & Staubert, R., 1996, A & A 311, 384.
23. Landt, H., Perlman, E. S., Padovani, P., & Giommi, P., 1999, MNRAS, submitted.
24. Laurent-Muehleisen, S. A., Kollgaard, R. I., Ciardullo, R., Feigelson, E. D., Brinkmann, W., & Siebert, J., 1998, ApJS 118, 127.
25. Laurent-Muehleisen, S. A., Kollgaard, R. I., Feigelson, E. D., Brinkmann, W., & Siebert, J., 1999, ApJ in press (astro-ph/9905133).
26. Laurent-Muehleisen, S. A., Kollgaard, R. I., Moellenbrock, G. A., & Feigelson, E. D., 1993, AJ 106, 875.
27. Marchã, M. J. M., Browne, I. W. A., Impey, C. D., & Smith, P. S. 1996, MNRAS 281, 425.
28. Marchã, M. J. M., & Browne, I. W. A., 1996, MNRAS 279, 72.
29. Mastichiadis, A., 1999, these proceedings.
30. Morris, S. L., Stocke, J. T., Gioia, I. M., Schild, R. E., Wolter, A., Maccacaro, T., & Della Ceca, R., 1991, ApJ 380, 49.
31. Padovani, P., & Giommi, P., 1995, ApJ 444, 567.
32. Padovani, P., Giommi, P., & Fiore, F., 1997, in Mem. Astron. Soc. Italia 68, 147.
33. Padovani, P., Giommi, P., & Fiore, F., 1997 MNRAS 284, 569.
34. Padovani, P., Perlman, E., Landt, H., Sambruna, R., Giommi, P., 1999, MNRAS, submitted.
35. Perlman, E. S., Padovani, P., Giommi, P., Sambruna, R., Jones, L. R., Tzioumis, A., & Reynolds, J., 1998, AJ 115, 1253.
36. Perlman, E. S., Schachter, J. F., & Stocke, J. T., 1999, in preparation.
37. Perlman, E. S., & Stocke, J. T., 1993, ApJ 406, 430.
38. Perlman, E. S., & Stocke, J. T., 1994, AJ 108, 56.
39. Perlman, E. S., et al., 1996, ApJS 104, 251.
40. Perlman, E. S., Stocke, J. T., Wang, Q. D., & Morris, S. L., 1996, ApJ 456, 451.
41. Rachen, J., 1999, these proceedings.
42. Rector, T. A., Stocke, J. T., & Perlman, E. S., 1999, ApJ 516, 145.
43. Rector, T. A., Stocke, J. T., Perlman, E. S., & Morris, S. L., 1999, ApJ submitted.
44. Sambruna, R. M., these proceedings.
45. Sambruna, R. M., 1997, ApJ 487, 536.
46. Sambruna, R. M., Maraschi, L., & Urry, C. M., 1996, ApJ 463, 444.
47. Scarpa, R., & Falomo, R., 1997 A & A 325, 109.
48. Stecker, F. W., & Salamon, M. H., 1996, ApJ 464, 600.
49. Stickel, M., Fried, J. W., & Kühr, H., 1993, A & A S 268, 53.
50. Stickel, M., Fried, J. W., Kühr, H., Padovani, P., & Urry, C. M., 1991, ApJ 374, 431.
51. Stickel, M., Meisenheimer, K., & Kühr, H., 1994, A & A S 105, 211.

52. Stickel, M., & Kühr, H., 1994a, A & A S 103, 349.
53. Stickel, M., & Kühr, H., 1994a, A & A S 115, 1.
54. Stocke, J. T., Morris, S. L., Gioia, I. M., Maccacaro, T., Schild, R., Wolter, A., Fleming, T. A., & Henry, J. P., 1991 ApJS 76, 813.
55. Stocke, J. T., 1996, in Extragalactic Radio Sources, IAU 175, ed. R. D. Ekers, C. Fanti, and L. Padrielli (Kluwer: Dordrecht), p. 385
56. Urry, C. M., & Padovani, P., 1995, PASP 107, 803.
57. Urry, C. M., Sambruna, R. M.,Worrall, D. M., Kollgaard, R. I., Feigelson, E. D., Perlman, E. S., Stocke, J. T., 1996, ApJ 463, 424.
58. Urry, C. M., 1999, *Astroparticle Physics* 11, 159.
59. Wolter, A., Caccianiga, A., Della Ceca, R., & Maccacaro, T., 1994, ApJ 433, 29.

Spectral Variability in the blazar PKS 0528+134

R. Mukherjee[*] & M. Böttcher[†]

[*] Dept. of Physics & Astronomy, Barnard College & Columbia University, NY, NY 10027
[†] Space Physics & Astronomy, Rice University, Houston, TX 77005

Abstract.
We present recent modeling results on all available simultaneous broadband spectra of the flat-spectrum radio quasar (FSRQ) PKS 0528+134 during 6 years of EGRET observations. We find that the spectral energy distributions of the source are most suitably reproduced by a two-component model, in which the target photons are produced externally to the γ-ray emitting region, but also including an SSC component. We observe that during the higher γ-ray states, the bulk Lorentz factor of the jet increases, possibly causing an intensification of the external soft photon field. Our model calculations indicate the trend that the energies of the electrons giving rise to the synchrotron peak decreases, as the source goes from a low to a high γ-ray state, in contrast to that observed in high-frequency peaked BL Lac objects (HBLs).

INTRODUCTION

PKS 0528+134, a compact radio-loud quasar, is one of the 66 blazars detected by EGRET (Energetic Gamma Ray Experiment Telescope) [9]. It is one of the most luminous EGRET AGN (active galactic nuclei) and has been observed to have high fluxes from radio through infrared, exhibiting extreme variations in the observed γ-ray emission.

The high γ-ray luminosity of PKS 0528+134 suggests that the emission is likely to be beamed and, therefore, Doppler-boosted along the line of sight. The jet models explain the radio to UV continuum from blazars as synchrotron radiation from high energy electrons in a relativistically outflowing jet which has been ejected from an accreting supermassive blackhole. The emission in the MeV-GeV range is believed to be due to the inverse Compton scattering of low-energy photons by the same relativistic electrons in the jet. However, the source of the soft photons that are inverse Compton scattered remains unresolved. The soft photons can originate either as synchrotron emission from within the jet ("Synchrotron self-Compton" (SSC) mechanism, e. g. [11,2]), or from a nearby accretion disk ("External radiation Compton" (ERC) mechanism e. g. [6]), or they can be disk radiation reprocessed in broad line region (BLR) clouds (e. g. [13,1,7]). An alternative model proposes that the synchrotron emission is due to the population of both electrons and electron-positron pairs arising from shock-accelerated electrons and protons [10]. Recently, a model combining the ERC and SSC scenarios has been used to fit the simultaneous COMPTEL and EGRET spectra of PKS 0528+134 [4] measured in Phases 1-3 of the CGRO observations. This model suggests that in the case of PKS 0528+134, the SSC process dominates the γ-ray spectrum in the γ-ray low state, while in the γ-ray high state an additional ERC component becomes dominant at energies above 10 MeV.

Multiwavelength studies of blazars play a very important role in the study of emission mechanisms in these objects. A large amount of archival γ-ray and lower energy data exist on PKS 0528+134. In this article we examine, collectively, the multiwavelength data on PKS 0528+134 after six years of observations of the source with EGRET, and study the spectral trends of the source during its various "high," "intermediate," or "low" states.

I OBSERVATIONS

The γ-ray (above 100 MeV) light curve of PKS 0528+134 from 1991 to 1997 is shown in Fig. 1. The source is highly variable in γ-rays, with a probability less than 10^{-15} that the flux variations are consistent with a constant flux. A description of the γ-ray data and further details of the observations is given elsewhere [12]. Of the data presented in the figure, we have analyzed the spectra of PKS 0528+134 in nine separate viewing periods (VPs), based on the availability of simultaneous multiwavelength data. The selected viewing periods, dates of observation, γ-ray fluxes, and the photon spectral indices are shown in Table 1.

Contemporaneous multiwavelength data on PKS 0528+134 for the observations listed in Table 1 were compiled from the literature or from archival data sets. A complete description of these data is given elsewhere [12].

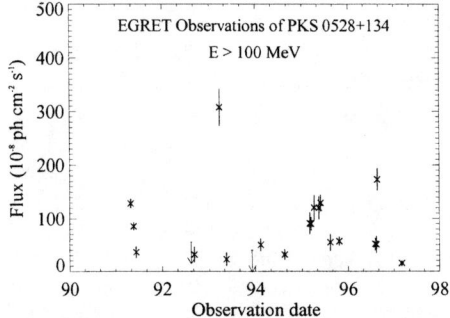

FIGURE 1. Summary of six years of EGRET observations of PKS 0528+134. The source has exhibited strong variability, flaring to nearly three times its mean flux during 1993 March (VP 213). The source was detected at it lowest state in 1997 March (VP 616.1).

II THE MODEL

The model for the geometry of the relativistic AGN jet used in this analysis is based on the one by Böttcher, Mause, & Schlickeiser (1997) [3]. The mass of the central black hole is assumed to be $M_{BH} = 5 \times 10^{10}~M_\odot$ and the luminosity of the accretion disk is assumed to be $L = 5 \times 10^{46}$ ergs s^{-1}. The model assumes a spherical blob filled with ultrarelativistic pair plasma which is moving with a bulk Lorentz factor Γ along an existing jet structure perpendicular to the accretion disk. The particles inside the blob are distributed isotropically according to the power-law $n_e(\gamma) \propto \gamma^{-s}$, where γ is the Lorentz factor in the rest frame of the blob, and s is the spectral index. The low and high energy cutoffs in the Lorentz factors of the electrons are given by γ_1 and γ_2. The value of the magnetic field B is chosen to be close to the equipartition value. The model uses a combination of SSC and ERC mechanisms to explain the spectral states of PKS 0528+134 during its various high and low states. The model follows the evolution of the pair distribution and the photon spectra as the blob moves out along the jet. A more detailed description of the model is given elsewhere [12].

III RESULTS AND DISCUSSION

Figure 2 shows the fit results to two extreme states of PKS 0528+134: the highest γ-ray state in 1993 March, and a low state in 1997 March. The model calculations for the other observations are given in Table 2.

These results reveal an important prediction of the model adopted in the analysis presented here: During the high γ-ray state, the synchrotron spectrum appears to peak at lower frequencies than in the low states. The results obtained from the spectral analysis are summarized below.

A pure SSC model does not reproduce the spectra well. Instead, a two-component model, in which the target photons are produced externally to the γ-ray emitting

TABLE 1. EGRET Spectral Analysis for PKS 0528+134 in Selected Observations

Viewing Period	Observing Date	Photon Spectral Index	Flux ($\times 10^{-7}$) photons cm^{-2} s^{-1}
0.2-0.5	1991 Apr	2.27 ± 0.07	12.9 ± 0.9
39.0	1992 Sep	2.39 ± 0.78	3.2 ± 1.4
213.0	1993 Mar	2.21 ± 0.11	30.8 ± 3.5
337.0	1994 Aug	2.68 ± 0.44	3.2 ± 1.0
413.0	1995 Mar	2.21 ± 0.16	9.0 ± 1.3
420.0	1995 May	2.37 ± 0.13	13.0 ± 1.6
502.0	1995 Oct	2.32 ± 0.16	5.7 ± 0.8
528.0	1996 Aug	2.44 ± 0.44	17.3 ± 2.1
616.1	1997 Mar	2.51 ± 0.47	1.1 ± 0.5

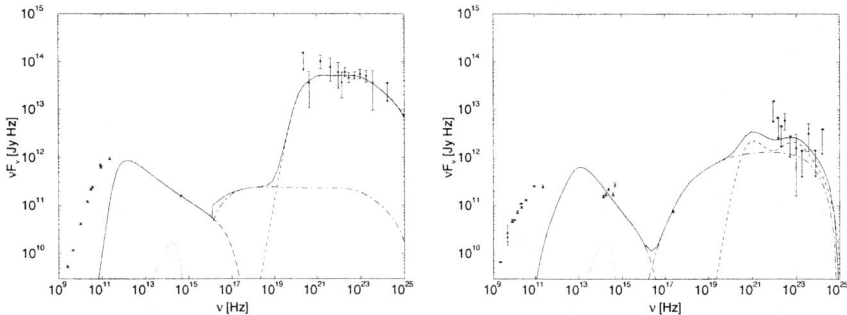

FIGURE 2. Fit to the SED of PKS 0528+134 during a (left) high γ-ray state (1993 March, VP 213) and (right) low γ-ray state (1992 September, VP 39). The long-dashed line represents the synchrotron spectrum, the short-dashed line is the ERC component, the dot-dashed line is the SSC component, and the solid line represents the combined SSC + ERC model. The dotted line is the accretion disk spectrum. Fit parameters are given in Table 2.

TABLE 2. EGRET Spectral Analysis for PKS 0528+134 in Selected Observations

VP	γ_1	γ_2	s	Γ	n_e cm^{-3}	B G	Gamma-ray State*
0.2-0.5	180	10^5	2.5	15	290	2.5	High
39.0	1000	10^5	2.5	5	150	2.5	Low
213.0	120	6×10^4	2.6	20	150	1.0	High
337.0	900	10^5	2.5	10	150	3.0	Low
413.0	1000	10^5	2.5	10	180	2.0	Intermediate
420.0	500	7×10^4	2.5	20	180	2.0	High
502.0	1000	10^5	2.5	7	180	3.2	Intermediate
528.0	600	10^5	2.2	10	180	2.5	High
616.1	1000	10^5	2.5	5	50	2.5	Low

* The high energy γ-ray state of the source [12].

region, but also including an SSC component, is required to suitably reproduce the SEDs of the source.

The spectral variability in PKS 0528+134 appears to arise from the different Doppler boosting patterns of the SSC and the ERC radiations. The SSC mechanism plays a larger role when the source is in a low flux state. The ERC mechanism is the dominant cooling mechanism when the source is in a high γ-ray state. There appears to be a trend in the observed properties of PKS 0528+134, as the source goes from a γ-ray low state to a flaring state. During the higher γ-ray states, the bulk Lorentz factor of the jet increases and the ERC component dominates the high-energy emission.

The model calculations indicate the trend that the energies of the electrons giving rise to the synchrotron peak decreases, and the power-ratio of the γ-ray and low energy spectral components increases, as the source goes from a low to a high γ-ray state. This is opposite to the change in the synchrotron peak observed in HBLs. Recently, Böttcher (1999) [5] concluded that such a shift in the synchrotron peak during a γ-ray flare may be typical of all flat-spectrum radio quasars that are γ-ray bright.

It has generally been seen that sources with higher bolometric luminosity, with the dominant component at γ-ray energies, have spectra that are better explained by the ERC models than pure SSC models. Ghisellini et al. (1998) [8] find that the properties of HBL (high frequency peaked BL Lacs), LBL (low frequency peaked BL Lacs) and FSRQs are located along a sequence, with the HBL characterized by the lowest intrinsic power and weakest external radiation field. Our results on PKS 0528+134 indicates that a similar trend might be exhibited even within a single source such as 0528+134 during the flaring and quiescent spectral states.

The main source of uncertainty in our results comes from the incomplete spectral coverage of PKS 0528+134, leading to the fits being poorly constrained. The results, however, are consistent with the predictions of the synchrotron peak shift in flaring blazars made by Böttcher (1999) [5].

R. Mukherjee was supported by NASA Grant NAG5-3696. M. Böttcher was supported by NASA Grant NAG5-4055.

REFERENCES

1. Blandford, R. D. & Levison, A. 1995, ApJ, 441, 79
2. Bloom, S. D. & Marscher, A. P. 1996, ApJ, 461, 657
3. Böttcher, M., Mause, H., & Schlickeiser, R. 1997, A&A, 324, 395.
4. Böttcher, M. & Collmar, W. 1998, A&A, 329, L57.
5. Böttcher, M. 1999, 515, L21.
6. Dermer, C. D. & Schlickeiser, R. 1994, ApJS, 90, 945
7. Ghisellini, G. & Madau, P. 1996, MNRAS, 280, 67
8. Ghisellini, G. et al. 1998, MNRAS, 301, 451.
9. Hartman, R. C., et al. 1999, ApJS, 123, 79.
10. Mannheim, K., 1993, A&A, 269, 67.
11. Maraschi, L., Ghisellini, G., & Celotti, A. 1992, ApJ, 397, L5
12. Mukherjee, R., et al. 1999, ApJ, December issue.
13. Sikora, M., Begelman, M. C., & Rees, M. J. 1994, ApJ, 421, 153

Observation of BL Lac Type AGNs With the Equatorial Mounted HEGRA Cherenkov Telescope

M. Schilling, O. Mang, G. Rauterberg, for the HEGRA Collaboration

Universität Kiel, Institut für Experimentelle und Angewandte Physik, Leibnizstraße 17-19, D-24118 Kiel, Germany

Abstract. The first Imaging Atmospheric Cherenkov Telescope of the HEGRA collaboration, CT1, has been used to observe BL Lac type objects. This equatorial mounted telescope is on account of the mirror configuration and the camera equipment especially suited for the observation of point sources, such as AGNs. In this paper we present the studies of the two objects MS 0116 and 3C 66A, which have been monitored by CT1 between August and December 1998. No evidence for TeV γ-ray emission was found.

INTRODUCTION

The study of BL Lac objects as possible TeV γ-sources is one of the major scopes of the HEGRA experiment. The two BL Lac objects Mkn 501 and Mkn 421 are well known TeV sources and under regular observation by our telescopes. During 1997 Mkn 501 showed a noticeable outburst with diurnal flux levels reaching ten Crab units [1]. The temporal coverage of the 1997 lightcurve of Mkn 501 measured with CT1 is unrivalled [2].

According to the widely accepted model of AGNs, which explains the difference between their subtypes by the orientation of the relativistic jet with respect to the observer, an object appears as a blazar if the jets are orientated along the observer's line of sight. The relativistic jets are supposed to be the source of inverse Compton scattered γ-ray beams. In high frequency BL Lacs (HBLs), which form a subset of the blazar class, the non-thermal emission components, widely believed to be synchrotron and inverse Compton radiation of the relativistic electrons, have their peaks at high energies, and thus are likely candidates for detection at energies above 500 GeV with Cherenkov Telescopes. For the better understanding of these kinds of sources, the detection of further BL Lac objects is very important.

EXPERIMENTAL SETUP

The HEGRA experiment is located on the Canary Island La Palma at the Observatorio Roque de los Muchachos at 2200 m above sea level, at 28.75°N, 17.89°W. It consists of several arrays of particle and Cherenkov detectors dedicated to cosmic ray research [8]. A total of six Cherenkov Telescopes is operated, five of them are alt-azimuth mounted and are operating as a stereoscopic system of telescopes.

The equatorial mounted Cherenkov Telescope CT1 has undergone significant modifications since its establishment as a prototype telescope in August 1992. In October 1997 it has been equipped with a new mirror dish of 33 aluminiumoxide coated hexagonal aluminium mirrors with a total reflecting area of 10 m^2. The camera consists of 127 photomultiplier tubes, which are arranged in a hexagonal tight packed structure, with a total field of view of 3.25°. The tracking accuracy of CT1 is better than 0.1°. For details on the hardware see eg. Rauterberg et al. [11].

OBSERVATION

Between August and December 1998 we observed 3C 66A (RA: 02 22 39.68, DEC: +43 02 07.8, redshift $z = 0.444$) for 36 hours at zenith angles between 14° and 35°, and MS 0116 (RA: 01 19 35.3, DEC: +32 10 47, redshift $z = 0.0592$) for 63 hours between 3° and 45° zenith angle. The objects have been observed during each moonless night for at least 1 h exposure time, whenever observations were possible, regarding weather and atmospherical conditions. The lightcurve on a diurnal basis allows us to detect high states and short range flares of the monitored object.

Background data were recorded for a total of 51 h in the considered time period. In order to maximize ON-source observation time for different objects, the background data were not taken in ON/OFF cycles, but at times when no other source of major interest was visible, taking advantage of the isotropic nature of the hadronic background, see Petry and Kranich [10] for details.

The absolute calibration and determination of the energy scale is difficult, as there is no artificial TeV γ-ray beam. Though the uncertainties concerning the energy spectrum of the Crab Nebula are about 50 % on the absolute scale [7], this source is the accepted standard candle of TeV gamma-ray emission.

In the following analysis, measured data are presented in Crab units. Results in Crab units have the advantage to rely only on measured data and to be free from systematic uncertainties due to the Monte Carlo (MC) simulations of the air showers and the detector response. For this new mirror configuration, up to now MC data are not available in sufficient amount. 52 hours of CT1 data from the Crab Nebula of the considered time period have been taken into account. The energy threshold is estimated to be 700 GeV near the zenith.

DATA ANALYSIS

We applied the standard gamma/hadron separation with dynamic supercuts on the Hillas parameters [10]. Due to the lack of MC Data, the cuts have been optimized with Crab data from 1998, and were applied to all three sets of data. Only events with more than 60 photoelectrons have been accepted for further analysis.

For each source, the number of ON-source events N_{ON}, the number of background events N_b, and the observation time T were determined. In the same way, the values N_{ON}^c, N_b^c, and T_c for Crab data were determined. From these values N_{ON}, N_b we calculated the upper (resp. lower) limit of the number of counts MAX (resp. MIN_c) at 99 % confidence level, using the probability density function of the number of source events [5]. The upper limits UL in Crab units are calculated from:
$UL = \frac{MAX}{MIN_c} \times \frac{T_c}{T}$.

RESULTS

The data from the 1998 observation period did not reveal evidence for TeV γ-ray emission from the examined BL Lac type sources. The upper flux limits are 18.6 % Crab for MS 0116 and 11.6 % Crab for 3C 66A, at 99 % confidence level.

As BL Lac sources are known to be highly variable, (eg. Mkn 501 was nearly not visible in the TeV range during 1996 [3],but showed high rates in a flare in the following year), we performed an analysis on a diurnal basis, in order to find possible flares and high states on a short time scale. No night of significantly higher excess rate could be found.

The obtained values are summarized in Table 1. Please note that the values MAX (resp. MIN_c) are upper (resp. lower) limits at 99% confidence level. The significance and flux in Crab units are upper limits for the two BL Lac objects.

Source	MS 0116	3C66A	Crab
obs. time/h	63.5	35.9	52.3
N_{ON}	1431	685	3776
N_b	1311	690	2370
Excess	120	-5	1406
Significance	1.9 σ	-0.14 σ	15.2 σ
MIN_c	-	-	1190
Significance$_{min}$	-	-	>12.8σ
MAX	269	95	-
Significance$_{max}$	<4.2σ	<2.5σ	-
Flux/Crab	<0.186	<0.116	1

TABLE 1. Results of the data analysis

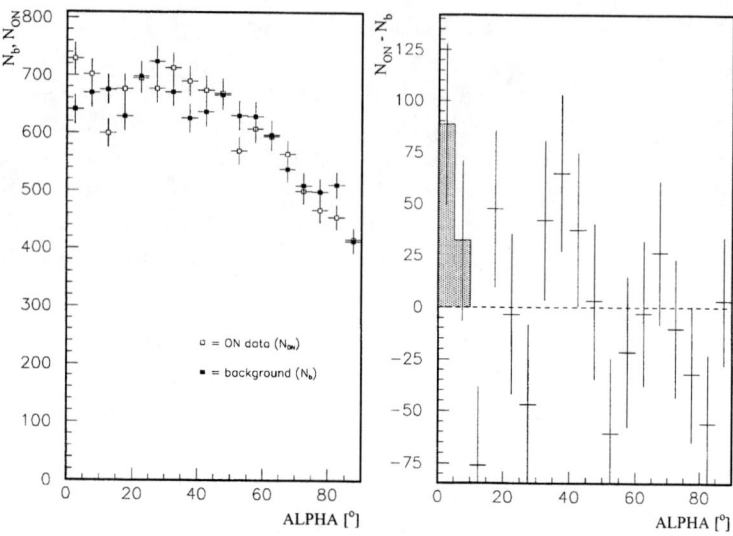

FIGURE 1. The left plot shows the number of events for ON-souce (N_{ON}) and background (N_b) measurements. The right plot shows the excess $N_{ON} - N_b$. 120 excess events are found in the two signal bins ALPHA < 10 °. This is a significance of $1.9\,\sigma$.

Figure 1 shows the second moment parameter ALPHA for the ON-source and background data of MS 0116, a $1.9\,\sigma$ excess is visible. Figure 2 shows the excess of the ALPHA parameter for 3C 66A. In total a $-0.14\,\sigma$ excess is to be seen.

CONCLUSION

The 1998 observations of 3C 66A did not reveal significance for TeV γ−ray emission. The Whipple Group gives an upper limit for 1993 [4,6], which is in good agreement with our data. We cannot confirm the $5\,\sigma$ excess of Neshpor et al. [9] for our 1998 data.

The observation of MS 0116 revealed only very small indication for a TeV γ−ray signal, despite of its rather long observation time.

ACKNOWLEDGEMENTS

The support of the German Ministry for Research and Technology BMBF and of the Spanish Research Council CYCIT is gratefully acknowledged. We thank the Instituto de Astrofísica de Canarias (IAC) for the excellent working conditions at La Palma.

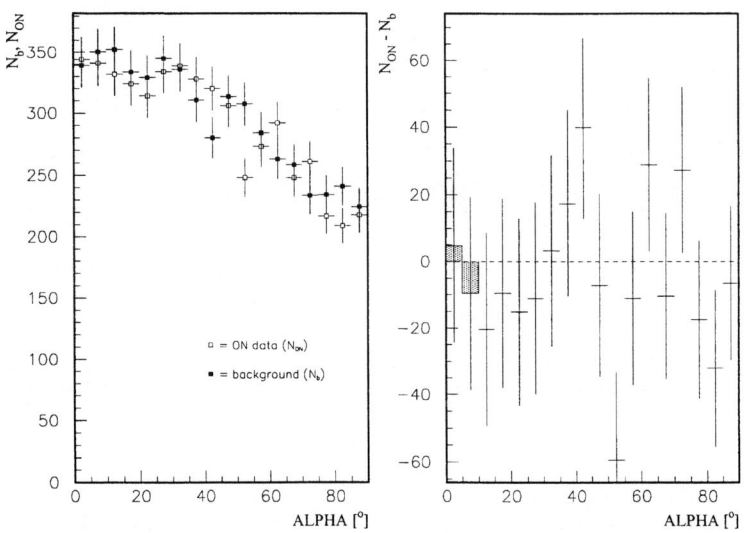

FIGURE 2. The left plot shows the number of events for ON-souce (N_{ON}) and background (N_b) measurements, the right plot shows the excess $N_{ON} - N_b$. The deficiency of five events in the two signal-bins, ALPHA $< 10°$, correspond to $-0.14\,\sigma$.

REFERENCES

1. Aharonian, F.A., et al., 1999, *A&A* **342**, 69.
2. Aharonian, F.A., et al., 1999, *submitted to A&A* (astro-ph/9901284).
3. Bradbury, S.M., Deckers, Th., Petry, D., et al., 1997, *A&A* **320**, L5.
4. Buckley, J.H., 1999, *Astroparticle Physics* **11**, 119.
5. Helene, O., 1983, *NIM* **212**, 319.
6. Kerrick A.D., 1995, *ApJ* **452**, 588.
7. Konopelko, A., 1999, *Proc. 26th ICRC*, Salt Lake City, OG.2.2.01.
8. Lindner A., 1997,*Proc. 25th ICRC* , Durban, **5**, 113.
9. Neshpor, Yu.I., Stepanyan, A.A., et al., 1998, *Astronomy Letters* **24**, 134.
10. Petry, D., Kranich, D., 1997, *O.C. de Jager (ed.) Towards a major atmospheric Cherenkov detector V*, Kruger Park, 368.
11. Rauterberg, G., et al., 1995, *Proc. 24th ICRC*, Rome, **3**, 460.

Search for TeV Gamma-Emissions from BL Lacertae with the HEGRA equatorial mount Cherenkov telescope

O. Mang[1], G. Rauterberg[1], M. Schilling[1] for the HEGRA Collaboration

[1] *Institut für Experimentelle und Angewandte Physik, Universität Kiel, 24098 Kiel, Germany*

Abstract. BL Lacertae is a nearby active galaxy of the blazar class with a redshift $z = 0.069$ and therefore, within the unified model of AGN, a candidate for the observation of gamma emissions in the TeV range. During 1998, BL Lacertae has been observed by the HEGRA equatorial mount Cherenkov telescope. The reconstructed shower images have been subjected to Gamma-Hadron separation. The sample has been used to derive an upper limit of the energy flux from BL Lacertae carried by Gamma-Rays with energies greater than 1 TeV.

I INTRODUCTION

The production of γ-rays in the TeV-domain by inverse Compton-scattering is assumed to originate in the relativistic jets of AGN. The unified model of AGN explains the differences between AGN subtypes by the orientation of the relativistic jets with respect to the observer. If the jets are oriented along the observer's line of sight, the object appears as a blazar. Blazars are therefore candidates for the emission of TeV γ-radiation. BL Lac objects form a subset of the Blazar class, which makes them interesting for the study of high energy γ-emissions. Furthermore, there are BL Lac objects that have already been established as sources of TeV γ-radiation, e.g. Mrk-421 [10] and Mrk-501 [11].

Ultra high energy photons can interact with the Cosmic Infrared Background (CIRB) in a pair production process. This limits the mean free path of TeV photons and causes considerable extinction of the photons, depending on the distance of the source. To determine this extinction and with it the hitherto unknown photon density of the CIRB, it is important to detect further BL Lac objects in the TeV regime.

The HEGRA experiment, located on the Canary Island of La Palma, has observed the object BL Lacertae with the equatorial mount Cherenkov Telescope CT1 for 46.9 hours in 1998. The data set was subjected to a standard Gamma-Hadron separation and compared to data of the Crab Nebula taken in 1998. From this an upper limit in units of the Crab flux was derived for the flux from BL Lacertae above ≈ 1 TeV.

II THE ACTIVE GALAXY BL LACERTAE

BL Lacertae is an active galaxy of the HBL-type (high frequency BL Lac). The two nonthermal components of the spectrum, which are believed to be the synchrotron and inverse Compton emissions of a population of relativistic electrons in the jets, peak at relatively high energies [4] and thus make HBLs candidates for the emission of TeV gamma radiation. In addition, the object BL Lacertae has a redshift $z = 0.069$, so the attenuation of UHE radiation by the CIRB is not considered to be a limiting factor. The degree of extinction by the CIRB is still under investigation, therefore the influence of the CIRB is not considered in this analysis.

Observations of BL Lacertae in the GeV and optical regimes showed a strong flare of the source around July 19th 1997 (MJD 50648) [3]. The established TeV-sources Mrk-421 and Mrk-501 also show irregular activity, most remarkably the outburst of Mrk-501 during 1997, reaching diurnal flux levels of 10 Crab units [2]. Therefore the monitoring of BL Lacertae is of high interest.

III EXPERIMENTAL SETUP

The HEGRA experiment is located at the Observatorio Roque de los Muchachos on the Canary Island *La Palma* (2200 m a.s.l., 28.75°N, 17.89°W). The experiment consists of several arrays of particle and Cherenkov detectors dedicated to cosmic ray research, and of six Cherenkov Telescopes [7]. Five of these telescopes have an Alt/Az-mount and are operated as a stereoscopic system of telescopes.

The other telescope, the HEGRA equatorial mount Cherenkov Telescope *CT1* is equipped with a camera consisting of 127 photomultipliers, with a field of view of 3.25°. Each pixel has an angular diameter of 0.25°. Further details of the Telescope hardware can be found in [12]. The tracking of the telescope is accurate to 0.1°. The telescope has been equipped with new hexagonal mirrors in October 1997. This has increased the reflective area to 10 m^2 and thus the sensitivity of the telescope, but since Monte Carlo statistics for the new configuration are not yet sufficient, the energy threshold can only be estimated to ≈ 700 GeV close to the zenith and, using $E_{thresh} \sim \cos^{-2.5}(\theta)$ with θ representing the zenith angle, ≈ 1.0 TeV at $\theta = 30°$. The limited Monte Carlo statistics imply that an absolute measurement of the flux is not possible. The method of determining the flux in Crab units has the advantage of being free from systematic uncertainties due to Monte Carlo simulations.

IV DATA SAMPLE

BL Lacertae has been observed by CT1 from May 29th to September 24th, 1998 for 46.9 hours. The data were obtained during moonless nights, with zenith angles up to 50°. The Crab Nebula has been monitored in Feb/Mar 1998 and in the fall/winter of 1998, yielding 52.3 hours of observation time. Both sources have been observed in ON-ON and ON-OFF mode. For the determination of background we used a data sample covering the same range of zenith angles, consisting of the OFF-runs and other measurements from regions where no source of gamma-rays is expected. First steps in the data processing include flatfielding of the camera, finding and excluding defective pixels, application of dynamical and topological tailcuts and computation of the Hillas parameters.

The analysis was done using the method of dynamic Supercuts [9] imposed on the Hillas parameters of the shower image. For the parameters WIDTH, LENGTH, CONC and DIST cuts were calculated from the parameters θ, SIZE and MDIST [8]. In addition a cut of SIZE > 60 photoelectrons was applied. Data from each source were combined with the background data to calculate the significance of a possible gamma excess and upper and lower limits [5].

V RESULTS

The analysis of the Crab Nebula data showed an excess of TeV-photons with a significance of $15.16\,\sigma$, with a total observation time of 52.3 hours. The data from BL Lac did not show a significant excess in the lower ALPHA bins (ALPHA < 10°). The significances and limits are shown in table 1. Since BL Lac type objects are known to be variable and an outburst of BL Lac in the optical and GeV regimes has been observed, the data were grouped by date to reveal a possible flare. No evidence for a signal on a one-day timescale could be found.

A comparison of the excesses gives an upper limit of the gamma ray flux from BL Lacertae of 6.9% of that of the Crab Nebula at a confidence level of 99%. The upper limit has been obtained using a normalization w.r.t. observation times using

$$\mathrm{UL}_{BL\ Lac} = \frac{\mathrm{MAX}_{BL\ Lac}}{T_{BL\ Lac}} \times \frac{T_{Crab}}{\mathrm{MAX}_{Crab}}.$$

Figure 1 visualizes the results of this analysis: It shows the distribution of the Alpha-parameter for BL Lac events. No excess events are seen, instead, there is a deficiency in the ON data.

Another upper limit of 11% Crab on TeV emisson from BL Lacertae has been determined by the HEGRA System of Imaging Atmospheric Cherenkov Telescopes [1]. The respective observation windows overlap only in May 1998 and thus complement each other to give a consistent long-term picture.

	BL Lacertae	Crab Nebula
$N_{on} - N_b$	-54	1406
Significance	$-1.32\,\sigma$	$15.16\,\sigma$
Upper Limit (MAX)	73.3	–
Upper Limit (rate)	$1.56\ h^{-1}$	–
Lower Limit (MIN)	–	1190
Lower Limit (rate)	–	$22.76\ h^{-1}$
Observation time	46.9 h	52.3 h

TABLE 1. Measured excess, upper and lower limit for BL Lacertae and the Crab Nebula at 99% confidence level. The numbers result in an upper limit of 0.069 Crab-units for BL Lac.

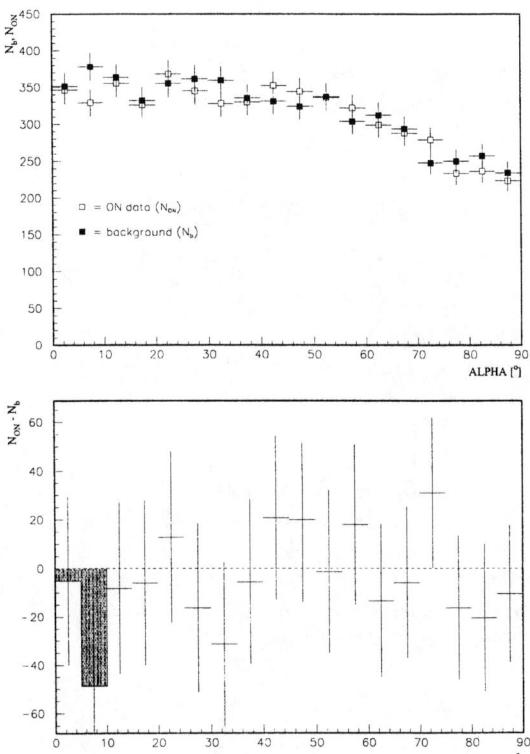

FIGURE 1. The distribution of Alpha including all zenith angles for BL Lac. The upper plot shows the superposition of ON and OFF data after cuts, the lower one shows the difference ON-OFF. This corresponds to an excess of $-1.32\,\sigma$. Binning the data by zenith angle did not reveal new features in the alpha distribution.

VI CONCLUSION

The analysis of 46.9 hours of data from BL Lacertae, taken in 1998, did not show a significant excess of gamma-like events from the direction of the observed object. A data sample of the Crab Nebula taken during the same year yielded an excess of gamma-like events with a significance of $15.16\,\sigma$ and was used as a flux unit. For the flux from BL Lacertae an upper limit of 0.069 Crab-Units was derived at a confidence level of 99%.

Since it was shown that BL Lacertae is a variable source in the GeV and optical regime [3], it is important to continue the monitoring of BL Lacertae in the future.

VII ACKNOWLEDGEMENTS

The support of the German Ministry of Research and Technology (BMBF) and of the Spanish Research Council CYCIT is gratefully acknowledged. We thank the Instituto de Astrofísica de Canarias (IAC) for hosting the HEGRA experiment and for supplying excellent working conditions at the Observatorio Roque de los Muchachos (ORM), La Palma.

REFERENCES

1. Aharonian, F.A., Akhperjanian, A.G., Barrio, J.A. et al. 1999a, *submitted to A&A* (astro/ph 9903455).
2. Aharonian, F.A., Akhperjanian, A.G., Barrio, J.A. et al. 1999b, *A&A* **342**, 69.
3. Bloom, S.D., Bertsch, D.L., Hartmann, R.C. et al. , *ApJL* **490**, L145 (1997).
4. Ghisellini, G., Celotti, A., Fossati, G., Maraschi, L., Comastri, A., *MNRAS* **301**, 451 (1998).
5. Helene, O., *NIM* **212**, 319 (1983).
6. Hillas, A.M., *Proc. 19th ICRC, La Jolla*, **3**, 445 (1985).
7. Lindner, A., *Proc. 25th ICRC, Durban*, **5**, 113 (1997).
8. Petry, D., PhD thesis, TU Munich (1997).
9. Petry, D. & Kranich, D., O.C. de Jager (ed.), *Towards a Major Atmospheric Cherenkov Detector V, Kruger Park*, 368 (1997).
10. Punch, M. et al., *Nature*, **358**, 477 (1992)
11. Quinn, J. et al., *ApJ*, **456**, L83
12. Rauterberg, G. et al. *Proc. 24th ICRC, Rome*, **3**, 460 (1995).

Upper Limits on low redshift AGN

H. Bojahr for the HEGRA Collaboration

Fachbereich Physik, Universität Wuppertal, Gaußstr. 20, D-42097 Wuppertal, Germany

Abstract. The HEGRA IACT stereoscopic system, with an energy threshold of about 0.5 TeV, was used to search for γ-ray emission of several low redshift AGN ($z < 0.2$), mostly BL Lacs. Here we present the results of observations that were made since June 1997 at zenith angles up to 40°. No evidence for TeV emission is detected and the corresponding upper limits in Crab units are given.

INTRODUCTION

The HEGRA Imaging Atmospheric Cherenkov Telescope system (IACT) was used for an extensive search of potential TeV-γ-ray sources like supernova remnants, gamma ray bursts, the galactic plane and other promising objects. The observations have focussed on blazars, since all EGRET detected Active Galactic Nuclei (AGN) at energies > 100 MeV are members of that subclass of AGN [8,13]. Blazars include both emission line radio quasars with a relatively flat spectrum (FSRQ) and lineless BL Lac objects.

The famous and well-studied blazars Mkn 421 and Mkn 501 are well known TeV-γ-ray sources and regularly monitored by the HEGRA experiment. Since explanation and unification of AGN characteristics are still a subject of discussion more informations about TeV-γ-ray sources are needed to complete the theoretical concept.

At HEGRA an observation program with special attention to low redshift AGN ($z < 0.2$), mostly BL Lacs, started in July 1997 and will be continued monitoring the established TeV sources on a regulary basis and searching for new ones gathering about 10 hours observation time on each object.

THE HEGRA CHERENKOV TELESCOPE SYSTEM

The HEGRA collaboration operates six IACTs and an array of several particle and Cherenkov detectors dedicated to cosmic ray research which are located on the Roque de los Muchachos (2200 m a.s.l., 28.75° N, 17.89° W) on La Palma, Canary Islands, Spain [7]. The HEGRA IACT stereoscopic system started to take data in 1996 as a system of three CTs and was completed to a five telescope system in September 1998. Each system telescope has a $8.5\,m^2$ reflector area focussing the light onto a high resolution camera with 271 pixels (PMTs), covering a field of view of $4.3°$ (pixel diameter $0.25°$). The system has an energy threshold of ≈ 500 GeV and a flux sensitivity νF_ν at 1 TeV of $10^{-11} erg\, cm^{-2}\, s^{-1}$ (S/N = 5σ) for one hour observation time. Detailed informations on the IACT system are given in [1] and [2].

THE DATA SAMPLE

The low redshift blazar sample was compiled from several catalogues of quasars, BL Lac objects and other galaxies [9,11,12,14] by taking into account the following selection criteria:

- for FSRQ the radio spectral index is $\alpha \leq +0.5$ $(S_\nu \propto \nu^{-\alpha})$

- restricting the zenith angle on $ZA \leq 30°$ leads to a declination range of $-2° \leq \delta \leq 58°$ for the HEGRA observation site.

- due to the extragalactic infrared background the constraint for the redshift was $z \leq 0.2$, which was the γ-ray horizon limit for 1 TeV source photons at this time [3].

This results in a low redshift blazar sample of about 48 objects (see table 4), which we continue to observe until 10 hours observation time on each object will be collected.

The here presented analysed data are a small subsample of objects that had been observed from September 1998 until November 1998.

TABLE 1. Summary of the presented analyzed low redshift data sample. The AGN are classified by Flat Spectrum Radio Quasars (FSRQ), high energy peaked BL Lacs (HBL) and low energy peaked BL Lacs (LBL) [10].

name	source	z	AGN class	number of CTs	obs. period	obs. hours
NGC 0315	0055+300	.017	FSRQ	5 CTs	14.Oct 98 - 29.Oct 98	12.2
NGC 1275	0316+413	.018	FSRQ	5 CTs	13.Nov 98 - 27.Nov 98	17.4
1ES 0323+022	0323+022	.147	HBL	5 CTs	20.Nov 98 - 26.Nov 98	7.0
1ES 0927+500	0927+500	.188	HBL	5 CTs	19.Nov 98 - 29.Nov 98	11.2
1ES 1741+196	1741+196	.083	HBL	4 CTs	22.Sep 98 - 23.Sep 98	2.3
[HB89] 2201+044	2201+044	.027	HBL	5 CTs	15.Oct 98 - 24.Oct 98	11.7
[HB89] 2254+074	2254+074	.190	LBL	5 CTs	09.Nov 98 - 20.Nov 98	9.0

UPPER LIMIT CALCULATION

To avoid systematic errors due to Monte Carlo simulations of the airshowers and the detector response in the data analysis the *upper flux limits* are determined *in Crab units*.

TABLE 2. Summary of the Crab reference sample which was used to calculate the upper limits of the presented data sample shown in table 1.

source name	number of CTs in system	obs. period	obs. hours
Crab Nebular	5 CTs	14.Oct 98 - 29.Oct 98	23.6
Crab Nebular	4 CTs	24.Jan 98	1.5

Therefore the number of events in the ON-source region (ON_S) and the number of events in the OFF-source region (OFF_S) and the observation time T_S for each single zenith-angle intervall (I_{ZA}) of the data sample of each observed source need to be known, and also for the Crab reference data sample (ON_C, OFF_C, T_C). Due to technical problems the CT-system partly operated only as a four or even three telescope system thus different experimental setups had to be considered in the analysis. For the γ-hadron-separation the following cuts were applied: the impact distance from the center of the telescope array $r \leq 250\,m$, the mean scaled width $MSW < 1.2$ (definition see [5]), the squared angular distance of the reconstructed shower direction to the source direction $\theta^2 < 0.05\,deg^2$.

The upper (lower) limits of the number of counts $MAX_S(MIN_C)$ from the source (Crab) on a 99% confidence level were calculated by using the probability density function of the number of source events [4]. The upper limits in Crab units UL_S are given by

$$UL_S = \frac{MAX_S}{MIN_C} \times \frac{T_C}{T_S}$$

These upper limits in Crab units are free from systematic errors. From 0.5 TeV to 20 TeV the integral Crab flux measured with the HEGRA-CT-System is $F_{Crab} = 5 \cdot 10^{-11}\,cm^{-2}\,s^{-1}$ [6].

RESULTS

TABLE 3. Results of the presented data sample: (a) source, (b) significance in σ, (c) expected excess rate per hour, (d) upper limits in Crab units based on a 99% C.L.

source name (a)	significance in σ (b)	excess rate in h^{-1} (c)	UL_{Crab} in $Crab\,units$ (d)
NGC 0315	-1.73	1.3	0.05
NGC 1275	1.73	4.9	0.19
1ES 0323+022	0.71	6.5	0.30
1ES 0927+500	0.75	4.7	0.17
1ES 1741+196	-0.42	5.0	0.13
[HB89] 2201+044	1.70	7.9	0.36
[HB89] 2254+074	0.56	5.2	0.22

TABLE 4. The low redshift blazar sample with 48 objects, marked is the analysed data sample.

name	RA (J 2000.0)			DEC (J 2000.0)			source	z	5 GHz flux in Jy	spect. index	opt. mag
NGC0315	00	57	48.8	+30	21	09	*0055+300*	*.017*	*1.180*	*-0.10*	*12.5*
4C+31.04	01	19	35.0	+32	10	50	0116+319	.060	1.450	-0.47	14.5
1ES0145+138	01	48	29.7	+14	02	18	0145+138	.125	0.006		17.9
UGC01651	02	09	38.6	+35	47	50	0206+355	.037	0.894	+0.77	13.0
1ES0229+200	02	32	48.6	+20	17	17	0229+200	.140	0.049		14.7
S40309+411	03	13	01.9	+41	20	01	0309+411	.134	0.514	+0.11	18.0
NGC1275	03	19	48.1	+41	30	42	*0316+413*	*.018*	*47.200*	*+1.02*	*11.9*
HB89-0317+185	03	19	51.8	+18	45	35	0317+183	.190	0.017		18.1
1ES0323+022	03	26	14.0	+02	25	15	*0323+022*	*.147*	*0.042*		*16.5*
4C+37.11	04	05	49.2	+38	03	32	0402+379	.055	1.160	-0.39	18.5
3C120.0	04	33	11.1	+05	21	16	0430+052	.033	8.440	+0.31	15.1
EXO0706.1+5913	07	10	30.8	+59	08	17	0706+591	.125	0.041		18.4
S40733+597	07	37	30.1	+59	41	03	0733+597	.041	0.518	+0.03	14.9
HB89-0736+017	07	39	18.0	+01	37	05	0736+017	.191	1.990	-0.13	16.5
3C197.1	08	21	33.7	+47	02	37	0818+472	.128	0.860	-0.49	16.5
BL0829+046	08	31	48.9	+04	29	39	0829+046	.180	2.110		16.0
1ES0927+500	09	30	37.6	+49	50	26	*0927+500*	*.188*	*0.018*		*17.2*
BL1011+496	10	15	04.2	+49	26	01	1013+498	.200	0.286		16.1
MS1019.0+5139	10	22	11.2	+51	24	15	1019+514	.141	0.003		18.1
MRK421	11	04	27.3	+38	12	32	1101+384	.031	0.722	-0.09	14.4
EXO1118+4228	11	20	48.1	+42	12	12	1118+423	.124	0.034		17.3
NGC3894	11	48	50.3	+59	24	56	1146+596	.011	0.606	+0.52	13.0
1ES1212+078	12	15	11.0	+07	32	04	1212+078	.136	0.117	-0.04	16.0
MS1214.3+3811	12	16	51.8	+37	54	39	1214+381	.062	0.002		16.7
PG1219+301	12	21	21.9	+30	10	37	1219+301	.130	0.056		16.4
WCOMAE	12	21	31.7	+28	13	59	1219+285	.102	0.981		16.5
MESSIER084	12	25	03.1	+12	53	11	1222+131	.003	3.090	-0.46	08.7
3C273	12	29	06.7	+02	03	09	1226+023	.158	42.850	+0.15	12.8
1ES1239+069	12	41	48.3	+06	36	01	1239+069	.150	0.010		19.4
1ES1255+244	12	57	31.9	+24	12	40	1255+244	.141	0.007		15.4
PG1418+546	14	19	46.6	+54	23	15	1418+546	.151	1.090	+0.38	15.3
1ES1426+428	14	28	32.6	+42	40	21	1426+428	.129	0.038		16.4
1ES1440+122	14	42	48.3	+12	00	40	1440+122	.162	0.410		17.0
CGCG021-063	15	16	40.2	+00	15	02	1514+004	.052	1.360	-0.48	16.5
RXJ16247+3726	16	24	43.4	+37	26	42	1625+373	.200	0.042		18.2
MRK501	16	53	52.2	+39	45	37	1652+398	.034	1.420	+0.06	13.7
V71-1721-026	17	24	37.8	-02	43	06	1721-026	.033	1.190	-0.41	15.0
1H1720+117	17	25	04.4	+11	52	16	1722+119	.018	0.741		15.2
IZw187	17	28	18.6	+50	13	10	1727+502	.059	0.155		16.7
1ES1741+196	17	43	57.8	+19	35	09	*1741+196*	*.083*	*0.223*		*16.6*
NGC6454	17	44	56.6	+55	42	17	1743+557	.031	0.521	-0.28	13.0
EXO1811+3143	18	13	35.3	+31	44	20	1812+314	.117	0.127	+0.17	17.4
BLLAC	22	02	43.3	+42	16	40	2200+420	.069	4.470	-0.13	14.9
HB89-2201	22	04	17.6	+04	40	02	*2201+044*	*.027*	*0.741*		*15.2*
HB89-2254	22	57	17.3	+07	43	12	*2254+074*	*.190*	*1.190*	*+1.19*	*16.4*
1ES2321+419	23	23	52.1	+42	10	59	2321+419	.059	0.019		17.0
1ES2344+514	23	47	04.8	+51	42	18	2344+513	.044	0.215		15.5
1ES0120+340	01	23	08.8	+34	20	50	0120+340	.272	0.034		15.2

The analysis of this data sample did not reveal positive evidence for TeV-γ-ray emission from any object of this sample. Based on a few hours of observation upper limits in Crab units with a confidence level of 99% could be obtained (see table 3). Here the absorption due to the extragalactical diffuse infrared background emission (DIRBE) was not taken into consideration.

ACKNOWLEGDEMENTS

The support of the German Ministry for Research and Technology BMBF and of the Spanish Research Council CYCIT is grateful acknowledged. We thank the Instituto Astrofisico de Canarias (IAC) for the use of the HEGRA site at the Roque de los Muchachos (ORM) and its facilities at La Palma.

REFERENCES

1. Aharonian, F. A. et al., A&A **342**, 69 (1999)
2. Daum, A. et al., Astrop. Phys. **8**, 1 (1997)
3. Funk, B., Magnussen, N. et al., Astrop. Phys.**9**, 97 (1998)
4. Helene, O., Nucl. Instr. Meth. **212**, 319 (1983)
5. Konopelko, A. et al., Conf.Proc. *Towards a Major Atmos. Cherencov Det. IV*, Padova (1995)
6. Konopelko, A. et al., 16th European Cosmic Ray Symposium Conf. Proc., 523 (1998)
7. Lindner, A., et al., Proc. 25th ICRC, Durban, **5**, 113 (1997)
8. Mukherjee, R. et al., ApJ **490**, 116 (1997)
9. Padovani, P. & Giommi, P., MNRS **277**, 1477 (1995)
10. Padovani, P. & Giommi, P., ApJ **444**, 567 (1995)
11. Stickel, M. & Kühr, H., A&AS **103**, 349 (1994)
12. Stickel, M., Meisenheimer, K. & Kühr, H., A&AS **105**, 211 (1994)
13. Thompson, D.J. et al., ApJS **101**, 259 (1995)
14. Véron-Cetty, Véron-Catalogue 7th edtion (1995)

TeV Emission from PKS 2155−304

P. M. Chadwick, K. Lyons, T. J. L. McComb, K. J. Orford,
J. L. Osborne, S. M. Rayner, S. E. Shaw, and K. E. Turver

Department of Physics, Rochester Building, Science Laboratories, University of Durham, Durham, DH1 3LE, U.K.

Abstract. The X-ray selected BL Lac PKS 2155−304 has been observed using the University of Durham Mark 6 very high energy gamma ray telescope during 1998. We find no evidence for TeV emission during these recent observations when the X-ray flux was observed to be low. We have reconsidered our measurements made in 1997 November when PKS 2155−304 was in a bright X-ray state and extended X-ray and GeV gamma ray observations were made as part of a multiwavelength campaign. Comparisons are made of the VHE emission during this time with the available data from other wavelengths.

INTRODUCTION

Evidence exists that at least four close X-ray selected BL Lacs are sources of episodic TeV gamma ray emission (Mrk 421 [1], Mrk 501 [2], 1ES 2344+514 [3] and PKS 2155−304 [4]). The discovery of PKS 2155−304 as a VHE gamma ray source was made with the University of Durham Mark 6 telescope during observations lasting 40 hrs in 1996 and 1997. These results suggested a time variable emitter with the strongest emission in 1997 November at the time of a successful multiwavelength campaign [4]. The details of the measurements at X-ray energies in 1997 November are now available from both *RXTE* [5] and *BeppoSAX* [6], together with data from *CGRO*/EGRET [5]. During the 36 hour observation with *BeppoSAX* a short interval (2 hrs) of simultaneous X-ray and TeV observations occurred.

We here report the results of our 1998 measurements and reconsider our 1997 November data in the light of the recently available X-ray results from the multi-wavelength campaign. All measurements reported here have been made with the University of Durham Mark 6 gamma ray telescope operating at Narrabri, NSW, Australia. The telescope has been described in detail by Armstrong et al. [7].

TABLE 1. Observing log for our observations of PKS 2155–304 during 1998.

Date	No. of scans ON source
1998 July 22	1
1998 August 18	5
1998 August 19	7
1998 August 20	4
1998 September 15	5
1998 September 16	2
1998 September 17	4
1998 September 19	3
1998 October 11	2
1998 October 13	2
1998 October 16	1

TABLE 2. The results of various event selections for the PKS 2155–304 data recorded in 1998.

	On	Off	Difference	Significance
Number of events	171723	173203	−1480	−2.5 σ
Number of size and distance selected events	97279	97493	−214	−0.48 σ
Number of shape selected events	6260	6053	207	1.9 σ
Number of shape and ALPHA selected events	950	992	−42	−0.10 σ

NEW MEASUREMENTS IN 1998

Observations in 1998 have involved 9.5 hrs of exposure ON source and an equal amount OFF source during 1998 July, August, September and October; the observing log is in Table 1. All analysis reported here used the background suppression employed in our previous anaysis of PKS 2155–304 [4]. We find no evidence for emission of TeV gamma rays throughout the 1998 observations — see Table 2. We calculate a time-averaged 3 σ flux limit of 4.0×10^{-11} cm^{-2} s^{-1} at an energy threshold of 300 GeV for 1998 July – October.

We show in Figure 1 the time averaged VHE gamma ray fluxes (normalised to the cosmic ray counting rate [4]) for all dark periods during which we have observed PKS 2155–304. We note that PKS 2155–304 was in an X-ray low state during our measurements in 1998 July – October. Our failure to detect VHE emission during 1998 is thus consistent with the hypothesis that the X-ray and VHE gamma-ray emission from PKS 2155–304 are correlated [4].

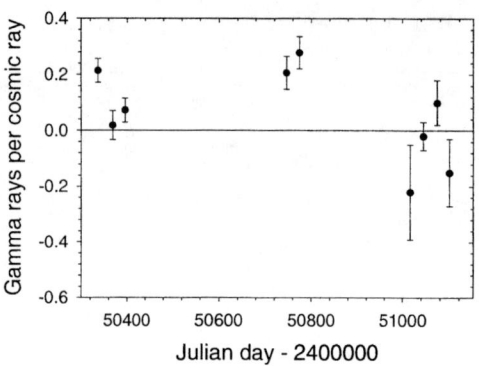

FIGURE 1. The measured VHE gamma ray flux above 300 GeV from PKS 2155–304 averaged over observing periods of approximately 10 days.

MULTIWAVELENGTH OBSERVATIONS IN 1997 NOVEMBER

TeV data are available between 1997 November 17–25 [4]. These observations were made as part of a multiwavelength campaign involving GeV gamma ray and X-ray measurements. The GeV gamma ray measurements were made using *CGRO*/EGRET from 1997 November 10 – 23 and showed strong emission during the first half of the period [5]. X-ray observations using PCA and HEXTE on board *RXTE* were made during 1997 November 20 – 22 [5] and ASM observations are available throughout. *BeppoSAX* observed this object for about 1.5 days during 1997 November 22 – 24 [6]. The X-ray and gamma ray observations clearly show that PKS 2155–304 was in an active flaring state in the middle of November 1997, with X-ray and gamma ray fluxes being as high as ever previously detected — see Figure 2. Recently the CANGAROO group have published data taken between 1997 November 24 and 1997 December 1 [8]. They fail to detect any TeV emission during this period, quoting a 2 σ flux limit of 9.5×10^{-12} cm^{-2} s^{-1} at an energy threshold of 1.5 TeV. Given the non-overlapping observation periods, the possibility of time variation in the emission and the different thresholds of the telescopes, this null result is not in conflict with our detection.

Our TeV observations, averaged over the total dataset for 1997 November, indicated the strongest emission during any of the dark periods to date [4]. The time averaged flux of VHE gamma rays for our observations in 1997 November is $(6.0 \pm 2.0_{stat} \pm 3.0_{sys}) \times 10^{-11}$ cm^{-2} s^{-1} for an energy threshold of 300 GeV. The EGRET and X-ray observations suggest that a large outburst occurred in early November prior to the TeV observations [5]. The TeV observations do not contradict this idea.

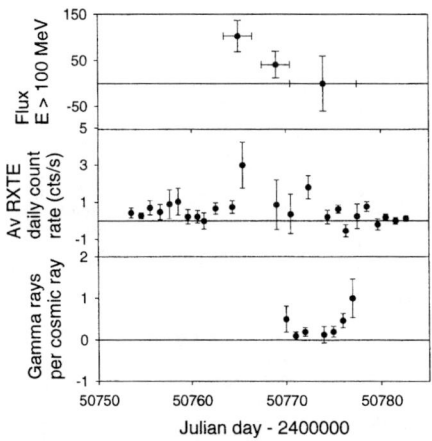

FIGURE 2. The GeV gamma rays recorded with EGRET ([5] – upper panel), X-rays recorded with ASM on *RXTE* (centre panel) and VHE gamma ray (present work – bottom panel) during the 1997 November observations of PKS 2155–304.

We have considered the information available on a day by day basis from ground based Cerenkov telescopes, *CGRO*/EGRET, *RXTE* and *BeppoSAX*. The only truly contemporaneous data were recorded by *BeppoSAX* and the Mark 6 telescope on 1997 November 23 between 1100 and 1300 hrs UTC. We reproduce the X-ray data from the paper of Chiappetti et al. [6] — see Figure 3(a).

Our VHE observations occur at the time which Chiappetti et al. [6] define as a region of low X-ray intensity defined on the basis of the MECS (medium energy) count rate, beginning about 2 hours after the peak of the second X-ray flare detected by *BeppoSAX*. We show in Figure 3(b) the results of our VHE observations for individual 15 min scans on 1997 November 23, along with data taken on 1997 November 22 which were obtained about three hours before the *BeppoSAX* observation started. We have no evidence for strong flaring activity within the VHE data taken during the *BeppoSAX* observations, consistent with the low X-ray state. The VHE data taken on 1997 November 22 are at the same activity level as on November 23. The X-ray data show that an X-ray flare peaked about 2 hrs after our VHE observation finished and that the typical time scale for X-ray flaring is such that the flare is likely to have commenced after our observation terminated. The VHE data yield a flux of $(2.0 \pm 5.0_{stat} \pm 1.0_{sys}) \times 10^{-11}$ cm^{-2} s^{-1} for the observation on 1997 November 22 and $(7.0 \pm 4.5_{stat} \pm 3.5_{sys}) \times 10^{-11}$ cm^{-2} s^{-1} for the observation on 1997 November 23, both at an energy threshold of 300 GeV.

We are grateful to the UK Particle Physics and Astronomy Research Council for support of the project. The Mark 6 telescope was constructed with the assistance of the staff of the Physics Department, University of Durham, and the efforts of Mr.

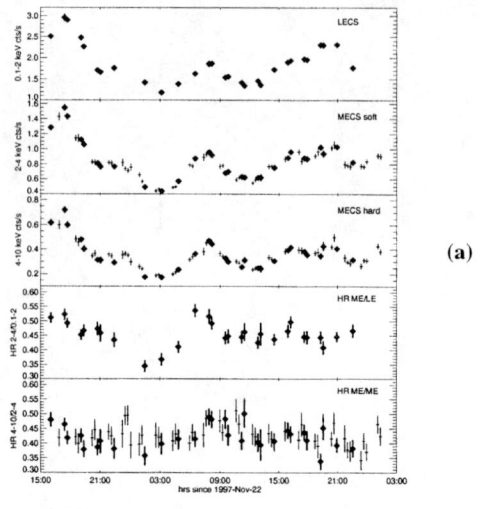

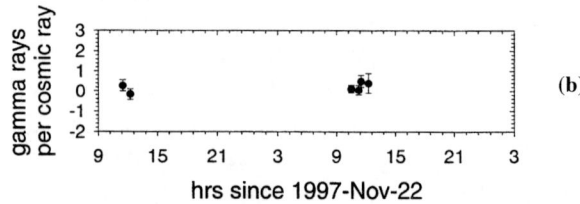

FIGURE 3. (a) The X-ray light curves (upper three panels) and hardness ratios (lower two panels) recorded with *BeppoSAX* during the 1997 November observations of PKS 2155–304 (taken from [6]). Also shown (b) are the VHE gamma ray results from the present work.

P. Cottle, Mrs. E. S. Hilton and Mr. K. Tindale are acknowledged with gratitude. This paper uses quick look results provided by the ASM/*RXTE* team.

REFERENCES

1. Punch, M., et al., *Nature*, **358**, 477 (1992).
2. Quinn, J., et al., *Astrophys. J.*, **456**, L83 (1996).
3. Catanese, M., et al., *Astrophys. J.*, **487**, L143 (1997).
4. Chadwick, P. M., et al., *Astrophys. J.*, **513**, 161 (1999).
5. Vestrand, W. T. & Sreekumar, P., *Astropart. Phys.*, **11**, 197 (1999).
6. Chiappetti, L., et al., *Astrophys. J.*, **521**, 552 (1999).
7. Armstrong, P., et al., *Exp. Astron.*, **9**, 51 (1999).
8. Roberts, M. D., et al., *Astron. Astrophys.*, in press (1999).

Flux Limits for TeV Emission from AGNs

P. M. Chadwick, K. Lyons, T. J. L. McComb, K. J. Orford,
J. L. Osborne, S. M. Rayner, S. E. Shaw, and K. E. Turver

Department of Physics, Rochester Building, Science Laboratories, University of Durham, Durham, DH1 3LE, U.K.

Abstract. The University of Durham Mark 6 telescope has been used to make observations of a number of AGNs visible from the Southern Hemisphere. Limits to VHE gamma ray emission from 1ES 0323+022, PKS 0829+046, 1ES 1101–232, Cen A, PKS 1514–24, RXJ 10578–275 and 1ES 2316–423 are presented, both for steady emission and for outbursts on timescales of ~ 1 day.

INTRODUCTION

The Durham AGN dataset consists of observations of 10 AGNs made with the Mark 6 telescope from 1996 to 1998. Results from the two close X-ray selected BL Lacs PKS 0548–322 and PKS 2005–489 will be reported elsewhere [1]. Here we describe observations of 1ES 0323+022, PKS 0829+046, RXJ 10578–275, 1ES 1101–232, Cen A, PKS 1514–24, and 1ES 2316–423, covering a range of classes of AGN. The typical energy threshold for these observations is ~ 300 to 400 GeV.

Observations were made with the University of Durham Mark 6 atmospheric Čerenkov telescope which has been in operation at Narrabri, NSW, Australia since July 1995. The telescope is described in detail elsewhere [2].

OBSERVATIONS

The seven AGNs which are discussed in this paper comprise three XBLs, two RBLs, one intermediate class object and one close radio galaxy (Cen A) which has been reported previously as a VHE γ-ray source. VHE γ-ray observations of BL Lacs have, in general, concentrated on the closest objects, but we have sought to extend the current redshift limit of $z = 0.117$ by observing more distant AGNs. In the case of one XBL, 1ES 1101–232, the VHE γ-ray observations were made shortly after *BeppoSAX* observations.

Data were taken in 15-minute segments. Off-source observations were taken by alternately observing regions of sky which differ by ± 15 minutes in right ascension from the position of the object. This off-source – on-source – on-source – off-source observing pattern is routinely used to eliminate any first order changes in count rate due to any residual secular changes in atmospheric clarity, temperature etc.

Data were accepted for analysis only if the sky was clear and stable as determined by an infra-red radiometer, and the gross counting rates in each on-off pair were consistent at the 2.5 σ level. A total of 54 hours of on-source observations under clear skies of 7 objects was completed, and an observing log is shown in Table 1.

Routine reduction and analysis of accepted data follow the procedures developed for the analysis of our PKS 2155–304 observations [3].

RESULTS

The dataset for each source has been tested for the presence of gamma ray signals. Typical results of the application of the cuts described above to one object (1ES 2316–423) are shown in Figure 1. The flux limits from the seven AGNs are summarised in Table 2. They are all 3σ flux limits, based on the maximum likelihood ratio test [4,5]. The threshold energy for the observations has been estimated on the basis of preliminary simulations, and is in the range 300 to 400 GeV for these objects, depending on the object's elevation. The effective collecting areas which have been assumed, again from simulations, are 5.5×10^8 cm^2 at an energy threshold of 300 GeV and 1.0×10^9 cm^2 at an energy threshold of 400 GeV. These are subject to systematic errors estimated to be $\sim 50\%$. We have assumed that our current selection procedures retain $\sim 20\%$ of the γ-ray signal, which is subject to a systematic error of $\sim 60\%$.

We have also searched our dataset for γ-ray emission on timescales of ~ 1 day. The search for enhanced emission has been conducted by calculating the on-source excess after the application of our selection criteria for the pairs of on/off observations recorded during an individual night. A typical observation comprising 6 on/off pairs of observations (1.5 hours of on-source observations) yields a flux limit of $\sim 1 \times 10^{-10}$ cm^{-2} s^{-1} at 300 GeV. Conversely, had any of the objects on which we report here produced a 15-minute flare similar to that seen from Mrk 421 with the Whipple telescope on 1996 May 7 [6], it would be detected with the Mark 6 telescope at a significance of around 7 σ. There is no evidence for any flaring activity.

DISCUSSION

Whilst the interpretation of VHE upper limits from BL Lacs is complicated by the lack of a complete theory of VHE γ-ray emission from AGNs, Stecker et al. [7] have predicted the TeV fluxes from a range of objects, one of which (1ES 0323+022), is included in the present work. The expected fluxes from the other XBLs included

TABLE 1. Observing log for observations of active galactic nuclei made with the University of Durham Mark 6 Telescope.

Object	Date	No. of ON source scans
Cen A	1997 March 08	5
Cen A	1997 March 10	5
Cen A	1997 March 11	6
Cen A	1997 March 12	6
Cen A	1997 March 13	5
PKS 0829+046	1996 March 15	3
PKS 0829+046	1996 March 17	6
PKS 0829+046	1996 March 18	7
PKS 1514−24	1996 April 14	2
PKS 1514−24	1996 April 15	6
PKS 1514−24	1996 April 17	6
PKS 1514−24	1996 April 18	7
PKS 1514−24	1996 April 19	10
PKS 1514−24	1996 April 20	6
PKS 1514−24	1996 April 21	6
PKS 1514−24	1996 April 22	8
1ES 2316−423	1997 August 26	2
1ES 2316−423	1997 August 27	2
1ES 2316−423	1997 August 29	8
1ES 2316−423	1997 August 30	13
1ES 2316−423	1997 August 03	11
1ES 2316−423	1997 September 06	4
1ES 1101−232	1998 May 19	6
1ES 1101−232	1998 May 20	3
1ES 1101−232	1998 May 21	7
1ES 1101−232	1998 May 22	6
1ES 1101−232	1998 May 23	4
1ES 1101−232	1998 May 24	4
1ES 1101−232	1998 May 25	8
1ES 1101−232	1998 May 26	8
1ES 1101−232	1998 May 27	6
RXJ 10578−2753	1996 March 20	7
RXJ 10578−2753	1996 March 21	6
RXJ 10578−2753	1996 March 22	2
1ES 0323+022	1996 September 14	4
1ES 0323+022	1996 September 15	4
1ES 0323+022	1996 September 17	7

TABLE 2. Flux limits (3 σ) for observations of active galactic nuclei made with the University of Durham Mark 6 Telescope.

Object	Estimated Threshold (GeV)	Flux Limit ($\times 10^{-11}$ cm^{-2} s^{-1})
Cen A	300	5.2
PKS 0829+046	400	4.7
PKS 1514−24	300	3.7
1ES 2316−423	300	4.5
1ES 1101−232	300	3.7
RXJ 10578−275	300	8.2
1ES 0323+022	400	3.7

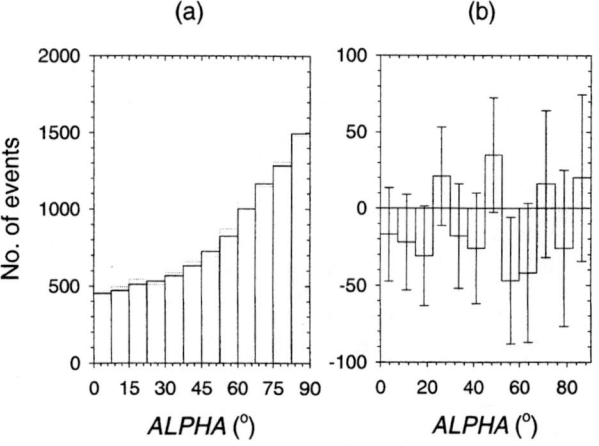

FIGURE 1. (a) The *ALPHA* distributions ON and OFF source for 1ES 2316–423. The dotted line refers to the OFF source data. (b) The difference in the *ALPHA* distributions for ON and OFF source events.

in this paper may be estimated on the basis of the work of Stecker et al. [7,8] using the simple relation $\nu_x F_x \sim \nu_\gamma F_\gamma$ and the published X-ray fluxes. We estimate that the 300 GeV fluxes of 1ES 1101–232, 1ES 2316–423 and RXJ 10578–275 would be 2.0×10^{-11} cm^{-2} s^{-1}, 1.5×10^{-12} cm^{-2} s^{-1}, and 3.3×10^{-12} cm^{-2} s^{-1} respectively, taking into account photon-photon absorption using the recent determination of γ-ray opacity by Stecker [8]. All these suggested fluxes are lower than the flux limits reported here. However, the lack of contemporaneous X-ray measurements in the case of most of our observations limits the usefulness of these predictions and emphasises the importance of simultaneous X-ray and γ-ray observations and multiwavelength campaigns. In the case of the RBLs, an extended observation of PKS 1514–24, a close RBL, lends support to the suggestion that RBLs are not strong VHE γ-ray emitters.

Our observations of Cen A were made when it was in an X-ray low state, in contrast to the earlier VHE detection of Cen A reported by Grindlay et al. [9], which was made when Cen A was in X-ray outburst.

We are grateful to the UK Particle Physics and Astronomy Research Council for support of the project and the University of Sydney for the lease of the Narrabri site. We would like to thank Anna Wolter for providing us with information about *BeppoSAX* observations of 1ES 1011–232 in advance of publication. This paper uses quick look results provided by the ASM/*RXTE* team and uses the NASA/IPAC Extragalactic database (NED), which is operated by the Jet Propulsion Laboratory, Caltech, under contract with the National Aeronautics and Space Administration.

REFERENCES

1. Chadwick, P. M., et al., this proceedings.
2. Armstrong, P., et al., *Exp. Astron.*, **9**, 51 (1999).
3. Chadwick, P. M., et al., *Astrophys. J.*, **513**, 161(1999).
4. Gibson, A. I., et al., *Proc. Intl. Workshop on Very High Energy Gamma Ray Astro.*, Bombay: Tata Institute, ed. P. V. Ramana Murthy & T. C. Weekes, p. 97 (1982).
5. Li, T. P., & Ma, Y. Q., *Astrophys. J.*, **272**, 317 (1983).
6. Gaidos, J. A., et al., *Nature*, **383**, 319 (1996).
7. Stecker, F. W., de Jager, O. C., & Salamon, M. H., *Astrophys. J.*, **473**, L75 (1996).
8. Stecker, F. W., *astro-ph/9812286*, (1998).
9. Grindlay, J. E., et al., *Astrophys. J.*, **197**, L9 (1975).

Observations of TeV Emission from PKS 2005–489 and PKS 0548–322

P. M. Chadwick, K. Lyons, T. J. L. McComb, K. J. Orford,
J. L. Osborne, S. M. Rayner, S. E. Shaw, and K. E. Turver

Department of Physics, Rochester Building, Science Laboratories, University of Durham, Durham, DH1 3LE, U.K.

Abstract. Observations have been made using the University of Durham Mark 6 telescope of the nearby X-ray selected BL Lacs PKS 2005–489 and PKS 0548–322 which should be strong candidates as sources of TeV emission. We find no evidence for long-term emission of VHE gamma rays from these sources in extended measurements covering $\sim$ 2 years. We find 3 σ limits to the VHE gamma ray flux of 0.79×10^{-11} cm^{-2} s^{-1} above 400 GeV for PKS 2005–489 and 2.4×10^{-11} cm^{-2} s^{-1} above 300 GeV for PKS 0548–322.

INTRODUCION

The detection of VHE gamma rays from at least 5 X-ray selected BL Lacs (XBLs) has emphasised the importance of such studies, both to help understand the physics of jets in AGNs and as a probe of the infra-red background. PKS 0548–322 is an XBL at a distance of $z = 0.069$ while PKS 2005–489, again an XBL, is at $z = 0.071$. They are the closest Southern hemisphere XBLs, and, since they are at similar distances, have the potential to help disentangle the effects of production processes and absorption on the energy spectrum.

Stecker et al. [1] have predicted VHE gamma ray fluxes from these objects, based on a simple model relating the X-ray flux to the gamma ray flux and taking into account the expected absorption on the IR background. These predictions confirm that PKS 0548–322 and PKS 2005–489 are strong candidates for detection as VHE emitters with the current generation of telescopes.

The CANGAROO group [2] have reported limits to VHE gamma ray emission above ~ 1.5 TeV. We have extended the VHE coverage of these objects down to ~ 300 GeV using extensive observations made with the University of Durham Mark 6 gamma ray telescope at Narrabri, NSW, Australia from 1996 – 1998. The results of these observations are reported.

TABLE 1. Observing log for observations of PKS 0548–322 and PKS 2005–489 made with the University of Durham Mark 6 Telescope.

PKS 2005–489		PKS 0548–322	
Date	No. of ON source scans	Date	No. of ON source scans
1997 July 5	8	1996 February 18	6
1997 July 6	8	1996 February 18	7
1997 July 7	4	1996 October 9	4
1997 July 28	8	1996 October 10	5
1997 July 29	8	1996 October 12	7
1997 July 30	10	1996 October 13	7
1997 August 5	9	1996 October 15	2
1997 August 7	13	1996 October 16	4
1997 August 26	10	1996 November 7	6
1997 August 27	9	1996 November 10	6
1997 August 28	10	1996 November 11	9
1997 August 29	10	1996 November 12	8
1997 September 20	2	1996 November 13	6
1997 September 21	3	1997 September 25	1
1997 September 24	8	1997 September 26	3
1997 September 25	10	1997 September 27	2
1997 September 26	6	1997 September 30	6
1997 September 27	4	1997 October 23	2
1997 September 30	9	1997 October 25	2
1998 May 28	7	1997 November 6	7
1998 May 29	4	1998 January 22	4
1998 August 11	2	1998 January 28	8
1998 August 13	3	1998 January 29	6
1998 August 17	7		
1998 August 18	10		
1998 August 19	10		
1998 August 20	7		
1998 August 23	5		
1998 August 24	7		
1998 August 25	3		
1998 September 13	9		
1998 September 15	6		
1998 September 16	7		
1998 September 17	4		
1998 September 19	6		
1998 September 21	5		
1998 Octobber 11	1		
1998 Octobber 14	2		
1998 Octobber 15	8		
1998 Octobber 17	4		
1998 Octobber 21	5		
1998 Octobber 22	4		

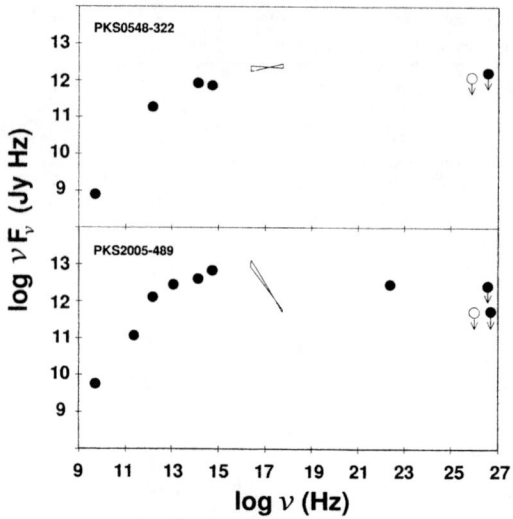

FIGURE 1. The broadband spectral energy distributions for (a) PKS 0548–322 and (b) PKS 2005–489. The datapoints denoted by • are from the compilation of Fossati et al. [7] with the addition of VHE data from Roberts et al. [2]; points from the present work are denoted by ○. All flux limits shown in this figure are 2 σ.

OBSERVATIONS

Observations were made with the Mark 6 telescope [3] on clear moonless nights. Data were taken in our normal off-on-on-off mode where a half-hour on-source observation is bracketed by two 15-minute off-source observations which are at the source postion displaced by ±15 minutes in RA. Data were subjected to our normal quality control procedures and processed following the criteria used for our detection of VHE emission from PKS 2155–304 [4].

After quality control, our dataset for PKS 0548–322 consists of 118 15 minute on-source segments spread over 23 nights of observation. The dataset for PKS 2005–489 consists of 275 on-source 15 minute data segments spread over 42 nights. An observation log is given in Table 1. The PKS 2005–489 data in 1998 October were taken during a multiwavelength campaign.

RESULTS

The dataset for each source has been tested for the presence of gamma ray signals. We find 3 σ flux limits, based on the maximum likelihood ratio test [5,6], of 0.79×10^{-11} cm^{-2} s^{-1} above 400 GeV for PKS 2005–489 and 2.4×10^{-11} cm^{-2} s^{-1} above 300 GeV for PKS 0548–322. The threshold energy for the observations has been estimated on the basis of preliminary simulations, and is in the range 300 to 400 GeV for these objects, depending on the object's elevation. The effective collecting areas which have been assumed, again from simulations, are 5.5×10^8 cm^2 at an energy threshold of 300 GeV and 1.0×10^9 cm^2 at an energy threshold of 400 GeV. These are subject to systematic errors estimated to be $\sim 50\%$. We have assumed that our current selection procedures retain $\sim 20\%$ of the γ-ray signal, which is subject to a systematic error of ~ 60 %.

Stecker et al. [1] have made predictions of the VHE flux from these objects, using a simple model for the VHE emission and taking into account absorption on the IR background. They predict values of 0.51×10^{-11} cm^{-2} s^{-1} for PKS 2005–489 and 1.3×10^{-11} cm^{-2} s^{-1} for PKS 0548–322. Our experimental upper limits are not in conflict with these predictions.

In Figure 1 we show the broadband spectral energy distributions (SEDs) for PKS 2005–489 and PKS 0548–322. Data points at other wavelengths are taken from the compilation of Fossati et al. [7] which are obtained by averaging all available data in wavelength bands. In plotting points from the present work we assume a spectral index of -1.6. Since our VHE points are derived from exposures over a period of ~ 2 years, it is appropriate to add our data to these compilations. We note that our new limits to VHE emission provide a much stronger constraint to the SEDs than previously reported results.

The results presented here are limits to the longterm emission from these objects. We will address the question as to whether there is any evidence for short term outbursts of VHE emission correlated with the activity at other wavelengths in a subsequent paper.

We are grateful to the UK Particle Physics and Astronomy Research Council for support of the project. The Mark 6 telescope was constructed with the assistance of the staff of the Physics Department, University of Durham, and the efforts of Mr. P. Cottle, Mrs. E. S. Hilton and Mr. K. Tindale are acknowledged with gratitude.

REFERENCES

1. Stecker, F. W., et al., *Astrophys. J.*, **473**, L75 (1996).
2. Roberts, M. D., et al., *astro-ph/9902008* (1999).
3. Armstrong, P., et al., *Experimental Astron.*, **9**, 51 (1999).
4. Chadwick, P. M., et al., *Astrophys. J.*, **513**, 161 (1999).
5. Gibson, A. I., et al., *Proc. Intl. Workshop on Very High Energy Gamma Ray Astro.*, Bombay: Tata Institute, ed. P. V. Ramana Murthy & T. C. Weekes, p. 97 (1982).
6. Li, T. P., & Ma, Y. Q., *Astrophys. J.*, **272**, 317 (1983).
7. Fossati, G., et al., *MNRAS*, **299**, 443 (1998).

The Implications of Galaxy Formation Models for the TeV Observations of Current Detectors

L.M. Boone[†], J.S. Bullock[‡], J.R. Primack[†], D.A. Williams[†]

[†] *Santa Cruz Institute for Particle Physics, University of California, Santa Cruz, CA 95064*
[‡] *Astronomy Department, Ohio State University, Columbus, OH 43215*

Abstract. This paper represents a step toward constraining galaxy formation models via TeV gamma ray observations. We use semi-analytic models of galaxy formation to predict a spectral distribution for the intergalactic infrared photon field, which in turn yields information about the absorption of TeV gamma rays from extra-galactic sources. By making predictions for integral flux observations at >200 GeV for several known EGRET sources, we directly compare our models with current observational upper limits obtained by Whipple. In addition, our predictions may offer a guide to the observing programs for the current population of TeV gamma ray observatories.

INTRODUCTION

As shown previously [8,11], measurements of the extra-galactic background light (EBL) may be used to probe models of galaxy formation. The model predictions for the EBL can be probed indirectly via the attenuation of high energy gamma rays, due to pair production with the EBL photon field. The $\gamma\gamma \to e^+e^-$ cross section [5] is maximized when $E_\gamma E_{EBL} \sim 2m_e^2$. From this, we expect TeV gamma rays to be primarily absorbed by EBL photons in the infrared (IR) region of the spectrum. Galaxy formation models which differ in their predicted amount of infrared EBL should be distinguishable by their predicted TeV gamma ray absorption. Here we present results of simulations for the observed spectra of six candidate blazars, including the absorption corrected integral flux we would expect from two plausible galaxy formation models. Other work calculating EBL absorption of TeV spectra under somewhat different assumptions is presented in [12] and [9].

SEMI ANALYTIC MODELS

We have modeled the EBL using the semi-analytic models (SAMs) of galaxy formation discussed in [16], and a similar ΛCDM cosmology where $\Omega_\Lambda = .7$, $\Omega_m =$

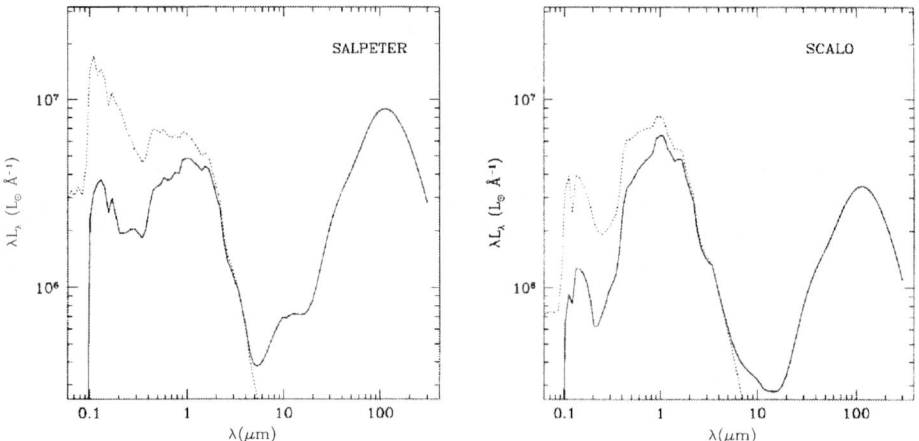

FIGURE 1. These figures represent the spectrum of an average "Milky Way" sized galaxy for each of the two IMFs considered here. The dotted lines indicate the starlight spectra without the effects of dust, while the solid lines represent the spectra with the effects of extinction and emission by dust.

0.3, and $h = 0.7$, normalized such that the rms mass variance on the scale 8 Mpc/h is 1.

We have modeled the EBL for our ΛCDM cosmology using two popular models of the stellar initial mass function (IMF): the Salpeter IMF [13] and the Scalo IMF [14]. The IMF, which describes the stellar mass distribution, affects the wavelength distribution of the starlight produced, and hence the wavelength distribution of the EBL.

The main difference between the two models is that the Salpeter IMF has a larger fraction of high-mass stars than Scalo, provided both are normalized to the same total mass of stars. Sample spectra for both models appear in Fig. 1. Note that the Salpeter IMF has more ultraviolet light than the corresponding Scalo model. This is because high-mass stars produce more ultraviolet light than low-mass stars. Furthermore, because dust absorbs ultraviolet light and emits around 100 μm, the additional ultraviolet light produced by the Salpeter IMF results in a greater amount of $\sim$100 μm EBL. Due to this enhanced 100 μm bump, the Salpeter IMF should produce stronger attenuation for corresponding gamma rays between about 10 GeV and 10 TeV.

IMPLICATIONS FOR TeV ASTRONOMY

We chose our candidate TeV sources from the third EGRET catalog [6]. Of the 67 AGN listed in that catalog, we considered a subset of 60 sources for which there was complete data. These sources are shown in Fig. 2a. Of the 60 sources

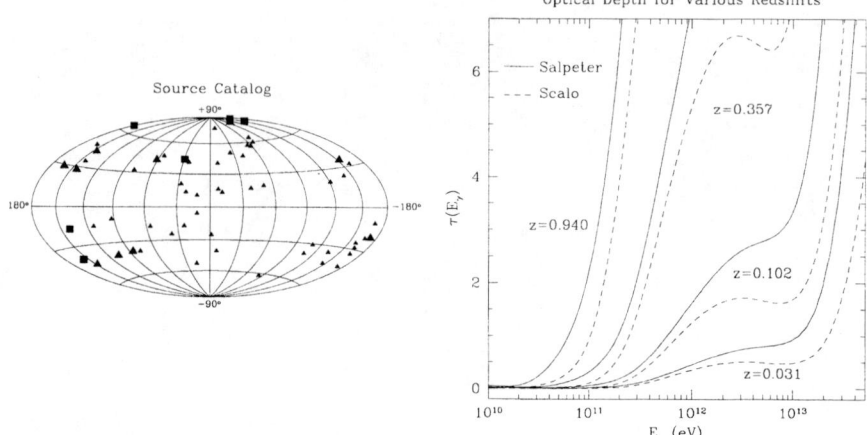

FIGURE 2. a) Source catalog in galactic coordinates. Small points denote all considered EGRET AGN, of which 16 (large points) were selected as "candidate" sources. We have plotted absorbed spectra for six of these candidate sources (large squares) **b)** Optical depth for selected redshifts as a function of gamma ray energy.

considered, 16 were chosen as candidate sources for TeV observations based on the hardness of their spectra and their integral flux above 100 MeV. Candidate sources appear as large points in Fig. 2a. We then simulated the observed integral flux for six of these sources (large squares), both with and without absorption corrections. Of our six sources, one (Mrk421) is an X-ray selected BL Lac, one (4C+29.45) is a flat spectrum radio quasar, and the rest are radio selected BL Lacs.

For the absorption simulations, we assumed a simple power law for the intrinsic spectrum of each source. We did not assume any source absorption effects. Spectral indices and pre-factors for the differential spectra were obtained from the third EGRET catalog [6], and calculations of these differential spectra followed [17]. We then modified each of these intrinsic spectra with an absorption factor. We assumed the functional form of this factor to be an exponential decay, whose exponent (τ) is derived from the calculations of the EBL mentioned earlier. Plots of τ as a function of energy for a sample of the red-shifts considered appear in Fig. 2b. The functional form for the simulated differential spectrum is then:

$$\left(\frac{d\phi}{dE}\right)_{abs} = \left(\frac{d\phi}{dE}\right)_{int} \exp[-\tau(E)] \qquad (1)$$

Integral fluxes were calculated numerically from this absorption corrected differential flux. Results are plotted in Fig. 3 for the two different stellar IMFs under consideration here. Also included for reference on the plots in Fig. 3 are sensitivity curves for a representative sample of ground based gamma ray detectors.

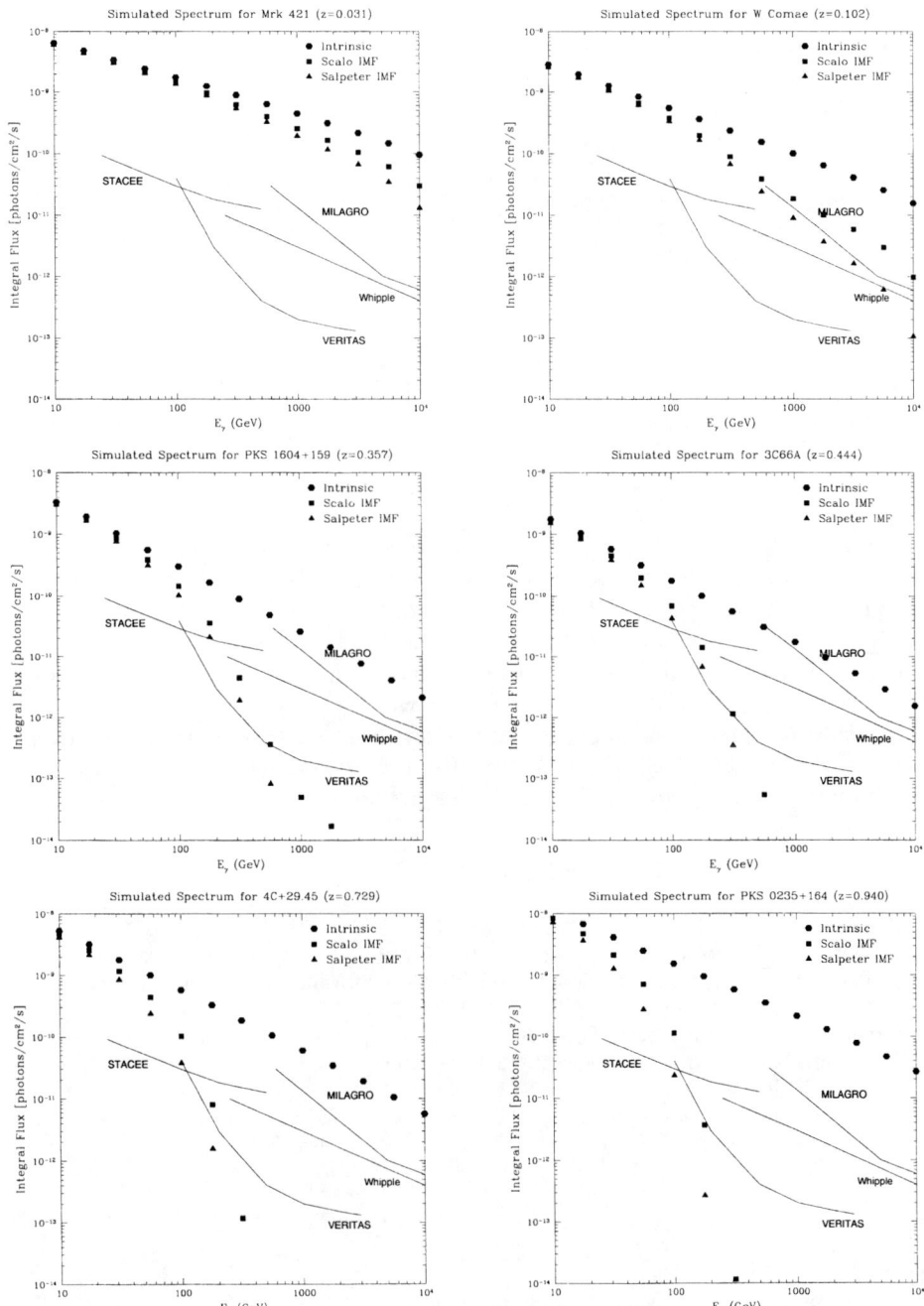

FIGURE 3. Calculation of expected integral flux for six EGRET sources at various redshifts.

Four of the six sources we considered had been previously selected for observations by the Whipple group [2], and each resulted in a non-detection. These sources are W Comae, PKS 1604+159, 3C66A, and PKS 0235+164. The upper limits placed on the latter three are consistent with our model. However, Whipple's upper limit on W Comae is significantly below both the Scalo and Salpeter corrections to a simple EGRET extrapolated power law spectrum. In addition to this discrepancy for W Comae, we are unable to reproduce Whipple's observed integral flux for Mrk 421 during the same epoch [15]. Comparison of our predicted differential

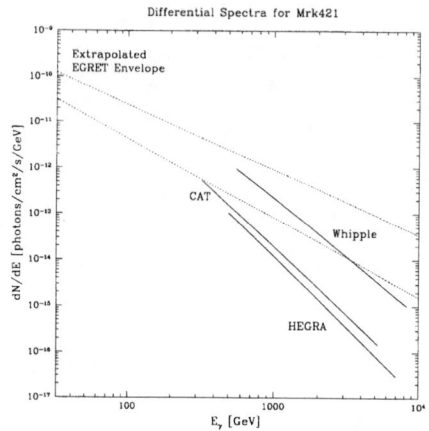

FIGURE 4. TeV spectra for Mrk421 and extrapolated EGRET spectrum.

spectrum and the current observed differential spectra from Whipple, HEGRA, and CAT [1,7,10] also yielded disparate results, as illustrated in Fig. 4. The poor correlation between our extrapolated EGRET data and the TeV observations make it clear that we cannot yet make a strong statement about the EBL, or the corresponding IMF, due mainly to the uncertainty of our simple power-law model of the intrinsic source spectrum. In future efforts, we will exchange our simple power law model for a more realistic simulation of the intrinsic source spectrum. However, while the plots in Fig. 3 may not be realistic, they do demonstrate the effects of absorption on the spectrum, and how this feature varies with redshift.

REFERENCES

1. Aharonian, F. et al., *Astron. & Astrophys.* **350**, 757 (1999).
2. Buckley, J.H., *Astropart. Phys.* **11**, 119 (1999).
3. Bullock, J.S., Ph.D. dissertation, University of California, Santa Cruz (1999).
4. Bullock, J.S. et al., *Astropart. Phys.* **11**, 111 (1999).
5. Gould R.J., Schreder, G.P., *Phys. Rev.* **155**, 1404 (1967).
6. Hartman, R.C. et al., *Astrophys. J.* **123**, 79 (1999).
7. Krennrich, F. et al., *Astrophys. J.* **511**, 149 (1999).
8. MacMinn, D., Primack, J.R., *Space Science Reviews* **75**, 413 (1996).
9. Mukherjee, R. et al., *Proc. 26th ICRC* (Salt Lake City) **3**, 362 (1999).
10. Piron, F. et al., *Proc. 26th ICRC* (Salt Lake City) **3**, 326 (1999).
11. Primack, J.R. et al., *Astropart. Phys.* **11**, 93 (1999).
12. Salamon, M.H., Stecker, F.W., *Astrophys. J.* **493**, 547 (1998).
13. Salpeter, E. *Astrophys. J.* **121**, 61 (1955).
14. Scalo, J.M., *Fund. Cosmic Phys* **11**, 1 (1986).
15. Schubnell, M.S. et al., *Astrophys. J.* **460**, 644 (1996).
16. Somerville, R.S., Primack, J.R., *MNRAS*, in press, astro-ph/9802268 (1999).
17. Thompson, D.J. et al., *Astrophys. J. Supp.* **107**, 227 (1996).

Extragalactic background light absorption signal in the 0.26 − 10 TeV spectra of blazars

Vladimir V. Vassiliev[1]

Whipple Observatory, Harvard-Smithsonian CfA, P.O. Box 97, Amado, AZ 85645, USA

Abstract. Recent observations of the TeV γ-ray spectra of the two closest active galactic nuclei (AGNs), Markarian 501 (Mrk 501) and Markarian 421 (Mrk 421), by the Whipple and HEGRA collaborations have stimulated efforts to estimate or limit the spectral energy density (SED) of extragalactic background light (EBL) which causes attenuation of TeV photons via pair-production when they travel cosmological distances. In spite of the lack of any distinct cutoff-like feature in the spectra of Mrk 501 and Mrk 421 (in the interval 0.26 − 10 TeV) which could clearly indicate the presence of such a photon absorption mechanism, we demonstrate that strong EBL attenuation signal (survival probability of 10 TeV photon $\sim 10^{-2}$) may still be present in the spectra of these AGNs. This attenuation could escape detection due to a special form of SED of EBL and unknown intrinsic spectra of these blazars. Here we show how the proposed and existing experiments, VERITAS, HESS, MAGIC, STACEE and CELESTE may be able to detect or severely limit the EBL SED by extension of spectral measurements into the critical 100 − 300 GeV regime.

Introduction

It has long been thought [1] that the detection of attenuation effect in the TeV spectra of extragalactic sources caused by pair production $\gamma + \gamma \to e^+ + e^-$ with the EBL would be of great value for the understanding of cosmology and many aspects of the astrophysics of the Universe. Finding the cutoff feature in the high energy end of the Markarian 501 (Mrk 501) spectrum (> 10 TeV), reported during this workshop [2], may well be a long awaited signature of such extinction of the highest energy photons. In the presentation [3] during this workshop, the claim has been made that the found cutoff is well explained by a semi-empirically derived EBL prediction [4] and by the simple power-law spectrum intrinsic to the source with an exponent close to two spanning from 0.2 to 20 TeV. The latter implies that the EBL attenuation mechanism below 10 TeV should rather be weak which seems to be confirmed intuitively by the absence of a distinguishable feature in this part of the spectra for both Mrk 501 ($z = 0.03$) and Mrk 421 ($z = 0.03$) blazars (Fig. 1, [5]).

---

[1] Corresponding author: vvassiliev@cfa.harvard.edu

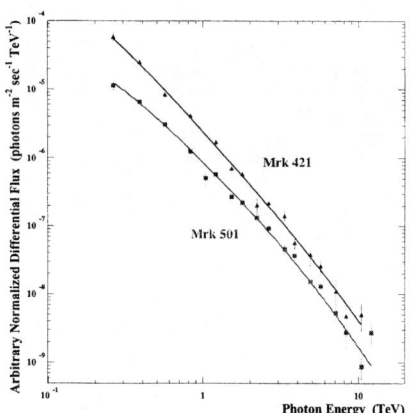
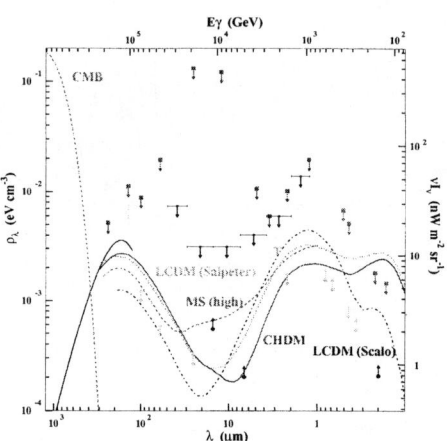

FIGURE 1. Spectra of Mrk 421 and Mrk 501 below 10 TeV during high state of activity as measured by the Whipple γ-ray telescope [5].

FIGURE 2. Compilation of various EBL detections, upper and lower limits, and EBL theoretical and semi-empirical models (data from [13]).

We might now ask, whether the proposed SED of EBL is uniquely consistent with experiment, in order to begin its interpretation in terms of astrophysical constraints, or even if the lack of the feature in the low energy part of the TeV AGN spectra does prove an absence of the EBL absorption. Here I argue that we are not yet ready to make such statements neither on theoretical nor experimental basis, and due to ironic coincidence there is still a substantial degree of freedom in the definition of the SED of EBL. In this talk I consider a peculiar degeneracy which allows a certain type of SED of EBL to avoid "apparent" detection in currently available experimental data below 10 TeV due to the unknown properties of intrinsic spectra of the sources. The conclusions which I draw at the end of this talk will show how we can narrow down the existing possibilities even if we use spectral data of only these two AGN which may become available in the near future with the introduction of new γ-ray instruments, such as VERITAS or STACEE.

The EBL

Any contemporary theoretical consideration of the SED of EBL predicts two well pronounced peaks (Fig. 2). One at ∼ 1 μm due to the starlight emitted and redshifted through the history of the Universe, the other at ∼ 100 μm generated by re-processing of the starlight by dust, its extinction and reemission. Theoretical modeling of the spectral evolution of the EBL field involve complex astrophysics with many unknown input parameters which specify cosmology, number density and evolution of dark matter halos, distribution of galaxies in them, mechanisms of converting cold gas into stars, the star formation rate (SFR), stellar initial mass function (IMF), supernovae feedback, and the mechanisms by which light is absorbed by dust and reemitted at longer wavelengths [6]. Semi-Analytical modeling of these processes show that the dominant factors shaping SED of EBL in the region $1-10$ μm are IMF, which provides a source of the UV light, generated mostly by the high mass stars and is therefore dependent on their fraction, and dust extinction which functions as a sink of UV light [7]. The region from 1 to 10 μm is primarily determined by the type of cosmology and the SFR. Allowing partial degeneracy between these two factors leads to an ambiguity in the interpretation of the actual SED of EBL [8]. It is also possible that a non-negligible contribution from a number of pregalactic and protogalactic sources may exist in this interval [9]. This EBL fraction is usually not considered in the EBL evolution models nor in semi-empirical EBL estimates. The long wavelength region, from approximately a few μm to 100 μm, is currently predicted with the largest uncertainty, due to poorly defined dust extinction and re-radiation mechanisms, which are crucial ingredients for modeling EBL in this band. In addition, it has been suggested recently [10], that a substantial energy in this wavelength interval may come from quasars. Most of their radiation should be absorbed in the dust and gas of the accretion disk and re-emitted later in the far-infrared. Up to now this contribution has been considered as negligible, but failure to explain existing X-ray background suggests a presence of a large population of the faint quasars generating this diffuse radiation field [11]. These sources, visible only in X-ray and far-infrared, are expected to be probed by the Chandra mission. Energy ejected into the surrounding media by supernovae and re-radiated later may also be concealed in this wavelength interval [12].

At present, the degree of the uncertainty of various theoretical considerations of the SED of EBL is about the same as the distance between current upper and lower experimental limits. Fig 2 shows a compilation [13] of various EBL detections and limits as well as several theoretical [12] and semi-empirical [4] estimates of the SED of EBL. In the current situation a preferential choice of a particular prediction based solely on a theoretical background seems unjustified. We do expect, however, that the SED of EBL is likely to be a function with complex behavior. If we take into account that attenuation of extragalactic TeV γ-rays is an exponential effect, one would intuitively expect appearance of the structures in the observable spectra of AGNs, such as cutoff, for example. Non-existence of any peculiar features in $0.25-10$ TeV spectra of AGNs should then indicate a very weak absorption effect. The problem, however, is more subtle than it first appears. There is a whole class of non-trivial solutions for the SED of EBL (see [13]), shown in Fig. 3, which may well describe starlight peak expected in the $0.1-10$ μm region. The important property of such SEDs is that they do not produce any peculiar change in the observable AGN spectrum. The only effect to be seen is change of the overall attenuation factor, change of the power-law spectral index, and slight change of the spectral curvature. All three of these potentially detectable EBL indicators are perfectly masked by the unknown intrinsic spectrum of the source. The existence of such SED solutions, which have been hinted in [14], is due to slow, power-law-like, change of an attenuation coefficient when the SED is proportional to energy

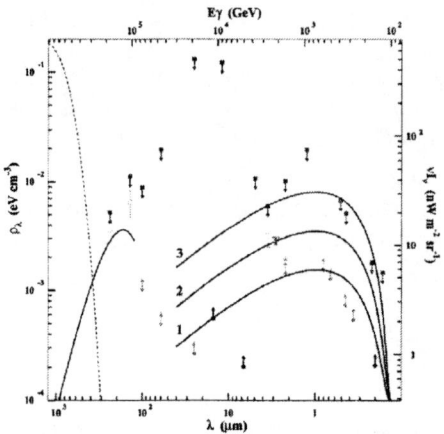
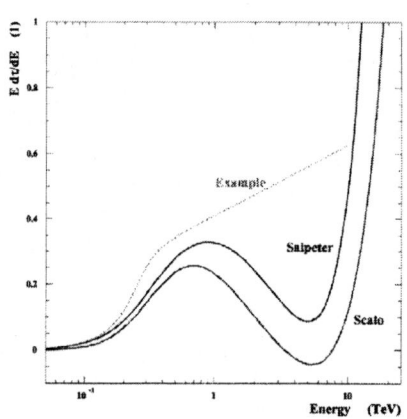

FIGURE 3. Three examples of "invisible" SED of EBL are marked 1, 2, and 3 [13].

FIGURE 4. Predicted change of the Mrk 421 and Mrk 501 power-law spectral index due to attenuation of TeV photons on EBL (data from [12]).

of the infrared photon with logarithmic accuracy. Such a case seems to take place in the $1 - 10$ μm region to which observations in $0.25 - 10$ TeV interval are most sensitive. It is an ironic coincidence that current observational window of TeV γ-ray astronomy coincided with the region of a special behavior of SED which makes EBL attenuation effect "invisible." If we were to move γ-ray observational window to lower or higher energies we would be sensitive to the bands in the EBL spectrum where the SED is rapidly falling. This would produce an exponential effect on the observable spectra of AGNs, such as one detected by the HEGRA collaboration in the region $10 - 25$ GeV [2].

Since there are a number of upper and lower limits for the SED of EBL in the $0.1 - 10$ μm interval, established by various experiments, it becomes possible to constrain the EBL attenuation effect in the spectra of Mrk 421 and Mrk 501 using an explicit form of "invisible" SEDs. Such considerations [13] lead to the conclusions that the optical depth for 10 TeV photons is bounded within the interval $0.85 - 4.43$, the EBL contribution to the power-law spectral index at photon energy 1 TeV can be in the range $0.19 - 0.94$, and spectral curvature does not exceed 0.22 (the Hubble constant used is 65 km s^{-1} Mpc^{-1}). Analogous constraints derived from the EGRET measurements of the spectral indices of these sources [15,16] provide similar upper limit for spectral index change due to the EBL attenuation effect (< 1.0). Source luminosity arguments [2] suggest that if the optical depth at 1 TeV exceeds ~ 3 then the intrinsic γ-ray luminosity of Mrk 501 is an order of magnitude larger than the luminosity in all other wavelengths which is difficult to explain with realistic parameters of the jet. This places approximately the same upper bound on the absolute value of the γ-ray absorption. Finally, a lack of the curvature in the spectrum of Mrk 421 provides upper limit of 0.3 which is no stronger at the moment than the direct EBL constraints.

The large degree of the uncertainty which still exists in an experimental detection of EBL signal via observations of γ-ray attenuation is due to the, as yet, undetermined initial conditions in the region approximately from 100 to 300 GeV. Although EBL induced attenuation of γ-rays in this interval is rather small for z=0.03, the absorption effect must produce a jump in the spectral index and curvature here. In Fig. 4 I show a derivative of the optical depth with respect to the logarithm of energy, which characterizes the change of the spectral index. Two examples shown correspond to the predictions made in [12] for Salpeter and Scalo IMFs. Note that the region from 100 GeV to 500 GeV is characterized by a rapid change in the spectral index. This result is, of course, hardly surprising since it is produced by a rapid fall of the SED of EBL in UV band. The subsequent dip above 1 TeV is due to a very low prediction of the models for EBL field in the region around 10 μm. Such a low estimate is not currently supported by the lower bound on EBL from ISO measurements at 15 μm [17]. Absence of the EBL signature in the spectra of Mrk 421 and Mrk 501 in the 0.3 − 10 TeV interval suggests that the behavior of the derivative of optical depth should likely be linear function of the logarithm of photon energy (curve marked "Example" in Fig. 4). The measurements of the AGN spectra in the region 100 − 300 GeV are of crucial importance then to experimentally determine two parameters of this curve and therefore unfold a unique SED "invisible" solution [13]. The expected change of spectral index, 0.19 − 0.94, of Mrk 501 and Mrk 421 between 100 GeV and 1 TeV due to the EBL absorption is a measurable effect. If the opacity of the Universe to TeV γ-rays is large, most of this change should occur below 300 GeV since such an effect has not been detected in the AGN spectra above this energy [5]. This would produce a "knee-like" feature in the spectra of these blazars in the 100 − 300 GeV interval. If the attenuation effect is small though, it is possible that the derivative of optical depth remains the same linear function in this energy band generating only a logarithmically small EBL footprint in the AGN spectra. In the latter case, a small curvature, $1/2\ d^2\tau/d\ ln(E)^2 \sim 0.1$, in 0.1 − 1 TeV energy range would be the only indicator of EBL presence.

Conclusions

1. The featureless spectra of the two closest AGNs in the 0.25 − 10 TeV energy band does not guarantee a low attenuation of TeV γ-rays via pair-production with EBL. The large class of "invisible" SED solutions exists [13] which change only overall attenuation factor, spectral index, and spectral curvature of the observable AGN spectrum. Behavior of these SEDs is consistent with the theoretically expected starlight EBL peak at 1 μm, but due to the unknown intrinsic properties of the sources such an attenuation effect cannot be unambiguously isolated based only on the data from this energy interval. The EBL spectral density suggested in [3] for explanation of spectral properties of Mrk 501 is, therefore, only one of many possibilities.

2. Even by observing only two known extragalactic TeV sources, Mrk 501 and Mrk 421, we can hope to constrain or possibly detect SED of EBL if spectral measurements are extended into the 100 − 300 GeV region where change of the spectral index would indicate a turn on of the absorption effect. Detection of this feature by future γ-ray observatories, for example VERITAS or STACEE, will provide a missing piece of information for proper unfolding of the SED of EBL in the wavelength interval above 0.1 μm. Of course, a certain ambiguity of spectra interpretation due to unknown properties of the sources will remain, the presence of a similar feature, which is static in time, in both spectra may allow to disentangle or severely limit intergalactic $\gamma - \gamma$ extinction effect.

3. It turns out that the most important constraints for limiting or unfolding the SED of EBL may be provided by accurate measurements of the curvature (better than $\sim$ 0.1 per decade in energy) in the spectrum of Mrk 421 through the interval 250 − 10 TeV. At the moment the spectrum of this source during high flaring state has been found consistent with a pure power-law [5]. If the source cooperates, an improved statistics may give EBL upper limits in the region around a few micrometers lower than the currently available *DIRBE* results.

Acknowledgments

I thank Whipple collaboration, J. Bullock and J. Primack for providing data, T. Weekes and S. Fegan for valuable discussions and invaluable help. This work was supported by grants from the U.S. Department of Energy.

REFERENCES

1. Gould R.P., & Schréder G.P.,*Phys. Rev.* **155**, 1408 (1967).
2. Aharonian F.A. et al., *Accepted for publication in A&A*, astro-ph/9903386 (1999).
3. Konopelko A.K. et al., *ApJ*, **518**, L13 (1999).
4. Malkan M.A., & Stecker F.W., *ApJ*, **496**, 13 (1998).
5. Krennrich F. et al., *ApJ*, **511**, 149 (1999).
6. Somerville R.S. & Primack J.R. *in press MNRAS*, astro-ph/9802268 (1999).
7. Bullock J.S. et al., *Astropart.Phys.*, **11**, 111 (1999).
8. MacMinn D. & Primack J.R. *Space Science Reviews*, **75**, 413 (1996).
9. Bond J.R., Carr B.J., & Hogan C.J. *ApJ*, **306**, 428 (1986).
10. Fabian A.C. *accepted for publication in MNRAS*, astro-ph/9908064 (1999).
11. Wilman R.J & Fabian A.C. *to appear in MNRAS*, astro-ph/9907204 (1999).
12. Primack J.R. et al., *Astropart.Phys.*, **11**, 93 (1999).
13. Vassiliev V.V. *accepted for publication in Astropart.Phys.*, astro-ph/9908088 (1999).
14. Dwek E. & Slavin J. *ApJ*, **436**, 696 (1994).
15. Hartman R.C. et al. *in press ApJ* (1999).
16. Kataoka J. et al., *in press ApJ*, astro-ph/9811014 (1999).
17. Oliver S.J. et al., *MNRAS*, **289** , 471 (1997).

ON MARKARIAN 421 AND 501

VHE γ-ray spectral properties of the blazars Mrk 501 and Mrk 421 from CAT observations in 1997 and 1998

Frédéric Piron, for the CAT collaboration

Laboratoire de Physique Nucléaire des Hautes Energies
Ecole Polytechnique, route de Saclay, 91128 Palaiseau Cedex, France

Abstract. The Very High Energy (VHE) γ-ray emission of the BL Lacertæ objects Markarian 501 and Markarian 421 has been observed by the CAT Imaging Atmospheric Cherenkov Telescope in 1997 and 1998. The spectrum extraction method is presented, and the spectral properties of both sources are compared in different activity states. Theoretical implications for jet astrophysics are discussed.

SPECTRAL ANALYSIS METHOD

The CAT (Cherenkov Array at Thémis) telescope records Cherenkov flashes due to VHE atmospheric showers through its 17.8m^2 mirror. The fine grain of its camera, combined with fast electronics, allows a relatively low γ-ray detection threshold energy of 250GeV (at Zenith) as well as an accurate analysis of the resulting images. The experiment is fully described elsewhere [3]. The analysis method is based on the fit of each individual shower image with theoretical mean γ-ray images coming from a semi-analytical model [7]. This permits the extraction of a γ-ray signal from the background due to cosmic-ray induced atmospheric showers; the optimization of the event selection using the Crab nebula data gives a hadron-rejection factor of ~200 for a γ-ray efficiency of ~40%. The fit also yields the energy of each shower in the γ hypothesis, with a resolution of ~20%, independent of energy, when restricting to showers with a small impact parameter (< 130 m). However, the spectrum reconstruction procedure presented below involves the exact energy-resolution function $\Psi(E \to \tilde{E}, cos\theta)$ [1] in order to use the entire available statistics. These functions have been determined by detailed Monte-Carlo simulations of the telescope response, as has the effective detection area $\mathcal{A}_{\text{eff}}$, which includes the effect of event-selection efficiency. The simulations have been checked and calibrated on the basis of several observables by using muons rings and the nearly-pure γ-ray signal from the highest flare of Mrk 501 in 1997 [9].

---

[1] Probability density, for fixed zenith angle θ and *real* energy E, to get an *estimated* energy $\tilde{E}$.

With typical statistics of ~1000 γ-ray events and a signal-to-background ratio of 0.4 (as obtained on the Crab nebula), a spectrum can be determined with reasonable accuracy as follows. First we define a set $\{\Delta_{i_e}\} = \{[\tilde{E}_{i_e}^{\min}, \tilde{E}_{i_e}^{\max}]\}_{i_e=1,n_e}$ of energy bins; since the threshold energy of the telescope increases with zenith angle, each of these is divided into a subset $\{\Delta_{i_e,i_z}\} = \{[\theta_{i_z}^{\min}, \theta_{i_z}^{\max}]\}_{i_z=1,n_z(i_e)}$ of zenith angle bins [2] with a width of $\delta(\cos\theta) = 0.05$. Then, for all ON and OFF-source runs [3], the number of events passing the selection cuts is determined within each Δ_{i_e,i_z} 2D-bin. Finally, assuming a given spectral shape, a maximum-likelihood estimation of the spectral parameters is performed, which takes into account the effective detection area and the energy-resolution function. At this stage, two hypotheses $\mathcal{H}^{\rm pl}$ and $\mathcal{H}^{\rm cv}$ are successively considered, which are a simple power law $\phi^{\rm pl}(E) = \phi_0 E_{\rm TeV}^{-\gamma}$ and a curved shape $\phi^{\rm cv}(E) = \phi_0 E_{\rm TeV}^{-(\gamma+\beta\log_{10}E_{\rm TeV})}$, respectively. Power laws often account for spectral properties in high energy astrophysics, at least because they often constitute a good approximation when considered over a restricted energy range. On the other hand, a curved shape is suggested by general considerations on emission processes within blazar jets (see the next sections); the latter parametrization, previously used by the Whipple group [11], corresponds to a parabolic law in a $\log(\phi)$ vs. $\log(E)$ representation and allows simple comparisons. The relevance of $\mathcal{H}^{\rm pl}$ with respect to $\mathcal{H}^{\rm cv}$ is estimated through the likelihood ratio of the two hypotheses, which is defined as $\lambda = -2 \times \log(\frac{\mathcal{L}^{\rm pl}}{\mathcal{L}^{\rm cv}})$: it behaves (asymptotically) like a χ^2 with one d.o.f. and permits a search for the presence of spectral curvature. *All spectral curves shown in the next sections come from the best set of fitted parameters obtained by this procedure. It is also customary to superimpose on them "experimental" points $\{E_{i_e}^{\rm exp}, \phi_{i_e}^{\rm exp}\}_{i_e=1,n_e}$, which are only indicative of the statistics used in each Δ_{i_e} energy bin.* The definition of these points is the following: if we note $\phi^{\rm best}(E)$ as the best fitted spectrum under the hypothesis $\phi^{\rm pl}$ or $\phi^{\rm cv}$ according to λ, $N_{i_e,i_z}^{\rm obs}$ as the number of γ-ray events observed in the Δ_{i_e,i_z} 2D-bin, and σ_{i_e,i_z} as its error, and $N_{i_e,i_z}^{\rm pred} = \int_{\tilde{E}_{i_e}^{\min}}^{\tilde{E}_{i_e}^{\max}} d\tilde{E} \int_0^\infty dE\, \phi^{\rm best}(E) \times \mathcal{A}_{\rm eff}(E, \langle\cos\theta\rangle_{i_z}) \times \Psi(E \to \tilde{E}, \langle\cos\theta\rangle_{i_z})$ as the corresponding predicted number, then $E_{i_e}^{\rm exp}$ is the unique energy within the bin for which $\phi^{\rm best}(E_{i_e}^{\rm exp}) = \langle\phi^{\rm best}\rangle_{i_e}$ holds [4], and $\phi_{i_e}^{\rm exp} = \phi^{\rm best}(E_{i_e}^{\rm exp}) \times \left[\sum_{i_z=1}^{n_z(i_e)} \left(N_{i_e,i_z}^{\rm obs}/N_{i_e,i_z}^{\rm pred}\right)/\sigma_{i_e,i_z}^2\right] / \left[\sum_{i_z=1}^{n_z(i_e)} 1/\sigma_{i_e,i_z}^2\right]$. It should be noted that this way of representing spectra cannot replace the likelihood method, which provides the only relevant physical results, i.e. the values of the most probable spectral parameters as well as their errors and covariance matrix. *In particular, one should not use the $\{E_{i_e}^{\rm exp}, \phi_{i_e}^{\rm exp}\}$ points in any kind of minimization, since the σ_{i_e,i_z} are correlated in a very complex manner.*

---

[2] The $i_z = 1$ bin corresponds to the transit of the source.
[3] Each run lasts ~30min, OFF-source runs being used to estimate the hadronic background.
[4] The average is taken over the bin according to the physical meaning of the flux, i.e. by integrating over E or $\log(E)$ in a dN/dE or in a $\nu F_\nu (\equiv E^2 \times dN/dE)$ representation, respectively.

CAT OBSERVATIONS OF MRK 501 AND MRK 421

Spectral properties of Mrk 501 in 1997 and 1998

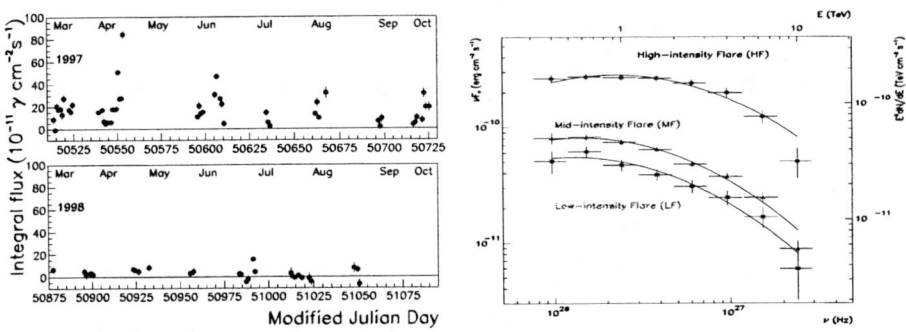

FIGURE 1. *(a)* Mrk 501 nightly integral flux above 250 GeV in 1997 and 1998; *(b)* Mrk 501 VHE SED between 330 GeV and 13 TeV for three independent data-subsets, corresponding to different activity states in 1997 (see [4] for details).

Mrk 501 exhibited a remarkable series of flares during the whole year 1997, going down to a much lower mean flux in 1998 (Fig. 1a). The detailed spectral analysis of the 1997 data can be found in [4]; here we extend it to the 1998 results.

The VHE spectral energy distribution (SED) of Mrk 501, derived for different flaring-activity states in 1997 (Fig. 1b), shows a significant curvature which is now well confirmed by different ground-based experiments [1,6]. The peak γ-energy is found to lie just above the CAT threshold, and it seems to shift towards higher energies as the flux increases. To check this spectral variability by a more robust method, the hardness ratio has been computed for five different-level intensities: the correlation observed in Fig. 2a confirms the hardening of the VHE SED during flaring periods. Fig. 2b shows the broad-band SED of Mrk 501 for two flaring

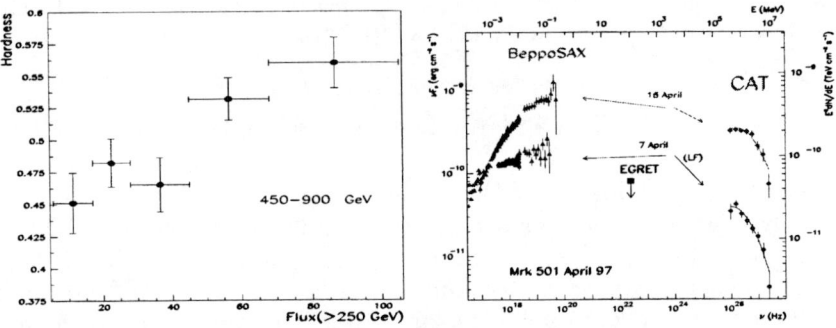

FIGURE 2. *(a)* Hardness-ratio ($HR = \frac{N_{E>900\,\text{GeV}}}{N_{E>450\,\text{GeV}}}$) vs. source intensity ($\Phi_{>250\,\text{GeV}}$ in units of $10^{-11}\,\text{cm}^{-2}\text{s}^{-1}$); *(b)* Mrk 501 X-ray and VHE spectra for April 7th and 16th. The EGRET upper limit corresponds to observations between April 9th and 15th.

dates in April 1997: it exhibits the two-bump structure typical of blazars, with a dip indicated by the contemporary EGRET upper-limit point in the GeV energy range [11]. The obvious correlation of the X-ray emission from BeppoSAX data [8] with TeV emission strongly suggests that the same particle population is responsible for emission in both energy-ranges, i.e. it supports the picture given by leptonic models [5], in which an energetic electron beam propagating in the magnetized plasma jet produces X-rays through synchrotron radiation as well as VHE γ-rays through inverse Compton scattering of low-energy photons.

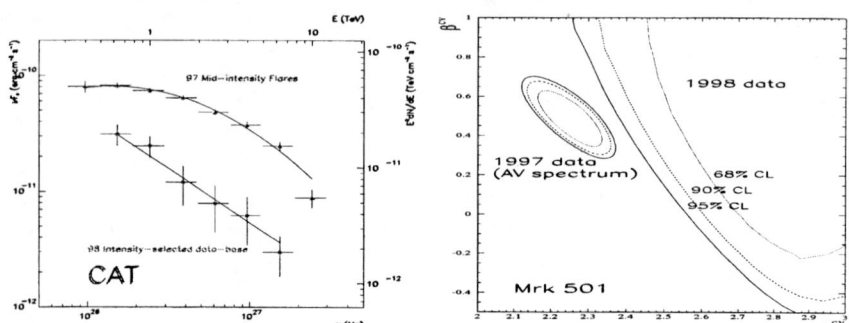

FIGURE 3. (a) Mrk 501 VHE SED from 330 GeV to 5.2 TeV in 1998, as compared to the MF data subset in 1997 (see Table 1); (b) 68%, 90% and 95% CL contours in the $\{\gamma^{cv}, \beta^{cv}\}$ plane for the 1997 AV and 1998 data sets (see Table 1); each contour is obtained by projecting the 3-dimension ellipsoid along the ϕ_0^{cv} axis.

Unlike the 1997 spectra, the spectrum of Mrk 501 in 1998 does not show any curvature (Fig. 3a), indicating a VHE peak energy well below the CAT threshold at that time. Fig. 3b confirms in particular that, in spite of larger statistical errors, the 1998 data cannot be fitted by the average spectrum shape found in 1997 (AV data set, see Table 1), as would be expected in the absence of any spectral variability. With a mean flux much lower than that of 1997, the power-law shape found in 1998 is therefore consistent with the previously suggested scenario, in which the peak γ-energy seems to be correlated with the VHE flux.

Mrk 421's spectrum in 1998

As reported in [10], Mrk 421 is the second extragalactic source detected by CAT: almost quiet in 1996-97, the source showed small bursts in 1997-98, together with a higher mean activity (Fig. 4a). The energy spectrum derived for the 1998 flaring periods is well represented by a simple power law (Fig. 4b) with a differential spectral index $\gamma = 2.96 \pm 0.13^{stat} \pm 0.05^{syst}$ [10]. This agrees with a recent result of the HEGRA group concerning the 1997-98 period [2]. It is however in contrast with the former behaviour of the source at the time of the 1995-96 flaring period, as observed by the Whipple group, who found $\gamma = 2.54 \pm 0.03^{stat} \pm 0.10^{sys}$ [6]. Thus,

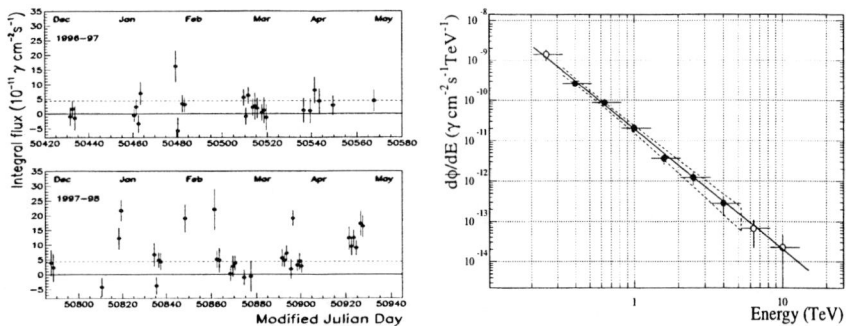

FIGURE 4. *(a)* Mrk 421 nightly integral flux above 250 GeV between December 1996 and May 1998. The dashed line represents the mean flux over the two years; *(b)* Differential flux of Mrk 421 for the flaring periods in 1998. Only the energy bins shown with fullfilled circles (from 330 GeV to 5.2 TeV) were used in the likelihood method (see text).

the higher value of the differential spectral index found in this work could come from a correlation between the intensity level and the spectral hardness, like that observed for Mrk 501 in 1997 (see [4] and above). A similar evidence for Mrk 421 was in fact suggested in [13] on the basis of the Whipple 1995-96 low-flux data.

In the framework of leptonic models [5], which succesfully explain the Mrk 501 broad-band SED in 1997, the absence of any obvious spectral curvature reported by all experiments implies in any case that the peak energy of the inverse-Compton contribution to Mrk 421's spectrum is significantly lower than the CAT detection threshold, as for Mrk 501 in 1998. This is not surprising since the corresponding synchrotron peak is lower than that of Mrk 501 in 1997, and since we have seen that leptonic models predict a strong correlation between X-rays and γ-rays. In fact, such a correlation was directly observed on Mrk 421 in Spring 1998, during a coordinated observation campaign involving ground-based Cherenkov imaging telescopes (Whipple, HEGRA, and CAT) and the ASCA X-ray satellite [12].

CONCLUSION

Since Mrk 501 and Mrk 421 lie at the the same redshift ($\sim$0.03), spectral differences between them must be intrinsic and not due, in particular, to absorption by the diffuse infrared background radiation. This allows direct and relevant comparison of their spectral properties. The correlation between X-ray and γ-ray emissions is now proven for both sources, supporting the simple and most natural scenario given by leptonic models [5] for the origin of blazar TeV flares, in which a single leptonic population is injected into the radio jets and produces correlated X-ray synchrotron and VHE γ-ray inverse Compton radiations. CAT observations complete this picture with some evidence of spectral variability of Mrk 501 in 1997, suggesting that this leptonic population is responsible for the hardening of the en-

TABLE 1. Best fitted spectral parameters of Mrk 501 and Mrk 421 in 1997 and 1998 from the $\mathcal{H}^{pl}$ and $\mathcal{H}^{cv}$ shape assumptions. The data-subsets of Mrk 501 in 1997 are defined in Fig. 1b, while the average set (AV) contains the data of the whole year. Fluxes are given in units of 10^{-11} cm^{-2}s^{-1}TeV^{-1}. If, according to λ, the $\mathcal{H}^{cv}$ hypothesis is favored, then the last two columns gives the curvature term β and the peak-emission energy $E_{\text{GeV}}^{\text{peak}} = 10^{\frac{2-\gamma}{2\beta}}$. If not, β is quoted between brackets.

Data set	T [a]	λ	ϕ_0	γ	β	$E_{\text{GeV}}^{\text{peak}}$
Mrk 501'97 LF	13.6	10.7	3.13 ± 0.19	2.32 ± 0.09	0.41 ± 0.17	410 ± 201
MF	40.5	47.1	4.72 ± 0.14	2.25 ± 0.05	0.52 ± 0.08	583 ± 104
HF	3.1	29.1	17.60 ± 0.61	2.07 ± 0.04	0.45 ± 0.09	840 ± 108
AV	57.2	61.5	5.19 ± 0.13	2.24 ± 0.04	0.50 ± 0.07	578 ± 98
Mrk 501 '98	3.3	0.09	1.25 ± 0.16	2.97 ± 0.20	(0.21 ± 0.73)	–
Mrk 421 '98	5.1	0.34	1.96 ± 0.20	2.96 ± 0.13	(0.28 ± 0.49)	–

[a] total time (h) ON-source.

tire high-energy part of the electromagnetic spectrum during flares. To date, the peak γ-energy of Mrk 421 has always remained well below the CAT threshold, precluding a more accurate spectral study. Therefore, testing VHE spectral variability as a general feature of blazars requires more multiwavelength observations with a large dynamic range in intensity. In any case, the comparison of the spectral properties of Mrk 501 and Mrk 421 in 1997 and 1998 fits the picture in which the VHE peak γ-energy shifts with increasing VHE flux: as shown in Table 1, spectral curvature seems to be characteristic of high flaring-activity states (i.e. Mrk 501 in 1997), whereas low-activity spectra (i.e. Mrk 501 and Mrk 421 in 1998) are always compatible with pure power-laws.

REFERENCES

1. Aharonian, F.A., et al, *A&A* **349**, 11 (1999).
2. Aharonian, F.A., et al, *A&A* **350**, 757 (1999).
3. Barrau, A., et al, *Nucl. Instr. Meth.* A **416**, 278 (1998).
4. Djannati-Ataï, A., et al, *A&A* **350**, 17 (1999).
5. Ghisellini, G., Maraschi, L., and Dondi, L., *A&AS* **120**, 503 (1996).
6. Krennrich, F., et al, *ApJ* **511**, 149 (1999).
7. Le Bohec, S., et al, *Nucl. Instr. Meth.* A **416**, 425 (1998).
8. Pian, E., et al, *ApJL* **492**, 17 (1998).
9. Piron, F., et al, these Proceedings.
10. Piron, F., et al, *Proc. XXVI ICRC* **3**, 326 (Salt-Lake City, 1999).
11. Samuelson, F., et al, *ApJL* **501**, 17 (1998).
12. Takahashi, T., et al, *APh* **11**, 177 (1999).
13. Zweerink, J.A., et al, *ApJL* **490**, 141 (1997).

Multiwavelength Observations of Mrk 501 in 1997

Markus Böttcher*, Dirk Petry†, and Valerie Connaughton×

*Space Physics and Astronomy Department
Rice University, MS 108, 6100 S. Main Street
Houston, TX 77005 - 1892, USA
†IFAE, Universitat Autonoma de Barcelona
08193 Bellaterra, Spain
×Marshall Space Flight Center, Huntsville, AL, USA

Abstract. We present the results of a long-term monitoring campaign of the TeV gamma-ray emitting blazar Mrk 501 during its active high-state in 1997. Observations at radio and optical frequencies, and in X-rays and TeV gamma-rays are included. We study the source's intermediate-term variability at all wavelengths and discuss implications for synchrotron and synchrotron-self-Compton dominated jet models. We find that the variability on weekly time scales is consistent with an SSC model where flux variations are primarily caused by a hardening of the electron spectra (change in spectral index) and an increasing density of ultrarelativistic electrons.

INTRODUCTION

During 1997, Mrk 501 was found to be in an extreme high state with a TeV flux on average 20 times higher than in 1996 [1]. The source exhibited strong variability on timescales of days and reached, in some of its flares, fluxes of more than 10^{-10} photons (> 1.5 TeV) cm^{-2} s^{-1} — the most intense TeV emission ever measured from any astronomical object. All available TeV observatories monitored the event for several months. The most complete dataset was produced by the HEGRA Cherenkov Telescope "1" (CT1) [2] which was even able to observe Mrk 501 under the presence of moonlight, thereby filling many gaps in the light curve. Largely concurrent with the CT1 measurements, the HEGRA system of (at the time 4) Cherenkov telescopes also provided very accurate spectral measurements in the TeV regime [3].

The origin of the TeV gamma-ray emission and the reasons for its variability are still under debate. The most popular models explain the TeV emission as near-infrared to UV photons which have been Compton-upscattered by very high energy electrons in a jet directed at a small angle with respect to our line of sight (for a review see, e. g., [4]). Possible sources of the seed photons for Compton scattering are the synchrotron radiation produced within the jet (SSC; [5]), or radiation from outside the jet (EIC), which could be the quasi-thermal radiation field of an accretion disk, either entering the jet directly from behind [6] or after being rescattered by circumnuclear material [7].

In this paper, we explore the multi-wavelength variability of Mrk 501 over intermediate timescales, and relate its behaviour to the physical parameters of a leptonic jet model. For this purpose we construct weekly SEDs using the HEGRA CT1 flux data and HEGRA CT System spectral data together with data from radio, optical, soft X-ray and hard X-rays. The instruments contributing to this campaign were the Metsähovi Radio Telescope, NOT, Tuorla Observatory, Osservatorio di Torino, Osservatorio di Perugia, RXTE ASM, BATSE, HEGRA CT1, and the HEGRA CT System. For a detailed description of this campaign and its results, see [8].

OBSERVATIONS AND DATA ANALYSIS

The observations used to construct the multiwavelength spectra of Mrk 501 cover mostly the synchrotron part of the SED. Only the TeV data explore what is believed to be the inverse-Compton emission, although OSSE and EGRET observations cover a few days in 1996 and 1997. As a guideline for our fitting procedure, we take into account the highest ever observed EGRET flux from Mrk 501, measured during 1996 Mar 25 – 28 [9], as an upper limit. All data sets are rebinned in time into 28 approximately weekly time bins from March through October 1997 (MJD 50517 – MJD 50707) defined on the basis of the HEGRA CT1 lightcurve. This weekly temporal resolution is an order of magnitude larger than the longest observed TeV intraday variability timescale of 15 h [3] and is believed appropriate, given the nature of the available data and the medium scale variability timescale which we have chosen to explore.

The optical observations were made by the groups of the Tuorla Oservatory, the Osservatorio di Torino and the Osservatorio di Perugia using their local telescopes, and the Nordic Optical Telescope on La Palma (see [8] and references therein). The observations were made in the V-band and partially also in the B-, R-, and I-bands and cover 22 of the 28 time bins. To bridge these gaps for our fits, we assume that the optical emission during these periods was within the total variability range observed over the whole duration of the campaign.

The radio observations were obtained by the Metsähovi Radio Observatory group at 22 GHz as part of the ongoing quasar monitoring program [10]. The observations cover 50 % of our time bins. Since the beginning of 1996, the ASM has been observing Mrk 501 in the 2 – 12 keV energy band. The data used in this analysis is taken from the publicly available 'quick look results'.

The hard X-ray fluxes were measured using BATSE with the Earth Occultation method [11]. Up to 32 independent flux measurements can be made per day. There exists, however, a wide variation in this number because of the passage of the spacecraft through the South Atlantic Anomaly, telemetry gaps from loss of

TDRSS contact, and other events, which occur randomly relative to the Mrk 501 steps. For the data shown here, between 61 and 211 measurements went into an individual weekly point. The fluxes between 20 and 200 keV are calculated by folding the measured counts through the BATSE detector response assuming a differential source power-law spectrum of photon index 2.0, which is the best fit to the flare measured on MJD 50550 – 51 between 20 and 1000 keV. Spectra for other time intervals were also calculated and were consistent with this power law.

For the TeV points, we use the data from the CT1 light curve of Mrk 501 (integral flux above 1.5 TeV) published in [2]. We group these points according to our weekly time bins and calculate an average flux from the up to seven values weighing each daily point by its observation time. In [12], the average spectral shape in the range between 0.5 and ≈ 25 TeV was determined with high accuracy as $dF/dE = N_0 E^{-\alpha} \exp(-E/E_0)$, where $N_0 = (10.8 \pm 0.2 \pm 2.1) \times 10^{-11}$ cm^{-2} s^{-1} TeV^{-1}, $\alpha = 1.92 \pm 0.03 \pm 0.20$, and $E_0 = (6.2 \pm 0.4 \pm 2.2)$ TeV (statistical and systematic errors, respectively). Furthermore, they find no spectral variability up to their sensitivity of $\delta\alpha \leq 0.1$ on all relevant time scales. In order to include this important spectral information in our model, we make the assumption that there is indeed no spectral variability and extrapolate the points measured at 1.5 TeV by CT1 up to 10 TeV and down to 0.8 TeV.

MODEL FITS TO THE SEDS

To each weekly SED we fit the combined SSC + EIC blazar jet model described in [13]. The model assumes that isolated components (blobs) of relativistic pair plasma are injected instantaneously into the jet, and follows the self-consistent evolution of the particle and radiation spectra as the blob moves outward along the jet, taking into account all radiation and cooling mechanisms mentioned in the introduction, as well as $\gamma\gamma$ absorption and pair production intrinsic to the source. Since we are interested in weekly averages, the emission from single blobs is time-averaged over the jet evolution. The emerging, time-averaged spectra are corrected for $\gamma\gamma$ absorption by the IIBR using the lower model spectrum given in [14].

It has been found previously that extreme HBLs are generally well described by a pure SSC model (e. g., [15,16]). However, in order not to introduce a bias towards such a choice of parameters, we first use the full simulation code to fit several weekly averaged broadband spectra. The extremely high peak frequencies of the synchrotron and γ-ray components and in particular the limit implied by the low EGRET flux force us to assume parameters in a way that the contribution from EIC is negligible. Subsequent fits to the same SEDs with a simplified version of our simulation code, neglecting external Comptonization, yield virtually the same parameters as obtained using the full simulations.

We construct a 3-dimensional mesh of simulations in parameter space, with the electron density n_e, the high-energy cutoff γ_2 and the spectral index p of the injected electron spectrum as free parameters on the grid points. Our simulated, time-

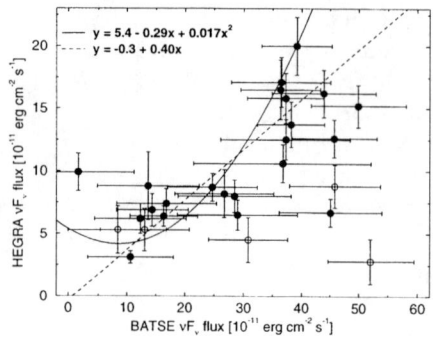

FIGURE 1. Correlation between the weekly averaged BATSE and HEGRA fluxes, fitted by a second-order (solid curve) and a first-order (dashed curve) polynomial. Open circles indicate points which were excluded from the fit because of poor time coverage (less than 3 days) by the TeV observations.

averaged spectra are only weakly dependent on the value of γ_2. All other parameters (in particular $B = 0.1$ G, $\gamma_1 = 300$, $D = 40$) are fixed to values known to be allowed for this object from independent arguments (e. g., [9]), and allowing good fits to the observed weekly averaged SEDs using our simulation code. Having constructed the three-dimensional mesh of simulations, we compare all weekly SEDs with the simulated spectra and find the simulation with the smallest χ^2.

RESULTS

The SSC model predicts a strong correlation between the emission in synchrotron and inverse-Compton radiation. We have derived an analytic estimate of the expected correlation, in particular for variations of n_e, p, and the average electron energy, assuming that the time-averaged (cooling) electron spectrum can be approximated by a broken power-law with index p below, and $p+1$ above the break energy. We use a delta-function approximation for the synchrotron spectrum, and calculate the expected power output in the SSC component in the Thomson regime up to the Klein-Nishina cut-off, beyond which we neglect Compton scattering events.

Trajectories in the (F_{sy}, F_{SSC}) plane of this approximation, varying a single model parameter, while all other parameters are fixed, generally have a non-linear shape. A variation in the electron density obviously yields a relation $F_{SSC} \propto F_{sy}^2$. A variation of p results in a relation which may be approximated by a second-order polynomial, where the linear term may be negative. If the break energy γ_b varies, an asymptotic behavior to less than quadratic order for high flux values is expected.

Fig. 1 shows the correlation between the HEGRA and BATSE points, fitted with a second-order polynomial, as expected if the spectral variability is dominated by variations of p. The fit has a χ^2_ν of 1.1. However, due to the large error bars, we cannot confidently distinguish between this and similar correlations.

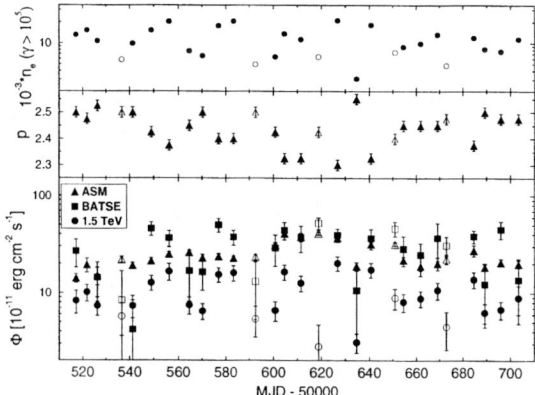

FIGURE 2. Temporal variation of the SSC fit parameters $n_e(\gamma > 10^5)$ and p compared to the weekly averaged light curves from the ASM, BATSE, and HEGRA. Open symbols indicate poor time coverage as in Fig. 1.

The results of our fitting procedure are represented in Fig. 2, where the temporal variation of the best-fit SSC parameters is compared to the variations of the soft and hard X-ray fluxes and the TeV γ-ray flux. There is a clear correlation between the hard X-ray and TeV fluxes with the injection spectral index and with the density of high-energy particles. This indicates that medium-timescale high activity states in X-rays and high-energy γ-rays are well explained by a hardening of the electron spectrum injected at the base of the jet. This is in good agreement with the observed correlation between the X-ray and TeV fluxes plotted in Fig. 1.

REFERENCES

1. Breslin, A. C., et al., *IAU Circular* **6592** (1997).
2. Aharonian, F., et al., *A&A*, in press (1999b).
3. Aharonian, F., et al., *A&A*, **342**, 69 (1999a).
4. Böttcher, M., these proceedings (1999).
5. Marscher, A. P. & Gear, W. K., *ApJ*, **298**, 114 (1985).
6. Dermer, C. D., Schlickeiser, R., & Mastichiadis, A., *A&A*, **256**, L27
7. Sikora, M., Begelman, M. C., & Rees, M. J., *ApJ*, **421**, 153 (1994).
8. Petry, D., et al., *ApJ*, submitted (1999).
9. Kataoka, J., et al., *ApJ*, **514**, 138 (1999).
10. Teräsranta, H., et al., *A&AS*, **132**, 305 (1998).
11. Harmon, A., et al., *AIP Conf. Proc.*, **280**, 314 (1992).
12. Aharonian, F., et al., *A&A*, in press (1999c).
13. Böttcher, M., Mause, H., & Schlickeiser, R., *A&A*, **324**, 395 (1997).
14. Malkan, M. A., & Stecker, F. W., *ApJ*, **469**, L33 (1998).
15. Mastichiadis, A., & Kirk, J. G., *A&A*, **320**, 19 (1997).
16. Pian, E., et al., *ApJ*, **492**, L17 (1998).

Comparison between HEGRA and RXTE ASM data from Mkn 501

J.C. González, D. Kranich

Max-Planck-Institute for Physics, Foehringer Ring 6, D-80805, Munich, Germany

Abstract. In this paper we make a comparison of the data collected from the blazar Mkn 501 with the Cherenkov telescope CT1 of the HEGRA Collaboration and the All Sky Monitor (ASM), on board of the RXTE (X-ray Timing Explorer) satellite. We analyze data from the observation seasons 1996 till 1999. In these periods the source Mkn 501 has shown quite different behaviors: quiescent low-state and a high-state showing quasi-periodic oscillations and episodic bursts. The study of correlation between the gamma-ray and the X-ray regime is of crucial importance for the development of unified scenarios.

INTRODUCTION

The Cherenkov telescope CT1 of the HEGRA Collaboration has been used to observe Mkn 501. It has a camera with 127 pixels in a 3.25° field of view (pixel diameter 0.25°). The mirror area has been enlarged from 5 m^2 to 10 m^2, and presently it has an energy threshold of about 700 GeV (before its refurbishing in 1998 the threshold was 1.2 TeV).

The XTE All Sky Monitor provides a nearly continuous monitoring of the X-ray sky, and also acts as an alarm for transient phenomena. The ASM consists of three Scanning Shadow Cameras (SSCs) mounted on a rotating drive assembly. Each camera has a field of view (full width at half maximum, FWHM) of 6°×90°. Two of the cameras share the same look direction but are canted by ±12° from each other, while the third camera looks in a direction parallel to the ASM drive axis. Each camera reports the total source intensity in the 1–12 keV band and the intensity in each of 3 energy, namely 1.3–3.0 keV, 3.0–5.0 keV and 5.0–12.1 keV.

ANALYSIS

The data collected and analyzed corresponds to the observation periods March–August 1996, March–September 1997, February–October 1998, and February–July of this year, 1999 (the observation still goes on at the moment of the presentation) (see [1], [2]). Although most IACTs take data in moon-less nights, we also observed Mkn 501 in the presence of moonlight (see [3]). After the reduction and analysis of the raw data we end up with a total amount of data of around 950 hours (about

10% is Moon data — see [3]). We applied for the gamma-hadron separation process the usual cuts (see for example [4]). Corresponding to the observation periods of CT1, we have used the data from ASM. They are obtained in the form of light curves. These data from the light curves are quoted as nominal 2–10 keV rates in ASM counts per second (ASM c/s), where Crab nebula flux is about 75 ASM c/s.

In Fig.1 the light curve obtained for CT1 in the analyzed periods, and the total light curve for ASM are shown.

Study of correlations

The correlation of two variables x and y is defined as $r = \text{cov}(x, y)/\sqrt{s_x^2 s_y^2}$, where $\text{cov}(x,y)$, s_x^2 and s_y^2 are the covariance between x and y, and the variances of x and y, respectively, and r is an estimator for the true population correlation coefficient ρ. Under the assumption of the null hypothesis $H_0 : \rho = 0$ (being the alternative hypothesis $H_1 : \rho \neq 0$), we define the significance of a given (calculated, observed) correlation r_{obs} as $t = (r_{\text{obs}} - \rho)/s_r = r_{\text{obs}}\sqrt{N-2}/\sqrt{1 - r_{\text{obs}}^2}$ (N being the number of data points).

Following this scheme, we have calculated the correlations for the observation periods in study. The results are shown in Table 1. The first thing we see is that in the low state (years 1996, 1997 and 1999) the correlations obtained are compatible with 0 within the errors. However, in 1997 when episodic bursts and significant quasi-periodic oscillations have been found (see [5]) the correlation is 0.62 with a significance of 9.0σ. In order to test whether this relatively high correlation could be due to statistical effects, we have studied the change of the correlation coefficient when applying time lags of an integer number of days. The result, plotted in the Fig. 2, shows that there is no evidence of such high correlations for time lags different than zero (no time lag). The same procedure applied to the data from 1996, 1998 and 1999 didn't show any significant correlation, as expected.

In view of these results, we can study the probability that such a difference in observed correlations would happen merely as a function of the sampling error, assuming that there is not intrinsic difference in the source of data from year to year. That is, we are interested in a possible difference of behavior of Mkn 501 from year to year. In order to do this we make the statistical hypothesis that two correlations for different periods are actually the same, and the observed difference comes from statistical errors in the observations. Our null hypothesis will be $H_0 : \rho_1 = \rho_2$

TABLE 1. Calculated correlations between CT1 and ASM data for Mkn 501.

Year	r_{obs}	Chance probability	Significance
1996	0.21 ± 0.29	$1.3 \cdot 10^{-1}$	1.5
1997	0.62 ± 0.08	$2.2 \cdot 10^{-19}$	9.0
1998	0.34 ± 0.38	$4.6 \cdot 10^{-2}$	2.0
1999	0.16 ± 0.17	$2.3 \cdot 10^{-1}$	1.2

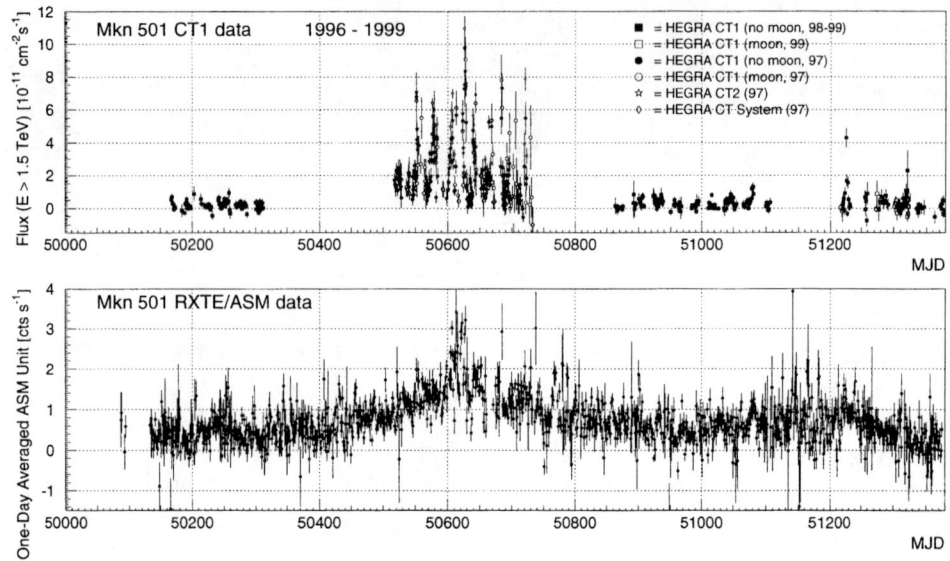

FIGURE 1. Light curve for Mkn 501 obtained with CT1 data (observation period 1996–1999) and with the XTE ASM. The data plotted for the ASM correspond to one-day average values.

(being the alternative hypothesis $H_1 : \rho_1 \neq \rho_2$). Since the difference of two correlation coefficients doesn't behave as a normal variable, we have to transform these variables to new ones, $z_r = \log[(1+r)/(1-r)]/2$, which is approximately normal. Then our difference in correlations relative to the standard error of such difference

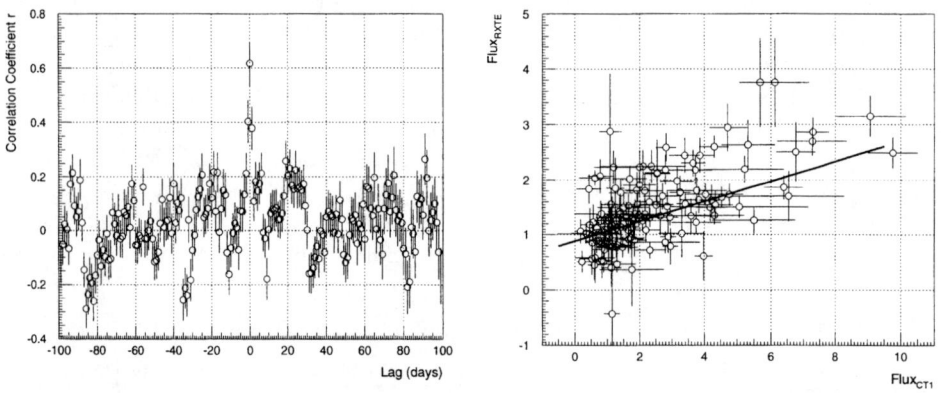

FIGURE 2. (a) Change of the correlation coefficient when applying 1-day time lags for 1997 data; (b) Regression applied to the data $\langle Flux \rangle_{\text{CT}}$ vs. $\langle Flux \rangle_{\text{RXTE}}$. The model used in a linear fit in the form $\langle Flux \rangle_{\text{RXTE}} = a + b \cdot \langle Flux \rangle_{\text{CT}}$.

TABLE 2. Results of comparison of correlation coefficients for different years for Mkn 501.

z_{obs}	1996	1997	1998	1999
1996	-	3.01	0.62	0.27
1997	3.01	-	1.76	3.43
1998	0.62	1.76	-	0.86
1999	0.27	3.43	0.86	-

is given by:

$$z_{obs} = \frac{z_{r_{y_1}} - z_{r_{y_2}}}{\sqrt{\frac{1}{N_{y_1}-3} + \frac{1}{N_{y_2}-3}}} \quad (1)$$

being N_{y_1} and N_{y_2} the number of data points for the years y_1 and y_2. The results from comparing two correlation coefficients from different years are shown in Table 2.

For a non-directional test at the 99% confidence level (p=0.01) the critical value for rejecting the null hypothesis H$_0$ is $|z_{obs}| < 2.58$. This is the case of the year 1997 compared with 1996 and 1999 (for the pair (1997,1998) the null hypothesis would be rejected at the 92% C.L.). Therefore, our analysis concludes that the differences in the correlation coefficients between CT1 and ASM data for Mkn 501 for 1997 compared with 1996 and 1999 (and 1998 at the 92% C.L.) come from an intrinsic, different behavior of the data in 1997.

Comparison of fluxes

If we compare the observed fluxes for Mkn 501 measured with CT1 in the GeV–TeV domain and the ASM in the keV domain, we get the results of the Table 3. This table shows the ratios $R_{TeV/keV}^{YEAR} = \langle Flux \rangle_{CT}/\langle Flux \rangle_{RXTE}$ (in units of $10^{11} cm^{-2} s^{-1}/Hz$) for the different periods. These ratios are quite different for the high state period 1997 and the low state periods 1996, 1998 and 1999.

While the keV flux rises by about a factor 3.5 from the 1996 to 1997 periods (and decreases afterwards by the same factor), the TeV flux increases by a factor 11, i.e., about in quadrature of the keV flux. The results from a linear fit to the data from CT1 and ASM are shown in Table 3. There's an obvious different correlation between different pairs of data. In Fig. 2 the correlation of the measured fluxes for Mkn 501 with CT1 and ASM is shown.

CONCLUSIONS

We have studied the correlation in variability of Mkn 501 from the telescope CT1 of the HEGRA Collaboration and the All Sky Monitor on board the RXTE. We obtain a clear correlation (0.62 ± 0.08, 9.0σ at 99% C.L.) between these data for the year 1997, and no correlation for 1996, 1998 and 1999 data. We conclude from

TABLE 3. Coefficients of the regression applied to the data $\langle Flux \rangle_{\text{CT}}$ vs. $\langle Flux \rangle_{\text{RXTE}}$. The model used in a linear fit in the form $\langle Flux \rangle_{\text{RXTE}} = a + b \cdot \langle Flux \rangle_{\text{CT}}$.

Year	a	b
1996	0.33 ± 0.03	0.23 ± 0.13
1997	0.88 ± 0.03	0.18 ± 0.01
1998	0.44 ± 0.06	0.33 ± 0.15
1999	0.39 ± 0.04	0.19 ± 0.05

our analysis of the difference in the correlation coefficients that these differences for the pairs of periods (1997,1996) and (1997,1999) (and (1997,1998) at the 92% C.L.) resolve in an intrinsic, distinct behavior of the Mkn 501 data for 1997. The increase in flux for the year 1997 in the TeV domain seems to go in quadrature with the increase in the keV domain.

Besides the inevitable reduction of the correlation due to uncorrelated experimental effects, it should be noted that there are two additional reasons for getting a correlation of only 0.62 in 1997, namely the light curve in keV might be caused by a two electron components, or the X-ray component becomes significantly harder during intense flares. This would result in a smaller value of r_{obs} due to the limited X-ray range of XTE. Also, one might observe a TeV γ-ray component caused by hadrons (hadrons induced γ's are not expected to show a strong keV peak.

Another confirmation of the correlation in 1997 comes from the QPO analysis of both keV and TeV data [5], which shows a significant period of 23 days in both light curves and a similar though not identical phase diagram.

ACKNOWLEDGMENTS

We thank the Instituto de Astrofísica de Canarias for the use of the site and the excellent working conditions. We also acknowledge the rapid availability of the RXTE data. This work is supported by the German Ministry of Education and Research, BMBF, the Deutsche Forschungsgemeinschaft, DFG, and the Spanish Research Foundation, CICYT.

REFERENCES

1. Aharonian F.A., et al. *A&A* **342**, 69 (1999).
2. Aharonian F.A., et al. *A&A* **349**, 29 (1999).
3. Kranich D., et al. *Ap.Phys.* **12**, 65 (1999).
4. Kranich D., *Diploma Thesis*, MPI-PhE/97-11.
5. Kranich D., et al. *Proc. XXVI ICRC* OG 2.1.18, **3**, pp. 358–361 (1999).

Periodicity analysis of Markarian 501 flaring activity in 1997

S.J. Fegan[1,2], I.H. Bond[3], S.M. Bradbury[3], A.C. Breslin[4],
J.H. Buckley[5], A.M. Burdett[1,3], M. Carson[4] D.A. Carter-Lewis[6],
M. Catanese[6], M.F. Cawley[7], S. Dunlea[7], M. D'Vali[3], D.J. Fegan[4],
J.P. Finley[8], J.A. Gaidos[8], T.A. Hall[8], A.M. Hillas[3], D. Horan[4],
J. Knapp[3], F. Krennrich[6], S. Le Bohec[6], R.W. Lessard[8],
C. Masterson[4], B. McKernan[4], J. Quinn[4], H.J. Rose[3],
F.W. Samuelson[6], G.H. Sembroski[8], V.V. Vassiliev[1], T.C. Weekes[1]

[1] *Fred Lawrence Whipple Observatory, Harvard-Smithsonian CfA, Amado, AZ 85645, USA*
[2] *Physics Department, University of Arizona, Tucson, AZ 85721, USA*
[3] *Department of Physics, Leeds University, Leeds, LS2 9JT, UK*
[4] *Experimental Physics Department, University College, Belfield, Dublin 4, Ireland*
[5] *Department of Physics, Washington University, St. Louis, MO 63130, USA*
[6] *Department of Physics and Astronomy, Iowa State University, Ames, IA 50011, USA*
[7] *Physics Department, St. Patrick's College, Maynooth, County Kildare, Ireland*
[8] *Department of Physics, Purdue University, West Lafayette, IN 47907, USA*

Abstract. We report on a periodicity analysis of γ-ray observations of very high energy (VHE, $E > 250$GeV) γ-rays from Markarian 501 (Mrk 501) taken with the Whipple telescope between March and July of 1997. The data correspond to a period of bright flaring of the source, quasi-periodic signals were detected by the Telescope Array Project (TAP) and HEGRA collaboration. Periodic analysis of these data is complicated because the source observations are not continuous and the reported period (≈ 24 days) is very close to the sampling period. We use the Lomb method, a technique for extracting spectral information from unevenly sampled data, in an attempt to overcome this difficulty. Similar results are shown to exist in data from the strongly correlated X-ray band sampled by the RXTE ASM instrument. No periodicity in the data could be identified with a source frequency.

INTRODUCTION

Mrk 501 is a BL Lacertae object at redshift of $z = 0.033$, discovered by the Whipple collaboration in 1995 to be a γ-ray source at $E > 300$GeV [8] and subsequently confirmed by the HEGRA collaboration [1]. In 1997, Mrk 501 was in an active state, characterised by an unprecedentedly high VHE flux, detected by several independent Čerenkov groups [7] .

Using a Rayleigh test on observations taken between March to July 1997, the TAP group reported evidence of a 12.7 day quasi-periodic signal from Mrk 501 [3]. However, it was not ruled out that the signal could be an artifact of the irregular sampling of the light curve which results from Čerenkov telescopes only being able to observer during dark periods of the moon. These results were confirmed more recently by the HEGRA collaboration, [4] and TAP group, who detected a 23.9 day periodicity in the TeV and X-ray light curves. Periodic emission from an AGN on such a short time scale would be very difficult to explain in the framework of relativistic jet models and, if confirmed, would impact significantly our understanding of AGN physics.

We present a periodic analysis of data from observations of Mrk501 in 1997 with the Whipple telescope and the All Sky Monitor (ASM) on board RXTE.

The VHE observations reported here were made with the atmospheric Čerenkov imaging technique using the Whipple Observatory 10m telescope [2]. Data taken with the instrument are analysed off line with γ-ray selection based on the Supercuts method. All data presented in this paper were analysed with a tracking analysis to give an average γ-ray rate for each 28 minute observation. The VHE dataset examined consists of 149 observations during the period February to June 1997 (50489 < MJD < 50610). The Whipple 10m telescope is limited to operation during the moon-less periods of the night and no observations are taken during the 10 night period around each full moon.

The X-ray data set consists of 4052 observations of 90 second duration. The RXTE has a 96 minute orbit and sees 80% of the sky per orbit. The satellite is moved frequently during the day to facilitate observations with the other instruments resulting in periods when Mrk501 is not visible to ASM.

ANALYSIS

Periodic analysis is performed by fitting single sine waves at different frequencies to the data using least squares minimisation. This technique is a variant of the Lomb method [5]. The reduction in the sum of the residuals is then plotted against frequency, giving the *Lomb normalised periodogram*, an indicator of spectral power as a function of frequency. Given a set of flux measurements $\hat{y}_i$ with known uncertainties σ_i taken at times t_i, a simple sine wave of frequency f with Gaussian noise is tested as a model of the data, $\hat{y}_i = a\cos 2\pi f(t_i - \tau) + b\sin 2\pi f(t_i - \tau) + \sigma_i \hat{\xi}_i$, where $\hat{\xi}_i$ is a random Gaussian distributed variable with zero mean and unit variance. τ is conveniently introduced by Lomb to make the periodogram independent of shifting the time points (t_i's) by a constant and to represent the residual as a sum of two positively defined terms. A best fit for a and b to the data is obtained by minimising the standard χ^2 function. The Lomb normalised periodogram at frequency ω is then defined as,

$$P_N(\omega) = \frac{1}{2} \left\{ \frac{\left[\sum_{i=1}^{N} \hat{y}_i \cos \omega (t_i - \tau)/\sigma_i^2\right]^2}{\sum_{i=1}^{N} \cos^2 \omega (t_i - \tau)/\sigma_i^2} + \frac{\left[\sum_{i=1}^{N} \hat{y}_i \sin \omega (t_i - \tau)/\sigma_i^2\right]^2}{\sum_{i=1}^{N} \sin^2 \omega (t_i - \tau)/\sigma_i^2} \right\}. \quad (1)$$

This definition is slightly different from that given in [6] in that it does not assume equal variances, σ_i^2, on the flux points, y_i, nor does it require that the noise component of the measurements dominate the signal ($\sigma_i \gg \sqrt{a^2 + b^2}$).

When the signal is absent in the data ($a = b = 0$), $P_N(\omega)$ is exponentially distributed with unit mean. The probability that $P_N(\omega) > z$ for some positive z is given by $P\{> z\} = e^{-z}$. The *false alarm probability*, $P_{FA}(z)$, or probability that at least one of the M independent frequencies tested in the periodogram will exceed a given z, $P_{FA}(z, M) = 1 - (1 - e^{-z})^M$. M can be evaluated by Monte Carlo simulation, for a given distribution of t_i's [6].

The extraction of a spectrum from unevenly sampled data is a more complex task than that from uniform data. For example, the orthogonality condition, customary for Fourier analysis, is no longer valid and it is possible for a true peak in the periodogram to give rise to any number of other false peaks at different periods, an effect called aliasing. It is therefore not possible to draw any immediate conclusion from a periodogram which contains more than one clear peak at any frequencies. To detect more than one period in the signal an approximate process of subtraction of previously found periods is usually employed, with the periodogram being recalculated at each step. It is common to find that subtracting one frequency from the data will remove all significant peaks from the periodogram. If the spectrum is complex and the subtraction process is repeated for many frequencies it becomes increasingly difficult to prove the existence of a particular mode in the spectrum.

However, the ASM observations during the period are sufficiently dense to allow the data to be binned equidistantly (2.5 days) and a standard Fourier Transform calculated under the assumption that no frequencies higher than the Nyquist frequency are present in the signal. This is done to show the validity of the Lomb method when applied to the ASM data and to calculate the probability that any periodicity found is not produced by chance due to inaccurate data measurements.

Finally, the windowing function, $W(\omega)$, is calculated for each of the data sets,

$$W(\omega) = \left| \mathcal{F}\left[\sum_{i=1}^{N} \delta(t - t_i)\right](w) \right|^2 = \sum_{i=1}^{N} \sin^2 \omega t_i + \cos^2 \omega t_i, \quad (2)$$

where and $\mathcal{F}$ is the Fourier transform and t_i are the times of the observations. $W(\omega)$ characterizes periodicities that arise from the sampling of the data. It is possible for these periodicities to show up in the results of the Lomb analysis of the dataset, a result of *aliasing* of power from other frequencies as discussed above.

RESULTS

The windowing function for both data sets are shown in figures 1(a) and 2(a), respectively. In both cases, there are significant periodicities present due to the irregular sampling of the lightcurve. In the RXTE dataset, the orbital period of 96 minutes is prominent as are some signals at longer periods. In the Whipple data, large peaks are seen at 29 minutes, 1 day and 28.2 days. In each of these datasets, any candidate signals detected at these periods with the Lomb method must be viewed as potentially arising from the aliasing of power from other frequencies.

Figure 1(b) shows the results of the Lomb analysis of the Whipple dataset. The most significant peaks present are at 25.4, 13.4 and 1 day. The windowing function suggests that the 1 day peak is due to aliasing. To determine the frequencies which might be actually present in the data from artifacts due to alising a signal subtraction proceedure is employed. Subtraction of the 25.4 day period results in the periodogram in figure 3; the 1 day and 13.4 day signals are suppressed, suggesting that they are due to aliasing of power from the 25.4 day signal. Similarly, starting with figure 1(b) and subtracting a 13.4 day signal results in a similar periodogram with suppression of the 25.4 and 1 day periods. We cannot, therefore, unambiguously identify the prominent periodicity found in the Whipple data with any particular source frequency. Whipple observations of Mrk 501 from 1996 do not show any significant periodicities.

The Lomb analysis of the RXTE ASM data is shown in figure 2. The periodogram is characterised mainly by a peaks at 200, 85 and 24 days and one at 96 minutes which is certainly due to aliasing. Removing the three lower frequencies results in the periodogram shown in figure 3(b). The Lomb power at $0 < f < 0.2$ day^{-1} is suppressed. The peaks left in this region are likely remnants due to aliasing and do not reflect the true spectrum of the source. $0.2 < f < 1$ day^{-1} has no prominent peaks (aliased peak corresponding to 96 minutes is suppressed). The Lomb periodogram, strongly affected by aliasing in this region, does not reflect the true source spectrum. Fourier analysis of the RXTE data shows a complex spectrum with several prominent ($> 4\sigma$) modes. Recovery of such a spectrum from an irregularly sampled data set is an ill-defined mathematical problem.

It seems that unless a similar pattern of flaring is observed from Mrk 501, a definitive conclusion about the presence of a 23.5 day periodicity in the time spectra of this source can not be firmly established.

REFERENCES

1. Bradbury, S. M., et al., *A&A* **320** L5 (1997)
2. Cawley, M.F., et al. *Exp. Astron.*, **1**, 173 (1990)
3. Hayashida, N., et al., *ApJ* **504**, L71 (1998)
4. Kranich, D., et al., *Proc. 26th Int. Cosmic Ray Conf., Utah* **3**, 358-361 (1999)
5. Lomb, N.R., *Ap. Space Sci.* **39**, 447 (1976)
6. Press, W.H., et al., *Numerical Recipies*, Cambridge: Cambridge University Press, 1993
7. Protheroe, R.J., et al., *Proc. 25th Int. Cosmic Ray Conf., Durban* **8**, 317 (1997)
8. Quinn, J., et al., *ApJ* **456** L83 (1996)

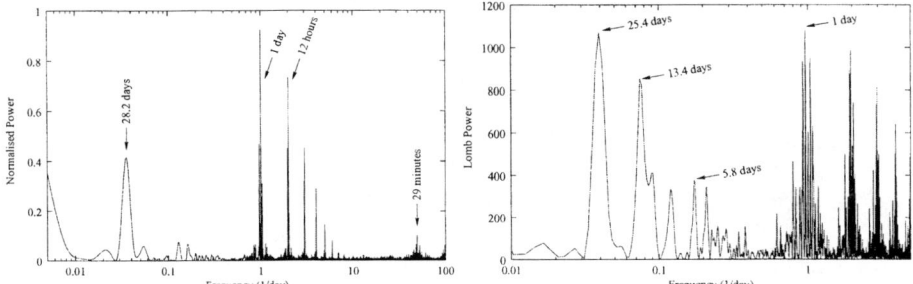

FIGURE 1. (a) Windowing function for Whipple data set. This VHE dataset is characterised by three peridicities, the lunar cycle (28 days), the diurnal cycle (1 day) and the data run length (29 minutes), all of which are visible in the windowing function. (b) Lomb analysis for the Whipple data set. The periodogram indicates possible periods of 25.4 days, 13.4 days and 1 day.

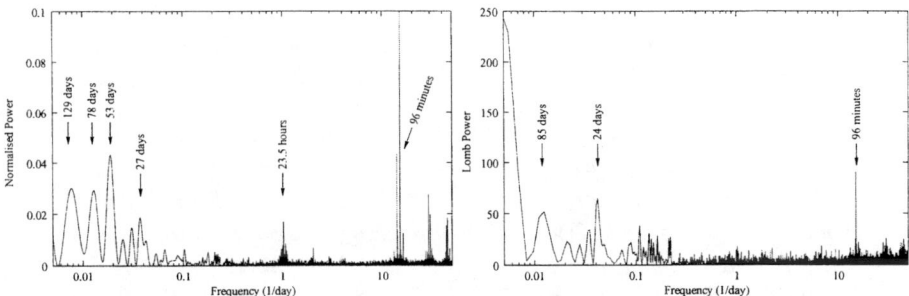

FIGURE 2. (a) Windowing function for RXTE ASM data set. A prominent periodicity, at the orbital time of 96 minutes, is evident in the data. (b) Lomb analysis for the RXTE ASM data set. Periodicites are present at 200, 85 and 24 days. Aliasing is visible at 96 minutes.

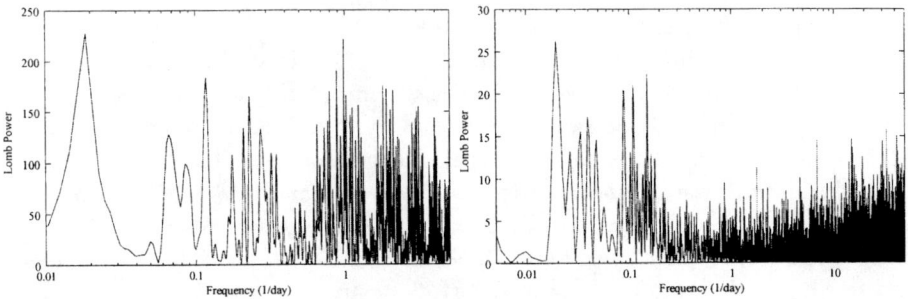

FIGURE 3. (a) Periodogram for Whipple dataset after subtraction of 25.4 day signal. (b) Periodogram for RXTE dataset after subtracting 200, 85 and 24 day signals.

Periodicity of TeV gamma-ray emissions from Mrk501

S.Hayashi[2], D.Nishikawa[1], N.Chamoto[2], M.Chikawa[3], Y.Hayashi[4], N.Hayashida[1], K.Hibino[5], H.Hirasawa[1], K.Honda[6], N.Hotta[7], N.Inoue[8], F.Ishikawa[1], N.Ito[8], S.Kabe[9], F.Kajino[2], T.Kashiwagi[5], S.kakizawa[19], S.Kawakami[4], Y.Kawasaki[4], N.Kawasumi[6], H.Kitamura[16], K.Kuramochi[11], E.Kusano[9], H.Lafoux[1], E.C.Loh[12], K.Mase[1], T.Matsuyama[4], K.Mizutani[8], Y.Morizane[3], M.Nagano[18], J.Nishimura[13], T.Nishiyama[2], M.Nishizawa[14], T.Ouchi[1], H.Ohoka[1], M.Ohnishi[1], S.Osone[1], To.Saito[15], N.Sakaki[1], M.Sakata[2], M.Sasano[1], H.Shimodaira[1], A.Shiomi[8], P.Sokolsky[12], T.Takahashi[4], S.F.Taylor[12], M.Takeda[1], M.Teshima[1], R.Torii[1], M.Tsukiji[2], Y.Uchihori[16], T.Yamamoto[1], Y.Yamamoto[2], K.Yasui[3], S.Yoshida[1], H.Yoshii[17], and T.Yuda[1]

[1] *Institute for Cosmic Ray Research, University of Tokyo, Tokyo 188-8502, Japan*
[2] *Department of Physics, Konan University, Kobe 658-8501, Japan*
[3] *Department of Physics, Kinki University, Osaka 577-8502, Japan*
[4] *Department of Physics, Osaka City University, Osaka 558-8585, Japan*
[5] *Faculty of Engineering, Kanagawa University, Yokohama 221-8686, Japan*
[6] *Faculty of Education, Yamanashi University, Kofu 400-8510, Japan*
[7] *Faculty of Education, Utsunomiya University, Utsunomiya 320-8538, Japan*
[8] *Department of Physics, Saitama University, Urawa 338-8570, Japan*
[9] *High Energy Accelerator Research Organization (KEK), Tsukuba 305-0801, Japan*
[10] *Department of Physics, Kobe University, Kobe 657-8501, Japan*
[11] *Faculty of Science and Technology, Meisei University, Tokyo 191-8506, Japan*
[12] *Department of Physics, University of Utah, Utah 84112, USA.*
[13] *Yamagata Academy of Technology, Yamagata 993-0021, Japan*
[14] *National Center for Science Information System, Tokyo 112-8640, Japan*
[15] *Tokyo Metropolitan College of Aeronautical Engineering, Tokyo 116-0003, Japan*
[16] *National Institute of Radiological Sciences, Chiba 263-8555, Japan*
[17] *Department of Physics, Ehime University, Matsuyama 790-8577, Japan*
[18] *Department of Applied Physics and Chemistry, Fukui University of Technology, Fukui 910-8505, Japan*
[19] *Department of Physics, Shinshu University, Matsumoto 390-8621, Japan*

Abstract. The BL Lac object Markarian 501 showed very high emission in TeV gamma rays from March to October in 1997. During this period the source was observed by the Whipple, HEGRA, CAT and the Utah Seven Telescope Array groups. Periodicity of TeV gamma-ray emissions for these data are analyzed using the Lomb method. Obtained periodicities for independent data are consistent with having a periodicity of about 24 days. Such a periodicity may give a key of an emission mechanism of VHE gamma rays and internal structure of super massive AGNs.

INTRODUCTION

Markarian 501 is a classical Bl Lacertae object and a sub-classification of the Blazar class of AGN with the redshift z=0.034. This source was first discovered as a TeV gamma-ray source by the Whipple group in 1996.[1] At the discovery stage the TeV emission level was only 8% of the Crab Nebula. The emission increased and showed intensity variability in 1996 [2], and this source was confirmed as a TeV gamma-ray source by the HEGRA group.[3]

In 1997 remarkable flaring activity of the gamma rays from Mrk 501 was observed by at least 7 groups, 7TA(the Utah Seven Telescope Array), Whipple, HEGRA, CAT, TACTIC, Mark6 and Tibet ASγ.[4][5] The TeV gamma-ray flux was highly variable and increased up to 10 times of that of the Crab Nebula.

The 7TA group suggested the possibility of the existence of the periodicity of 12-14 days or 24-26 days in the light curve of the TeV emission.[6]

In this paper, we report the periodicity analysis for the TeV gamma-ray flux data observed by four groups in 1997 by using the Lomb method.

OBSERVATION DATA

We have used the observation data of Cherenkov telescopes of four groups; 7TA, Whipple, HEGRA, and CAT.[4] Used light curves of TeV gamma rays from Mrk501 observed by these groups in 1997 are shown in Fig 1. For 7TA, we have selected data obtained under good weather condition. Eleven data points out of 47 are omitted in this process. For HEGRA, we used CT2 data only. Original datasets were used for Whipple and CAT.

ANALYSIS METHOD AND RESULTS

In order to test the periodicity in the TeV gamma-ray light curves, we used the Lomb method.[7][8] This method can be applied to unevenly sampled data such as observations using the air Cherenkov detector.

The Lomb normalized periodogram (spectral power as a function of angular frequency $\omega \equiv 2\pi f > 0$) is defined by

$$P_N(\omega) \equiv \frac{1}{2\sigma^2}\{\frac{[\sum(h_j - \bar{h})\cos\omega(t_j - \tau)]^2}{\sum \cos^2\omega(t_j - \tau)} + \frac{[\sum(h_j - \bar{h})\sin\omega(t_j - \tau)]^2}{\sum \sin^2\omega(t_j - \tau)}\}$$

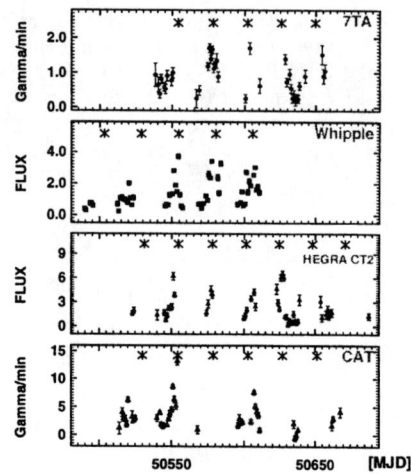

Figure 1: Light curves of TeV gamma rays from Mrk501 observed by four groups in 1997.

where, τ is defined by
$$\tan 2\omega\tau = \frac{\sum \sin 2\omega t_j}{\sum \cos 2\omega\, t_j}$$

The constant τ is a kind of time offset that makes $P_N(\omega)$ independent of shifting all the $t'_j s$ by any constant. h_j corresponds to individual gamma-ray flux observed at the time t_j, $\bar{h}$ is the mean value of the h_j and σ^2 is the variance.
The probability to be $z < P_N(\omega) < z + dz$ is expressed as
$$\mathcal{P}(z < P_N(\omega) < z + dz) = \exp(-z)dz.$$
False-alarm probability of the null hypothesis with M *independent* frequency is , therefore, expressed as
$$\mathcal{P}(P_N(\omega) > z) = 1 - (1 - e^{-z})^M.$$

The spectral power and the false-alarm probability as a function of period for the four experiments are obtained and are shown in Fig. 2.

Error bars of the data points are taking into account for the analysis by the Monte Carlo method. The effect of the error bars are found to be small.

The maximum power values, the minimum probability values and corresponding periods in Fig. 2 are listed in Table 1 for respective experimental groups.

Maximum power values, minimum probability values and corresponding periods in Fig. 2 are listed in Table 1 for respective experimental groups.

Phase diagrams by using obtained period in Table 1 for the four experimental

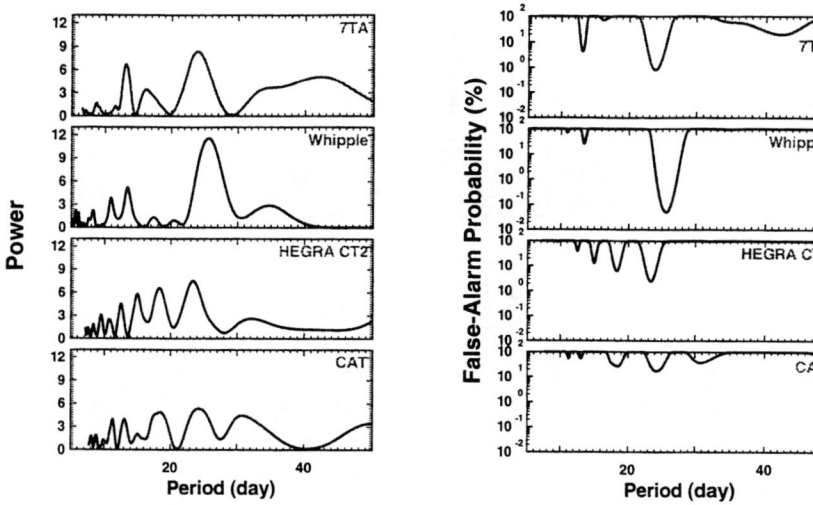

Figure 2: Left : Power as a function of period for four experiments. Right : False-alarm probability as a function of period for four experiments.

groups are shown in Fig. 3. In these figures flux data points are superimposed using the obtained period.

CONCLUSIONS AND DISCUSSION

Mrk 501 has been a very active phase in 1997. Gamma-ray flares at TeV energies has been observed by several groups. We have tried to examine the existence of the periodicity in the light curves of TeV emissions for four experimental groups observed by the Cherenkov detectors by using the Lomb method. Analyzed results show obtained values of period for respective experiments are very similar, which are around 24 days. Obtained chance probabilities are very small. Phases of the sinusoidal functions fitted to respective experimental data are plotted with * in Fig. 1. It is seen that the phases among the light curves also agree well.

We have also pointed out a good correlation of the periodicity between the TeV

Table 1: Obtained power, probability and corresponding period.

Experimental Group	Period(day)	Power	Probability(%)
Utah Seven Telescope Array (7TA)	23.9	8.4	0.82
Whipple	25.6	11.6	0.05
HEGRA CT2	23.4	7.5	2.4
CAT	24.3	5.4	17

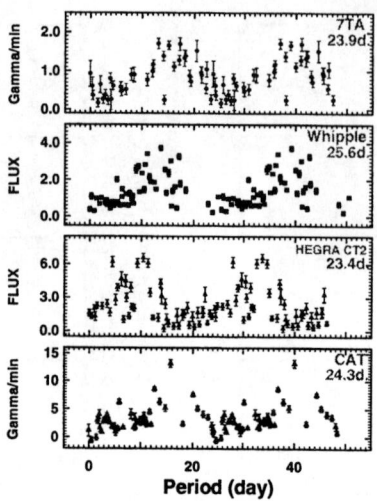

Figure 3: Phase diagrams of TeV gamma rays for four groups for respective periods.

and X-ray light curves elsewhere.[9]

Existence of such periodicity may suggest the interaction of the shock wave and the helical structure of the jet[10], or internal structure of the core of the AGN.

Acknowledgments : The authors would like to thank the Hires group and people at Dugway for the help of observations. This work is supported in part by the Grants-in-Aid for Scientific Research (Grants #0724102 and #08041096) from the Ministry of Education, Science and Culture and is also partly supported by the Promotion and Mutual Aid Corporation for Private Schools of Japan.

REFERENCES

1. Quinn, J. et et al., 1996, ApJ **456**, L83.
2. Quinn, J. et et al., 1997, 25th ICRC, **3**, 249.
3. Bradbury, S.M., et et al. 1997, A&A **320**, L5.
4. Protheroe, R.J., et al., 25th ICRC, **8**, 317 (1997).
5. Amenomori, M., et al., 26th ICRC, **3**, 382 (1999).
6. Hayashida, N., et al., 1998, ApJ **504**, L71.
7. Lomb, N.R., 1976, Astrophysics and Space Science, **39**, 447.
8. Press, W.H. et al., 1992, Numerical Recipes in C, Second edition.
9. Nishikawa, D., et al., 26th ICRC, **3**, 354 (1999).
10. Bednarek, W. and Protheroe, R.J 1997, MNRAS, **292**, 646.

Detection of TeV Gamma Rays from Mrk 501 at High Flaring State of Activity in 1997 with the Tibet Air Shower Array

The Tibet ASγ Collaboration

M.Amenomori[1], S.Ayabe[2], P.Y.Cao[3],
Danzengluobu[4], L.K.Ding[5], Z.Y.Fen[6], Y.Fu[3], H.W.Guo[4],
M.He[3], K.Hibino[7], N.Hotta[8], Q.Huang[6], A.X.Huo[5], K.Izu[9],
H.Y.Jia[6], F.Kajino[10], K.Kasahara[11], Y.Katayose[9], Labaciren[4], J.Y.Li[3],
H.Lu[5], S.L.Lu[5], G.X.Luo[5], X.R.Meng[4], K.Mizutani[2], J.Mu[12], H.Nanjo[1],
M.Nishizawa[13], M.Ohnishi[9], I.Ohta[8], T.Ouchi[7], J.R.Ren[5], T.Saito[14], M.Sakata[10],
T.Sasaki[10], Z.Z.Shi[5], M.Shibata[15], A.Shiomi[9], T.Shirai[7], H.Sugimoto[16],
K.Taira[16], Y.H.Tan[5], N.Tateyama[7], S.Torii[7], T.Utsugi[2], C.R.Wang[3],
H.Wang[5], X.W.Xu[5], Y.Yamamoto[10], G.C.Yu[6], A.F.Yuan[4], T.Yuda[9],
C.S.Zhang[5], H.M.Zhang[5], J.L.Zhang[5], N.J.Zhang[3], X.Y.Zhang[3],
Zhaxisanzhu[4], Zhaxiciren[4], W.D.Zhou[12]

[1] *Department of Physics, Hirosaki University, Hirosaki, Japan*
[2] *Department of Physics, Saitama University, Urawa, Japan*
[3] *Department of Physics, Shangdong University, Jinan, China*
[4] *Department of Mathematics and Physics, Tibet University, Lhasa, China*
[5] *Institute of High Energy Physics, Academia Sinica, Beijing, China*
[6] *Department of Physics, South West Jiaotong University, Chengdu, China*
[7] *Faculty of Engineering, Kanagawa University, Yokohama, Japan*
[8] *Faculty of Education, Utsunomiya University, Utsunomiya, Japan*
[9] *Institute for Cosmic Ray Research, University of Tokyo, Tanashi, Japan*
[10] *Department of Physics, Konan University, Kobe, Japan*
[11] *Faculty of Systems Engineering, Shibaura Institute of Technology, Omiya, Japan*
[12] *Department of Physics, Yunnan University, Kunming, China*
[13] *National Center for Science Information Systems, Tokyo, Japan*
[14] *Tokyo Metropolitan College of Aeronautical Engineering, Tokyo, Japan*
[15] *Faculty of Engineering, Yokohama National University, Yokohama, Japan*
[16] *Shonan Institute of Technology, Fujisawa, Japan*

Abstract. The BL Lac object Mrk 501 came into a high flaring state in 1997, brightest in the sky at TeV energies. Using the air shower data obtained with a high density part of the Tibet II array, being operating at Yangbajing since 1996, we searched for γ-ray signals from this source during the period from February to August in 1997. The result shows a 3.7σ excess for showers with its mode energy 3 TeV for the whole period and 4.7σ excess for the period from April 7 to June 16 . The estimated energy spectrum of γ rays is consistent with those obtained by imaging atmospheric Cherenkov detectors.

INTRODUCTION

Mrk 501 and Mrk 421 have been well established as extra-galactic TeV γ-ray sources by Whipple and subsequent ground-based Cherenkov detectors [1]. They are the so-called BL Lac objects, which are radio-loud AGNs whose relativistic jets are considered to be aligned along our line of sight.

Flux variability of radiation on various time scales is a common feature of BL Lac objects, and spectral variations of γ rays arriving from these sources are considered to give us a powerful key to understand the physics of BL Lac objects. When Mrk 501 was first detected by the Whipple Collaboration in 1995, it showed low fluxes significantly below the level of the Crab. In March of 1997, however, this source began a remarkably high flaring activity and lasted for almost half a year with highly variable and strong γ-ray emission and reached the maximum flux roughly 10 times of the Crab.

During this period, several groups [2] observed strong γ-ray emission from this source with imaging atmospheric Cherenkov detectors. Independent measurements of the γ-ray spectrum seem to show a gradual softening towards higher energy. The energy spectrum and its shape are very important to clarify the mechanism of γ-ray production at the source, and eventually to lead to the actual measurement of the intergalactic infrared background photon field. Hence, confirmation of the detection of γ rays with a different technique is strongly required, especially with advantages of air shower arrays of such as wide aperture and high duty cycle for monitoring such variable sources.

The Tibet air shower array, operating since 1990 at Yangbajing (4,300 m above sea level) in Tibet [3], has a capability of detecting γ rays in the TeV energy region with high efficiency and high angular resolution. With this array, we succeeded to detect TeV γ-ray flares from Mrk 501 in 1997. The result obtained with traditional and well established air shower technique is important to compare with those by atmospheric Cherenkov technique.

EXPERIMENT

The high density (HD) part of Tibet II air shower array consists of 109 scintillation counters with 7.5 m spacing covering an area of 5,175 m^2 inside the Tibet II array as described elsewhere [4]. The Tibet II array comprises 221 scintillation counters of 0.5 m^2 each placed on a 15 m square grid with an enclosed area of 36,900 m^2, and the HD array is operating inside the Tibet II array to detect cosmic ray showers lower than 10 TeV (some detectors are commonly used).

The HD array was constructed in 1996 and from November the events have been accumulated at a trigger rate of about 115 Hz under any 4-fold coincidence of detectors, while the Tibet II array has triggered the events at about 200 Hz under the same condition. In this paper, we used the data obtained during the period from February to August in 1997. The event selection was done by imposing the

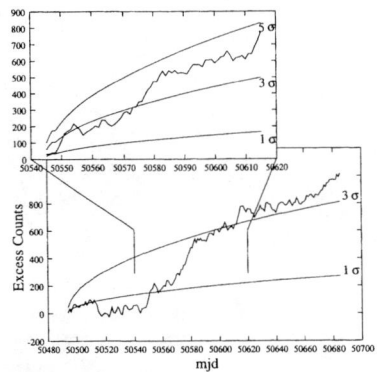
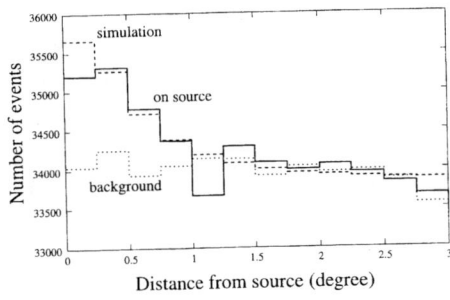

FIGURE 1. Cumulative excess of the events with $\sum \rho \geq 15$.

FIGURE 2. Opening angle distribution of the events with $\sum \rho \geq 15$.

following two conditions ; 1) Each of any four FT detectors should give a response more than 1.25 particles. 2) Among the four detectors giving the highest responses, two or more should be within the inner array. After data processing and quality cuts, the total number of events selected are 6.0×10^8 for the HD array and 1.1×10^9 for the Tibet II array, respectively, with effective running time of 155.3 days.

Since the background cosmic rays are isotropic and γ rays from a source must be centered on the source direction, a bin size for collecting on-source data should be determined based on the array's angular resolution so as to optimize the S/N ratio. The angular resolution of the Tibet array can be easily examined by observing the Moon shadow casting in the cosmic-ray intensity distribution. Using the HD array, the Moon shadow was observed with the significance of 15σ at the maximum deficit position for all events.

The Moon shadow was also observed in the direction deflected to the west by about $0.°4$. The mean energy of protons responsible for casting the shadow is estimated to be about 4 TeV by a Monte Carlo simulation. The observed deflection of the Moon shadow is consistent with the expected one for protons by the geomagnetic field, in which primary energy deduced from the shower size measured by the array can be calibrated by the experiment itself.

RESULTS

We use a circular search bin whose radius adjusted to the estimated angular resolution of the array so as to maximize the ratio $N_S/N_B^{1/2}$, where N_S is the number of signals and N_B the number of background events, and to contain about 50% of the signals. The used angular resolutions are $0.°9$ for showers with $\sum \rho \geq 15$ ($E_\gamma^{mode} \approx 3$ TeV), $0.°8$ for $\sum \rho \geq 50$ and $0.°5$ for $\sum \rho \geq 100$. The signal, N_S, is given by excess number of the ON-source events from the OFF-source events. The signals were searched for by shifting the ON-source window by $0.°125$ step around the object.

The background was estimated by averaging over events falling in neighboring ten OFF-source windows with the same zenith angle as the ON-source, but excluding the neighborest two windows. The signals were searched for by shifting the ON-source window by 0.°125 step around the object. The background was estimated by averaging over events falling in neighboring ten OFF-source windows with the same zenith angle as the ON-source, but excluding the neighborest two windows. Figure 1 shows the cumulative excess for all events as a function of mjd. There is seen no excess until the end of March 1997, the excess events rapidly increase in the period from April 7 to June 16 which is shown by an inserted enlarged figure.

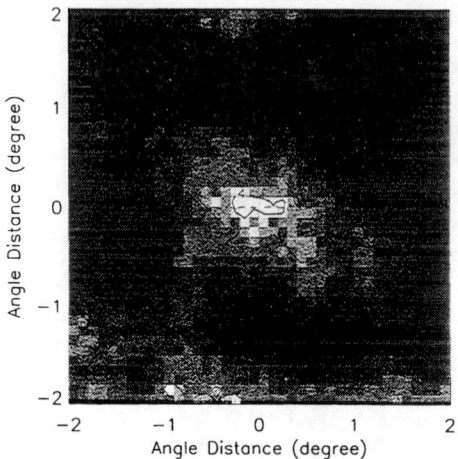

Figure 3. Two dimensional significance map of excess centered at Mrk 501 in the period April 7 to June 16.

The statistical significance reached the maximum 4.7σ in this period. These features are consistent with other observations by air Cherenkov detectors. Figure 2 shows the distribution of the opening angles relative to the Mrk 501 direction for all events with $\sum \rho \geq 15$ ($E_\gamma^{\text{mode}} \approx 3$ TeV) in this period. The excess in the small opening angle region less than 0.°5 would correspond to the γ rays from Mrk 501. Figure 3 shows the two dimensional significance map of excess events in this period. For the whole period from February to August, on the other hand, the statistical significances showed 3.7σ, 2.3σ and 1.6σ for showers with $\sum \rho \geq 15$, 30 and 50, respectively.

We also searched for γ-ray emission using the Tibet II array, but no excess was found and the flux upper limits are obtained at the 90 % confidence level.

We tried to estimate the γ-ray spectrum from Mrk 501 by a Monte Carlo simulation, assuming a differential power-law spectrum with the index $-(\beta + 1)$ and the cut-off at a certain energy, E_c, where the cut-off means that the spectral slope steepens by -1.0 at E_c. The value of β was varied in 1.4 ~ 1.7 and also the effect of E_c was examined in 7 ~ 50 TeV. The primary γ rays are generated between 0.2 TeV and 30 TeV from the direction of Mrk 501. The detection condition of simulated events at Yangbajing level was same as the experiment. The number of simulated events with respective $\sum \rho$ - value was compared with the observed one. The energy of γ rays was defined as the energy of the maximum flux of simulated events for respective $\sum \rho$. These steps were repeated until the observed results are well reproduced. A combination of $\beta \cong 1.6$ and $E_c \sim 20$ - 30 TeV seems to reproduce the data well, while the absolute flux around 3 TeV almost unchanged. Figures 4 and 5 show the energy spectra averaged in the period from 1997 February 15 to

1997 August 25 and from 1997 February 15 to 1997 June 8, respectively. The latter observation time corresponds to that of the Whipple Collaboration [5]. It is seen that the observed results by other experiments [5,6,7] are not inconsistent with ours.

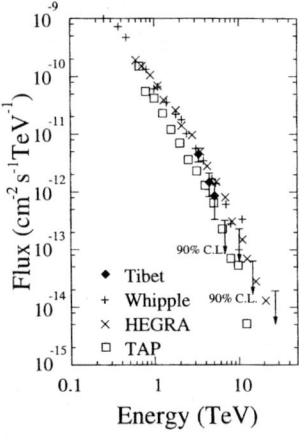 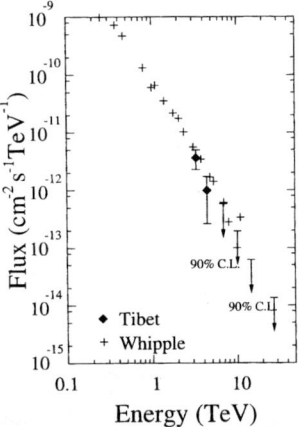

FIGURE 4. Energy spectrum of γ rays from Mrk 501 averaged over the whole period from February 15 to August 25, compared with other results [5,6,7]. Upper limits are at the 2σ confidence level.

FIGURE 5. Energy spectrum of γ rays from Mrk 501 averaged in the period from February 15 to June 9, to compare with the Whipple results. Upper limits are at the 2σ confidence level.

SUMMARY

Mrk 501 suddenly came into a very high active phase from March in 1997, with several large flares and lasted for about half a year. The maximum γ-ray flux during this period reached about 10 times as high as the Crab Nebula. The high resolution Tibet air shower array succeeded to detect γ rays during this very active phase. This is the first observation of γ-ray signals from AGNs by well-established air shower technique.

Acknowledgements : This work is supported in part by Grants-in-Aid for ScientificResearch and also for International Science Research of Monbusho in Japan and the Academy of Sciences in China.

REFERENCES

1. R.A. Ong, Phys. Rep., **305**, 93 (1998).
2. R.J. Protheroe *et al.*, 25th ICRC, **8**, 317 (1997).
3. M. Amenomori *et al.*, Phys. Rev. Let., **69**, 2468 (1992).
4. T. Yuda, Proc. Inter. Symp. on EHECR (Univ. of Tokyo, 1996), 175 (1996).
5. F.W. Samuelson *et al.*, ApJ, **501**, L17 (1998).
6. N. Hayashida *et al.*, ApJ, **504**, L71 (1998).
7. A. Konopelko *et al.*, astro-ph/9901093

Results from Milagrito on TeV Emission by Active Galactic Nuclei

David A. Williams
for the Milagro Collaboration[1]

*Santa Cruz Institute for Particle Physics
University of California
Santa Cruz, California 95064*

Abstract. Milagrito was a preliminary phase of the Milagro experiment and operated from February 1997 to May 1998. Milagrito was a water-Cherenkov detector for air showers produced by gamma rays (and background cosmic rays). The data have been studied for emission from several nearby blazars. Emission from Markarian 501 is detected coincident with the large flare observed in 1997 by several atmospheric Cherenkov telescopes.

INTRODUCTION

Milagro is an experiment to study gamma-rays with energy near one TeV using a large water-Cherenkov detector. Very high-energy particles interacting in the atmosphere produce extensive air showers. The small relativistic portion of the shower that reaches the ground can be detected by the flash of Cherenkov light produced in the detector, and the time of arrival of the light at different photomultiplier tubes in the detector can be used to reconstruct the arrival direction of the shower to about 1° or better.

---

[1] The Milagro Collaboration: R.W. Ellsworth, *George Mason University*; C. Espinoza, G. Gisler, T.J. Haines, C.M. Hoffman, R.S. Miller, M.M. Murray, C. Sinnis and T.N. Thompson, *Los Alamos National Laboratory*; L. Fleysher, R. Fleysher, A.I. Mincer and P. Nemethy, *New York University*; R.S. Delay, S. Hugenberger, I. Leonor, A. Shoup and G.B. Yodh, *University of California, Irvine*; B. Shen, A.J. Smith, T. Tumer, K. Wang and M.O. Wascko, *University of California, Riverside*; W. Benbow, D.G. Coyne, D.E. Dorfan, L.A. Kelley, J.F. McCullough, M.F. Morales, M. Schneider, S. Westerhoff, D.A. Williams and T. Yang, *University of California, Santa Cruz*; D. Berley, M.L. Chen, D. Evans, J.A. Goodman and G.W. Sullivan, *University of Maryland, College Park*; A. Falcone, M. McConnell and J.M. Ryan, *University of New Hampshire*; R. Atkins, B.L. Dingus and J.E. McEnery, *University of Utah*.

FIGURE 1. The reservoir used by the Milagrito and Milagro experiments is 80 m by 60 m at the surface and 8 m deep. In the photograph above the reservoir is covered and empty; the sloping sides can be seen.

THE MILAGRITO DETECTOR

The Milagrito detector [4] used a man-made water reservoir, shown in Figure 1, in the Jemez Mountains of New Mexico, at an elevation of 2650 m and location 35.9° N, 106.7° W. The same reservoir is now used by the Milagro detector. Cosmic gamma-rays and charged particles interact in the atmosphere, producing a shower of low-energy, but still relativistic, secondary particles. The secondary particles in a shower were detected by their interactions in the water of the Milagrito detector, where they produced flashes of Cherenkov light that were detected by photomultiplier tubes. An opaque, light-tight cover keeps the water clean (the reservoir also has a man-made liner) and allows operation regardless of daylight and weather conditions. Such a detector is hence able to operate around the clock, and is sensitive to showers from a large portion of the overhead sky.

Whereas the Milagro detector operates with the reservoir full of water and phototubes deployed in two layers, Milagrito used a single layer of 228 phototubes in a single layer at the bottom of the reservoir. The reservoir was partially filled to cover the phototubes by between 1 m and 2 m of purified water. The tubes are arranged on a square grid with 2.8 m spacing. Events having more than 100 phototubes firing within a 300 ns coincidence window triggered readout by the data acquisition system. The trigger rate was about 300 Hz when the tubes were covered by 1 m of water, and increased to about 400 Hz with the addition of another meter of water above the tubes.

The direction of the incoming primary responsible for each shower was recontructed by precise measurement of the time of arrival of the Cherenkov light at each struck phototube. The secondary particles are relativistic and form a shower front which is nearly planar impinging on the pond. Recording the arrival time of the phototube signals with subnanosecond accuracy allows the shower plane to be reconstructed, and the normal to the plane gives the primary direction. Shower directions are determined in this manner with a resolution of about 1°.

ANALYSIS METHOD

Gamma-ray emission is identified as an excess of events, above the cosmic ray background, from a particular direction. A background map is constructed from the data, which is predominantly cosmic-ray events. A distribution of the arrival direction of all events is made in local coordinates for each 2 hour interval. For each event, entries are made in the background map by choosing the local coordinates at random from the accumulated distribution, and using those coordinates together with the time of the actual event to determine the celestial coordinates of the background entry. This is done at least ten times, so that the statistical fluctuations in the background map are small compared to the data. See Ref. [2] for more details on this method.

The data are compared to the background map by counting the total number of events within some angular separation from the candidate source and comparing that number to the number predicted by the background map. There is an optimal search region, or angular bin size, which is related to the pointspread function of the detector. For Milagrito, the quality of the direction reconstruction is related to the number of good tube hits which are used in the fit to the shower plane. We have studied how the expected sensitivity to a source varies with the choice of bin size and with the requirement of a certain number of good hits in the directional fit. The results are shown in Figure 2. The significance is maximized by the use of a 1° bin and a requirement of >40 tubes in the fit. These values are

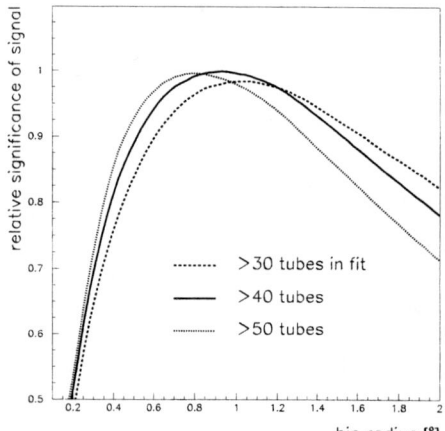

FIGURE 2. The expected significance of a gamma-ray signal as a function of the radius of the search bin and the number of tube hits kept in the fit to the shower plane. (©1999. The American Astronomical Society. [3])

used in the subsequent analysis. The optimum is sufficiently broad that the results are not sensitive in detail to these criteria. (In the case of much larger expected signal to background, the optimum is obtained by relaxing the tube requirement and opening up the search bin, as has been done in the gamma-ray burst analysis from Milagrito [5].)

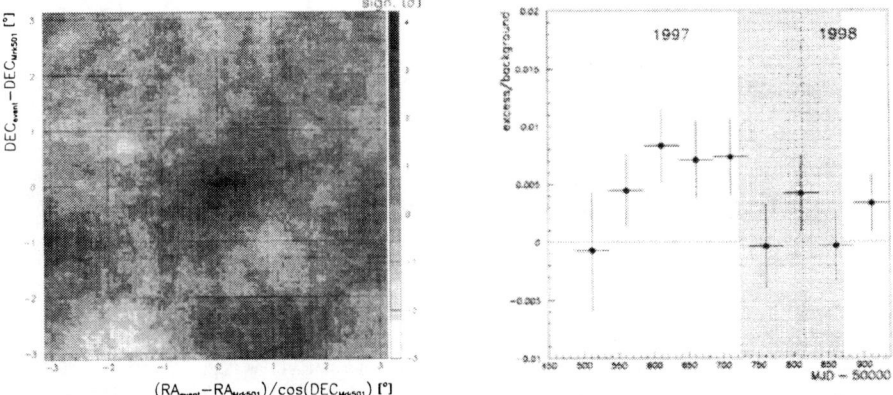

FIGURE 3. On the left, the significance for an event excess as a function of right ascension (RA) and declination (DEC) in a $6° \times 6°$ region with the Mrk 501 position in the center. On the right, the fractional event excess from Mrk 501 as a function of time. (©1999. The American Astronomical Society. [3])

MARKARIAN 501

Figure 3 shows the significance of the excess over background in the vicinity of Mrk 501. For each point the significance is calculated for the area of the circle with radius 1° and the bin center as the central point; thus neighboring bins are highly correlated. The position of Mrk 501 is at the center of the plot. The excess at the position of Mrk 501 is 3624 ± 990 events, or 3.7σ. The flux between February and October 1997 has been reported by atmospheric Cherenkov telescopes [1,6]. The number of Milagrito events from that period is in good agreement with the reported flux and calculations of Milagrito sensitivity. More details appear in Ref. [3]. The agreement indicates that the water-Cherenkov technique is working as anticipated.

Figure 3 also shows the fractional excess (the excess divided by the background) as a function of time in 50 day bins. The shaded area indicates the time period for which no data from atmospheric Cherenkov telescopes is available because Mrk 501 is a daytime object. While the lightcurve suggests a higher overall flux for the first half of the data set, at Milagrito's level of sensitivity, the flux is consistent with being constant in time.

OTHER BLAZARS

We have performed a similar search for continuous emission from 10 other nearby blazars: 0310+4104, 0430+052, 0554+5328, Mrk 421, 1133+704, 1727+502, 1807+698, 1959+650, 2321+419, 2344+513. No significant excess was seen in the Milagrito data set from any of these candidates [7], consistent with calculations of

Milagrito sensitivity and prior observations of many of these objects by atmospheric Cherenkov telescopes.

REFERENCES

1. Aharonian, F., et al., *Astron. & Astrophys.* **342**, 69 (1999).
2. Alexandreas, D. E., et al., *Nucl. Instr. Meth. A* **328**, 570 (1993).
3. Atkins, R., et al., *Astrophys. J. Lett.* **525**, L25 (1999).
4. Atkins, R., et al., submitted to *Nucl. Instr. Meth. A*.
5. McEnery, J. E., these proceedings.
6. Samuelson, F. W., et al., *Astrophys. J. Lett.* **501**, L17 (1998).
7. Westerhoff, S., et al., Proc. 19th Texas Symposium on Relativistic Astrophysics, December 14-18, 1998, Paris, France (in press).

Modeling the April 1997 flare of Mkn 501

A. Mücke and R.J. Protheroe

*The University of Adelaide
Dept. of Physics & Mathematical Physics
Adelaide, SA 5005, Australia*

Abstract. The April 1997 giant flare of Mkn 501 is modelled by a stationary Synchrotron-Proton-Blazar model. Our derived model parameters are consistent with X-ray-to-TeV-data in the flare state and diffusive shock acceleration of e^- and p in a Kolmogorov/Kraichnan turbulence sprectrum. While the emerging pair-synchrotron cascade spectra initiated by photons from π^0-decay and electrons from $\pi^\pm \to \mu^\pm \to e^\pm$-decay turn out to be relatively featureless, $\mu^\pm$ and p synchrotron radiation and their cascade radiation produce a double-humped spectral energy distribution. For the present model we find p synchrotron radiation to dominate the TeV emission, while the contribution from the synchrotron radiation of the pairs, produced by the high energy hump, is only minor.

I THE CO-ACCELERATION SCENARIO

With its giant outburst in 1997, emitting photons up to 24 TeV and 0.5 MeV in the γ-ray and X-ray bands, Mkn 501 has proved to be the most extreme TeV-blazar observed so far (e.g. Catanese et al 1997, Pian et al 1997, Aharonian et al 1999).

In this paper, we consider the April 1997 flare of Mkn 501 in the light of a modified version of the Synchrotron Proton Blazar model (SPB) (Mannheim 1993), and present a preliminary model fit.

In the model, shock accelerated protons (p) interact in the synchrotron photon field generated by the electrons (e^-) co-accelerated at the same shock. This scenario may put constraints on the maximum achievable particle energies.

The usual process considered for accelerating charged particles in the plasma jet is diffusive shock acceleration (see e.g. Drury 1983, Biermann & Strittmatter 1987). If the particle spectra are cut off due to synchrotron losses, the ratio of the maximum particle energies $\gamma_{p,max}/\gamma_{e,max}$ can be derived by equating $t_{acc,p}/t_{acc,e} = t_{syn,p}/t_{syn,e}$, with $t_{syn,p}$ and $t_{syn,e}$ being the synchrotron loss time scales for p and e^-, respectively. We find that for shocks of compression ratio 4 (see Mücke & Protheroe 1999 for a detailed derivation)

$$\frac{\gamma_{p,\max}}{\gamma_{e,\max}} \leq \frac{m_p}{m_e}\left(\frac{m_p}{m_e}\right)^{\frac{2(\delta-1)}{3-\delta}} \sqrt{\frac{F(\theta,\eta_{e,\max})}{F(\theta,\eta_{p,\max})}} = \frac{m_p}{m_e}\sqrt{\frac{\eta_{e,\max}F(\theta,\eta_{e,\max})}{\eta_{p,\max}F(\theta,\eta_{p,\max})}} \quad (1)$$

where the "="-sign corresponds to synchrotron loss, and the "<"-sign to adiabatic loss determining the maximum energies. δ is the power law index of the magnetic turbulence spectrum ($\delta = 5/3$: Kolmogorov turbulence, $\delta = 3/2$: Kraichnan turbulence, and $\delta = 1$ corresponds to Bohm diffusion). $\eta_{e,\max}$ is the mean free path at maximum energy in units of the particle's gyroradius and $F(\theta,\eta_{e,\max})$ takes account of the shock angle θ (Jokipii 1987). The ratio $F(\theta,\eta_{e,\max})\eta_{e,\max}/F(\theta,\eta_{p,\max})\eta_{p,\max}$ can be constrained by the variability time scale t_{var}, requiring $t_{var}D \geq t_{acc,p,max}$ (D = Doppler factor, $t_{acc,p,max}$ = acceleration time scale at maximum particle energy) for a given parameter combination. As an example, we adopt $D = 10$, $B = 20$G and $t_{\text{var}} = 2$ days. Eq. 1 then restricts for these parameters the ratio of the allowed maximum particle energies to the range below the solid lines shown in Fig. 1. Points exactly on this line correspond to synchrotron-loss limited particle spectra which are accelerated with exactly the variability time scale.

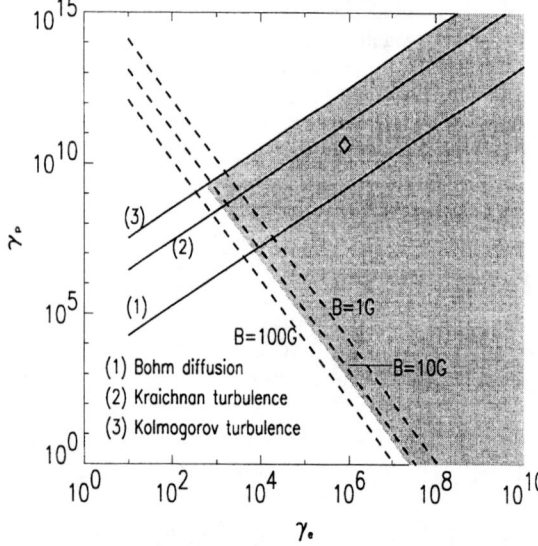

FIGURE 1. Allowed parameter space (shaded area) for $\gamma_{p,\max}$, $\gamma_{e,\max}$ in the SBP-model for typical TeV-blazar parameters (B=20 G, D=10, $u_1 = 0.5c$, $\beta = 1$, $t_{\text{var}} = 2$days) and for different magnetic turbulence spectra $I(k) \propto k^{-\delta}$. The diamond symbol corresponds to the Mkn 501-model presented below.

In hadronic models π photoproduction is essential for γ-ray production. The threshold of this process is given by $\epsilon_{\max}\gamma_{p,\max} = 0.0745$ GeV where $\epsilon_{\max}$ is the maximum photon energy of the synchrotron target field. Inserting $\epsilon_{\max} = 3/8\gamma_{e,\max}^2 B/(4.414 \times 10^{13}\text{G}) \, m_e c^2$ into the threshold condition, we find

$$\gamma_{p,\max} \geq 1.72 \cdot 10^{16}\left(\frac{B}{\text{Gauss}}\right)^{-1}\gamma_{e,\max}^{-2}$$

which is shown in Fig. 1 as dashed line for various magnetic field strengths. Together with Eq. 1, the allowed range of maximum particle energies is then restricted to the shaded area in Fig. 1.

II THE MKN 501 FLARE IN THE SYNCHROTRON PROTON BLAZAR (SPB) MODEL

We assume the parameters used in Fig. 1, and that the co-accelerated e^- produce the observed synchrotron spectrum, unlike in previous SPB models, and this is the target radiation field for the $p\gamma$-interactions. This synchrotron spectrum, and its hardening with rising flux, has recently been convincingly reproduced by a shock model with escape and synchrotron losses (Kirk et al 1998). We use the Monte-Carlo technique for particle production/cascade development, which allows us to use exact cross sections.

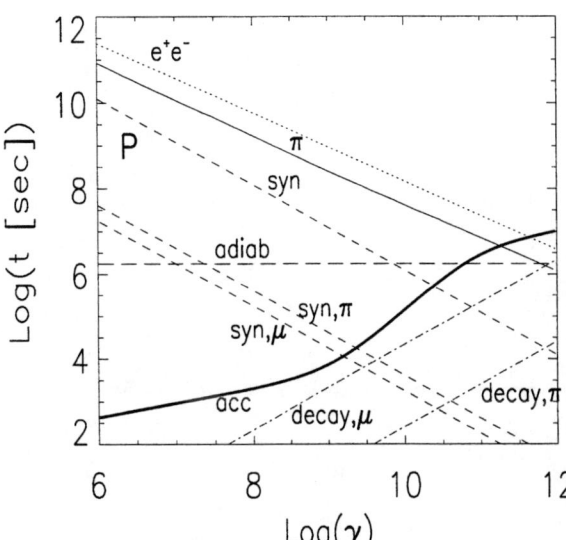

FIGURE 2. Mean energy loss time of p for synchrotron radiation (syn), π-photoproduction (π) and Bethe-Heitler pair production ($e^\pm$), and for $\pi^\pm$- and $\mu^\pm$-synchrotron radiation (syn π, syn μ) for $B = 19.6$ G with their mean decay time scales (decay π, decay μ) in the SPB-model. The acceleration time scale (acc), based on Kolmogorov turbulence, is calculated for $u_1 = 0.5c$, $\eta_p = 40$ and shock angle $\theta = 85°$. Its curvature reflects the influence of the shock angle. The adiabatic loss time (adiab) is assumed to be $R/u_1 \approx Dt_{var}$. All quantities are in the jet frame.

For simplicity we represent the observed synchrotron spectrum (target photon field for the $p\gamma$-collisions) as a broken power law in the jet frame with photon power law index 1.4 below the break energy of 0.2 keV, and index 1.8 up to 50 keV.

The variability time scale restricts the radius R of the emission region. For our model we use $t_{var,x} \approx 2$ days (Catanese et al 1997), and find $R \approx 2.6 \times 10^{16}$cm for $D = 10$, $B = 19.6$ G. With these parameters the $\gamma\gamma$-pair production optical depth reaches unity for ≈ 25 TeV photons.

Our model considers photomeson production (simulated using SOPHIA, Mücke et al 1999), Bethe-Heitler pair production (simulated using the code of Protheroe & Johnson 1996), p synchrotron radiation and adiabatic losses due to jet expansion. The mean energy loss and acceleration time scales are presented in Fig. 2.

Synchrotron losses, which turn out to be at least as important as losses due to π photoproduction for the assumed 2 day variability, limit the injected p spectrum $\propto \gamma_p^{-2}$ to $2 \leq \gamma_p \leq 4.4 \times 10^{10}$. This leads to a p energy density $u_p \approx 0.2$ TeV/cm^3, which is bracketed by the photon energy density $u_{\text{target}} \approx 0.01$ TeV/cm^3, and a magnetic field energy density $u_B \approx 9.5$ TeV/cm^3. With $u_B \gg u_{\text{target}}$ significant Inverse Compton radiation from the co-accelerated e^- is not expected.

Rachen & Meszaros (1998) noted the importance of synchrotron losses of $\mu^{\pm}$- (and $\pi^{\pm}$-) prior to their decay in AGN jets and GRBs. For the present model, the critical Lorentz factors $\gamma_{\mu} \approx 3 \times 10^9$ and $\gamma_{\pi} \approx 4 \times 10^{10}$, above which synchrotron losses dominate above decay, lie well below the maximum particle energy for $\mu^{\pm}$, while $\pi^{\pm}$-synchrotron losses can be neglected due to the shorter decay time.

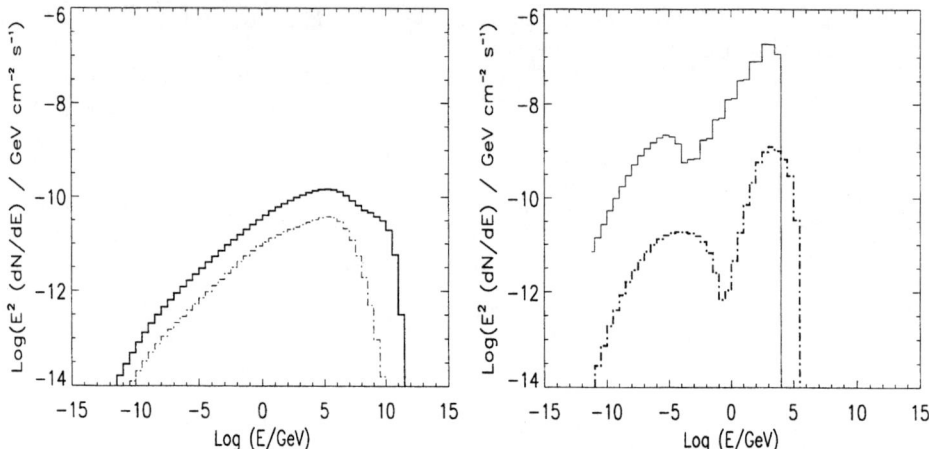

FIGURE 3. Left: Average emerging cascade spectra initiated by π^0-decay (solid line) and $\pi^{\pm}$-decay synchrotron photons (dashed-dotted line). Right: Average emerging cascade spectra initiated by p- (solid line) and $\mu^{\pm}$-synchrotron photons (dashed-dotted line).

The matrix method (e.g. Protheroe & Johnson 1996) is used to follow the pair-synchrotron cascade in the ambient synchrotron radiation field and magnetic field, developing as a result of photon-photon pair production. The cascade can be initiated by photons from π^0-decay ("π^0-cascade"), electrons from the $\pi^{\pm} \to \mu^{\pm} \to e^{\pm}$-decay ("$\pi^{\pm}$-cascade"), $e^{\pm}$ from the proton-photon Bethe-Heitler pair production ("Bethe-Heitler-cascade") and p and μ-synchrotron photons ("p-synchrotron cascade" and "$\mu^{\pm}$-synchrotron cascade"). In this model, the cascades develop linearly.

Fig. 3 shows an example of cascade spectra initiated by photons of different origin, and for the parameter combination given above. π^0- and $\pi^{\pm}$-cascades obviously produce featureless spectra whereas p- and $\mu^{\pm}$-synchrotron cascades cause the typical double hump shaped SED as observed in γ-ray blazars (see also Rachen, these proceedings). The contribution from Bethe-Heitler cascades turns out to be negligible. Direct p- and $\mu^{\pm}$-synchrotron radiation is responsible for the high energy peak, whereas the low energy hump may be either synchrotron radiation from the directly accelerated e^- and/or by pairs produced by the "low energy hump".

Adding the four components of the cascade spectrum in Fig. 3 and normalizing to an ambient, accelerated p density of $n_{tot,p} = 7$ cm^{-3}, we obtain the SED shown in Fig. 4 where it is compared with the multifrequency observations of the 16 April 1997 flare of Mkn 501.

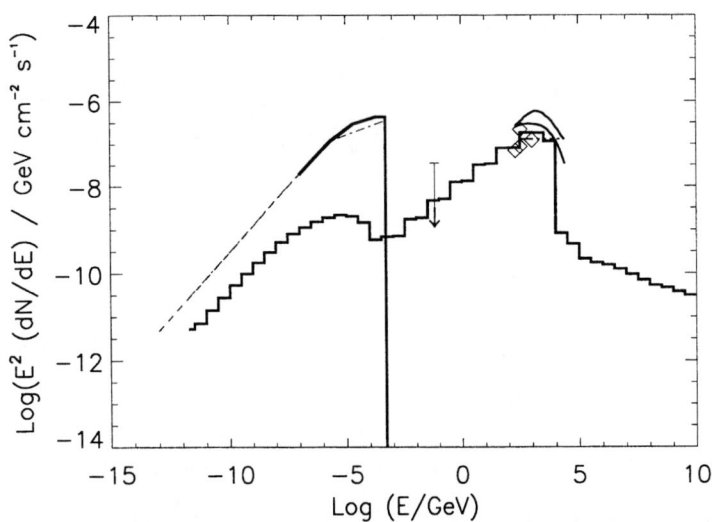

FIGURE 4. Present model (histogram) in comparison with the data of the 16 April 1997-flare of Mkn 501. Photon absorption on the cosmic diffuse background radiation field is not included in the model. Straight solid lines: parametrization of the observed, curved synchrotron spectrum (BeppoSAX & OSSE) by Bednarek & Protheroe (1999) and observed TeV-emission corrected for cosmic background absorption (Bednarek & Protheroe 1999) with peak energy output ~ 2 TeV; the 100 MeV upper limit is from Catanese et al 1997 (observed 9-15 April 1997), diamonds: nearly simultaneous (uncorrected) Whipple data (Catanese et al 1997); dashed-dotted line: synchrotron target spectrum.

REFERENCES

1. Aharonian F.A. et al, *A&A* **342**, 69 (1999).
2. Bednarek W. & Protheroe R.J., to appear in *MNRAS* (1999).
3. Biermann, P.L. & Strittmatter, P.A., *ApJ* **322**, 643 (1987).
4. Catanese M., et al, *ApJ* **487**, L143 (1997).
5. Drury L.O'C, *Rep. Prog. Phys.* **46**, 973 (1983).
6. Jokipii J.R., *ApJ* **313**, 842 (1987).
7. Kirk J.R., Rieger, F.M. & Mastichiadis, A. *A&A* **333**, 452 (1998).
8. Mannheim K., *A&A* **269**, 67 (1993).
9. Mücke A. et al, to appear in *Comm.Phys.Comp.* (1999).
10. Mücke A. & Protheroe R.J., in preparation (1999).
11. Pian E. et al, *ApJ* **492**, L17 (1998).
12. Protheroe R.J. & Johnson P., *Astropart.Phys.* **4** 253, & erratum **5**, 215 (1996).
13. Rachen J.P. & Meszaros P., *Phys.Rev.D* **58**, 123005 (1998).

Spectral Measurements of TeV γ-ray Emission from Mkn501 and Mkn421 Using the HEGRA Stereoscopic System of IACTs

Michael Panter and Henric Krawczynski
for the HEGRA Collaboration

Max - Planck - Institut für Kernphysik, Saupfercheckweg 1, D-69117 Heidelberg

Abstract. Since October'96 the HEGRA collaboration is operating the first stereoscopic system world wide. It consists of five imaging atmospheric Cherenkov telescopes (IACTs) installed at La Palma, Canary Islands. The performance of the system shows its unique capabilities for the study of TeV γ-ray sources with an energy threshold of 500 GeV, an angular resolution of 0.1°, and an energy resolution of better than 20%. The stereoscopic system is able to detect within an hour a γ-ray flux of νF_ν $10^{-11} ergs/cm^2 sec$ at 1 TeV with a signal-to-noise ratio of 5σ. The recently developed technique of spectrum evaluation from the stereoscopic data was used in the observations of two BL Lac objects - Mkn501 and Mkn421. The Mkn501 flared with up to ten times the Crab flux in 1997 with an average flux of about three Crabs. Recording of several Cherenkov light images from an individual air shower and the unprecedented statistics of about 38,000 TeV γ-rays allowed us to test each step of the spectrum evaluation procedure in great detail and to extend the energy measurements far beyond 10 TeV. Despite the low state of TeV emission of Mkn421 during 1997 and 1998 the spectrum was measured over the energy range from 500 GeV to $\sim$ 7 TeV.

Introduction The BL Lac objects Mkn421 and Mkn501 were the first two extragalactic sources established as TeV-emitters [1] [2] [3] [4] and have been subject to intense studies ever since. They are located at similar redshifts (Mkn501 $z = 0.034$, Mkn421 $z = 0.031$), and show strong TeV-flux variability. The object Mkn501, after showing moderate flux levels of one third of the flux of the Crab Nebula during the first years after its discovery as a TeV γ-ray source, went during 1997 into a phase of high activity and dramatic variability, outshining during several nights the brightest known source in the TeV sky, the Crab Nebula, by factors as large as $\sim$ 10. The high activity allowed us to study the temporal characteristics on time scales of several minutes and to perform detailed spectral studies on diurnal basis. In this contribution we summarize the results of the stereoscopic IACT system of HEGRA on the emission of Mkn501 in 1997 and of Mkn421 in 1997 and 1998. We compare the TeV characteristics of the two sources and discuss the implications of the results. Detailed further information on the Mkn501 and Mkn421 observations as well as on the analysis tools can be found in [5] [6] [7].

Detector, Data Sample, and Analysis Method The HEGRA system [8] of five (until August'98 four) IACTs is located on the Roque de los Muchachos on the Canary Island of La Palma, (28.8° N, 17.9° W, 2200m a.s.l.). Each telescope is equipped with a segmented 8.5 m^2 mirror and a high resolution camera consisting of 271 pixels of 0.25° diameter which add up to a 4.3° field of view. The stereoscopic observation technique allows the reconstruction of the shower impact point with an accuracy of ± 10 m. The knowledge of the impact point improves in particular the energy determination and the γ/hadron-separation. The simultaneous observation of air showers under widely differing viewing angles with two or more Cherenkov telescopes results in an energy threshold of 500 GeV, an angular resolution of 0.1° and an energy resolution better than 20% for an individual photon.

The Mkn501 analysis is based on 110 hours of observations acquired between March 16th, and October 1st 1997. The Mkn421 analysis is based on 165 hours between January 1st, 1997 and May 27th, 1998. Only data taken under optimal weather conditions, with the optimal detector performance, and with the source being more than 45° above the horizon were used for the analysis. For minimizing the systematic uncertainties in the energy dependent cut-efficiencies, the analysis uses "loose" γ/hadron-separation cuts which accept a large fraction of $\sim$ 80% of the γ-rays at all energies above 1 TeV. The cut optimization and the calculation of effective detection areas and cut efficiencies, is based on detailed Monte Carlo simulations [9] and was verified with experimental data. Extensive studies, only possible due to redundant shower information provided by a stereoscopic IACT system, were carried out to estimate and to reduce the systematic error of the energy spectra. The pointing accuracy of the IACTs was checked by verifying the Mkn501 location to an accuracy of 35 arcsec [10].

Mkn501 Results from 1997 For fluxes comparable to the Crab flux the IACT system of HEGRA is not only sensitive enough to assess within a fraction of an hour the TeV flux, but also to determine differential spectra on a diurnal basis. One of 63 diurnal Mkn501 energy spectra is shown in Figure 1 (left side). The systematic error of the spectrum can be divided into two contributions: a 15% uncertainty on the absolute energy scale, and an error on the curvature of the spectrum, shown by the hatched region. An integration time of only 1.9 hours was sufficient to determine the differential spectrum over the broad energy region from 500 GeV to above 10 TeV, and to prove a significant deviation from a pure power law. The diurnal 1 TeV to 5 TeV photon index has been determined to be 2.24 with a statistical accuracy of 0.11 and a systematic accuracy of 0.05.

The 1997 light curve [5] showed flux variations from a fraction of the Crab flux to more than 10 Crab with flux increases and decreases by more than a factor of two within 24 hours. The 1 TeV to 5 TeV photon indices, determined with typical accuracies between 0.1 and 0.3 remained stable and no highly significant deviation from the mean index of 2.28 could be established. Furthermore, accumulating the data according to the diurnal 2 TeV flux levels, or according to a "rising" or "falling" flux behavior yielded time averaged spectra with the same curvature within statistical errors.

The time averaged Mkn501 spectrum is shown in Figure 1. From 500 GeV to 24 TeV, the spectrum can be described by a power law with an exponential cut off: $dN/dE = N_0 (E/1\,\mathrm{TeV})^{-\alpha} \exp(-E/E_0)$, with $N_0 = (10.8 \pm 0.2_{\mathrm{stat}} \pm 2.1_{\mathrm{syst}}) \cdot 10^{-11}\,\mathrm{cm}^{-2}\mathrm{s}^{-1}\mathrm{TeV}^{-1}$, $\alpha = 1.92 \pm 0.03_{\mathrm{stat}} \pm 0.20_{\mathrm{syst}}$, and $E_0 = (6.2 \pm 0.4_{\mathrm{stat}} (-1.5 + 2.9)_{\mathrm{syst}})$ TeV. The systematic errors on the fit parameters result from worst case assumptions on the systematic errors of the data points and their correlations, and include the error caused by the 15% uncertainty in the energy scale. Note that the errors on E_0 and α are strongly correlated. A dedicated χ^2-analysis yields a very conservative lower limit on the maximum γ-ray energy of 16 TeV.

Intensive Mkn501 multiwavelength campaigns have been performed with the aim to unambiguously identify the emission mechanism. We found a significant correlation of the TeV fluxes with the 2 keV-12 keV X-ray fluxes determined with the All Sky Monitor on board the RXTE *Rossi X-Ray Timing Explorer* [5]. The analysis of multiwavelength campaigns performed during 1998 together with more sensitive X-ray instruments, namely RXTE, and ASCA is underway.

Mkn421 Results from 1997 and 1998 The Mkn421 lightcurve is shown in Figure 2 (left side). During the observations, the source showed moderate flux

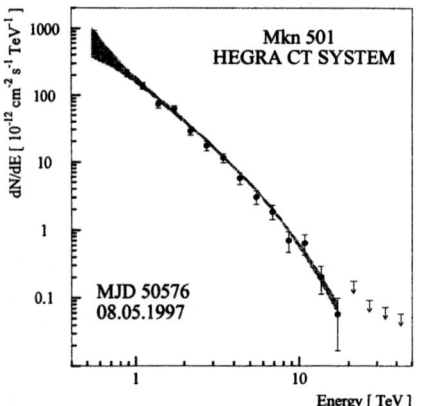 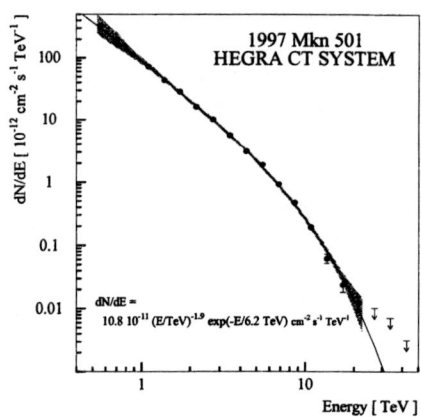

FIGURE 1. The left side shows a diurnal Mkn501 energy spectrum taken with an integration time of 1.9 hours (data with <30° zenith angle). The right side shows the 1997 time averaged Mkn501 spectrum. Upper limits are 2σ confidence level; vertical error bars show the statistical errors; the hatched region shows the systematic error on the curvature of the spectrum.

levels at 1 TeV of approximately one third of the Crab flux. During sporadic flares, the flux increased up to a level a little above one Crab unit. Also in the case of Mkn421 we do not find significant evidence for spectral shape variability, neither studying the diurnal spectra, nor by dividing the data in a high- and a low-flux data sample. A pure power law model satisfactorily describes the data over the energy range from 500 GeV to several TeV (Figure 2, right side, solid line): $dN/dE = (12.1 \pm 0.5_{stat} \pm 4.3_{syst})\, 10^{-12}\, (E/TeV)^{(-3.09 \pm 0.07_{stat} \pm 0.10_{syst})} cm^{-2} s^{-1} TeV^{-1}$. A power law with an exponential cutoff fits the data as well (see Figure 2, right side, dashed line): $dN/dE \propto (E/TeV)^{(-2.5 \pm 0.4_{stat})} \exp(-E/E_0)$ with $E_0 = 2.8 \left(^{+2.0}_{-0.9}\right)_{stat}$ TeV.

For the 1997 data we obtain a power law photon index of $3.28 \pm 0.20_{stat}$, for the 1998 data we get $3.00 \pm 0.05_{stat}$, and from the April 1998 data which was taken in the frame of a worldwide multiwavelength campaign [11], we obtain the photon index $3.03 \pm 0.08_{stat}$.

Discussion The IACT system of HEGRA has been used to obtain a wealth of detailed spectral and temporal information about the TeV-emission of the two BL-Lac objects Mkn501 and Mkn421. For both sources, we studied the spectra on diurnal basis but did not find strong evidence neither for spectral variability nor for a correlation of absolute flux and spectral shape. In the case of Mkn501, the high emission levels allowed us to assess the spectrum up to energies of $\sim$20 TeV. The spectrum is well described by a power law with an exponential cutoff at 6.2 TeV. This cutoff could certainly be caused by several effects: it could reflect e. g. the

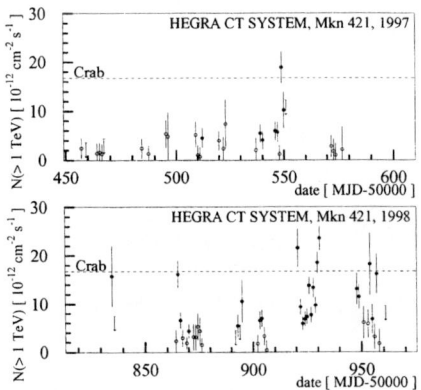
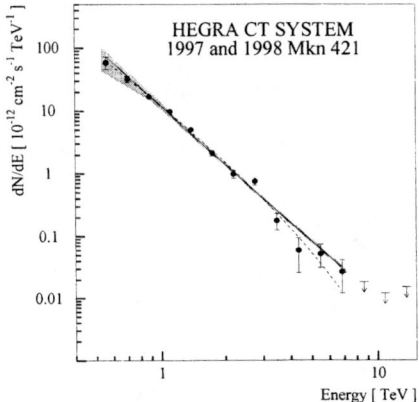

FIGURE 2. The left side shows the Mkn421 lightcurve. The right side shows the 1997–1998 time averaged Mkn421 energy spectrum. The solid line shows the power law fit, the dashed line the fit of a power law with an exponential cutoff, both as described in the text. Upper limits are 2σ confidence level; the error bars show the statistical errors; the hatched region show the systematic error on the curvature of the spectrum

maximum energy to which nonthermal particles are accelerated, it could be caused by absorption of the TeV-photons in pair production processes inside the source, or by extinction of the TeV photons by the infrared diffuse extragalactic background radiation (DEBRA). As pointed out already in [5], the rather stable TeV spectra indicate a time independent spectrum of accelerated nonthermal particles together with cooling times of the very energetic particles which are too short to yield spectral cooling observable with the typical HEGRA integration times in the order of hours. Since Mkn501 and Mkn421 are located at similar redshifts, our observations clearly show that the intrinsic source spectra of the two BL Lac objects are different. The DEBRA extinction would deform the Mkn501 and the Mkn421 spectra in the same way. Due to the low flux levels during the Mkn421 observations we cannot verify or exclude at this time that the Mkn421 spectrum cuts off at a similar energy as the Mkn501 spectrum.

Under rather general assumptions the pure fact of recording Mkn501 photons with energies well above 10 TeV yields sensitive upper limits on the DEBRA density in the wavelength region from 1 to 50 microns. Note that a constant DEBRA energy density of $\sim 1.1 nW/m^2 sr$ per logarithmic bandwidth would deform a power law of photon index 1.92 exactly into the observed power law spectrum with an exponential cutoff at 6.2 TeV.

Acknowledgments

We thank the Instituto de Astrofísica de Canarias (IAC) for supplying excellent working conditions at La Palma. HEGRA is supported by the BMBF (Germany) and CYCIT (Spain).

REFERENCES

1. Punch M., Akerlof C.W., Cawley M.F., et al., 1992, Nat 358, 477
2. Quinn J., Akerlof C.W., Biller S., et al., 1996, ApJ 456, L83
3. Petry D., Bradbury S.M., Konopelko A., et al., 1996, A&A 311, L13
4. Bradbury S.M., Deckers T., Petry D., et al., 1997, A&A 320, L5
5. Aharonian F.A., Akhperjanian A.G., Barrio J.A., et al., 1999a, A&A 342, 69
6. Aharonian F.A., Akhperjanian A.G., Barrio J.A., et al., 1999b, submitted to A&A, astro-ph/9903386
7. Aharonian F.A., Akhperjanian A.G., Andronache M., et al., 1999c, submitted to A&A, astro-ph/9905032
8. Daum A., Hermann G., Heß M., et al., 1997, Astropart. Phys. 8, 1
9. Konopelko A. et al., 1999, Astropart. Phys. 10, 275-289
10. Pühlhofer G., Daum A., Hermann G., et al., 1997, Astropart. Phys. 8, 101
11. Takahashi T., Madejski G., Kubo H., 1999. In: Proc. the Veritas Workshop on the TeV Astrophysics of Extragalactic Objects, ed. Weekes T.C., Catanese M., Astroparticle Physics in press.

Effect of intergalactic absorption in the TeV γ-ray spectrum of Mkn 501

Alexander K. Konopelko

Max-Planck-Institut für Kernphysik,
Heidelberg D-69029, Postfach 10 39 80, Germany
e-mail: alexander.konopelko@mpi-hd.mpg.de

Abstract. We discuss an effect of the intergalactic absorption of the TeV γ-rays in time-averaged spectrum of Mkn 501 measured by the HEGRA collaboration. Analysis of the spectral behavior, variability time scale and relevant calculations of TeV γ-ray emission allow to conclude the presence of a noticeable absorption of the TeV γ-rays in the Mkn 501 energy spectrum.

INTRODUCTION

The ground-based detectors, utilizing the so-called imaging air Čerenkov technique, offer an effective tool to study the cosmic TeV γ-rays. Recently, a number of celestial objects has been identified as TeV γ-ray emitters by use of such technique [1]. Among them there are two active galactic nuclei (AGN) – Mkn 421 and Mkn 501 – which for almost similar redshift of 0.031 and 0.034 respectively, have very different properties of a TeV γ-ray emission. In particular, TeV γ-ray fluxes from Mkn 421 and Mkn 501 differ in variability time scale and spectral behavior. Mkn 421 has shown significant flux variations within a time period as short as 15 minutes [2] whereas Mkn 501 may outburst during a period of 6 months with an extraordinary high γ-ray flux of more than 3 Crabs on average [3]. The energy spectrum of Mkn 501, as measured in the energy range from 0.5 TeV up to 20 TeV, shows evident curvature ($dJ_\gamma/dE \propto E^{-1.9} exp(-E/6.2)$) and the spectrum shape does not depend on the flux level [4]. At the same time the Mkn 421 energy spectrum is very steep and consistent with the pure power law ($dJ_\gamma/E \propto E^{-3.1}$) over the energy range 0.5-7 TeV, at least during low state of emission [5]. All that shows an apparent intrinsic difference in the mechanism of the TeV γ-ray emission which is widely believed to be an inverse Compton scattering of electrons within a relativistic jet directed along the observer line of site (for review see [1]). In addition the measured spectra of TeV γ-rays from such distant sources as Mkn 421 and Mkn 501 might be affected by the γ-ray absorption on the diffuse intergalactic infrared (IR) background. Here we discuss how important might be the effect of

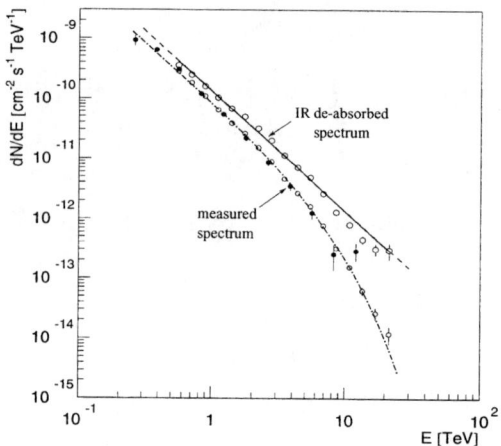

FIGURE 1. The energy spectrum of Mkn 501 as measured by the HEGRA IACT array (open circles) [4]). The combined fit (power law plus exponent) of the HEGRA data is shown by the dotted-dashed curve. The Mkn 501 spectrum measured by the Whipple group (filled circles) is from [6]. The "de-absorbed" HEGRA data and a power law fit (solid line) are shown also.

such absorption on the spectra of two observed AGNs in particular Mkn 501 which shows a spectacular shape of its spectrum.

OBSERVATIONS

During an extraordinary outburst of TeV γ-rays from Mkn 501 in 1997 observation period this object was monitored by several ground-based imaging air Čerenkov telescopes (IACTs) [3]. The HEGRA stereoscopic system of 4 IACTs has observed Mkn 501 for a total exposure time of 110 hours [4]. The unprecedented statistics of about 38,000 TeV photons, combined with the good energy resolution of $\sim 20\%$ allowed determination of a spectrum over the energy range from 500 GeV up to 24 TeV. The shape of the spectrum does not depend on intensity of the source. It justifies the determination of the time-averaged Mkn 501 spectrum. The energy spectrum of Mkn 501, as measured by the HEGRA group, shows apparent curvature over entire energy range. The shape of the spectrum may be well described by the power law with an exponential cutoff. A fit of the data gives:

$$dN/dE = 10.8 \cdot 10^{-11} E^{-1.92} \exp\left[-E/6.2\right], \, \text{cm}^{-2}\text{s}^{-1}\text{TeV}^{-1} \qquad (1)$$

The detailed systematic analysis of the fit parameters has been discussed in [4]. The HEGRA data are also shown in Figure 1.

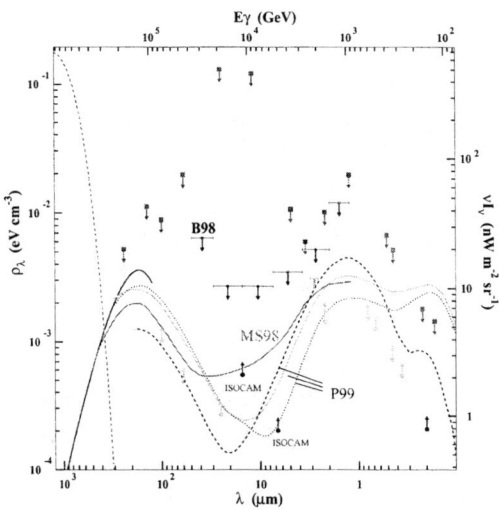

FIGURE 2. Compilation of data and models for a spectral energy distribution of the diffuse intergalactic background taken from [1] (adapted). The results of calculations using the model [14] for high IR photon field are shown by solid curve and denoted by MS99. Calculations from [15] for different model parameters are shown by dotted, dashed, thin solid curves and denoted by P99.

DISCUSSION ON SPECTRUM SHAPE

The curvature in Mkn 501 energy spectrum may be caused by several reasons. The curved energy spectrum of TeV γ-rays may be attributed to (i) the intrinsic spectrum of TeV γ-ray emission within the synchrotron self-Compton or external inverse Compton scenarios; (ii) the curvature might be due to the absorption of TeV γ-rays by the pair production inside the source, or (iii) in intergalactic medium; finally, the observed energy spectrum may be affected by a combination of several reasons noticed above.

The recent calculations based on the synchrotron self-Compton (SSC) and external Compton (EC) models could explain rather well currently established variability time scales of X-ray and TeV emission of Mkn 421 and Mkn 501 (see e.g., [7]). Thus the observation of the variability of TeV γ-ray flux at the level of $\simeq 1$ hr [8], at least, limits the Doppler boosting factor of the emitting jet as $\delta \geq 10$. For such big Doppler boosting factor the γ-ray absorption within the sources does not play an important role and γ-ray photons can easily escape from the emitting region [9].

The shape of Mkn 501 energy spectrum as measured by the HEGRA collaboration can not be easily fitted by pure SSC and EC models (see e.g., [9]). In particular, the shape of a spectrum appears to be very steep above 5 TeV. In addition, the observations with Rossi X-Ray Timing Explorer have shown that the spectrum of

Mkn 501 varies strongly with generally a very hard spectral index extending to much higher energies ($\geq$ 100 keV) [10]. On the contrary, the TeV γ-ray spectrum does not show any variations in the spectrum shape [4]. The simultaneous variations in the X-ray and TeV γ-ray fluxes may be well described by a change in the maximum energy to which electrons can be accelerated γ_{max} [11] (hereafter γ_{max} is a corresponding maximum Lorentz factor). Thus the energy spectrum of TeV γ-rays can be extremely soft, e.g., $\alpha \geq 3.0$ (α is an index of a power law energy spectrum), due to the cutoff in the spectrum of emitting electrons. However for that the variations in γ_{max} lead to significant change of the spectrum slope in TeV γ-rays which is not a case for the HEGRA observations. It is more likely that synchrotron photons of approximately 1-20 keV, emitted by the electrons accelerated within the jet, are up scattered by the same electrons to the TeV energies. For such scenario the X-ray variabilities caused by the hardening of the initial electron spectrum not necessarily lead to a variations of the spectrum slope in TeV γ-rays which is relatively flat $\alpha \simeq 2.0$ and remains constant. Interestingly, the TeV energy γ-ray spectrum as measured by HEGRA shows very similar spectrum slope in the energy range below 5 TeV whereas it deviates from the power law strongly in the high energy part. Such behavior might be easily explained by the effect of intergalactic absorption.

IR ABSORPTION IN TEV SPECTRUM OF MKN 501

While propagating in the interstellar medium the TeV γ-rays may attenuate through pair production process in the intergalactic infrared radiation field (IIRF) [12]. The corresponding opacity of intergalactic medium is determined by the spectral energy distribution (SED) of the IR photon field (see Figure 1). The absorption of γ-rays in the energy range from 0.5 to 20 TeV rely on IR SED of photon field between 1 to 50 mkm. Recently measurements as well as low upper limits of SED strongly constrain the shape of SED in the range relevant to the TeV γ-rays. Compilation of present data is shown in Figure 1. We also show two models of SED from [13] and [14]. Note that the recent tentative detection of IR photons at 3.5 mkm by COBE [15] is consistent with both models whereas the ISOCAM lower limit on IR photon field, if true, favour model from [13] with rather flat SED at mid IR region. An optical depth of γ-ray absorption as a function of energy and redshift, $\tau = \tau(E_\gamma, z)$, was calculated in [16] using the predictions on SED of intergalactic IR photon field according to [13]. As such these data may be used to unfold the Mkn 501 energy spectrum measured by the HEGRA group, $(dN_\gamma/dE)_m$, in order to get a "de-absorbed" intrinsic energy spectrum of Mkn 501, $(dN_\gamma/dE)_i$.

$$(dN_\gamma/dE)_i = (dN_\gamma/dE)_m \cdot e^{\tau(E,z)} \qquad (2)$$

The "de-absorbed" HEGRA data are shown in Figure 1 together with a power law fit. We find [17] that the data points can be well fitted by

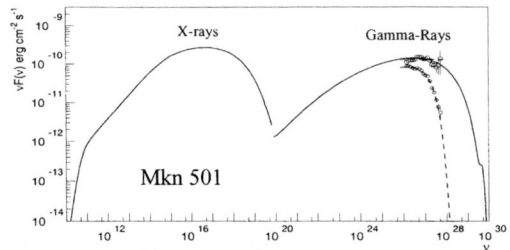

FIGURE 3. Spectral energy distribution of Mkn 501 computed using the homogeneous model [11]. Calculations have been done assuming the variability time scale of $t_{var} = 10^4$s, maximum Lorentz factor of the emitting electrons: $\gamma_{max} = 1.4 \cdot 10^6$, magnetic field: $B = 0.7 \cdot 10^{-3}$G, Doppler boosting factor: $\delta = 80$. The HEGRA data are from [4].

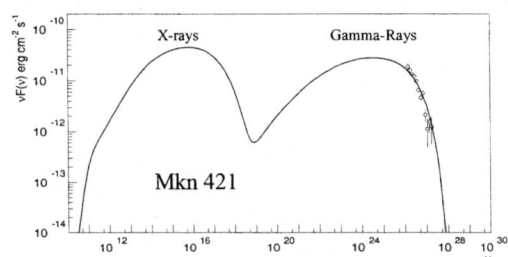

FIGURE 4. Spectral energy distribution of Mkn 421 calculated in [11] ($t_{var} = 10^3$s, $\gamma_{max} = 2 \cdot 10^5$, $B = 2.5 \cdot 10^{-3}$G, $\delta = 47$). The HEGRA data are from [5].

$$(dN_\gamma/dE)_i = (1.32 \pm 0.04) \cdot 10^{-10} (E/1\ TeV)^{-2.0 \pm 0.03}. \qquad (3)$$

Note that similar results have been shown at this Workshop by the Telescope Array group using their measurement of the Mkn 501 TeV γ-ray spectrum. We show in Figure 3 the large scale spectral energy distribution of Mkn 501 calculated assuming the absorption.

COMPARISON OF MKN 501 AND MKN 421 SPECTRA

As reported in [18] the spectra of Mkn 421 and Mkn 501, measured by the Whipple group in the high state of emission, show noticeable difference in their spectral shape over the energy range 0.3-10 TeV. The spectrum of Mkn 421 is a power law whereas the spectrum of Mkn 501 is apparently curved. Since two objects Mkn 421 and Mkn 501 have almost the same red shift one may conclude that these two objects have different intrinsic energy spectra of TeV γ-rays [18]. However the Whipple data for the Mkn 421 and Mkn 501 energy spectra at the energies above 1 TeV do not show prominent difference and both could be well fitted by power law. Apparent difference in two spectra is at energies less than 1 TeV, namely in the range where the absorption of TeV γ-rays in the intergalactic IR photon field does strongly affect the spectra. Similar behavior of both spectra in the energy range above 5 TeV does not contradict the effect of absorption at these energies as stated above.

The spectrum of Mkn 421 as measured by HEGRA collaboration in low state shows power law behavior $dN_\gamma/dE \propto E^{-3.1}$ [5]. The HEGRA data allow to extend the spectral measurements only up to 7 TeV. Such steep spectrum most likely can be attributed to the very soft intrinsic source spectrum and may not disprove the effect of absorption of TeV γ-rays (see Figure 3).

CONCLUSION

We propose a possible scenario explaining the spectral shape of the Mkn 501 energy spectrum as measured by HEGRA collaboration. Strong variations of X-ray emission argue in favor of rather flat intrinsic spectrum of TeV γ-rays. We conclude that absorption in the interstellar IR photon field plays an important role and produces an apparent curvature observed in Mkn 501 spectrum. The SSC fit of the spectral shape constrain rather high value of the Doppler boosting factor of a emitting jet, $\delta > 50$. Future multi-wavelength observations as well as detections of other BL Lac objects will help in future understanding of mechanisms of the TeV γ-ray emission and propagation processes.

REFERENCES

1. Catanese, M., Weekes, T.C. *Publ. Astron. Soc. Pac.*, **111**, 1193 (1998).
2. Gaidos, J.A., et al. *Nature*, **383**, 319 (1996).
3. Protheroe, R.J., et al. *Proc. 25th ICRC (Dublin, South Africa)*, ed. M.S. Potgeier, B.C. Raubenheimer, & D.J. van der Walt, **8**, 317 (1997).
4. Aharonian, F., et al. *A & A*, **349**, 11 (1999).
5. Aharonian, F., et al. *A & A*, in press, (1999); astro-ph/9905032
6. Samuelson, F., et al. *ApJ*, **501**, L17 (1998).
7. Ghiselline, G., et al. *MNRAS*, **301**, 451 (1998).
8. Quinn, J., et al. *ApJ*, **518**, 693 (1999).
9. Bednarek, W., Protheroe, R.J. *MNRAS*, in press (1999); astroph/9902050
10. Pian, E., et al. *ApJ*, **492**, L17 (1998).
11. Mastichiadis, A., Kirk, J.G. *A & A.* **320**, 19 (1997).
12. Gould, R.J., Schreder, G.P. *Phys. Rev.*, **155**, 1408 (1967).
13. Malkan, M.A., Stecker, F.W. *ApJ*, **496**, 13 (1998).
14. MacMinn, D., Primack, J.R., *Space Sci. Rev.*, **75**, 413 (1996).
15. Dwek, E., Arendt, R.G. *ApJ*, **508**, L9 (1998).
16. Stecker, F.W., De Jager, O.C. *A & A*, **334**, L85 (1998).
17. Konopelko, A.K., Kirk, J.G., Stecker, F.W., Mastichiadis, A. *ApJ*, **518**, L13 (1999).
18. Krennrich, F., et al. *ApJ*, **511**, 149 (1999).

The probable binary galaxy system MKN 421: kinematics & structure from optical observations

P. W. Gorham[*], L. van Zee[†], S. C. Unwin[*] and C. S. Jacobs[*]

[*] Jet Propulsion Laboratory, California Institute of Technology
4800 Oak Grove Drive, Pasadena, CA, 91109
[†] National Radio Astronomical Observatory
PO Box 0, Socorro, New Mexico

Abstract. We present Hubble Space Telescope (HST) imagery, and ground–based spectroscopy and CCD photometry of the active galaxy Markarian 421 and its companion galaxy 14 arcsec to the ENE. The HST images indicate that the companion is a morphological spiral rather than elliptical as previous ground–based imaging has concluded. The companion has a bright, compact nucleus, appearing unresolved in the HST images. This is suggestive of Seyfert activity, or possibly a highly luminous compact star cluster. We also report the results of high dynamic range long-slit spectroscopy with the slit placed to extend across both galaxies and nuclei. Velocities derived from a number of absorption lines visible in both galaxies indicate that the two systems are probably tidally bound and thus in close physical proximity. Using the measured relative velocities, we derive a lower limit on the MKN 421 mass within the companion orbit ($R \sim 10$ kpc) of 5.9×10^{11} solar masses, and a mass-to-light ratio of ≥ 17. Our spectroscopy also shows for the first time the presence of Hα and [NII] emission lines from the nucleus of MKN 421. We see both broad and narrow line emission, with a velocity dispersion of several thousand km s^{-1} evident in the broad lines. Based on the imagery and broad-line velocity dispersion we find evidence for a black hole with mass of order 10^9 $M_\odot$ at the center of MKN 421.

INTRODUCTION

The host galaxy of MKN 421 has been the subject of several spectroscopic and photometric studies. The first such study (Ulrich 1975) established the redshift (z=0.0308) based on weak stellar absorption lines, and also noted that a nearby galaxy 14 arcsec to the ENE had a similar redshift (z=0.0316), indicating that it was probably physically related, although if the velocity difference were due to the Hubble flow, the distance could be a few Mpc or more. The companion galaxy was classified as a normal elliptical (Hickson et al. 1982). Further work by Ulrich (1978) showed that MKN 421 was the brightest member of a group of 5–7 galaxies of similar redshift spread over a region of sky of order 10 arcmin in radius. The presence of this group increases the likelihood that the companion's proximity to MKN 421 is physical rather than a random alignment.

There is mounting evidence that AGN phenomena appear to be associated with galaxy mergers or encounters (cf. Shlosman, Begelman, and Frank 1990; Hernquist & Mihos 1995). In the case of BL Lac objects, a significant number have been found in the last decade or so to be associated with close companions or groups of nearby galaxies (cf. Falomo 1996; Heidt 1999), although MKN 421 has apparently been overlooked in this regard. Intrigued by the proximity of these two galaxies, we have analyzed previously unpublished HST imagery of the system, and performed Palomar 1.5 m photometric and Hale 5 m long-slit spectroscopic observations aimed at clarifying this association. Our results will show that the companion galaxy contains a Seyfert-like nucleus, and is likely to be tidally interacting with MKN 421. This association does appear to lend weight to the suggestions that galaxy encounters are an important factor in AGN evolution, and that close companions are associated in some way with the BL Lac phenomenon.

HST IMAGERY

MKN 421 was observed with the HST wide field/planetary camera (WF/PC2) on 1997 March 05, using the F702W filter. Five exposures, one of duration 2 s, two of duration 30 s, and two of duration 120 s were made [1].

Figure 1 shows a slightly smoothed grayscale of the summed image, with a logarithmic stretch and quantized levels chosen to show the details of the host galaxy of MKN 421 and MKN 421-5. Several features of MKN 421-5 are evident even from this image. First, its structure is not a simple elliptical. A suggestion of spiral arms is evident, and possible evidence of barlike structure appears at the outer edges of the galaxy. Second, the nucleus of the companion is clearly brightened relative to the galactic bulge, as is shown in the inset frame.

PALOMAR SPECTROSCOPY

Low resolution optical spectra of MKN 421 and its companion were obtained with the Double Spectrograph on the 5m Palomar[2] telescope during the night of 1999 February 19. The long slit (2′) with a 2″ aperture was centered on the companion galaxy and two 600 sec exposures were obtained. The slit was positioned at an angle of 53°, and passed through both the companion galaxy and the nucleus of MKN 421.

As seen in Figure 2a, the nucleus of MKN 421 is dominated by nonthermal emission with very weak absorption features. This new spectrum is similar to

---

[1] The HST observations used here are available as part of the Space Telescope Science Institute public archive, and were made originally as a result of a proposal by C. M. Urry.

[2] These observations at the Palomar Observatory were made as part of a continuing cooperative agreement between Cornell University and the California Institute of Technology.

other observations of the nucleus of MKN 421 (e.g., Marchã et al. 1996), with the exception that [OI], [NII], and Hα emission lines have been detected for the first time. Estimates of the flux of the [NII] line indicates that it is significantly stronger than the Hα feature. Large [NII]/Hα ratios are not uncommon in AGN, however (e.g., Veilleux & Osterbrock 1987). Both the broad and narrow emission lines appear to be associated only with the nucleus of MKN 421.

The instrinsic width of the broad-line emission has a full-width-at-half-maximum (FWHM) of order 80 Å with detectable emission that extends out to nearly twice this value. Interpretation of the width of the broad line emission is complicated by the fact that it is most likely a blend of [NII] and Hα, but even accounting for this the implied velocity dispersion is ~ 5000 km s^{-1} Implications of this are discussed in a separate paper (Gorham et al. 1999).

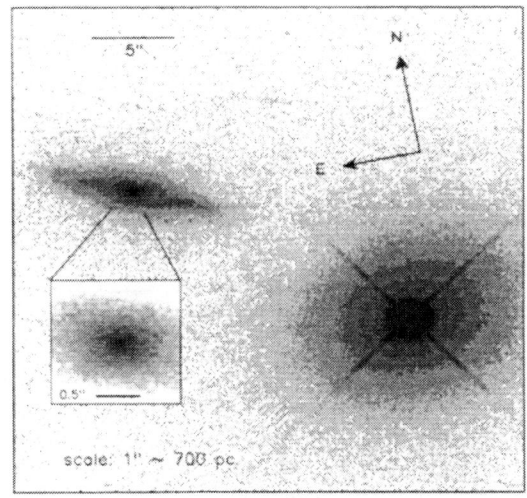

FIGURE 1. HST image of MKN 421 and companion MKN 421-5, through red F702W filter, 300 s total exposure. In the larger image the cores of MKN 421 and the companion are saturated to show detail of the companion disk structure. Levels are quantized to more clearly show the companion structure. The inset shows the nucleus of the companion plotted with a logarithmic stretch, showing the bright, unresolved nucleus. The scale in is approximately 700 pc/arcsec for $h = 0.65$.

Radial velocity gradient

Both MKN 421 and its companion galaxy have high signal–to–noise Na I absorption features which can be used to trace their gas kinematics. The mean velocity as a function of position along the slit for several of the absorption lines is shown in Figure 3. The errors in each individual measurement are large due to the relatively low signal–to–noise ratio and the somewhat coarse spectral resolution of the observations; nonetheless, the two systems are clearly offset in velocity, with a sense of velocity continuity between the two galaxies. [3]

This work was performed at the Jet Propulsion Laboratory, Calif. Inst. of Technology, under contract with NASA. The National Radio Astronomy Observatory is a facility of the National Science Foundation, operated under a cooperative agreement by Associated Universities Inc. This research is based in part on observations made with the NASA/ESA Hubble Space Telescope, obtained from the data archive at the Space Telescope Science Institute. STScI is operated by the Association of Universities for Research in Astronomy, Inc. under NASA contract NAS 5-26555.

---

[3] Further details of the results of this work are available in Gorham et al. (1999), including discussions of the derived masses and mass-to-light ratio of the MKN 421 host, and estimates of the central black hole mass.

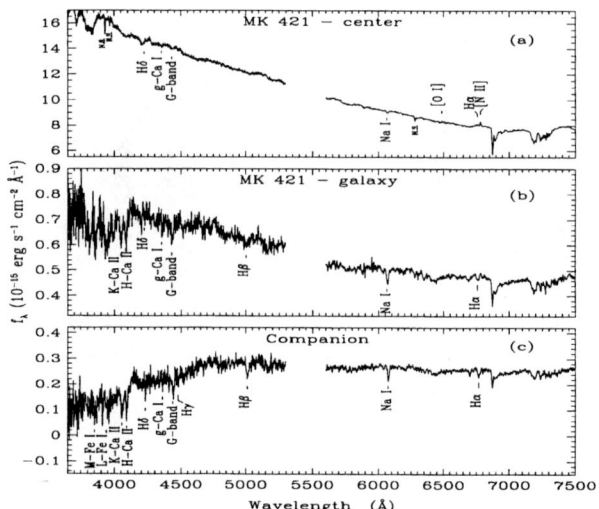

FIGURE 2. Palomar 5 m double spectrograph spectrum of MKN 421 and its companion. The slit was aligned with the two nuclei, and thus covered both galactic bulges as well. (a) MKN 421 shows weak emission at Hα and NII, the first time any emission lines have been noted for this usually almost featureless BL Lac object. (b),(c) MKN 421 and its companion galaxy have a number of well-defined absorption lines but no detectable emission.

REFERENCES

1. Falomo, R., 1996, MNRAS 283, 241.
2. Gorham, P.W., van Zee, L., Unwin, S. C., and Jacobs, C. S., 1999, AJ, in press (see LANL astro-ph 9908077).
3. Heidt, J. 1999, in Takalo L. O., and Sillanpaa, A. (eds.), Proc. BL Lac Phenomena, PASP 159, 367.
4. Hernquist, L., and Mihos, J. C., 1995, ApJ 448, 41.
5. Hickson, P., Fahlman, G. G., Auman, J. R., Walker, G. A. H., Menon, T. K., and Ninkov, Z., 1982, ApJ 258, 53.
6. Marchã, M. J. M., Browne, I. W. A., Impey, C. D., & Smith, P. S. 1996, MNRAS 281, 425
7. Shlosman, I., Begelman, M. C., Frank, J., 1990 Nature 345, 679.
8. Ulrich, M.-H. 1978, ApJ, 222, L3
9. Ulrich, M.-H., Kinman, T. D., Lynds, C. R., Rieke, G. J., & Ekers, R. D. 1975, ApJ, 198, 261
10. Veilleux, S., & Osterbrock, D. E. 1987, ApJ Suppl., 63, 295

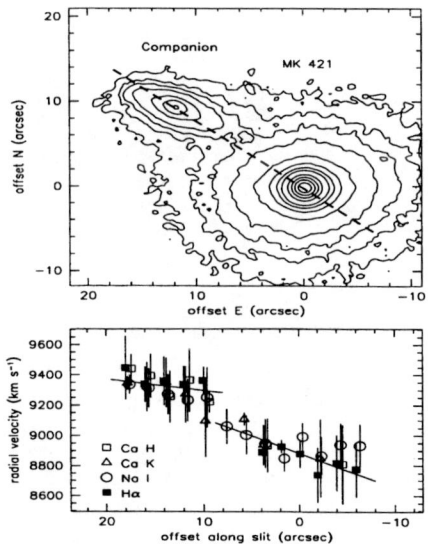

FIGURE 3. (a) Contour plot of Palomar 60" Gunn r-band image of MKN 421 and companion. The heavy dashed line marks the slit position for the absorption line spectroscopy. (b) Plot of line centroids as a function of slit position. The lines show the fitted velocity gradients across each galaxy separately. The velocity gradient of MKN 421 is consistent with a component of Keplerian rotation that is in the same sense as the companion's velocity relative to MKN 421. The two thus appear to be a bound pair.

SUPERNOVA REMNANTS

Modelling Hard Gamma-Ray Emission From Supernova Remnants

Matthew G. Baring[†]

Laboratory for High Energy Astrophysics, Code 661
NASA Goddard Space Flight Center, Greenbelt, MD 20771
baring@lheavx.gsfc.nasa.gov
[†] *Universities Space Research Association*

Abstract. The observation by the CANGAROO experiment of TeV emission from SN 1006, in conjunction with several instances of non-thermal X-ray emission from supernova remnants, has led to inferences of super-TeV electrons in these extended sources. While this is sufficient to propel the theoretical community in their modelling of particle acceleration and associated radiation, the anticipated emergence in the next decade of a number of new experiments probing the TeV and sub-TeV bands provides further substantial motivation for modellers. In particular, the quest for obtaining unambiguous gamma-ray signatures of cosmic ray ion acceleration defines a "Holy Grail" for observers and theorists alike. This review summarizes theoretical developments in the prediction of MeV–TeV gamma-rays from supernova remnants over the last five years, focusing on how global properties of models can impact, and be impacted by, hard gamma-ray observational programs, thereby probing the supernova remnant environment. Properties of central consideration include the maximum energy of accelerated particles, the density of the unshocked interstellar medium, the ambient magnetic field, and the relativistic electron-to-proton ratio. Criteria for determining good candidate remnants for observability in the TeV band are identified.

INTRODUCTION

It is widely believed that supernova remnants (SNRs) are the primary sources of cosmic-ray ions and electrons up to energies of at least $\sim 10^{15}$ eV, where the so-called *knee* in the spectrum marks a deviation from almost pure power-law behavior. Such cosmic rays are presumed to be generated by diffusive (also called first-order Fermi) acceleration at the remnants' forward shocks. These cosmic rays can generate gamma rays via interactions with the ambient interstellar medium, including nuclear interactions between relativistic and cold interstellar ions, by bremsstrahlung of energetic electrons colliding with the ambient gas, and inverse Compton (IC) emission off cosmic background radiation. Rudimentary models of gamma-ray production in remnants involving nuclear interactions date back to the late 1970s [1,2]. These preceded the first tentative associations of two COS-B gamma-ray sources [3] with the remnants γ Cygni and W28. Apart from the work of Dorfi [4], who provided the first model including a more sophisticated study of non-linear effects of shock acceleration to treat gamma-ray production, the study of gamma-ray SNRs remained quietly in the background until the observational program of the EGRET experiment aboard the Compton Gamma Ray Observatory. This provided a large number of unidentified sources above 50 MeV, a handful of which have interesting associations with relatively young SNRs [5].

Following the EGRET advances, the modelling of gamma-ray and other non-thermal emission from supernova remnants "burgeoned," beginning with the paper of Drury, Aharonian, & Völk [6] (hereafter DAV94), who computed the photon spectra expected from the decay of neutral pions generated in collisions of power-law shock-accelerated ions with those of the interstellar medium (ISM). This work spawned a number of subsequent papers that used different approaches, as discussed in the next section, and propelled the TeV gamma-ray astronomy community into a significant observational program given the prediction of substantial TeV fluxes from the DAV94 model. The initial expectations of TeV gamma-ray astronomers were dampened by the lack of success of the Whipple and HEGRA groups [7–9] in detecting emission from SNRs after a concerted campaign. While sectors of the community contended that the constraining TeV upper limits posed difficulties for SNR shock acceleration models, these observational results were naturally explained [10–12] by the maximum particle energies expected (in the 1–50 TeV range) in remnants and the concomitant anti-correlation between maximum energy of gamma-ray emission and the gamma-ray luminosity [13] (discussed below).

The observational breakthrough in this field came with the recent report of a spatially-resolved detection of SN1006 (not accessible by northern hemisphere atmospheric Čerenkov telescopes (ACTs) such as Whipple and HEGRA) by the CANGAROO experiment [14] at energies above 1.7 TeV. The interpretation (actually predicted for SN 1006 by [10,15]) that evolved was that this emission was due to energetic electrons accelerated in the low density environs of this high-latitude remnant, generating flat-spectrum inverse Compton radiation seeded by the cosmic microwave background. This suggestion was influenced, if not motivated by the earlier detection [16] of the steep non-thermal X-ray emission from SN 1006 that has been assumed to be the upper end of a broad synchrotron component, implying the presence of electrons in the 20–100 TeV range. Studies of gamma-ray emission from remnants have adapted to this discovery by suggesting (e.g. [11,13]) that galactic plane remnants such as Cas A that possess denser interstellar surroundings may have acceleration and emission properties distinct from high-latitude sources; the exploration of such a contention may be on the horizon, given the detection of Cas A by HEGRA announced at this meeting [17]. Given the complexity of recent shock acceleration/SNR emission models, the range of spectral possibilities is considerable, and a source of confusion for both theorists and observers. It is the aim of this paper to elucidate the study of gamma-ray remnants by pinpointing the key spectral variations/trends with changes in model parameters, and thereby identify the principal parameters that impact TeV astronomy programs.

MODELS: A BRIEF HISTORY

Reviews of recent models of gamma-ray emission from SNRs can be found in [11,13,18,19]; a brief exposition is given here. Drury, Aharonian, & Völk [6] provided impetus for recent developments when they calculated gamma-ray emission from protons using the time-dependent, two-fluid analysis (thermal ions plus cosmic rays) of [20], following on from the similar work of [4]. They assumed a power-law proton spectrum, so that no self-consistent determination of spectral curvature to the distributions [21-23] or temporal or spatial limits to the maximum energy of acceleration was made. The omission of environmentally-determined high energy cutoffs in their model was a critical driver for the interpretative discussion that ensued. [6] found that during much of Sedov evolution, maximal diffusion length scales are considerably less than a remnant's shock radius.

Gaisser, et al. [24] computed emission from bremsstrahlung, inverse Compton scattering, and pion-decay from proton interactions, but did not consider non-linear shock dynamics or time-dependence and assumed test-particle power-law distributions of protons and electrons with arbitrary e/p ratios. In order to suppress the flat inverse Compton component and thereby accommodate the EGRET observations of γ Cygni and IC443, [24] obtained approximate constraints on the ambient matter density and the primary e/p ratio.

A time-dependent model of gamma-ray emission from SNRs using the Sedov solution for the expansion was presented by Sturner, et al. [12]. They numerically solved equations for electron and proton distributions subject to cooling by inverse Compton scattering, bremsstrahlung, π^0 decay, and synchrotron radiation (to supply a radio flux). Expansion dynamics and non-linear acceleration effects were not treated, and power-law spectra were assumed. Sturner et al. (1997) introduced cutoffs in the distributions of the accelerated particles (following [10,25,26]), which are defined by the limits (discussed below) on the achievable energies in Fermi acceleration. Hence, given suitable model parameters, they were able to accommodate the constraints imposed by Whipple's upper limits [9] to γ Cygni and IC 443.

To date, the two most complete models coupling the time-dependent dynamics of the SNR to cosmic ray acceleration are those of Berezhko & Völk [27], based on the model of [28], and Baring et al. [13]. Berezhko & Völk numerically solve the gas dynamic equations including the cosmic ray pressure and Alfvén wave dissipation, following the evolution of a spherical remnant in a homogeneous medium. Originally only pion decay was considered, though this has now been extended [32] to include other components. Baring et al. simulate the diffusion of particles in the environs of steady-state planar shocks via a well-documented Monte Carlo technique [23,29] that has had considerable success in modelling particle acceleration at the Earth bow shock [30] and interplanetary shocks [31] in the heliosphere. They also solve the gas dynamics numerically, and incorporate the principal effects of time-dependence through constraints imposed by the Sedov solution.

These two refined models possess a number of similarities. Both generate upward spectral curvature (predicted by [21]; see the review in [11]), a signature that is a consequence of the higher energy particles diffusing on larger scales and therefore sampling larger effective compressions, and both obtain overall compression ratios r well above standard test-particle Rankine-Hugoniot values. Yet, there are two major differences between these two approaches. First Berezhko et al. [27,32] include time-dependent details of energy dilution near the maximum particle energy self-consistently, while Baring et al. [13] mimic this property by using the Sedov solution to constrain parametrically the maximum scale of diffusion (defining an escape energy). These two approaches merge in the Sedov phase [33], because particle escape from strong shocks is a fundamental part of the non-linear acceleration process and is determined primarily by energy and momentum conservation, not time-dependence or a particular geometry. Second, [13] injects ions into the non-linear acceleration process automatically from the thermal population, and so determine the dynamical feedback self-consistently, whereas [27] must specify the injection efficiency as a free parameter. Berezhko & Ellison [33] recently demonstrated that, for most cases of interest, the shock dynamics are relatively insensitive to the efficiency of injection, and that there is good agreement between the two approaches when the Monte Carlo simulation [13,29] specifies injection for the model of [27]. This convergence of results from two complimentary methods is reassuring to astronomers, and underpins the expected reliability of emission models to the point that a hybrid "simple model" has been developed [34] to describe the essential acceleration features of both techniques. This has been extended to a new and comprehensive parameter survey [35] of broad-band SNR emission that provides results that form the basis of much of the discussion below.

GLOBAL THEORETICAL PREDICTIONS

Since there is considerable agreement between the most developed acceleration/emission models, we are in the comfortable position of being able to identify the salient global properties that should be characteristics of any particular model. Clearly a treatment of non-linear dynamics and associated spectral curvature are an essential ingredient to more accurate predictions of emission fluxes, particularly in the X-ray and gamma-ray bands where large dynamic ranges in particle momenta are sampled, so that discrepancies of factors of a few or more arise when test-particle power-laws are used. Concomitantly, test-particle shock solutions considerably over-estimate [29,34,35] the dissipational heating of the downstream plasma in high Mach number shocks, thereby introducing errors that propagate into predictions of X-ray emission and substantially influence the overall normalization of hard X-ray to gamma-ray emission (which depends on the plasma temperature [13,35]). These points emphasize that a cohesive treatment of the entire particle distributions is requisite for the accuracy of a given model.

In addition, finite maximum energies of cosmic rays imposed by spatial and temporal acceleration constraints (e.g. [13,36]) must be integral to any model, influencing feedback that modifies the non-linear acceleration problem profoundly.

In SNR evolutionary scenarios, a natural scaling of this maximum energy $E_{\max}$ arises, defined approximately by the energy attained at the onset of the Sedov phase [13,36]:

$$E_{\max} \sim 60 \frac{Q}{\eta} \left(\frac{B_{\rm ISM}}{3\mu{\rm G}}\right) \left(\frac{n_{\rm ISM}}{1\,{\rm cm}^{-3}}\right)^{-1/3} \left(\frac{\mathcal{E}_{\rm SN}}{10^{51}{\rm erg}}\right)^{1/2} \left(\frac{M_{\rm ej}}{M_\odot}\right)^{-1/6} {\rm TeV} \,, \qquad (1)$$

where Q is the particle's charge, η (≥ 1) is the ratio between its scattering mean-free-path and its gyroradius, $\mathcal{E}_{\rm SN}$ is the supernova energy, $M_{\rm ej}$ is its ejecta mass, and other quantities are self-explanatory. At earlier epochs, the maximum energy scales approximately linearly with time, while in the Sedov phase, it slowly asymptotes [13,37] to a value a factor of a few above that in Eq. (1).

Three properties emerge as global signatures of models that impact observational programs. The first is that there is a strong anti-correlation of $E_{\max}$ (and therefore the maximum energy of gamma-ray emission) with gamma-ray luminosity, first highlighted by [13]. High ISM densities are conducive to brighter sources in the EGRET to sub-TeV band [13,27,35], but reduce $E_{\max}$ in Eq. (1) and accordingly act to inhibit detection by ACTs. Low ISM magnetic fields produce a similar trend, raising the gamma-ray flux by flattening the cosmic ray distribution (discussed below). Clearly, high density, low $B_{\rm ISM}$ remnants are the best candidates for producing cosmic rays up to the knee. Fig. 1 displays a sample model spectrum for Cas A, which has a high density, high $B_{\rm ISM}$ environment. In it the various spectral components are evident, and the lower $E_{\max}$ for electrons (relative to that for protons) that is generated by strong cooling is evident in the bremsstrahlung and inverse Compton spectra.

The other two global properties are of a temporal nature. The first is the approximate constancy of the observed gamma-ray flux (and $E_{\max}$ [37]) in time during Sedov phase, an insensitivity first predicted by [4] and confirmed in the analyses of [6,13,37]. The origin of this insensitivity to SNR age $t_{\rm SNR}$ is an approximate compensation between the SNR volume $\mathcal{V}$ that scales as $t_{\rm SNR}^{6/5}$ (radius $\propto t_{\rm SNR}^{2/5}$) in the Sedov phase, and the normalization coefficient $\mathcal{N}$ of the roughly E^{-2} particle distribution function: since the shock speed (and therefore also the square root of the temperature $T_{\rm pd}$) declines as $t_{\rm SNR}^{-3/5}$, it follows that $\mathcal{N} \propto T_{\rm pd} \propto t_{\rm SNR}^{-6/5}$ and flux $\propto \mathcal{N}\mathcal{V} \approx$ const. There is also a limb brightening with age [6] that follows from the constant maximum particle length scale concurrent with continuing expansion.

Key Parameters and Model Behavioural Trends

The principal aim here is to distill the complexity of non-linear acceleration models for time-dependent SNR expansions and discern the key parameters controlling spectral behaviour and simple reasons for behavioural trends. This should elucidate for theorist and experimentalist alike the scientific gains to be made by present and next generation experimental programs. Parameters are grouped according to them

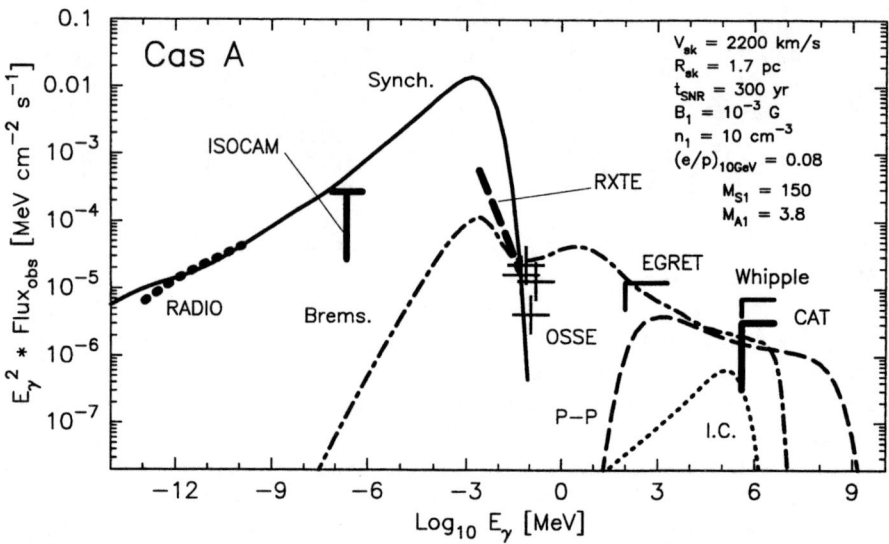

FIGURE 1. The Cassiopeia A spectrum from the Monte Carlo acceleration calculation of Ellison et al. ([38], see this for detailed referencing of the data sources). The model photons come from a single set of proton, helium, and electron spectra calculated with the upstream parameters shown in the figure. A single normalization factor has been applied to all components to match the radio flux. Note that the bremsstrahlung and inverse Compton (IC) emission cuts off at a much lower energy than the pion decay radiation due to the synchrotron losses the electrons experience. In these results, the IC component does not include a synchrotron self-Compton contribution.

being of model origin and environmental nature (trends associated with the age of a remnant were discussed just above), and details can be found in the comprehensive survey of Ellison, Berezhko & Baring [35].

There are three relevant *model* parameters in non-linear acceleration, the ratio of downstream electron and proton temperatures T_{ed}/T_{pd}, the injection efficiency η_{inj} (after [27,28]), and the electron-to-proton ratio $(e/p)_{rel}$ at relativistic energies (i.e. $\gtrsim 1-10$ GeV). The injection efficiency is the most crucial of these, since it controls the pressure contained in non-thermal ions, and therefore the non-linearity of the acceleration process. It mainly impacts the X-ray to soft gamma-ray bremsstrahlung contribution, a component that is generally dominated by pion decay emission in the hard gamma-ray band. The shape and normalization of the π^0 decay gamma-rays is only affected when η_{inj} drops below 10^{-4} and the shock solution becomes close to the test-particle one, i.e. an overall spectral steepening arises. Variations in $(e/p)_{rel}$ influence the strength of the inverse Compton and bremsstrahlung components, which modify the total gamma-ray flux only if $(e/p)_{rel} \gtrsim 0.1$, a high value relative to cosmic ray abundances, or the ambient field is strong.

The most interesting behavioural trends are elicited by the *environmental* parameters n_{ISM} and B_{ISM}, and the results adapted from [35] are illustrated in

Fig. 2. Naively, one expects that the radio-to-X-ray synchrotron and gamma-ray inverse Compton components should scale linearly with density increases, while the bremsstrahlung and pion decay contributions intuitively should be proportional to $n_{\rm ISM}^2$. However, global spectral properties are complicated by the non-linear acceleration mechanism and the evolution of the SNR. As $n_{\rm ISM}$ rises, the expanding supernova sweeps up its ejecta mass sooner, and therefore decelerates on shorter timescales, thereby reducing both the volume $\mathcal{V}$ of a remnant of given age, and lowering the shock speed and the associated downstream ion temperature $T_{\rm pd}$. Hence, the density increase is partially offset by the "shifting" of the particle distributions to lower energies (due to lower $T_{\rm pd}$) so that the normalization $\mathcal{N}$ of the non-thermal distributions at a given energy is a weakly increasing function of $n_{\rm ISM}$. Clearly $\mathcal{V}$ times this normalization controls the observed flux of the synchrotron and inverse Compton components, while the product of $\mathcal{N}$, the target density $n_{\rm ISM}$ and $\mathcal{V}$ determines the bremsstrahlung and $\pi^0 \to \gamma\gamma$ emission, with results shown in Fig. 2. Observe that the approximate constancy of the inverse Compton contribution effectively provides a lower bound to the gamma-ray flux in the 1 GeV–1 TeV band, a property that is of significant import in defining experimental goals.

The principal property in Fig. 2 pertaining to variations in $B_{\rm ISM}$ is the anticorrelation between radio and TeV fluxes: the higher the value of $B_{\rm ISM}$, the brighter the radio synchrotron, but the fainter the hard gamma-ray pion emission. This property is dictated largely by the influence of the field on the shock dynamics and total compression ratio r: the higher the value of $B_{\rm ISM}$, the more the field contributes to the overall pressure, reducing the Alfvénic Mach number and accordingly r, as the flow becomes less compressible. This weakening of the shock steepens the particle distributions and the overall photon spectrum. An immediate offshoot of this behaviour is the premise [35] that radio-selected SNRs may not provide the best targets for TeV observational programs. Case in point: Cas A is a very bright radio source while SN 1006 is not, and the latter was observed first.

Since $n_{\rm ISM}$ and $B_{\rm ISM}$ principally determine the gamma-ray spectral shape, flux normalization and whether or not the gamma-ray signatures indicating the presence of cosmic ray ions are apparent, they are the most salient parameters to current and future ACT programs and the GLAST experiment.

KEY ISSUES AND EXPERIMENTAL POTENTIAL

There are a handful of quickly-identifiable key issues that define goals for future experiments, and these can be broken down into two categories: spatial and spectral. First and foremost, the astronomy community needs to know whether the EGRET band gamma-ray emission is actually shell-related. While the associations of [5] were enticing, subsequent research [39–41] has suggested that perhaps compact objects like pulsars and plerions or concentrated regions of dense molecular material may be responsible for the EGRET unidentified sources. If a connection to the shell is eventually established, it is desirable to know if it is localized only

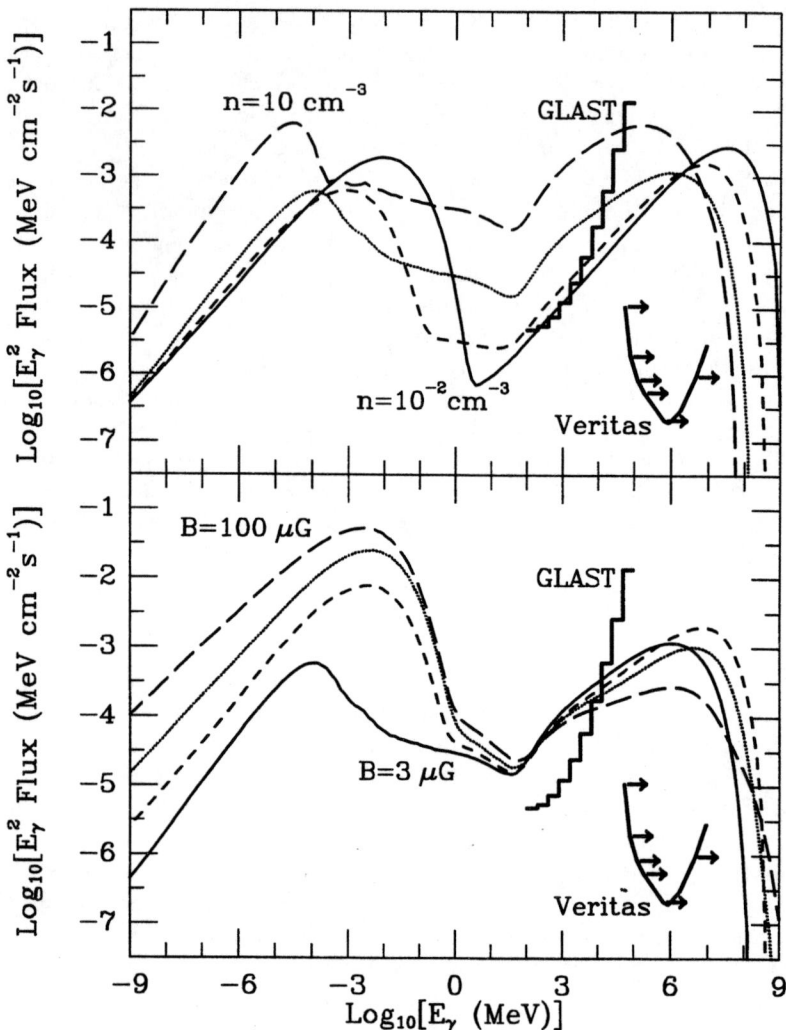

FIGURE 2. Trends of total photon emission for variations of ISM parameters $n \equiv n_{\rm ISM}$ and $B \equiv B_{\rm ISM}$, adapted from the simplified approximate description of non-linear acceleration in [35]. Top panel: the ISM field is fixed at $B = 3\mu G$, and the ambient number density is varied such that: $n = 0.01$ cm^{-3} (solid), $n = 0.1$ cm^{-3} (short dashes), $n = 1$ cm^{-3} (small dots), and $n = 10$ cm^{-3} (long dashes). Bottom panel: B is varied: $B = 3\mu G$ (solid), $B = 10\mu G$ (short dashes), $B = 30\mu G$ (small dots), and $B = 100\mu G$ (long dashes), with the density pinned to $n \doteq 1$ cm^{-3}. Here $(e/p)_{\rm rel} = 0.03$; consult [35] for other model parameters. Also depicted are the canonical integral flux sensitivity for Veritas [42] and the differential flux sensitivity for GLAST (Digel, private communication) to facilitate the discussion in the text.

to portions of the shell. One naturally expects that shock obliquity effects [29] can play an important role in determining the gamma-ray flux in "clean" systems like SN 1006, and that turbulent substructure within the remnant (e.g. Cas A) can complicate the picture dramatically. Such clumping issues impact radio/gamma-ray flux determinations, since the radio-emitting electrons diffuse on shorter length scales and therefore are more prone to trapping. Another contention that needs observational verification is whether or not limb-brightening increases with SNR age? Improvements in the angular resolution of ACTs can resolve these issues and discern variations in gamma-ray luminosities across SNR shells: the typical capability of planned experiments such as HESS, Veritas, MAGIC and CANGAROO-III is of the order of 2–3 arcminutes in the TeV band [42,43].

The principal gamma-ray spectral issue is whether or not there is evidence of cosmic ray *ions* near remnant shocks. The goal in answering this is obviously the detection of $\sim 70\,\mathrm{MeV}$ π^0 bump, the unambiguous signature of cosmic ray ions, and given the GLAST *differential* sensitivity (the measure of capability in performing spectral diagnostics as opposed to detection above a given energy) plotted in Fig. 2, GLAST will be sensitive to remnants with $n_{\mathrm{ISM}} \gtrsim 0.1$ cm^{-3}. Atmospheric Čerenkov experiments can also make progress on this issue, with the dominant component in the super-TeV band for moderately or highly magnetized remnant environs being that of pion decay emission (see Fig. 1). Such a circumstance may already be realized in the recent marginal detection [17] of Cas A by HEGRA. The most powerful diagnostic the sensitive TeV experiments will provide is the determination of the maximum energy (see Fig. 2) of emission (and therefore also that of cosmic ray ions or electrons), thereby constraining n_{ISM}, B_{ISM} and the e/p ratio. Furthermore, the next generation of ACTs should be able to discern the expected anti-correlation between E_{max} and γ-ray flux, and with the help of GLAST, search for spectral concavity, a principal signature of non-linear acceleration theory. In view of the anticipated increase in the number of TeV SNRs, a population classification may be possible, determining whether or not SN1006 and other out-of-the-plane remnants differ intrinsically in their gamma-ray and cosmic ray production from the Cas A-type SNRs. These potential probes augur well for exciting times in the next 5–10 years in the field of TeV gamma-ray astronomy.

Acknowledgments: I thank my collaborators Don Ellison, Steve Reynolds, Isabelle Grenier, Frank Jones and Philippe Goret for many insightful discussions, Seth Digel for providing results of simulations of GLAST spectral capabilities, and Rod Lessard for supplying the Veritas integral flux sensitivity data for Fig. 2.

REFERENCES

1. Higdon, J. C. & Lingenfelter, R. E. *Ap. J. Lett.* **198**, L17 (1975).
2. Chevalier, R. A. *Ap. J.* **213**, 52 (1977).
3. Pollock, A. M. T. *Astron. Astr.* **150**, 339 (1985).
4. Dorfi, E. A. *Astron. Astr.* **251**, 597 (1991).

5. Esposito, J. A., Hunter, S. D., Kanbach, G. & Sreekumar, P. *Ap. J.* **461**, 820 (1996).
6. Drury, L. O'C., Aharonian, F. A., & Völk, H. J. *Astron. Astr.* **287**, 959 (1994).
7. Lessard, R. W., et al. *Proc. 24th ICRC (Rome)* **2**, 475 (1995).
8. Prosch, C., et al. *Astron. Astr.* **314**, 275 (1996).
9. Buckley, J. H. et al. *Astron. Astr.* **329**, 639 (1997).
10. Mastichiadis, A., & de Jager, O. C. *Astron. Astr.* **311**, L5 (1996).
11. Baring, M. G. 1997, in *Very High Energy Phenomena in the Universe*, ed. Trân Thanh Vân, J., et al. (Éditions Frontières, Paris), p. 97 & p. 107.
12. Sturner, S. J., Skibo, J. G., Dermer, C. D., & Mattox, J. R. *Ap. J.* **490**, 619 (1997).
13. Baring, M. G., Ellison, D. C., Reynolds, S. P., Grenier, I. A., & Goret, P. *Ap. J.* **513**, 311 (1999).
14. Tanimori, T., et al. *Ap. J. Lett.* **497**, L25 (1998).
15. Pohl, M. *Astron. Astr.* **307**, L57 (1996).
16. Koyama, K. et al. *Nature* **378**, 255 (1995).
17. Völk, H. J., et al. (1999, these proceedings).
18. de Jager, O. C. & Baring, M. G. *Proc. 4th Compton Symposium*, ed. Dermer, C. D. & Kurfess, J. D. (AIP Conf. Proc. 410, New York), p. 171 (1997).
19. Völk, H. J. in *Towards a Major Atmospheric Čerenkov Detector*, ed. O. C. de Jager (Wesprint, Pochefstroom) p. 87 (1998).
20. Drury, L. O'C., Markiewicz, W. J. & Völk, H. J. *Astron. Astr.* **225**, 179 (1989).
21. Eichler, D. *Ap. J.* **277**, 429 (1984).
22. Ellison, D. C., & Eichler, D. *Ap. J.* **286**, 691 (1984).
23. Jones, F. C. & Ellison, D. C. *Space Sci. Rev.* **58**, 259 (1991).
24. Gaisser, T. K., Protheroe, R. J., & Stanev, T. *Ap. J.* **492**, 219 (1998).
25. Reynolds, S. P. *Ap. J. Lett.* **459**, L13 (1996).
26. de Jager, O. C., & Mastichiadis, A. *Ap. J.* **482**, 874 (1997).
27. Berezhko, E. G., & Völk, H. J. *Astroparticle Phys.* **7**, 183 (1997).
28. Berezhko, E. G., Yelshin, V., & Ksenofontov, L. *Sov. Phys. JETP* **82**, 1 (1996).
29. Ellison, D. C., Baring, M. G. & Jones, F. C. *Ap. J.* **473**, 1029 (1996).
30. Ellison, D. C., Möbius, E., & Paschmann, G. *Ap. J.* **352**, 376 (1990).
31. Baring, M. G., et al., *Ap. J.* **476**, 889 (1997).
32. Berezhko, E. G., Ksenofontov, L., & Petukhov, S. I. *Proc. 26th ICRC (Salt Lake City)*, **4**, 431 (1999).
33. Ellison, D. C. & Berezhko, E. G. *Proc. 26th ICRC (Salt Lake City)*, **4**, 446 (1999).
34. Berezhko, E. G. & Ellison, D. C. *Ap. J.* **526**, 385 (1999).
35. Ellison, D. C., Berezhko, E. G. & Baring, M. G. *Ap. J.* submitted (2000).
36. Berezhko, E. G. *Astroparticle Phys.* **5**, 367 (1996).
37. Berezhko, E. G., & Völk, H. J. *Proc. 26th ICRC (Salt Lake City)*, **4**, 377 (1999).
38. Ellison, D. C., et al. *Proc. 26th ICRC (Salt Lake City)*, **3**, 468 (1999).
39. Brazier, K. T. S., et al., *MNRAS* **281**, 1033 (1996).
40. Keohane, J. W., et al., *Ap. J.* **484**, 350 (1997).
41. Brazier, K. T. S., et al., *MNRAS* **295**, 819 (1998).
42. Weekes, T. C., et al., VERITAS proposal (1999).
43. Kohnle, A. et al. *Proc. 26th ICRC (Salt Lake City)*, **5**, 239 (1999).

Environmental and Age Limits on Particle Acceleration in Supernova Remnants

L O'C Drury*, J Kirk[†] and P Duffy[‡]

*Dublin Institute for Advanced Studies,
5 Merrion Square,
Dublin 2,
Ireland.
[†]Max-Planck-Institut für Kernphysik,
D69026 Heidelberg,
Germany.
[‡]Department of Mathematical Physics,
University College Dublin,
Dublin 4,
Ireland.

Abstract. Ion-neutral damping of resonantly excited Alfvén waves can be the critical factor determining the maximum energy to which particles are accelerated by diffusive shock acceleration if the upstream medium is only partially ionized. The implications of this for acceleration in Supernova remnants is discussed and it is shown that old remnants are unlikely to be observable TeV gamma-ray sources.

INTRODUCTION

In general the medium into which an astrophysical shock propagates is only partially ionized. In this case the resonantly excited Alfvén waves, which are required to provide the strong scattering needed for the diffusive shock acceleration to work rapidly, are damped by ion-neutral friction. The detailed analysis [1] shows that there are two cases.

If the excitation is sufficiently strong the damping can be overcome on all scales and does not, by itself, affect the accelerated particle spectrum. "Sufficiently strong" means that, modulo factors of order unity,

$$\left(\frac{U}{10^3 \text{ km s}^{-1}}\right)^3 > 8 \times 10^{-3} \left(\frac{n}{1 \text{ cm}^{-3}}\right)^{-1} \frac{1-x}{x^2} \left(\frac{B}{1 \,\mu\text{G}}\right)^2$$

where U is the shock speed, n the total upstream density, x the ionisation fraction and B the magnetic field strength. If, however, the above condition is not satisfied, the wave damping leads to a cut-off in the accelerated particle spectrum at an energy (for protons) of

$$E_{\max} \approx \left(\frac{U}{10^3 \text{ km s}^{-1}}\right)^3 \left(\frac{n}{1 \text{ cm}^{-3}}\right)^{-0.5} \frac{x^{0.5}}{1-x} \text{ TeV}.$$

In the strong excitation case, it has been pointed out by Tagger et al [2] that the ponderomotive force working on the neutrals may produce an interesting non-linear instability. Essentially the neutrals are expelled from regions of higher ionization and wave activity and the medium may spontaneously separate into ionized and neutral "filaments". The importance of this effect for cosmic ray acceleration in supernovae has not been examined in detail.

SUPERNOVA REMNANT SHOCKS

For practical application to SNRs it is more convenient to use the remnant age t rather than the shock speed U. During the approximately self-similar Sedov-like phase of the remnant evolution and assuming a canonical mechanical explosion energy of 10^{51} ergs

$$\frac{U}{10^3 \text{ km s}^{-1}} \approx 9 \left(\frac{n}{1 \text{ cm}^{-3}}\right)^{-0.2} \left(\frac{t}{10^2 \text{ yr}}\right)^{-0.6}.$$

This relation holds for $t \gg t_{\text{SW}}$ where the sweep-up time t_{SW} is that at which the mass of ambient material swept-up by the blast wave equals that ejected in the explosion (a few solar masses). At earlier times most of the explosion energy is in the form of kinetic energy of the expanding ejecta and is not available for particle acceleration. For canonical values

$$t_{\text{SW}} \approx 70 \left(\frac{n}{1 \text{ cm}^{-3}}\right)^{-1/3} \text{ yr}.$$

Thus, for remnants in the Sedov phase of their evolution, the condition that wave-damping be unimportant is that

$$\left(\frac{n}{1 \text{ cm}^{-3}}\right)^{2/5} > 10^{-5} \frac{1-x}{x^2} \left(\frac{B}{1 \text{ }\mu\text{G}}\right)^2 \cdot \left(\frac{t}{10^2 \text{ yr}}\right)^{9/5}$$

and if this condition is not satisfied the maximum particle energy is

$$E_{\max} \approx 700 \left(\frac{n}{1 \text{ cm}^{-3}}\right)^{-11/10} \left(\frac{t}{10^2 \text{ yr}}\right)^{-9/5} \frac{x^{1/2}}{1-x} \text{ TeV}.$$

OBSERVABILITY OF REMNANTS

What are the implications of these results for supernova remnants, and in particular their observability as gamma-ray sources at TeV energies? To produce TeV gamma-rays through hadronic processes protons will have to be accelerated to at least 10 TeV. By coincidence this is also the energy that electrons require to produce TeV gamma-rays by inverse Compton scattering of the CMB. Assuming the ionization fraction $x = 0.1$, the figure below indicates where, as a function of ambient density and remnant age, the acceleration is restricted to energies below about 10 TeV and where, in consequence, the remnants are not expected to be visible as TeV gamma-ray sources.

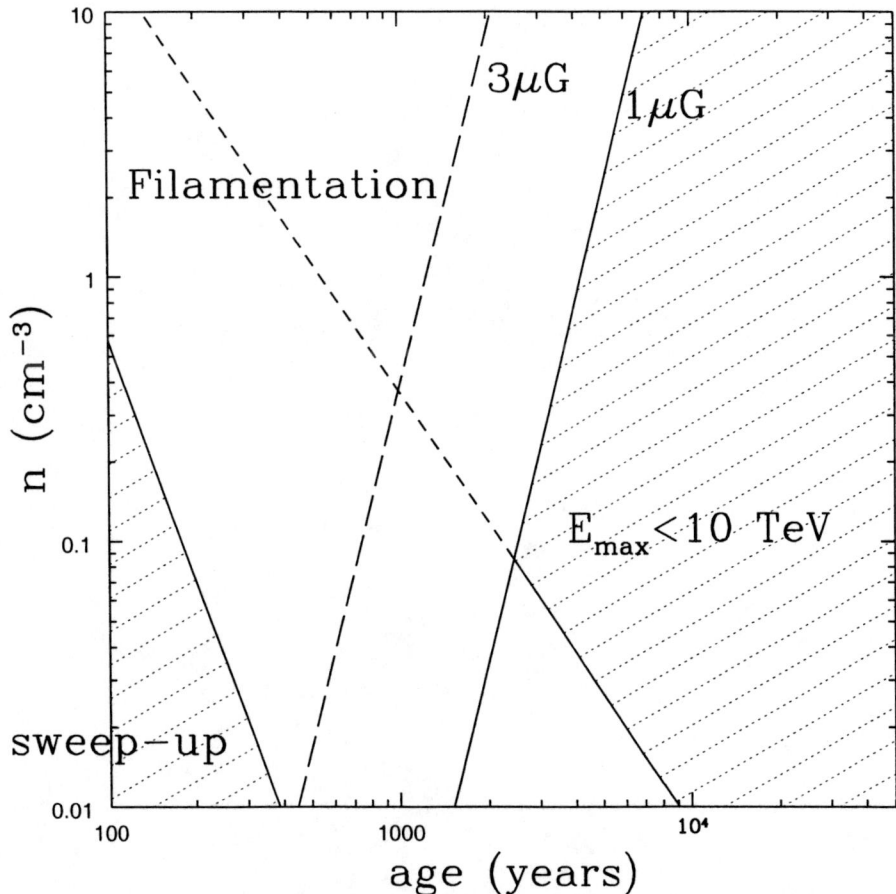

The shaded region on the left hand side corresponds to the very young remnants which are still in the free-expansion phase of their evolution and in consequence are not expected to be strong sources. More interesting is the region to the right where for remnants older than a few thousand years the shocks have slowed to the point that wave damping becomes important even at quite low ambient densities. What is particularly interesting is that at ambient densities of $1\,\mathrm{cm}^{-3}$ and above and for modest magnetic field strengths only young remnants are expected to be TeV sources.

ACKNOWLEDGMENTS

This work was supported by the TMR programme of the European Communities under contract FMRX-CT98-0168.

REFERENCES

1. Drury, L O'C, Kirk, J, and Duffy, P *Astron. Astrophys.* **309** 1002-1010 (1996).
2. Tagger, M, Falgarone, E and Shukurov A *Astron. Astrophys.* **299** 940-946 (1995).

The SNR W28 at TeV Energies

G.P. Rowell[1], T. Naito[2], S.A. Dazeley[3], P.G. Edwards[4], S. Gunji[5],
T. Hara[2], J. Holder[1], A. Kawachi[1], T. Kifune[1], Y. Matsubara[8],
Y. Mizumoto[9], M. Mori[1], H. Muraishi[10], Y. Muraki[8], K. Nishijima[7],
S. Ogio[6], J.R. Patterson[3], M.D. Roberts[1], T. Sako[8],
K. Sakurazawa[6], R. Susukita[11], T. Tamura[12], T. Tanimori[6],
G.J. Thornton[3], S. Yanagita[10], T. Yoshida[10] and T. Yoshikoshi[1]

[1] *Institute for Cosmic Ray Research, University of Tokyo, Tokyo 188-8502, Japan*
[2] *Fac. of Management Information, Yamanashi Gakuin University, Yamanashi 400-8575, Japan*
[3] *Dept. of Physics and Math. Physics, University of Adelaide 5005, Australia*
[4] *Institute of Space and Astronautical Science, Kanagawa 229-8510, Japan*
[5] *Dept. of Physics, Yamagata University, Yamagata 990-8560, Japan*
[6] *Dept. of Physics, Tokyo Institute of Technology, Tokyo 152-8551, Japan*
[7] *Dept. of Physics, Tokai University, Kanagawa 259-1292, Japan*
[8] *Solar-Terrestrial Environment Lab., Nagoya University, Aichi 464-8601, Japan*
[9] *National Astronomical Observatory of Japan, Tokyo 181-8588, Japan*
[10] *Faculty of Science, Ibaraki University, Ibaraki 310-8512, Japan*
[11] *Institute of Physical and Chemical Research, Saitama 351-0198, Japan*
[12] *Faculty of Engineering, Kanagawa University, Kanagawa 221-8686, Japan*

Abstract. The southern supernova remnant (SNR) W28 was observed in 1994 and 1995 by the CANGAROO 3.8m telescope in a search for multi-TeV gamma ray emission, using the Čerenkov imaging technique. We obtained upper limits for a variety of point-like and extended features within a $\pm 1°$ region and briefly discuss these results, together with that of EGRET within the framework of a shock acceleration model of the W28 SNR.

INTRODUCTION

W28 is a composite SNR (mixed or M-type) with centrally filled X-ray and optical emission and limb brightened or shell-like radio emission [10,6]. It lies at a distance of about 1.8 kpc (from Σ-D, although kinematic arguments place a higher figure of 4 kpc), has an age of between $3.5\text{-}15 \times 10^4$ yrs, and evolution consistent with the radiative or Sedov phases. The radio shell ($\sim 1°$ diameter) is dominated by the northern half and over 40 maser emission (1720 MHz) sites have been identified indicating strong interaction with a molecular cloud [2]. The

ROSAT X-ray emission is well explained by a thermal model, but recent ASCA data hint at non-thermal emission in the southwest region [13]. A flat spectrum (integral index −0.9) unidentified EGRET source, 3EG J1800-2338 (0.32° 95% error circle radius), [4] is centred on the southern radio edge. W28 and the EGRET source are a strong example of an EGRET source/SNR association [11]. The radio pulsar PSR J1801-23 at the northern SNR edge is not thought to be associated with W28 given the difference in distances of this and the SNR [5].

SNR are thought primarily responsible for the acceleration of galactic cosmic-rays (CR) and W28 is a good southern hemisphere example of such a site. Gamma-ray emission can be produced from one, or a combination of hadronic ($p+p \rightarrow \pi^\circ \rightarrow 2\gamma$) and electronic (inverse Compton boosting of ambient soft photons and bremsstrahlung) processes extending up to TeV energies. The emission at TeV energies (and non-thermal X-ray synchrotron emission) is therefore a tracer of CR acceleration.

DATA ANALYSIS AND RESULTS

We used the 3.8 metre telescope of CANGAROO [3] in a search for TeV gamma-ray emission from the W28 region over two observation seasons (1994 and 1995). The imaging camera on this telescope has a field of view $\sim 3°$ on a side and we have used an analysis that maintains a roughly constant gamma-ray selection power for sources located within a $\sim 1° \times 1°$ area [12] of the telescope tracking position. ON source data were complemented by a set of OFF source data (tracking position displaced in right ascension only) for background comparison. Following removal of data under the influence of weather and instrumental effects, a total (for 1994 and 1995) of 57.5 hours ON and 53.5 hours OFF source data were accepted for analysis.

The image cuts on data are based on a combination of the Hillas image orientation, location and size parameters (see [9] for a technique summary). W28, if a TeV emitter, may contain both point-like and extended features, requiring a detailed study of the off-axis performance of the CANGAROO 3.8m camera. Simulations of the telescope/camera combination reveal a decreasing gamma-ray selection efficiency of the cuts for off-axis point sources, due to camera-edge effects. It is possible to maintain an improved gamma-ray cut efficiency over the camera using a combination of cuts that are dependent on the location of the assumed source. One of these cuts, D, characterises the distance between the assumed and reconstructed source position for each event:

$$D = \sqrt{\left(\frac{miss}{\sigma_{miss}}\right)^2 + \left(\frac{dis - dis_{ex}}{\sigma_{dis}}\right)^2} \qquad (1)$$

where the expected dis of an image, $dis_{ex} = 1.25(1.0 - \frac{width}{length})$, is dependent on the image elongation. The standard deviations are given by $\sigma_{dis} = 0.21 + 0.09d$, ie.

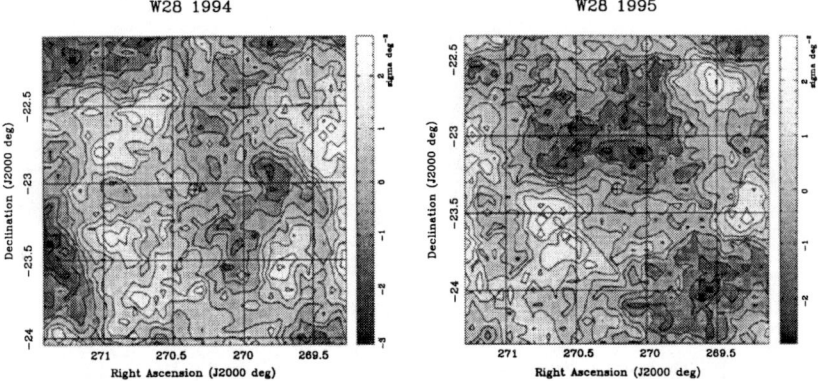

FIGURE 1. Skymaps of point source statistical significance (normalised) for the ON–OFF source excess over a ±1° field after application of all cuts (table 1). The tracking positions for each year's data differs by ~ 0.34°.

dependent on source distance from the camera centre d, and $\sigma_{miss} = 0.09$ for all d. A *length* cut dependent upon source position is also applied. These cuts are all selected *a priori* using Monte Carlo simulations of the CANGAROO 3.8 metre telescope and camera, and provide a quality factor of ~ 4 at ~40% gamma-ray cut efficiency for any point source located within a ±1° area. A full description of this analysis is given by [12].

A number of sites (point and extended, listed in table 1) within the W28 region were searched for TeV gamma-ray emission. The tracking position of 1994 data was PSR J1801−23 and that for 1995 was a radio position labelled 'A83', defined by [1]. Skymaps (normalised ON−OFF excess in sigma for a point source as a function of assumed source position) are presented in figure 1, and reveal no statistically significant excesses. Upper limits to the TeV gamma-ray flux at the 3σ level (listed in table 1) were calculated. For the extended source examples, the events satisfying the cuts at source positions within a radius of interest were summed for ON and OFF source data. A slightly different gamma-ray trigger efficiency was used for each year's data to account for a lower event rate in 1995 compared to 1994, and all results were normalised to a 1.5 TeV energy threshold (representing the energy at the half-maximum of the distribution of triggered energies).

MODEL COMPARISON

We make a comparison of the EGRET results (spectrum from [7]) and our upper limit for an extended source centred on A83 from 1994 data (the lowest of our extended source upper limits) with a model of the TeV gamma-ray flux due to the decay of neutral pions [8] in figure 2. The model flux will scale according to

TABLE 1. Summary of the 3σ flux upper limits from several sites/features within the W28 region.

Feature	Flux($\geq$ 1.5 TeV) ph cm^{-2} s^{-1}	
	1994 Data	1995 Data
Radio position A83[a]	$< 3.36 \times 10^{-12}$	$< 2.95 \times 10^{-12}$
Radio position A83[b]	$< 8.75 \times 10^{-12}$	$< 6.67 \times 10^{-12}$
PSR J1801$-$23[c]	$< 3.20 \times 10^{-12}$	$< 3.32 \times 10^{-12}$
Masers (E&F)[d]	$< 4.14 \times 10^{-12}$	$< 3.47 \times 10^{-12}$
3EGJ1800$-$2338[e]	$< 8.82 \times 10^{-12}$	$< 1.18 \times 10^{-11}$

a: Point source at radio position A83, defined by [1].
b: Extended source of radius 0.35° centred on A83.
c: Point source at pulsar position [5].
d: Point source at average position of the two strongest maser sites E and F [2].
e: Highest pointlike significance within EGRET 95% error circle (0.32°) [4].

$F_\gamma \propto \frac{E_t(10^{51}\text{erg})\ n(\text{cm}^{-3})}{d^2(\text{kpc}^2)}$, where values of the total energy available for CR production, $E_t = 0.01 \to 0.1$, the distance to the remnant, $d = 1.8 \to 4.0$ kpc, and the density of ambient matter, $n = 1.3$ cm^{-3}, are published ranges. The results of scaling the model flux according to these range of values are defined by the hashed region in figure 2. Results when assuming a much higher matter density of $n = 20$ cm^{-3} in combination with favourable values of E_t and d are indicated by the thick dot-dashed line. A proton injection spectrum of -2.1 (differential) and cutoff energy 10^{14}eV has been used in the model, ie. consistent with shock acceleration. When assuming a high value of n, the model flux is able to meet the EGRET data without violating our upper limit.

DISCUSSION

A search for TeV gamma ray emission from the W28 region was carried out on data taken in 1994 and 1995 with the CANGAROO 3.8m telescope. No evidence for point-like or extended TeV γ-ray emission from a number of sites in the W28 region was found. We compare the lowest of our extended source upper limits to the predicted TeV gamma ray emission from π° decay. From figure 2, the EGRET flux may be accounted by π° decay gamma-rays alone, if we assume a high ambient matter density ($n \sim 20$ cm^{-3}). Such a density is possible for W28, given the presense of a nearby molecular cloud and maser emission. However, an accelerated electron component from bremsstrahlung and inverse Compton scattering may also contribute. Further studies of results at X-ray energies (for e.g. [13]), may hint at the level of such components. A more detailed investigation of model parameters including spectral cutoffs is underway.

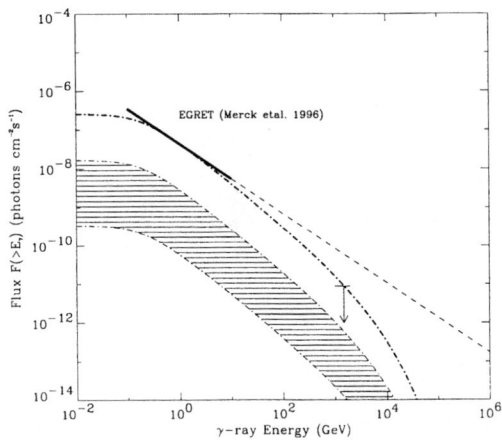

FIGURE 2. Comparison of our upper limit (extended source at A83 for 1994 data) and the EGRET flux of 3EG J1800−2338 [7] with a model predicting the TeV gamma-ray flux due to π° decay (hashed area and single dot-dashed line, [8]). See text for details.

ACKNOWLEDGEMENTS

This work is supported by a Grant-in-Aid in Scientific Research from the Japanese Ministry of Science, Sports and Culture, and also by the Australian Research Council. GR acknowledges the receipt of a JSPS postdoctoral fellowship.

REFERENCES

1. Andrews M.D. et al. 1983, *ApJ*, **266**, 684
2. Claussen M.J., Frail D.A., and Goss W.M. 1997, *ApJ* **489**, 143
3. Hara T. et al. 1993, *Nucl. Inst. Meth.*, **300**, A332
4. Hartman R.C. et al. 1999, *ApJ (Supp)*, **123**, 79
5. Kaspi A.G. et al. 1993, *ApJ*, **409**, L57
6. Long K.S., Blair W.P., White R.L. et al. 1991 *ApJ*, **373**, 567
7. Merck M. et al. 1996 *A & A*, **120**, 465
8. Naito T. and Takahara F. 1994 *J Phys G: Nucl Part. Phys.*, **20**, 477
9. Ong R.A. 1998 *Physics Reports*, **305**, 93
10. Rho J.H., Petre R., Pisarski R, and Jones L.R. 1996, *MPE Report*, **263**, 273
11. Romero G.E., Benaglia P. and Torres D.F., *A & A*, **348**, 868
12. Rowell G.P. et al. 1999, in preparation
13. Tomida H. et al. 1998, *Genshikaku Kenkyu*, **42**, 123, in japanese

Recent Results from the CANGAROO 3.8 m Telescope

Takanori Yoshikoshi*[1] for CANGAROO[2]

*Institute for Cosmic Ray Research, University of Tokyo, Tokyo 188-8501, Japan

Abstract. We observed the shell-type supernova remnant (SNR) RX J1713.7–3946 and the X-ray binaries Cen X-3 and Vela X-1 at TeV energies in 1998 using the CANGAROO 3.8 m telescope. Preliminary results of data analyses on these objects are summarized here. We also estimated a spectrum of the Vela pulsar region using 116 hr data taken by the 3.8 m telescope during 1994 and 1997. The two data sets having different mirror reflectivities give consistent results and the photon index of the best-fit power-law spectrum is -2.4 ± 0.2.

INTRODUCTION

The 3.8 m telescope of CANGAROO [1] started to observe southern TeV gamma-ray sources in 1992 from Woomera, South Australia and has detected gamma-ray signals from three pulsar nebulae (Crab [2], Vela [3], PSR 1706–44 [4]) and a SNR (SN 1006 [5]) so far. These results give direct evidence of particle acceleration up to at least multi-TeV energies in our galaxy and activated discussions on non-thermal phenomena in such objects. The important role of the 3.8 m telescope will be taken over by the new 7 m telescope of the CANGAROO II project [6–8], which was completed in March 1999. The 3.8 m telescope has not been in operation since October 1998 to concentrate to the construction of the 7 m telescope. In 1998, before stopping the 3.8 m telescope, we observed two important objects: RX J1713.7–3946 which is a very similar SNR to SN 1006, and Cen X-3 from which a 400 GeV gamma-ray signal was recently detected by the Durham group [9]. Preliminary results on these objects are summarized in the next section. We also present a new result on the spectrum of the Vela pulsar region obtained using all available data taken by the 3.8 m telescope so far as well as the details of the analysis.

---

[1] Present address: Department of Physics, Osaka City University, Osaka 558-8585, Japan; Email: tyoshiko@alpha.sci.osaka-cu.ac.jp
[2] URL: http://icrhp9.icrr.u-tokyo.ac.jp/

TABLE 1. Summary of new results from the CANGAROO 3.8 m telescope.

Object	T_{obs} (hr)	Flux ($\times 10^{-12}$ cm^{-2} s^{-1})
SNR		
RX J1713.7−3946	42	∼ 3 (> 2 TeV, preliminary)
W 28	58	< 8.8 (> 1.5 TeV^{a})
X-ray binary		
Cen X-3	17	< 5.2 (> 2 TeV, preliminary)
Vela X-1	12	< 5.5 (> 2 TeV, preliminary)
Pulsar nebula		
Vela pulsar region	116	2.6 $(E/2\ \mathrm{TeV})^{-2.4}$ TeV^{-1}

^{a} A different definition of the energy threshold is used (see [13]).

NEW RESULTS FROM THE 3.8 M TELESCOPE

In the recent CANGAROO observations, we have given the highest priority to RX J1713.7−3946, which is a shell-type SNR recently discovered by *ROSAT*. Subsequent observations of this SNR by *ASCA* revealed non-thermal hard X-ray emission from its northeastern rim, which is even brighter than that from SN 1006 [10]. We could accumulate about 42 hr on-source data on RX J1713.7−3946 and a preliminary analysis of the data shows a 5σ excess of the on-source events over the background [11]. The gamma-ray flux derived is shown in Table 1.

Another SNR, W 28, has also been thought to be a possible TeV gamma-ray emitter since it includes pulsars and an unidentified EGRET source. However, we have detected no significant signal from W 28 so far in spite of the long exposure time [12]. Rowell et al. recently reanalyzed all of our data on W 28 taking into account the positions of the possible sources in the field of view [13].

Following the detection of an unpulsed signal from Cen X-3 by the Durham group, we observed the X-ray binaries Cen X-3 and Vela X-1 in February and March 1998. Our preliminary analyses show no evidence of TeV emission from either source. The calculated 3σ upper limits to the TeV fluxes (> 2 TeV) are shown in Table 1. The upper limit for Cen X-3 does not conflict with the Durham flux if we assume a power-law model and the photon index is smaller than −2.0.

The CANGAROO result on the Vela pulsar region using the data taken during 1993 and 1995 has already been reported by Yoshikoshi et al. [3]. An unpulsed TeV gamma-ray signal from the Vela pulsar region was found at the 5.8σ level, with the TeV source being offset from the Vela pulsar to the southeast by about $0°.13$. We observed the same field in 1997 with the recoated 3.8 m mirror (the reflectivity was improved from about 45% to 75% on average) and again found a significant gamma-ray signal at a consistent position. The statistical significance of the total signal increased to 6.8σ and using these data we have estimated the spectrum of this source. Table 2 summarizes the data used in this analysis. The 1993 data were not used here since they are more contaminated by electronic noise. The image analysis procedure is almost the same as that described in [3] but the

TABLE 2. Summary of the data set of the Vela pulsar region used to estimate the spectrum.

Year	T_{ON} (hr)	N_{ON}	T_{OFF} (hr)	N_{OFF}	Reflectivity (%)
1994	41.9	194,795	41.9	162,029	45
1995	44.9	183,262	44.9	169,425	45
----	----	---- Mirror recoating ----	----	----	----
1997	28.8	291,567	23.3	189,792	75

gamma-ray selection criteria based on the image parameters were slightly changed so as to avoid the overcut of gamma-ray events in the higher energy region. The 1994 to 1995 data and the 1997 data were separately analyzed owing to the different reflectivities.

If we observe a gamma-ray source having an intrinsic source spectrum $f(E)$, the observed gamma-ray rate $r(E)$ is represented by the following convolution function:

$$r(E) = \int_0^\infty f(E')A(E')P(E;E')dE', \quad (1)$$

where $A(E)$ and $P(E;E')$ are the effective area including the efficiency of the gamma-ray selections and the response function of the detector including the atmosphere, respectively. The functions $A(E)$ and $P(E;E')$ were estimated using Monte Carlo simulations. The equation (1) has to be deconvolved to obtain $f(E)$. However, it is not easy if the functions being convolved are not simple. Instead of doing a deconvolution, we directly fit the function (1) to the observed rate histogram using numerical techniques and taking the binning effect into account. Figure 1 shows the results of power-law fits for the 1994 to 1995 and 1997 data sets. These two spectra are consistent with each other within their error regions. Fitting a power-law function to the both data together between 1 TeV and 100 TeV gives a best-fit spectrum of $f(E) = (2.6 \pm 0.6) \times 10^{-12}(E/2 \text{ TeV})^{-2.4\pm0.2}$ photons cm^{-2} s^{-1} TeV^{-1} with $\chi^2 = 7.2$ for 8 degrees of freedom, where the errors are statistical only.

As a method to estimate the systematic error of the spectrum, we tried to reconstruct a rate spectrum of cosmic-ray protons using the same technique. The result is shown in Figure 2. The rate spectrum calculated using a proton spectrum directly measured by a balloon experiment [14] is slightly higher than the observed rate since the observed rate also includes the heavier components of cosmic rays. However, the contribution of the heavier components to the total rate is smaller than that of protons [15]. Therefore, even if we conservatively understand this result, the systematic error of the spectrum must be smaller than a factor of 2.

The Durham group recently reported an upper limit to the unpulsed emission from the Vela pulsar with the threshold energy of 300 GeV [16]. If we compare it with our spectrum although their result is not for the position of our TeV source, the upper limit is very close to the extrapolation of our spectrum, but does not conflict taking the error region into account.

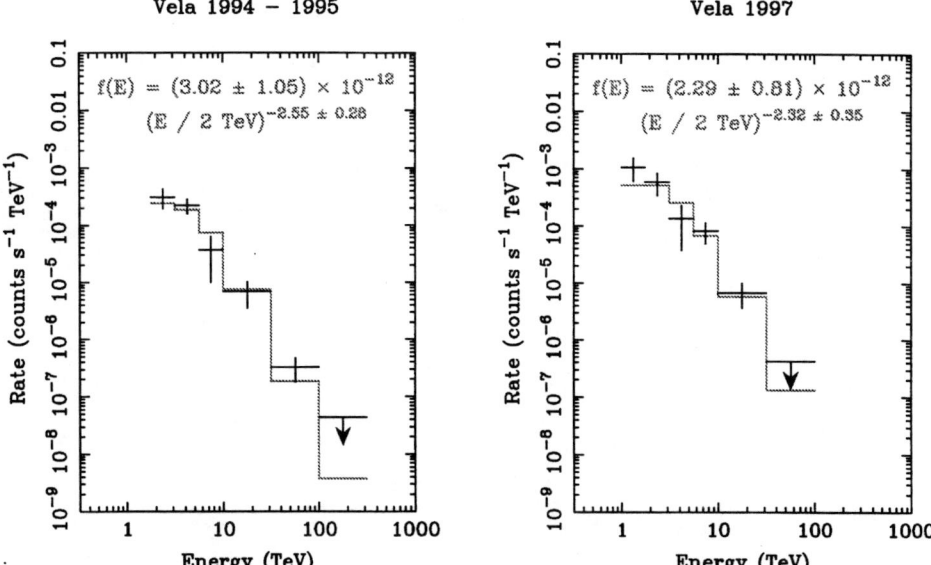

FIGURE 1. Rate spectra of the Vela pulsar region with the best-fit power-law spectra. The left and right graphs are for the 1994 to 1995 data and for the 1997 data, respectively.

CONCLUSIONS

We observed RX J1713.7–3946 and the X-ray binaries Cen X-3 and Vela X-1 in 1998 using the 3.8 m telescope. The preliminary analysis shows some evidence of TeV emission from RX J1713.7–3946. On the other hand, no signal was found from either of the X-ray binaries. We also estimated a spectrum of the Vela pulsar region using the 1994 to 1997 data taken by the 3.8 m telescope. The best-fit spectrum has the form of $(2.6 \pm 0.6) \times 10^{-12}(E/2 \text{ TeV})^{-2.4\pm0.2}$ photons cm^{-2} s^{-1} TeV^{-1} and the systematic error of the spectrum is smaller than a factor of 2.

ACKNOWLEDGMENTS

This work is supported by International Scientific Research Program of a Grant-in-Aid in Scientific Research of the Ministry of Education, Science, Sports and Culture, Japan, and by the Australian Research Council. The receipt of a JSPS Research Fellowship is also acknowledged by T.Y.

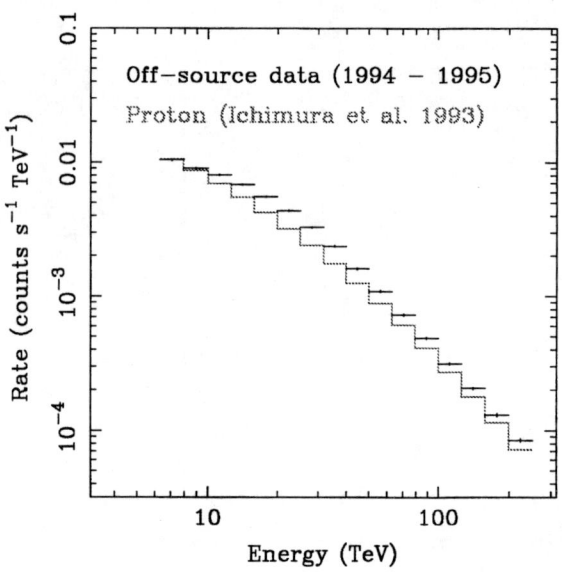

FIGURE 2. Rate spectra of cosmic-ray protons. The lower spectrum was calculated using a directly measured primary proton spectrum [14] and the upper spectrum is the observed off-source data, which includes not only protons but also helium and other components of cosmic rays.

REFERENCES

1. Hara, T., et al., *Nucl. Instrum. Methods Phys. Res. A* **332**, 300 (1993).
2. Tanimori, T., et al., *ApJ* **429**, L61 (1994).
3. Yoshikoshi, T., et al., *ApJ* **487**, L65 (1997).
4. Kifune, T., et al., *ApJ* **438**, L91 (1995).
5. Tanimori, T., et al., *ApJ* **497**, L25 (1998).
6. Tanimori, T., et al., *Proc. 26th Int. Cosmic-Ray Conf. (SLC)* **5**, pp. 203 (1999).
7. Mori, M., et al., *Proc. 26th Int. Cosmic-Ray Conf. (SLC)* **5**, pp. 287 (1999).
8. Kawachi, A., et al., *Proc. 26th Int. Cosmic-Ray Conf. (SLC)* **5**, pp. 207 (1999).
9. Chadwick, P.M., et al., *ApJ* **503**, 391 (1998).
10. Koyama, K., et al., *Publ. Astron. Soc. Japan* **49**, L7 (1997).
11. Muraishi, H., et al., *Proc. 26th Int. Cosmic-Ray Conf. (SLC)* **3**, pp. 500 (1999).
12. Mori, M., et al., *Proc. 24th Int. Cosmic-Ray Conf. (Rome)* **2**, pp. 487 (1995).
13. Rowell, G.P., et al., *this workshop*.
14. Ichimura, M., et al., *Phys. Rev. D* **48**, 5, 1949 (1993).
15. Aharonian, et al., *Phys. Rev. D* **59**, 9, 2003 (1999).
16. Chadwick, P.M., et al., *Proc. 26th Int. Cosmic-Ray Conf. (SLC)* **3**, pp. 504 (1999).

HEGRA Observations of Galactic Sources

Heinz Völk*, HEGRA Collaboration

*Max-Planck-Institut für Kernphysik
P.O. Box 103980, 69029 Heidelberg, Germany

Abstract. In this talk I will first give a summary of the observations of expected Galactic TeV γ-ray sources with the HEGRA CT-Sytem since the Kruger Park Workshop in 1997. Then I will go into some detail regarding the observations of Supernova Remnants (SNRs), especially those of Tycho's SNR and of Cas A. The emphasis will not be on all aspects of these published data. I will rather review the selection of these observational targets, and discuss some of the physical implications of the results.

SUMMARY OF OBSERVATIONS

The stereoscopic system of imaging atmospheric Cherenkov telescopes (IACTs) of HEGRA has been running since late 1996 with four telescopes. After a fire in the array which also damaged one of these telescopes late in the year 1997, the final configuration of five equal telescopes with identical cameras has become operational in August 1998. Apart from the IACT system, HEGRA successfully operates a stand-alone telescope, called CT1; it is also doing obervations during moon periods. However in this review, I will be concerned with the stereoscopic system alone.

Since 1997 a number of Galactic source candidates has been observed with the IACT system. The Galactic coordinates of the objects discussed in this paper are indicated in Figure 1. The objects analyzed are given in Table 1.

1. The observations of the Crab Nebula were done both at normal (ZA ≤ 30°) and at high zenith angles (ZA ~ 60°). They led to a (combined) energy spectrum up to 20 TeV (Konopelko et al. 1999). It is within the errors compatible with an extension of the power law spectrum inferred from measurements in the TeV energy range (Konopelko et al., these Proceedings). Thus no possible hard hadronic emission component has been identified up to these energies.

2. A search for a periodic signal from the Crab and Geminga Pulsars was also performed (Aharonian et al. 1999a). No evidence for pulsed emission was found. Even though we had expected Geminga to be a major contributor to the distribution of very energetic CR electrons in the neighborhood the Solar System, it showed up

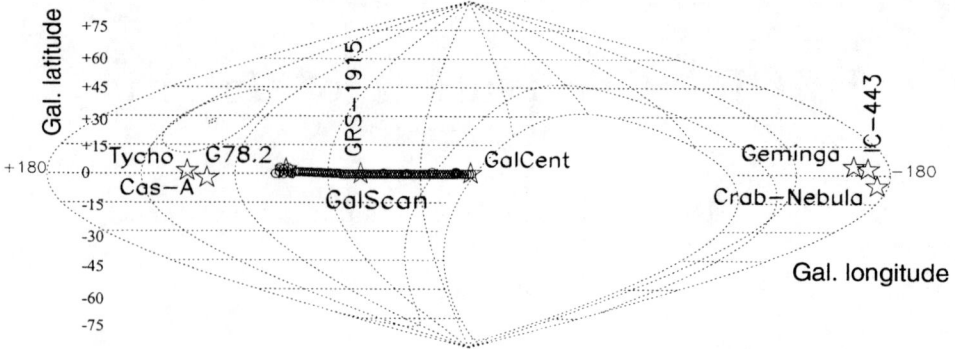

FIGURE 1. Part of the sky in Galactic coordinates. Shown are the positions of the objects discussed in this paper. The white part is not visible from La Palma; the band surrounding that region can only be observed under restricted conditions.

as a TeV-quiet object.

TABLE 1. Galactic objects observed 1997-1999 with the HEGRA IACT-System. The numbers given are the observation hours in the respective years; for the analysis of the Galactic Cosmic-Ray (CR) proton spectrum, CR background (bgr) events were used

Source	[1997]	[1998]	[1999]	References
Crab	92	138	31	Konopelko et al. 1999, Aharonian et al. 1999a
Geminga	–	23	–	Aharonian et al. 1999a
GRS 1915	50	12	11	Kettler 1999
Cas A	102	85	–	Pühlhofer et al. 1999a
Tycho	23	35	–	Pühlhofer et al. 1999a
Gal. Plane	111	66	–	Pühlhofer et al. 1999b
Diff. VHE	–	–	53	Lampeitl et al. 1999
CR Protons	bgr	bgr	bgr	Aharonian et al. 1999b

3. The observations of the Galactic Microquasar GRS 1915 have been analysed (Kettler 1999). No signals have been found during these observation periods. This is not too surprising since, unfortunately, the source had also been low in other wavelength ranges in those times.

4. The two SNRs Tycho and Cas A have been observed extensively with the stereoscopic system. This is especially true for Cas A, where a deep observation of 128 hr duration has been included in the present analysis (Pühlhofer et al. 1999a). The two objects will be discussed in some detail in section 2.

5. An extensive Galactic Plane scan (≥ 2 hrs of observation time for each point, plus some re-observations) in the TeV band covered the Galactic longitude region

from the Galactic Center $(l, -1.5°)$ to the Cygnus region $(l, 83.5°)$, see Figure 2 .

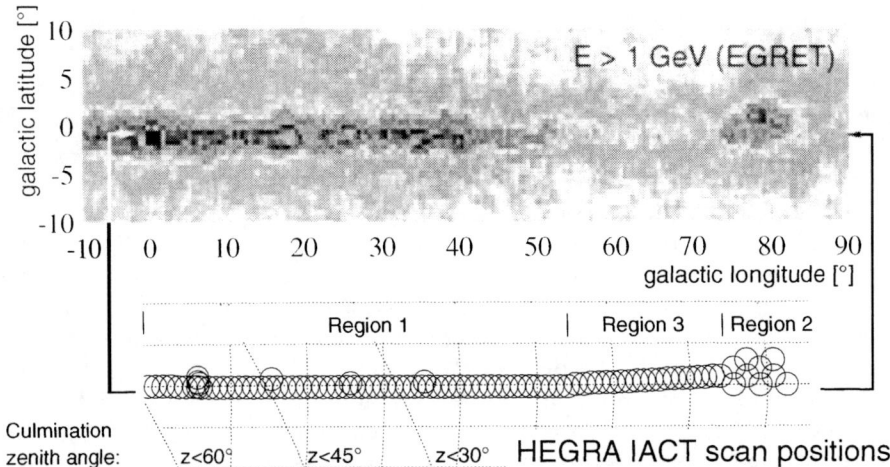

FIGURE 2. The HEGRA IACT scan positions in the Galactic Plane. For comparison also the corresponding part of the EGRET sky map for γ-ray energies $E > 1$ GeV is shown. Regions 1 and 2 were observed in 1997, region 3 in 1998.

Sources with a flux above 1/4 Crab units should have been detected, as indicated by Table 2 below. A first analysis reveals no hints for such strong TeV point sources (Pühlhofer et al. 1999b).

6. In a similar vein, a program regarding the search for diffuse VHE γ-ray emission from the Galactic Plane was started. The present analysis is largely of a technical nature (Lampeitl et al. 1999). The observations will be continued during this summer of 1999.

7. Finally, the imaging Cherenkov technique was applied for the first time to the determination of the flux and the TeV energy spectrum of the charged CR protons. For this work background events for the Mkn 501 observations from 1997 were used. Calibration is exclusively by Monte Carlo simulations that include a detailed detector simulation. For physical reasons, the proton detection rate strongly exceeds that for heavier CR nuclei near threshold, around 1.5 TeV. The stereoscopic detection of the air showers permits the effective suppression of air showers induced by heavier particles already at the trigger level, and in addition by software analysis cuts. The results are in good agreement with the recent results of satellite and balloone-borne experiments and reach similar accuracy (Aharonian et al. 1999b). Without any knowledge of the CR composition, the proton spectrum can only be determined with precision near threshold. However, it should be possible to obtain in addition an approximate CR composition, using further specific image cuts (Plyasheshnikov et al. 1998). This will allow an extension of the dynamical range of the spectrum into an energy region that is very costly to cover by direct

detection CR experiments. I think it would be important if in addition the large Zenith angle technique could be applied to this problem.

Of course, more than these objects have been observed in the Galaxy. However the data have not been analyzed yet, and are therefore not a subject of this summary.

Let me conclude this section with a general consideration.

As mentioned above, the HEGRA Galactic Plane scan has not yet led to the detection of new sources. In his excellent introductory review, Trevor Weekes (these Proceedings) described this result as "depressing".

We were also disappointed. On the other hand, the result is perhaps not too surprising, given the low sensitivity level with which this survey had to be done. The result should also prompt a new discussion about the aims and possibilities of ground-based γ-ray astronomy with imaging telescopes. Space is scarce in these proceedings. So, I will summarize my arguments only briefly in four points, and hope that they open a broader debate: (i) we should of course continue such surveys; any field of astronomy must do this (ii) however it is not too probable that we will find new sources that have not been seen as unusual objects in another wavelength range already, considering the enormous investments in ever more powerful instruments in the radio, infrared, optical, and X-ray domains that have been made over the last two decades (iii) thus, our main activity should perhaps be to look at sources also known in other wavelength-ranges; only then we can hope to obtain a physical understanding of the γ-ray results (iv) given the much higher physical complexity of the acceleration and transport processes for the nonthermal component than for the thermal component, the potential for discovery is one for strong nonthermal activity in known objects, and it is as important as the potential for discovery of previously unknown objects in the more conventional "thermal" astronomy.

I do not believe that the γ-ray bursts provide a counter argument to this point of view: they are explosive events in previously inconspicuous objects, and could not have been found in a survey with a narrow-FoV instrument like IACTs; an all-sky capability was needed to discover them, and they were difficult to understand for decades before they were detected also in other wavelength ranges. Also Geminga, originally an enigmatic Cos B source, is not really a counterexample, because Geminga could only be physically identified after many years, when ROSAT discovered that it was a long-period Pulsar and determined its period, which was subsequently confirmed by EGRET.

SUPERNOVA REMNANTS

Observations

Earlier observations of the SNRs G87.2+2.1 (γ Cygni) and IC 443 in 1996/97 gave consistent upper limits between Whipple (Buckley et al. 1998) and the HEGRA

CT-System (Heß 1998) at effective threshold energies, for the Zenith angles involved, of $E_\gamma > 300$ GeV and $E_\gamma > 800$ GeV, respectively. They were slightly above theoretical predictions regarding the π^0-decay γ-ray emission for a uniform ISM but well within astronomical uncertainties (Völk 1997). Both objects had originally been assumed to interact with interstellar clouds. Under ideal assumptions such an interaction could have increased the π^0-decay γ-ray luminosity significantly. These two SNRs are presumably the result of core collapse Supernovae, due to massive ($M > 8\,M_\odot$) progenitor stars. If they have masses exceeding roughly 15 $M_\odot$, these stars have stellar winds which significantly modify the circumstellar environment. For such "Wind-SNe" the time history of the π^0-decay γ-ray emission is much more complex: except within the wind zone, it is much lower than for a uniform ISM of the same density (Berezhko & Völk 1995, Berezhko & Völk 1997).

The recent HEGRA observations concern deep observations of Tycho's SNR, believed to be a SN Ia in a uniform ISM with strong X-ray lines and no or only a very weak nonthermal X-ray continuum, and of Cas A, assumed to be a SN Ib resulting from the core collapse of a very massive Wolf-Rayet star, with a strong nonthermal X-ray continuum - an archetypical Wind-SN (Pühlhofer et al. 1999a). Both SNRs are very young in an evolutionary sense, presumably still in the sweep-up phase, even though this might only be marginally true for Tycho's SNR.

The long observation times which we have reserved for these objects imply a change in our ideology: the emphasis is no more on SNR shocks that presumably interact with interstellar clouds, but on very young objects, either in a supposedly uniform ISM or in a strongly modified precursor wind structure.

Data analysis for Cas A and Tycho

The following data analysis for Tycho and Cas A has been done by G. Pühlhofer. For the angular resolution of the HEGRA CT-system of 0.05 to 0.1° Tycho is a γ-ray point source, and Cas A is marginally extended. Due to the available Zenith angles the instrument threshold is at 1 TeV or slightly above (see Table 2). The complete sensitivities of the IACT system are described in the article by M. Panter (these Proceedings). The observation times and significances are given in Table 2.

With ~ 38 hrs of observation no signal has yet been found from Tycho, whereas the full data sample of ~ 128 hrs for Cas A shows evidence for a signal above 1 TeV. For Cas A the event statistics as a function of (distance)^2 is shown in the left panel of Figure 3, both for a point source assumption (I), and for a slightly extended source (II). The position of the γ-ray source on the sky is given in the right panel, and is consistent with the radio astronomical position that corresponds to the center of the picture.

Figure 4 shows a model calculation for the energy spectrum of Cas A (Atoyan et al. 1999b). The full and the dashed lines correspond to the expected inverse Compton (IC) emission, as derived phenomenologically from the observed synchrotron

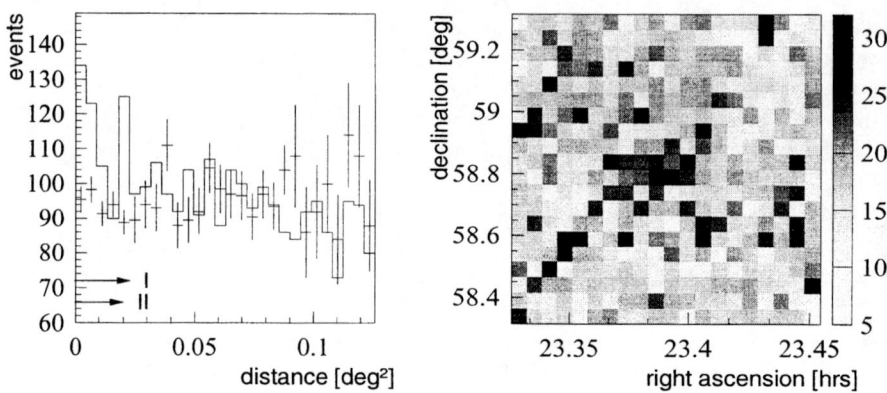

FIGURE 3. Event statistics and position determination of the γ-ray source Cas A. See text for details.

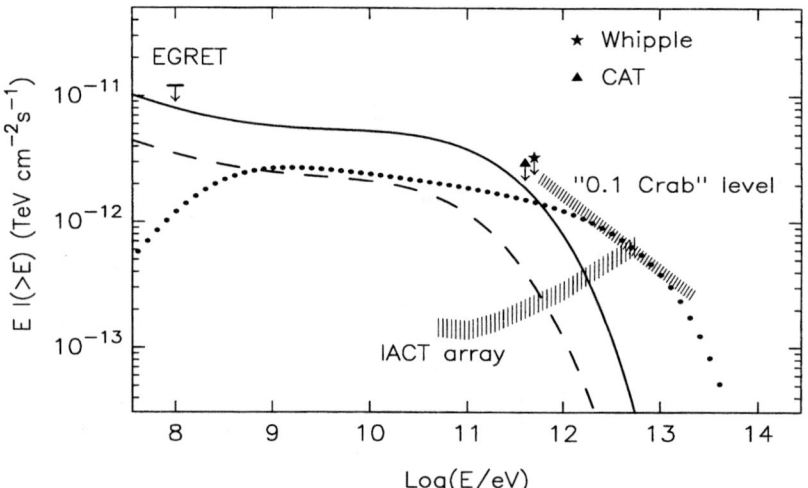

FIGURE 4. Energy flux spectrum of Cas A in γ-rays. The full and the dotted curve correspond to the inferred IC emission, for a magnetic field strength B_1 in the bright compact radio knots, given by $B_1 = 1$mG and $= 1.6$mG, respectively. The heavily dotted curve is an assumed π^0-decay spectrum thought to be appropriate for the present evolutionary state of the remnant (see text). Indicated are also the upper limits reported by the Whipple (Lessard et al. 1999) and CAT collaborations (Goret et al. 1999). The slantedly hatched curve denotes 1/10 of the Crab flux. The vertically hatched curve corresponds to the sensitivity of the future arrays Cangaroo III, H.E.S.S. and VERITAS.

	Tycho's SNR		Cas-A	
Configuration:	time	mean zenith angle / E_{thr}	time	mean zenith angle / E_{thr}
1: 1997 CT 3-6	20.8 hrs	$\vartheta = 37°$	49.3 hrs	$\vartheta = 31°$
2: 1997 CT 3,5,6	-		20.0 hrs	$\vartheta = 32°$
3: 1998 CT 3-6	16.8 hrs	$\vartheta = 36°$	58.6 hrs	$\vartheta = 33°$
Total	37.6 hrs	$E_{thr} \approx 1.2$ TeV	127.9 hrs	$E_{thr} \approx 1$ TeV
significance above background (probability cut)	-1.1 σ		4.5 σ	

TABLE 2. Observation times and significances for Tycho's SNR and Cas A; the 37.6 hrs for Tycho constitute only part of the total observation time available. The stereoscopic configuration of the four sytem telescopes CT 3-6, used for these observations, was different for the second 1997 period due to the fire on La Palma that hit CT 4. For the analysis of Cas A the source was assumed to be slighly extended.

spectrum from the radio to the hard X-ray region (Atoyan et al. 1999a), for the two mean magnetic field strengths $B_1 = $ 1mGand $= 1.6$mG, respectively. The heavily dotted curve corresponds to a π^0-decay spectrum, produced by an assumed power law spectrum of protons accelerated in the source, with a total energy content of $W_p = 2 \times 10^{49}$ergs $\simeq 1/5 W_p(t = \infty)$, where $W_p(t = \infty) = 10^{50}$erg corresponds to an assumed time-asymptotic nonthermal fraction of 10 percent of the total hydrodynamic energy of 10^{51}erg, generally assumed to be released in Cas A. The proton spectral index assumed is 2.15, with a rather high cutoff at 200 TeV for this Wind-SN already at very early times (Völk & Biermann 1988). The mean thermal gas density seen by the relativistic protons is taken as 15 cm^{-3}. The energetic electrons responsible for the IC emission are assumed to come from three different regions of the SNR interior: the bright compact radio components with magnetic field strength $B_1 = 1$ mG, where electrons are accelerated locally, an extended "plateau" of shocked circumstellar gas with $B_2 = B_1/4$ due to global acceleration at the forward SNR shock, and a low-field part of this "plateau" with $B_3 = 0.1$ mG. The electron spectral index is assumed to be uniformly 2.15, as for the protons. The electron cutoff energy, however, is only 17 TeV, corresponding to the steep drop-off of the observed hard X-ray spectrum with increasing energy.

Clearly, the magnitude of the γ-ray flux at about 1 TeV, if ultimately detected with a significance exceeding 5σ, could be equally due to electronic IC or hadronic π^0-decay emission. However the spectra would be very different for the two cases: an IC spectrum should fall off strongly with energy, in contrast to a π^0-decay spectrum. Therefore I believe that every effort should be made to obtain a TeV-spectrum of Cas A.

Acknowledgements I am grateful to Gerd Pühlhofer for providing several of the Figures in this paper.

REFERENCES

1. Konopelko, A.K., Pühlhofer, G., et al., *Proc. 26th ICRC, Salt Lake City* **3**, 444 (1999).
2. Aharonian, F.A., Akhperjanian, A.G., Barrio, J.A., et al., *A&A* **346**, 913 (1999a).
3. Kettler, J., private communication, (1999).
4. Pühlhofer, G., Völk, H.J., Wiedner, C.-A., et al., *Proc. 26th ICRC, Salt Lake City* **3**, 492 (1999).
5. Pühlhofer, G., Bernlöhr, K., Daum, A., et al., *Proc. 26th ICRC, Salt Lake City* **4**, 77 (1999).
6. Lampeitl, H., Konopelko, A.K., et al., *Proc. 26th ICRC, Salt Lake City* **4**, 81 (1999).
7. Aharonian, F.A., Akhperjanian, A.G., Barrio, J.A., et al., *A&A* **59**, 092003-1 (1999b).
8. Plyasheshnikov, A.V., Konopelko, A.K., Aharonian, F.A., et al., *J. Phys. G* **24**, 653 (1998).
9. Buckley, J.H., Akerlof, C.W., Carter-Lewis, D.A., et al., *A&A* **329**, 639 (1999).
10. Heß, M., *PhD Thesis Univ. Heidelberg* (1998).
11. Völk, H.J., *Proc. "Towards a Major Atmospheric Cherenkov Detector V", Kruger Park, S.A.*, 87 (1997).
12. Berezhko, E.G., Völk, H.J., *Proc. 24th ICRC, Rome* **3**, 380 (1995).
13. Berezhko, E.G., Völk, H.J., *Astroparticle Phys.* **7**, 183 (1997).
14. Atoyan, A.M., Aharonian, F.A., Tuffs, R.J., et al., *A&A* in press (1999a).
15. Atoyan, A.M., Aharonian, F.A., Tuffs, R.J., et al., *A&A* submitted to *A&A* (1999b).
16. Völk, H.J., Biermann, P.L., *ApJ* **333**, L65 (1988).
17. Lessard, R.W., Bond, I.H., Boyle, P.J., et al., *Proc. 26th ICRC, Salt Lake City* **3**, 488 (1999).
18. Goret, P., Guiffes, C., Nuss, E., et al., *Proc. 26th ICRC, Salt Lake City* **3**, 496 (1999).

THE SUPERNOVA REMNANTS AS THE MAIN SOURCES OF GAMMA-QUANTA WITH ENERGIES MORE THAN 1 TeV IN OUR GALACTIC

V.G. Sinitsyna

P.N. Lebedev Physical Institute, Russian Academy of Science,
Leninsky pr. 53, Moscow 117924, Russia.

Abstract. Four years of TeV observations by means of telescope SHALON on SHALON-ALATOO mountain observatory are received data on the fluxes of gamma-quanta with energy 10^{12}-10^{13} Galactic objects (Crab Nebula, Supernova Remnant Tycho and binary source in our Galactic Cygnus X-3). Observed fluxes from already known sources at about equal: $10^{-12} cm^{-2} s^{-1}$. It is the such intensity which observed from extragalactic gamma-sources, although the distances from Earth to Galactic and Extragalactic Sources differ for about 10^4 times, thant means into 10^8 times lager intensity then observed extragalactic sources. With allowance for the limited number of sources in our Galactic in comparison with Metagalactic it is necessary to suppose the predominantly extragalactic origin of the cosmic rays with energy more then 10^{13} eV.

More than twelve years ago one was proposed the plan of the mirror Cherenkov telescope SHALON (Sinitsyna, 1987) and the first observations were begun in 1992 on the ALATOO mountain observatory (Sinitsyna, 1992-1998). The distinctive feature of telescope is large full angle, that is achieved by for relatively large size of photomultypliers matrix (FIGURE 1). This allows the registration of extensive air showers that are coming at distance up to 120 m from the optical axis of telescope, that increase the obtaining of the statistics data from the sources of gamma-quanta of very high energies. In addition to these large angular dimensions of the imaging matrix will enable to research for isotropic

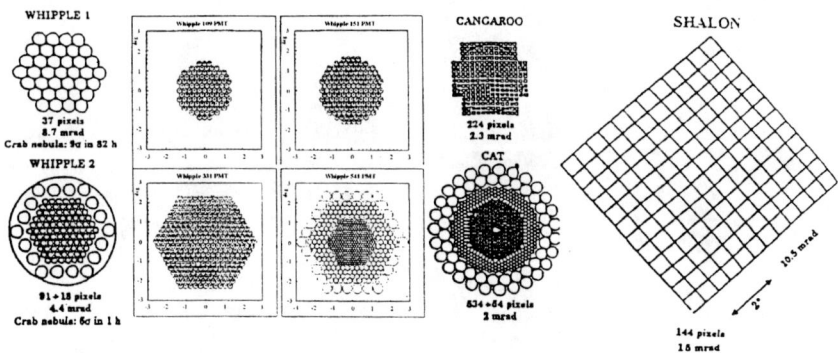

Pixel distribution the focal plan of the 10m reflector: top left: 109 pixels (1993-1996); top right: 151 pixels (Dec.,1996); 331 pixels (Oct., 1997); 541 pixels (planned).

FIGURE 1. *Evolution of photomultyplier arrays to record images seen in Cherenkov light*

background extensive air showers from charged particles of cosmic rays simultaneously with the observation of the local sources of gamma-quanta, that is at the same optical characteristics of atmosphere. It is particularly important because of in our research of gamma-sources the extensive air showers generating by the gamma-quanta are selected not only in accordance with the exceeding of flux of showers in small angle, but also as long as the differences of the evaluation in the depth atmospheres of the electron-photon cascades generated by protons and by nuclei of the cosmic rays.

Cherenkov imaging telescope SHALON equipped with a very high definition camera (144 pixels, full angle 8°) takes data from of 1993 at the mountain altitude 3338 m. We discuss some results of the observations of the indicated gamma-ray sources and the discrimination methods between gamma-rays and protons. Selection of showers produced by gamma-quanta from a background of showers produced by protons (FIGURE 2,3):

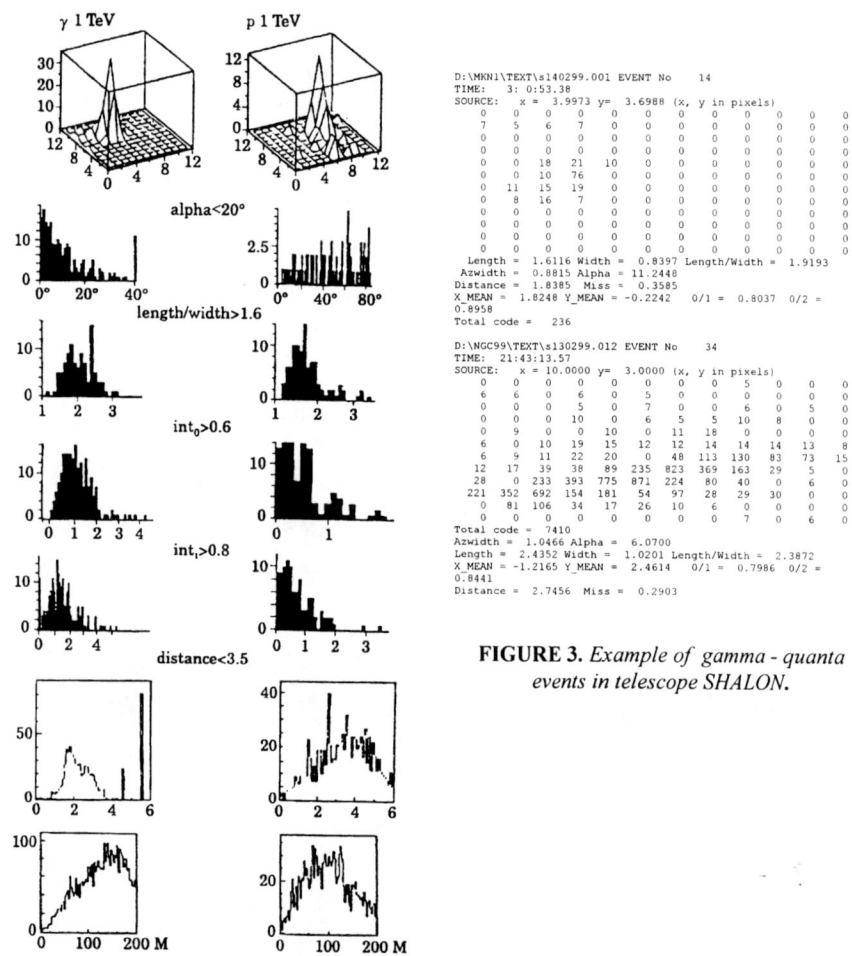

FIGURE 3. *Example of gamma - quanta events in telescope SHALON.*

FIGURE 2. *Monte Carlo distribution of image parameters for proton and gamma showers of 1TeV.*

1. α<20°
2. length/width>1.6 for γ
3. Cherenkov light intensity in pixels with max light to the light in the eight pixels near around is for
γ>0.6
4. Cherenkov light intensity in pixels with max light to all light in imaging except nine in center is for
γ>0.8
5. distance for γ<3.5 pixels.

The intensity of the different sources E>0.8 TeV is following:
$I_{Crab}=(1.1\pm0.30)\bullet10^{-12} cm^{-2}s^{-1}$ (FIGURE 5), $I_{Mrk421}=(1.09\pm0.41)\bullet10^{-12} cm^{-2}s^{-1}$,
$I_{Mrk501}=(1.32\pm0.30)\bullet10^{-12} cm^{-2}s^{-1}$, $I_{NGC1275}=(1.10\pm0.40)\bullet10^{-12} cm^{-2}s^{-1}$,
$I_{Cyg X-3}=(4.2\pm0.80)\bullet10^{-13} cm^{-2}s^{-1}$ (FIGURE 4), $I_{Geminga}=(5,7\pm4.0)\bullet10^{-13} cm^{-2}s^{-1}$ (FIGURE 6) and $I_{Tycho}<2\bullet10^{-13} cm^{-2}s^{-1}$ upper limit (FIGURE 6).

Observations Crab Nebula and Geminga (fig. 5, 6), Markarian 421 and Markarian 501 (FIGURE 3, 4, Sinitsyna, this book) compare with dates of other experiments including the data form observations on satellites in an energy range of 10^8-10^9 eV. Experimental data may be represented by energy spectrum $F(>E)\sim E^{-\gamma}$ in energy interval 10^8-10^{13} eV, but with different indexes: γ~1.5 for Crab Nebula and γ~1.1-1.3 for Markarian 421. May be Crab Nebula is typical source of Galactic cosmic rays and NGC 1275, Markarian 421 and Markarian 501 is a typical source of Extragalactic cosmic rays.

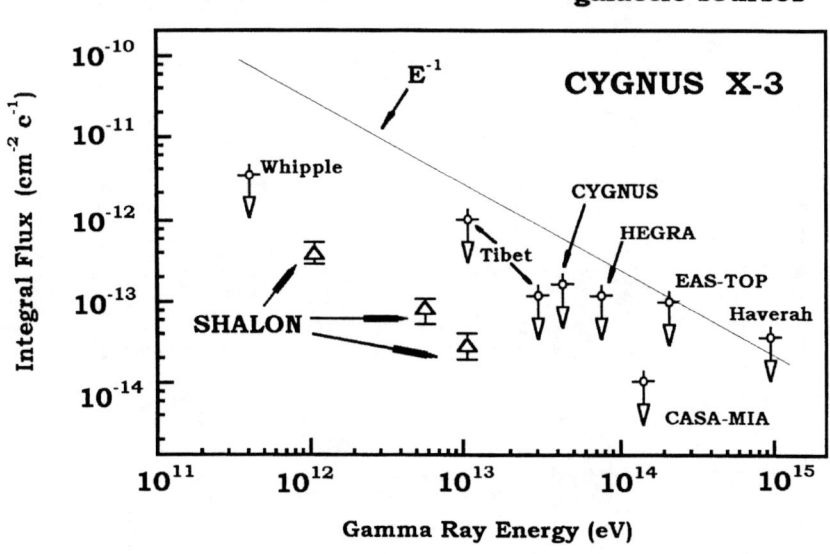

FIGURE 4. *Galactic high energy gamma-ray spectrum from CYGNUS X-3.*

CONCLUSION

The flux of gamma-quanta from sources in a high degree is parallel. The direction of Cherenkov radiation results to effective observation of sources up to distance ≤ 100 м from an axis of a shower. It means, that one telescope can observe showers on area 10^5 м^2. However, if with the objective selection of an observable shower with a shower from gamma-quanta from a source to increase angular resolution is lost width of a corner of the review, that provides the large square of registration of showers. If the angular resolution become rough is necessary to use the analysis of development of the cascade in atmosphere for distinguish of showers generated by gamma-quanta.

Among different methods of search of local sources of gamma-quanta most important is observations of Cherenkov radiation of electron - photon cascades generated in atmosphere in very high energy gamma - quanta. Thus the trend of Cherenkov radiation determines a direction on a source, that is coordinate of an observable source. The approximately equality observable of intensity of sources is connected to limitation of time observation of a assumed point source. Distance from Earth to Galactic and Extragalactic Sources differs for about 10^4 times of means, that detected in Extragalactic sources have large in 10^8 times a radiation possibilities.

This work was support by the Russian Foundation for Fundamental Research, project №98-02-16536.

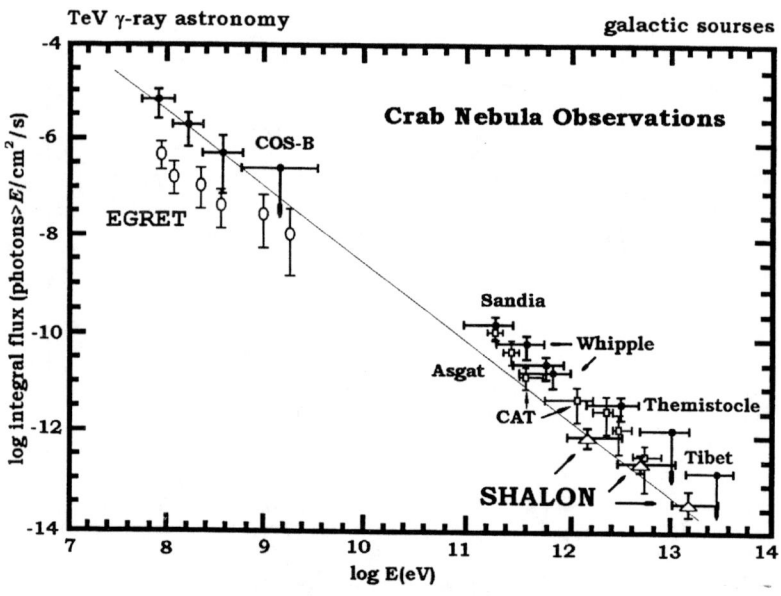

FIGURE. 5. *Galactic high energy gamma-ray spectrum from CRAB NEBULAR.*

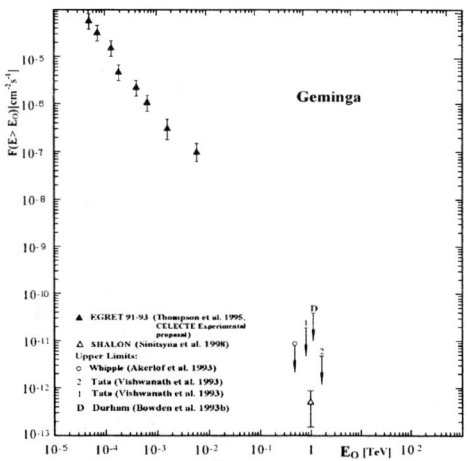
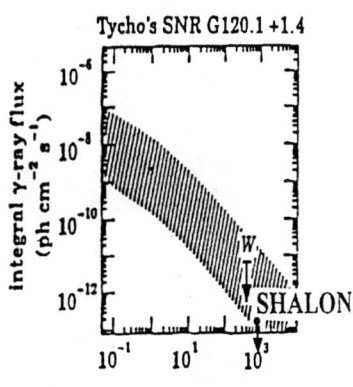

FFIGURE 6. *Galactic high energy gamma-ray spectrum from GIMINGA, and TYCHO.*

REFERENCES

Nikolsky S.I., Sinitsyna V.G., 1987, VANT, Ser.TFE (1331), 30.
Nikolsky S.I., Sinitsyna V.G., 1989, Proc. Workshop VHE Gamma-ray Astr., 11-21.
Weekes T.C., 1994, Preprint Series, N 3870, (Proceedings of the NATO School "The Gamma Ray Sky with Compton GRO and SIGMA" 1994), 1-33.
Lamb R.C., 1994, Proc.Workshop, Major Atm. Detector-III, ed. T.Kifune, 11-24.
Aglietta M., 1996, Alessandro B., Antonioli P., Preprint INFN/AE-96/21.
Hillas A.M., 1989, Proc. Workshop VHE Gamma Ray Astronomy, Crimea, 130-137.
Fegan D.J., 1992, Proc. Workshop, Towards a Major Atmospheric Cherenkov Detector - I, ed. P.Fleury, (Ecole Polytechnique Palaiseau, Paris), 3-42.
Weekes T.C., 1997, 25th ICRC, Dublin, v. 5, 257-260.
Hillas A.M., 1996, Nuovo Cimento, vol. 19C, №5, 701-712.
Weekes T.C. at al., 1996, Preprint Series less then galactic №4450 Harvard-Smithsonian Center for Astrophysics, "TeV Gamma Ray from Active Galactic Nuclei".
Iacoucci, L. and Nuss, E. (for the CAT collaboration), 1998, Proc. of the 16th European Cosmic Ray Symposium, 363-366.
Thompson, D.J. et al.,1995, Ap.J.S.101,209.CELESTEexperimental proposal,1996,1-78.
Sinitsyna V.G., Alaverdian A.Y., Arsov T.P., Mirzafatihov R.M., Nikolsky S.I., Striguin V.B., Vasilev V.D., Vorobiov S.P., 1997, 25th ICRC, Durban, v.3, 273-276.
Sinitsyna V.G., Alaverdian A.Y., Dashtojan A.A., Ivanov I.A., Mirzafatihov R.M., Nikolsky S.I., Platonov G.F., Sinitsyna V.Y., Striguin V.B., 1997, 25th ICRC, Durban, v. 5, 97-100.
Sinitsyna V.G., 1996, Nuovo Cimento, v. 19C, N6, 965-971.
Sinitsyna V.G., Alaverdian A.Y., Arsov T.P., Ivanov I.A., Nikolsky S.I, Mirzafatihov R.M., Platonov G.F., Sinitsyna V.Y., Vorobyev S.P., Strigin V.B., 16th European Cosmic Ray Symposium, ed. J.Medina, 1998, Spain, 367.
Sinitsyna V.G., 16th European Cosmic Ray Symposium, ed. J.Medina, 1998, Spain, 383.
Sinitsyna V.G., 1992, Proc. Workshop, Towards a Major Atmospheric Cherenkov Detector-I, ed. P.Fleury, (Paris), 299-304; 1993, Detector -II, ed. R.Lamb, (Calgary), 91-101; 1995, Detector-IV, ed. M.Cresti, 133-140; 1997, Detector-V, ed. O.De Jager, Kruger Park, 136-141, 190-195.

TeV Emission from Supernovae

P. M. Chadwick, K. Lyons, T. J. L. McComb, K. J. Orford,
J. L. Osborne, S. M. Rayner, S. E. Shaw, and K. E. Turver

Department of Physics, Rochester Building, Science Laboratories, University of Durham, Durham, DH1 3LE, U.K.

Abstract. Observations have been made with the University of Durham Mark 6 telescope of SN 1006 and SN 1998bw. Limits to VHE gamma ray emission from these objects have been derived. The upper limit from SN 1006 may be in conflict with the measurement made with the CANGAROO telescope.

INTRODUCTION

SN 1006

The supernova of 1006 A.D. was probably one of the brightest stellar events in recorded history. Despite its southerly declination, there are several records of contemporary sightings [1]. The remnant of SN 1006 was first observed in 1965, and it was detected as an X-ray source in 1976 [2,3]. Observations made in 1995 with the *ASCA* satellite indicated that emission from the edges of the remnant shell is non-thermal, and dominated by radiation from electrons accelerated to energies of ~ 100 TeV within the shock front [4]. If ions are accelerated to similar energies, then SN 1006 will be a source of very high energy cosmic rays. The presence of relativistic particles in the remnant also make it a strong candidate for VHE gamma ray emission.

In 1998, Tanimori et al. [5] reported the detection of VHE gamma rays with energy $> 1.7 \pm 0.5$ TeV at the 4.8 σ level from the NE rim of SN 1006, using the CANGAROO telescope. The flux from the NE rim is $(2.4 \pm 0.5_{\text{stat}} \pm 0.7_{\text{syst}}) \times 10^{-12}$ cm^{-2} s^{-1}; the authors also placed an upper limit of $\sim 1.1 \times 10^{-12}$ cm^{-2} s^{-1} for TeV gamma ray emission from the SW rim of the remnant, which is almost as X-ray bright as the NE rim.

The magnetic field strength in the emission region of SN 1006 was initially estimated to be 20 μG [6] on the basis of the minimum energy assumption; the magnetic field inferred from the CANGAROO measurements is 6.5 ± 2.0 μG since the flux detected is somewhat higher than might be expected if the magnetic field were 20 μG.

SN 1998bw

Soffitta et al. [7] detected a gamma ray burst (GRB 980425) of about 30 s duration in the *BeppoSAX* GRB Monitor and the Wide Field Camera (WFC). Galama et al. [8] discovered within the WFC error circle a supernova (SN1998bw) in the western spiral arm of the barred spiral galaxy ESO 184-G82. Spectra show that this is a type Ib/Ic SN. The redshift of the galaxy is $z = 0.0083$, giving a distance of 38 Mpc for $H = 65$ km s^{-1} Mpc^{-1}. Extensive radio measurements by Kulkarni et al. [9] showed what they describe as extraordinary radio emission from a most unusual supernova. Starting 3 days after the GRB the radio light curve showed a first peak about day 12 at 3 and 6 cm and a second lower peak at about day 30 at 3 and 6 cm and about day 45 at 20 cm.

OBSERVATIONS

Data were taken in 15-minute segments using the Durham Mark 6 telescope [10]. Off-source observations were taken by alternately viewing regions of sky which differ by ± 15 minutes in R.A. from the position of the object. This off-source/on-source/on-source/off-source observing pattern is used routinely to eliminate to first order any secular variation in count rate caused by changes in sky clarity etc.

The data were accepted for analysis only if the sky was clear and stable, as determined by an infra-red radiometer, and if the gross counting rates were consistent at the 2.5 σ level. A total of 41 hours of on-source observations were made on SN 1006, and a total of 6.25 hours of on-source observations were made on SN 1998bw. In the case of SN 1998bw, these observations were made during days 25, 27, 30 and 31 after the GRB when the radio light curve was around its second maximum. An observing log for both objects is shown in Table 1

The analysis and reduction of the accepted data follow the procedures developed from our analysis of our PKS 2155–304 observations [11].

RESULTS

SN 1006

The SN 1006 on-source field contains a 3rd magnitude star. In order to avoid any biases due to PMT gain changes which might arise due to the presence of this star, the Mark 6 telescope was offset steered by 1° in R.A. and 0.2° in declination. Treating the NE limb as a candidate point source and using the standard image cuts of Chadwick et al. [11], an on source excess of only 0.4 σ was found. The corresponding 3 σ upper limit to the flux is 1.7×10^{-11} cm^{-2} s^{-1} assuming an effective area of 5.5×10^8 cm^2 and that we retain 20% of the gamma ray flux. As SN 1006 is an extended object, we have also performed a false-source analysis, the results of which are shown in Figure 1. We see no evidence for VHE gamma ray

TABLE 1. Observing log for observations of supernovae made with the University of Durham Mark 6 Telescope.

Object	Date	No. of ON source scans
SN 1006	1997 March 15	4
SN 1006	1997 March 16	8
SN 1006	1998 April 28	5
SN 1006	1998 May 17	4
SN 1006	1998 May 29	9
SN 1006	1998 July 12	2
SN 1006	1998 July 14	4
SN 1006	1998 July 19	2
SN 1006	1998 July 21	3
SN 1006	1998 July 23	3
SN 1006	1999 April 11	4
SN 1006	1999 April 12	7
SN 1006	1999 April 13	9
SN 1006	1999 April 14	12
SN 1006	1999 April 15	11
SN 1006	1999 April 16	3
SN 1006	1999 April 18	13
SN 1006	1999 April 19	15
SN 1006	1999 April 20	14
SN 1006	1999 April 21	6
SN 1006	1999 April 22	7
SN 1006	1999 April 23	9
SN 1006	1999 April 11	6
SN 1998bw	1998 May 20	3
SN 1998bw	1998 May 22	4
SN 1998bw	1998 May 25	11
SN 1998bw	1998 May 26	7

emission from either the NE or the SW limb of SN 1006. This flux limit above 300 GeV when combined with the CANGAROO flux is in conflict with an inverse Compton spectrum.

We have produced a flux limit of 1.7×10^{-11} cm^{-2} s^{-1} at $E > 300$ GeV to emission from SN 1006. At first sight, this is in conflict with the CANGAROO measurement. However, it should be noted that this is a preliminary result based on initial simulations. It is possible that with further refinements produced by our ongoing programme of simulations any conflict may be resolved.

SN 1998bw

No on-source excess was seen. The 3 σ limit to the number of gammas is 142 which corresponds to 8.6×10^{-3} s^{-1}. If for the sake of argument there was an energy

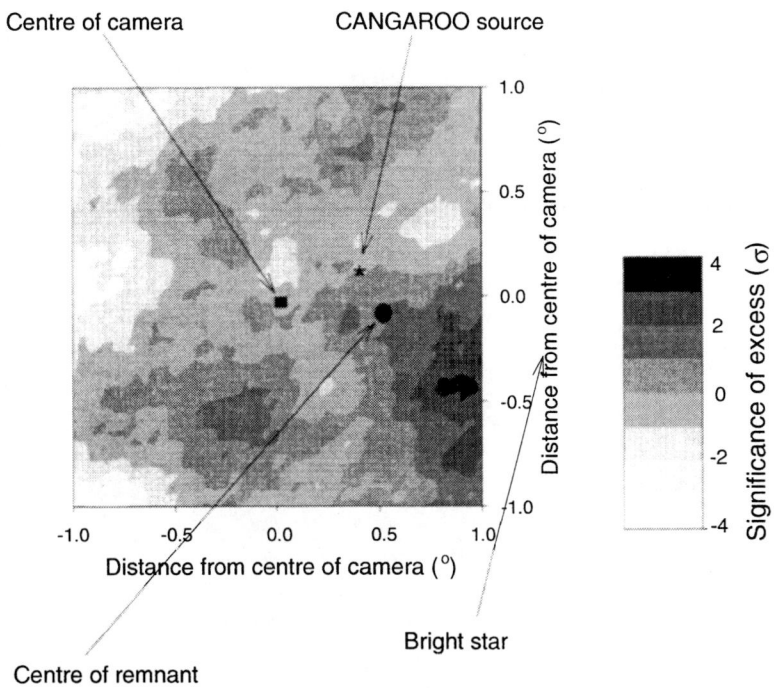

FIGURE 1. The significance of the number of any excess events as a function of assumed source position for the observation of the SN1006 region.

spectrum with spectral index 2.5 the upper limit to the total photon energy above 300 GeV would be 2×10^{40} J. This is to be compared with the 10^{42} to 10^{45} J in relativistic electrons. There is, however, no proposed mechanism for producing TeV gamma-rays in the order of 30 days after the explosion. The mechanism of Colgate [12] proposes the acceleration of the SN shock to relativistic speed down the density gradient of a massive stellar envelope. This method, favoured by Kulkarni et al. [9], could produce TeV energy gammas from a type I supernova but only for a few seconds. The diffusive shock mechanism in the expanding SNR might produce this total energy in TeV gammas but only when integrated over a much longer time.

We are grateful to the UK Particle Physics and Astronomy Research Council for support of the project. The Mark 6 telescope was constructed with the assistance of the staff of the Physics Department, University of Durham, and the efforts of Mr. P. Cottle, Mrs. E. S. Hilton and Mr. K. Tindale are acknowledged with gratitude.

REFERENCES

1. Stephenson, F.R., Clark, D.H., and Crawford, D.F., *MNRAS*, **180**, 567 (1997).
2. Gardner, F. F., and Milne, D. K., *Astron. J.*, **70**, 754 (1965).
3. Hill, R. W., et al., *Astrophys. J.*, **171**, 519 (1972).
4. Koyama, K. et al., *Nature*, **378**, 255 (1995).
5. Tanimori, T. et al., *Astrophys. J.*, **497**, L25 (1998).
6. Pohl, M., *Astron. Astrophys.*, **307**, L57 (1996).
7. Soffitta, P., et al., *IAU Circ. 6884* (1998).
8. Galama, T. J., et al., *IAU Circ. 6895* (1998).
9. Kulkarni S. R., et al., *Nature*, **395**, 663 (1998).
10. Armstrong P., et al., *Experimental Astron.*, **9**, 51 (1999).
11. Chadwick P. M., et al., *Astrophys. J.*, **513**, 161 (1999).
12. Colgate, S., *Astrophys. J.* **187**, 333 (1974).

Stereoscopic observations of the Crab Nebula with the HEGRA system of imaging air Čerenkov telescopes

A. Konopelko, G. Pühlhofer,

for the HEGRA Collaboration

Max-Planck-Institut für Kernphysik, Heidelberg D-69029, Postfach 10 39 80, Germany

Abstract. The HEGRA system of 5 imaging air Čerenkov telescopes was used for the extensive observations of the Crab Nebula. The observations have been extended to the zenith angles as large as 65 degree. Here we present the results of the data analysis in particular for the data taken at large zenith angles. LZA observations significantly improve the statistics for the γ-rays of energy above 5 TeV. Resulting spectrum of the Crab Nebula does not show apparent curvature in this energy range and may be fitted by pure power law in the energy range from 500 GeV to 20 TeV.

INTRODUCTION

The sensitivity of imaging atmospheric Čerenkov telescopes (IACTs) in TeV γ-ray observations reaches its maximum at small zenith angles ($\leq$30 degree) which permits the minimum attainable energy threshold of an instrument and detection of high quality two-dimensional angular images of Čerenkov light from air showers. However, for a specific telescope site a number of γ-ray sources, or source candidates, can only be observed at much larger zenith angles (LZAs). Extension of observations to LZAs might substantially widen the observational time window. Apparently that is very important for multi-wavelength campaigns involving ground-based and satellite-born instruments in simultaneous observations of variable γ-ray sources, e.g. BL Lac objects. In addition, observations at LZAs favor the detection of the high energy γ-rays (E > 10 TeV) which, at present, can be registered only using the ground-based Čerenkov technique.

We present here the results respecting the sensitivity and measurements of the energy spectra in stereoscopic observations of the Crab Nebula at LZAs ($\simeq$65 degree) with the HEGRA system of IACTs.

The HEGRA IACT system was primarily designed for detailed spectral measurements of TeV γ-rays, utilizing the advantages of the stereoscopic observations.

Stereo imaging allows (i) direct measurement of shower impact parameter with accuracy better than 10 m; (ii) good energy resolution of 18%; (iii) extended abilities for systematic studies using several images for an individual shower. The detailed systematic studies for the spectrum evaluation technique for observations at small zenith angles (<30 degree) have been recently completed using Mrk 501 1997 data [1]. Here we present the Crab Nebula data taken with the the HEGRA system of IACTs in two 97/98 and 98/99 observation campaigns, in particular data taken at LZAs ($\simeq 60°$). The LZA data were analyzed using new analysis technique for the stereoscopic observations [2].

OBSERVATIONS

The Crab Nebula was extensively observed with the HEGRA IACT system in two observational seasons from Sep'97 to Mar'98 and from Oct'98 to Apr'99. The HEGRA system of IACTs is located at La Palma, Canary Islands. Since Oct'98 HEGRA collaboration operates 5 telescopes. Each of the telescopes consists of a 8.5 m^2 reflector focusing onto a photo multiplier tube camera. The number of photo multipliers in the camera was 271, which were arranged in a hexagonal matrix covering a field of view with a radius of 2.°3. Any telescope camera was triggered when the signal in two neighbours of the 271 photo multiplier tubes exceeded a threshold of 8 photoelectrons, and the system readout started when at least two telescopes were triggered by Čerenkov light from air shower. The detection rate was 12.6 Hz near the zenith in Dec'97 and dropped down to about 10 Hz in Dec'99 for the 4 telescope system due to aging of the PMTs and reduced mirror reflectivity.

The Crab was observed in a wobble mode; i.e., the telescopes were pointed in Declination ±0.5° aside from the nominal Crab position (a sign of angular shift was altered from one run of 20 min to another). This is useful for continuous monitoring of the cosmic ray background taking the OFF-source region being symmetric about the camera center, and 1° apart relative to the ON-source region. Observations of the Crab at zenith angles up to 50 degree were made in a period from Sep'97 to Mar'98, for a total of 82.5 hrs. During the last observational period in addition to the observations at zenith angles less than 50° for 76.1 hrs we have carried out the observations at LZAs ($50° < \theta < 65°$) for a total of 24 hrs in order to study the performance of the telescope system at LZA and to extend measurements of the Crab Nebula energy spectrum beyond 10 TeV.

The collection areas, as a function of energy and zenith angle, for γ-ray showers has been inferred from Monte Carlo simulations as described in [3]. The energy threshold of the telescope system, defined as the energy at which the γ-ray detection rate reaches its maximum for the differential spectrum $dN_\gamma/dE \sim E^{-2.5}$, is ~ 0.5 TeV at small zenith angles, and increases up to ~ 5 TeV at 60° zenith angle. The simulations have been tuned for the different system configurations of 3,4 and 5 telescopes. The energy spectra derived for the different configurations have been gathered in the overall spectrum.

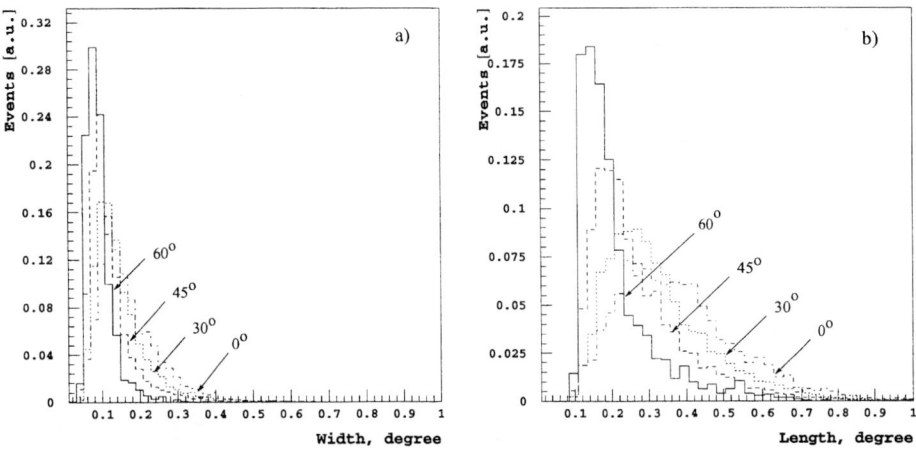

FIGURE 1. Distributions of image shape parameters for a $3 \div 5$ TeV γ-ray showers simulated at different zenith angles of $0°, 30°, 45°, 60°$.

ANALYSIS

The *stereoscopic imaging* analysis of the data is based on the geometrical reconstruction of shower arrival direction and shower core position in observation plane, as well as on the joint parameterization of a shape of the Čerenkov light images. The simultaneous registration of several (≥ 2) Čerenkov light images from air shower provided an angular resolution of $\sim 0.°1$ for γ-ray showers. For each individual shower stereoscopic observations permit the determination of the position of the shower axis in the observation plane. We selected only air showers within a certain impact distance R_0 from the center of the telescope system. The limiting upper radius for zenith angles less than 50 degrees was 200 m and a significantly larger radius of 400 m was used for LZAs. For the data taken at zenith angles up to 50° we applied an orientation cut $\theta^2 < 0.05 \,[\text{deg}^2]$, where θ^2 is the squared angular distance of the reconstructed source position from the true source position. In addition we analyzed the data by *mean scaled Width* parameter, $< \tilde{w} >$. We used relatively loose cut, $< \tilde{w} > < 1.2$.

For the air showers at LZAs, the height of shower maximum is far above the observation level, and the geometrical distance from the shower maximum to the observer is correspondingly very large. That basicly determines the topology of Čerenkov light images from such showers. In Figure 1 the distributions of the image shape parameters are shown for the different LZAs: $0°, 30°, 45°, 60°$. The calculations have been done for the primary $3 \div 5$ TeV γ-rays. One can see that both the transverse (Width) and longitudinal (Length) angular size of Čerenkov light images become smaller at LZAs. Air showers of the same primary energy generate less density of the Čerenkov light at the observation level resulting, on

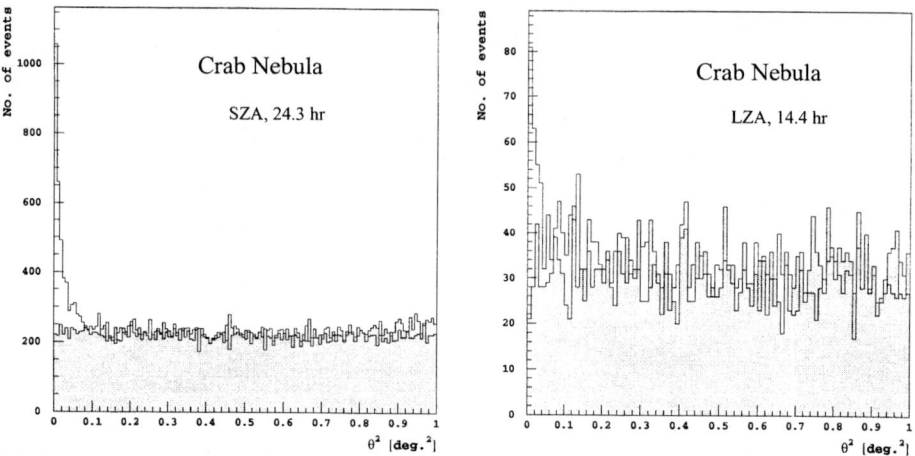

FIGURE 2. The event distribution as a function of θ^2 parameter for the ON and OFF (shaded histogram) regions for Crab Nebula observations at small ($\theta \leq 25°$) and LZAs ($55° \leq theta \leq 65°$).

average, in smaller number of Čerenkov photons in image. Calculations show that the images from LZA showers have almost circular shape and the image ellipticity, $1 - Width/Length$, noticeably decreases. The orientation of such images can be determined with large errors. That results in large uncertainties in the shower reconstruction, i.e., determination of the shower arrival direction, impact parameter, shower energy etc, and worse cosmic ray rejection. Thus the angular cut, $\theta^2 < 0.1\,[\text{deg}^2]$, gives a 90% acceptance of the γ-rays at SZAs and only 52% acceptance at LZAs. In present analysis we used the orientation cut of $\theta^2 < 0.1\,[\text{deg}^2]$ for LZA observations. Uncertainty in determination of the shower impact distance results in much broader distribution of mean scaled Width for the γ-rays and consequently significantly broader overlap with the cosmic ray events over this parameter. The shape cut $<\tilde{w}>\leq 1.0$ yields the enhancement factor (Q-factor) of 5.3 and 2.0 for the SZAs and LZAs, respectively.

At LZAs the angular size of the images becomes comparable with the pixel size (angular size of photo multiplier in the focal plane) and a small number of pixels is used (about 4) for the image parameterization. In order to improve the analysis at LZAs one can effectively use the fine granularity camera with the pixels of 0.15°. Calculations show that both orientation and shape of Čerenkov light images from air showers at LZAs can be better measured with the camera comprised of smaller pixels.

In order to improve the cosmic ray rejection in observations at LZA we introduced an additional parameter, mean scaled Length, $<\tilde{l}>$, defined by analogy with $<\tilde{w}>$. Two parameters, $<\tilde{w}>$ and $<\tilde{l}>$, can be used for calculating a

Mahalanobis distance, MD [4], in two-dimensional space as

$$\mathrm{MD} = ((1- <\tilde{w}>)^2/\sigma^2_{<\tilde{w}>} + (1- <\tilde{l}>)^2/\sigma^2_{<\tilde{l}>})^{1/2} \qquad (1)$$

where $\sigma_{<\tilde{w}>}$ and $\sigma_{<\tilde{l}>}$ are the standard deviations for the corresponding distributions of $<\tilde{w}>$ and $<\tilde{l}>$. We found that the optimum value of the MD cut for LZA is 1.5. Note that this analysis improves the cosmic ray rejection by a factor of 1.7.

The summary of the data is shown in Table 1. Data are shown after applying the orientation cut for the shape analysis by loose mean scaled Width cut of $<\tilde{w}> < 1.2$ (L. cut), MD cut (MD≤ 1.5) and without image shape cut (Raw).

TABLE 1. Summary of the data taken with the system of 4 IACTs during the observational period from Aug'98 to Dec'98.

	Z.A.	$0°-25°$	$25°-40°$	$40°-50°$	$55°-65°$
	Time, hrs	7.30	11.34	7.56	5.44
	Raw	592	972	327	143
γ	L. cut	577	670	244	124
	MD cut	450	499	157	101
	Raw	2964	4495	2772	3353
CR	L. cut	536	846	567	658
	MD cut	191	268	151	114
	Raw	7.3	9.7	4.3	1.8
S/N, σ	L. cut	14.2	13.8	6.6	3.3
	MD cut	15.6	15.5	7.3	5.6

The energy of a γ-ray shower is defined by interpolation over the "size" parameter S (total number of photoelectrons in Čerenkov light image) at the fixed impact distance R, as $E = f_{MC}(S, R, \theta)$, where θ is the zenith angle and f_{MC} is a function obtained from Monte Carlo simulations. The energy distribution for the ON- and OFF-source events, after the orientation and shape image cuts, were histogrammed over the energy range from 500 GeV to 30 TeV with 10 bins per decade. The γ-ray energy spectrum was obtained by subtracting ON- and OFF-histograms and dividing the resulting energy distribution by the corresponding collection area and the γ-ray acceptance. In the present Crab Nebula analysis we have extended the energy spectrum measurements up to LZAs (65°). The data were processed independently for each of four zenith angle bins: $(0°-25°), (25°-40°), (40°-50°), (50°-65°)$. The corresponding effective collection areas as well as the cut efficiencies were calculated as a function of the zenith angle. First, we derived the energy spectra for all zenith angle bins independently. Note that the spectra evaluated at different zenith angles are in a good agreement. For the final energy spectrum we joined the different zenith angle bins together

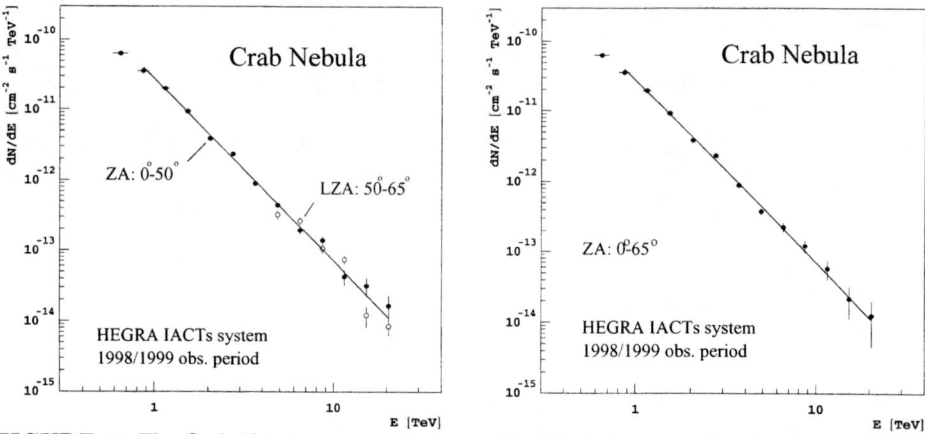

FIGURE 3. The Crab Nebula energy spectrum. The filled circles are for the observations at zenith angles up to 50°, the open circles are for the LZA data (60°).

$$dN_\gamma^i/dE = \sum_{j=1}^{4} w_j (dN_\gamma^i/dE)_j H(E^i - E_{th}^j),$$
$$w_j = t_j/t_0, \; i = 1, n; \; H = (0, E^i < E_{th}^j; 1, E^i > E_{th}^j) \qquad (2)$$

where dN_γ^i/dE, $(dN_\gamma^i)_j/dE$ are the differential energy spectra at energy E^i as measured over all zenith angle range and for the particular zenith angle bin (j), respectively. E_{th} is an estimated energy threshold for the zenith angle bin j. t_j is the observation time for j bin on the zenith angle, and t_0 is the total observation time. E_{th}^j is an energy threshold of the γ-rays observed within the zenith angle bin j. The Monte Carlo studies show that for the good energy resolution of 18% this approach does not distort the initial spectrum shape. The collection area for γ-rays rises very quickly in the energy range near the energy threshold of the telescope system, which is 500 GeV, whereas it is almost constant at the energy ≥ 3 TeV. Even slight variations of the trigger threshold could lead to noticeable systematic changes in the predicted spectral behavior in the energy range of $\sim 0.5 - 1$ TeV whereas such effect is negligible for energies above 3 TeV.

Statistics of the γ-rays from the Crab Nebula provides the measurement of the energy spectrum up to a few tens of TeV. However, detection of Čerenkov light images with extremely large amplitudes - several thousands of ph.e. - is complicated by the nonlinearity in the PMT response as well as by saturation in the 8 bit Flash-ADC readout. Such measurements need special treatment [5]. This effect becomes less important in observations at LZAs. Observations at LZAs permit the measurements of the energy spectrum far beyond 10 TeV. The images of the γ-ray air showers observed at LZAs have small *Size* and are not influenced by the saturation effect.

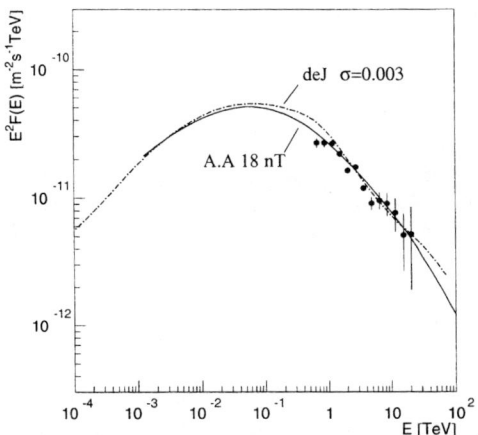

FIGURE 4. The Crab Nebula energy spectrum as measured by HEGRA group. The results of modeling the TeV γ-ray emission from Crab Nebula from [9] (A.A) and [10] (deJ) are shown by solid and dotted-dashed curve, respectively.

RESULTS

We have observed the Crab Nebula extensively in two observational seasons with the HEGRA IACT system. The differential energy spectrum of the Crab Nebula has been derived from the HEGRA data for both observational campaigns using recently developed advanced techniques for the measurements of the spectrum using *stereoscopic data*. The analysis for the different system configurations as well as, for different trigger threshold values give the resulting spectrum

$$dJ_\gamma/dE = (2.75 \pm 0.02 \pm 0.5) \cdot 10^{-7} \left(\frac{E}{1\,\text{TeV}}\right)^{-2.59 \pm 0.03 \pm 0.05} \text{ph}\,\text{m}^{-2}\,\text{s}^{-1}\,\text{TeV}^{-1} \quad (3)$$

as measured at zenith angles up to 60°. The statistical and systematic errors are also given.

Based on the Monte Carlo simulations for the γ-ray induced and cosmic ray induced air showers at LZAs, we have developed a specific analysis technique to be used for such data. The Crab Nebula differential energy spectrum derived from small zenith angle data matches quite well the spectrum derived at LZAs. The γ-ray rate measured at energies above 10 TeV in observations at LZAs exceeds the corresponding rate measured at small zenith angles. Our Crab Nebula spectrum is best fitted by a pure power law. To assess the contribution of π^0-produced γ-rays, measurements of the energy spectrum above 30 TeV are necessary.

DISCUSSION

The Crab Nebula has been observed and studied over an exclusively broad photon energy range embracing radio, optical, X-ray bands as well as high energy γ-rays up to hundreds of TeVs. The various theoretical scenarios of photon emission are primarily based on the Synchrotron Compton model [6,7], which combines the synchrotron and inverse Compton (IC) emissions from high-energy electrons, accelerated up to $\sim$ 100 TeV, which interact with magnetic field and seed photons within the nebula [8-10]. The predicted IC spectrum in the TeV energy domain appears to be very sensitive to the model parameters: the value of magnetic field, the nature of seed photons, the maximum energy of electrons, etc. Even IC scenarios of the photon emission are widely believed as most appropriate for the Crab Nebula one can not exclude the possible contribution of the γ-ray fluxes induced by π^0 decay [9]. The HEGRA data are shown in Figure 4 together with two SSC models of the TeV γ-ray emission [9,10]. One may conclude that IC modeling fits rather well the HEGRA data. According to [9] the estimated magnetic field in Crab Nebula is about 18 nT. To assess the contribution of π^0-produced γ-rays measurements of the Crab Nebula spectrum beyond 20 TeV are needed. The LZA technique could help to perform these future observations.

ACKNOWLEDGMENTS

The support of the German ministry for Research and technology BMBF and of the Spanish Research Council CYCIT is gratefully acknowledged. We thank the Instituto de Astrophysica de Canarias for the use of the site and for supplying excellent working conditions at La Palma. We gratefully acknowledge the technical support staff of the Heidelberg, Kiel, Munich, and Yerevan Institutes.

REFERENCES

1. Aharonian, F., et al., *A & A*, **342**, 69 (1999).
2. Konopelko, A., et al., *J. Phys. G: Nucl. Part. Phys.*, **25**, 1989-2000 (1999).
3. Konopelko, A., et al., *Astoparticle Physics*, **10**, 275-289 (1999).
4. Mahalanobis, H.C., *Proc. Nat. Inst. Sci. India*, **12**, 49 (1963).
5. Hess, M., et al.,*Astroparticle Physics*, **11**, 363-377 (1998).
6. Gould, R.J. *Phys. Rev. Lett.*, **15**, 511 (1965).
7. Kennel, C.F., Coroniti, F.V. *ApJ*, **283**, 694 (1984).
8. De Jager, O.C., Harding, A.K. *ApJ*, **396**, 161 (1992).
9. Atoyan, A.M., Aharonian, F.A. *MNRAS*, **278**, 525 (1996).
10. De Jager, O.C., et al. *ApJ*, **457**, 253, (1996).

GAMMA RAY BURSTS

High Energy Radiation from Gamma Ray Bursts

Charles D. Dermer[*][1] and James Chiang[†]

[*] *Naval Research Laboratory, Code 7653, Washington, DC 20375-5352*
[†] *JILA, University of Colorado, Campus Box 440, Boulder, CO 80309-0444*

Abstract. Gamma-ray burst (GRB) engines are probed most intimately during the prompt gamma-ray luminous phase when the expanding blast wave is closest to the explosion center. Using GRBs 990123 and 940217 as guides, we briefly review observations of high-energy emission from GRBs and summarize some problems in GRB physics. $\gamma\gamma$ transparency arguments imply relativistic beaming. The parameters that go into the external shock model are stated, and we show numerical simulation results of gamma-ray light curves from relativistic blast waves with different amounts of baryon loading. A distinct component due to the synchrotron self-Compton process produces significant emission at GeV and TeV energies. Predictions for spectral and temporal evolution at these energies are presented for a blast wave expanding into uniform surroundings. Observations of the slow decay of GeV-TeV radiation provides evidence for ultra-high energy cosmic ray acceleration in GRBs.

I INTRODUCTION

The cosmological hypothesis for the origin of GRBs has been favored ever since BATSE showed that GRB sources were isotropically distributed about us, yet were bounded in spatial extent [1]. Burst studies were revolutionized by the discovery of X-ray afterglows [2] with the Beppo-SAX satellite. The good X-ray imaging of the Narrow Field Instrument on Beppo-SAX quickly led to the identification of optical [3] and radio counterparts [4], permitting redshift measurements from optical transient counterparts and directionally coincident host galaxies. Once the redshift is known, the power and energy release follow modulo the collimation factor $\delta\Omega/4\pi$.

Table 1 lists soft gamma-ray luminosities and energies for a sample of the dozen GRBs now known with measured redshifts. In the case of GRB 990123, the directional γ-ray power and energy reach values as large as $\partial L_\gamma/\partial\Omega \sim 3 \times 10^{51}$ ergs s^{-1} sr^{-1} and $\partial E/\partial\Omega \sim 2 \times 10^{53}$ ergs sr^{-1} [5], respectively. Fig. 1 shows the 50-300 keV light curve (left) and νF_ν spectrum [6] (right) of GRB 990123. The spectrum is

---
[1]) Work supported by the Office of Naval Research.

TABLE 1. Inferred Isotropic 50-300 keV×(1 + z) Luminosities and Energy Releases from a Sample of GRBs with Redshifts

GRB	Redshift z	Peak Flux[a]	Fluence[b]	$L_\gamma{}^c$ (10^{51} ergs s^{-1})	E_γ (10^{52} ergs)
970228	0.695	3.5	6.1	1.1	0.65
970508	0.835	1.2	3.1	0.57	0.47
971214	3.418	1.95	10.9	24	17
980425	0.0085	0.96	4.0	6.2×10^{-5}	7.3×10^{-5}
980703	0.966	2.4	45.9	1.6	9.1
990123	1.60	16.4	509	35	240

[a] Units of photons cm^{-2} s^{-1} in 50-300 keV range.
[b] Units of 10^{-6} ergs cm^{-2} in 50-300 keV range, except for GRB 970228 (35-1000 keV) and GRB 971214 (> 20 keV).
[c] Assuming an $\Omega_m = 0.3$, $\Omega_\Lambda = 0.7$ cosmology with $H_0 = 65$ km s^{-1} Mpc^{-1}, and a mean photon energy of 107 keV implied by a flat νF_ν spectrum.

summed over the 12.3 - 45.1 s interval after the trigger time[2]. Given this luminosity and the intrinsic variability time scale $t_v \sim 5/(1+z)$ s, the compactness parameter $\ell = L\sigma_T/(4\pi m_e c^3 c t_v) \sim 10^{12}$ is enormous, so that gamma rays could not escape without invoking directed beams of photons and directed relativistic motions of the emitting particles.

The inference of bulk relativistic motion from the gamma-ray observations anticipated the expanding relativistic blast wave model that so readily explains temporal X-ray and optical power-law afterglow decays [7–9]. In the external shock model (ESM), a relativistic blast wave is energized as it passes through and captures material from the surrounding gas and dust [10,11]. In the colliding shell (internal shock) model [12,13], the engine's activity is prolonged and intermittent. The ESM

---

[2] Except for EGRET's Total Absorption Shower Counter (TASC) spectrum, which accumulated photons from -0.057 s to 64.5 s.

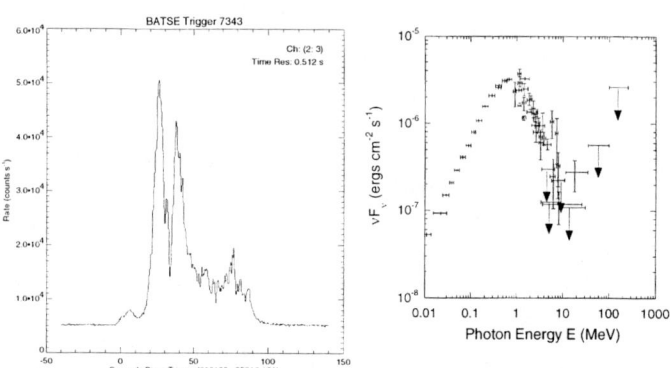

FIGURE 1. (left) The gamma-ray light curve of GRB 990123, observed with BATSE in the energy range 50-300 keV. (right) The time-averaged broadband spectrum of GRB 990123.

explains the long wavelength afterglow behavior and, as argued elsewhere [14,15], the phenomenology of the prompt gamma-ray luminous phase and short timescale variability [16].

Here we describe the potential of gamma-ray observations to characterize properties of GRB sources. In Section 2, a brief summary of GRB gamma-ray observations is given. Some unsolved GRB problems are mentioned in Section 3. Section 4 spotlights the method of inferring properties of the blast wave from $\gamma\gamma$ transparency arguments. This leads to a description of the standard fireball/blast-wave model for GRBs in Section 5, with its luminous GeV-TeV radiation from the SSC process. Calculations from the ESM in a uniform surrounding medium establish quantitative predictions for the MeV, GeV, and TeV behavior of GRBs in Section 6 – for one parameter set. Finally, in Section 7 we mention possible high energy gamma-ray signatures of ultrahigh energy hadrons accelerated by GRB blast waves. We summarize in Section 8.

II GAMMA RAYS FROM GAMMA RAY BURSTS

The largest and most complete GRB data set has been obtained with the BATSE detector on the *Compton Observatory* [17,18]. In its normal mode of operation, BATSE triggers when the 50-300 keV count rate in two detectors exceeds 5.5σ over background on time scales of 64, 256, and 1024 ms. The background is obtained from a commandable time interval, usually set at ≈ 17 seconds. Fig. 2 shows the t_{50} duration [17] and $E_{\rm pk}$ [19] distributions measured with BATSE. The t_{50} duration is the time interval over which the integrated counts range from 25% to 75% of the total counts over background. The value of $E_{\rm pk}$ is the photon energy of the peak of the time-averaged νF_ν GRB spectral energy distribution. As can be seen, the duration distribution shows two distinct components [20]; furthermore, there is a

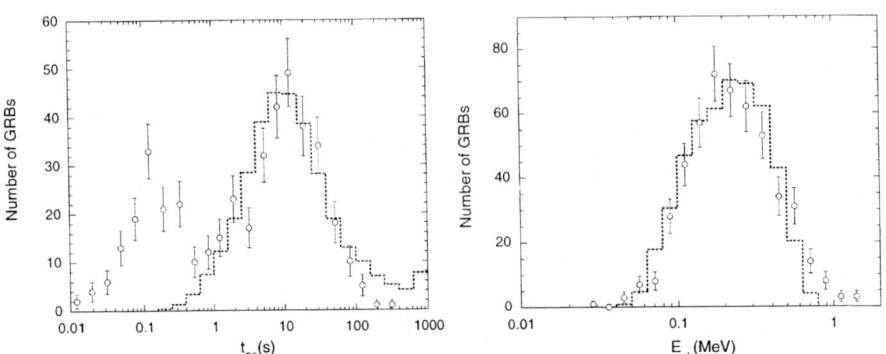

FIGURE 2. Data points give the t_{50} duration (left) and $E_{\rm pk}$ (right) distributions of GRBs measured with BATSE [17,19]. Dotted histograms give model fits from the external shock model [15].

clear correlation for shorter GRBs to have harder spectra. The range of E_{pk} is quite narrowly distributed in a range centered at ~ 200 keV, which is right in the middle of the triggering range of BATSE. According to the ESM [15], E_{pk} is primarily determined by the baryon loading, and the E_{pk} distribution is a consequence of the triggering properties of BATSE convolved with the flux behavior of GRB blast waves with different total energies and baryon-loading factors which explode in surroundings with a range of densities.

The generic spectral form of GRB emission in the BATSE energy range is

$$\frac{dN}{dE} \propto \begin{cases} E^{-\alpha_{ph}}, & \text{for } E < E_{pk} \\ E^{-\beta_{ph}} & \text{for } E > E_{pk} \end{cases} \quad (1)$$

where, typically, $\alpha_{ph} \cong 1$ and $\beta_{ph} = 2\text{-}2.5$. The Solar Maximum Mission satellite revealed that $\gtrsim 1$ MeV emission was a common property of GRBs [21], thus establishing that the radiation has a nonthermal origin. COMPTEL has detected over 30 GRBs at $E > 0.75$ MeV [22]. The spark chamber on EGRET detected $E \gtrsim 30$ MeV photons from 7 GRBs [23]. These GRBs are invariably among the brightest BATSE bursts. The average spectral index of four EGRET GRBs, consisting of 45 photons with energies > 30 MeV, is $\beta_{ph} = 1.95 \pm 0.25$ [23], consistent with this emission being an extension of the spectrum near E_{pk} observed with BATSE. EGRET's TASC, which measures $\sim 1\text{-}200$ MeV spectra and serves as a calorimeter to measure total photon energy for EGRET, has detected at least 16 GRBs [24].

Fig. 3 shows the light curve and spectra of the famous burst GRB 940217, which displayed an Earth-occulted ~ 100 MeV tail that lasted for ~ 95 minutes, two ~ 3 GeV photons during the BATSE burst, and an 18 GeV photon 90 minutes later [25]. The Interval 1 and 2 νF_ν spectra are shown in the inset, and the three EGRET photons detected during the brief interval 4 are shown in the upper right panel.

III UNSOLVED GRB PROBLEMS

If, as generally reasoned, GRB emissions originate from a fireball that ejects either a single blast wave into inhomogeneous surroundings or expels a long-lasting relativistic wind, then a central problem in GRB studies is to understand the nature of the central engine and how it powers the energy released into the blast wave. The favored, but by no means proven scenario is that GRBs are powered by the core collapse of a massive star to a black hole. The short events (i.e., $t_{50} \lesssim 1$ s in Fig. 2a) may have a separate origin, for example, through compact object coalescence. A massive star origin for GRBs is in accord with the vigorous star formation implied by the blue galaxy hosts, the evidence for large quantities of gas and dust in GRB environs, and the coincidence of GRB directions with the disk and central regions of host galaxies.

The degree of GRB blast-wave collimation remains a crucial unknown. Neither compact object coalescence scenarios nor collapsar/hypernova models invoking neutrino annihilation or poorly quantified mhd processes make sufficient fireball energy

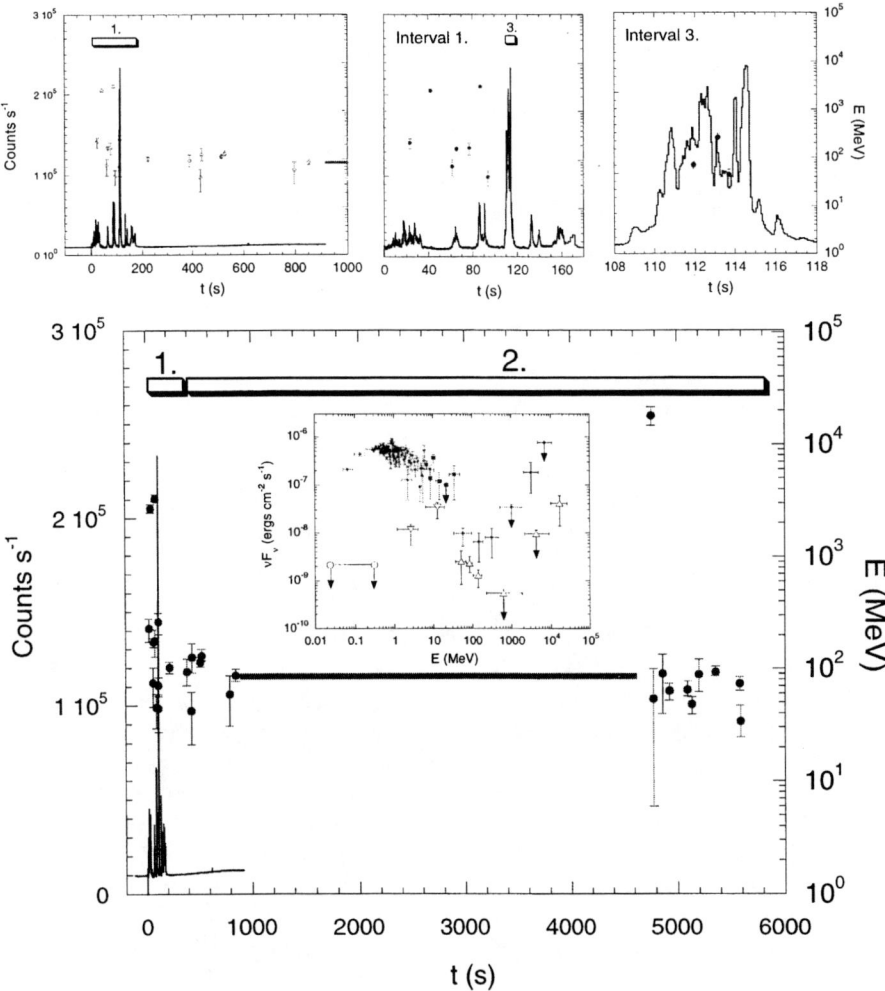

FIGURE 3. Central figure: Times and energies of EGRET-detected photons (data from [25]) and BATSE light curve of GRB 940217 (BATSE trigger # 2831). Inset shows composite νF_ν spectra during interval 1 and interval 2 (large symbols), naively obtained by multiplying photon spectrum in Ref. [25] by E^2. The BATSE light curve is summed 16 channel MER data (excluding channel 12 for which only the first 8 seconds of data exist). Successive blow-ups of the BATSE light curve are shown in the top panels. Note that deadtime effects from background vetos could have reduced EGRET's efficiency for detecting gamma-rays.

to account for the largest measured GRB energies without invoking opening half-angles $\psi \lesssim 10°$ (e.g., [26,27]). Easing the energy requirements is a great boon to these and other models.

A third open question is whether the prompt GRB emission results from collisions between a succession of shells ejected from the GRB engine [28,12] or is instead due to an ESM where a single impulsive relativistic blast wave interacts with inhomogeneities in the external medium [11,16]. The answer to this problem characterizes the accretion/collapse and coalescence activity taking place near GRB engines.

IV $\gamma\gamma$ TRANSPARENCY ARGUMENTS

Gamma-ray observations set important constraints on the location and speed of the blast wave shell through the requirement that the emission region be optically thin to $\gamma\gamma$ pair production attenuation [29–31]. We estimate the optical depth $\tau_{\gamma\gamma}(\epsilon')$ to pair production at dimensionless photon energy $\epsilon' = h\nu'/m_e c^2$. Primed quantities refer to the comoving blast wave frame and unprimed quantities refer to the observer frame. We have

$$\tau_{\gamma\gamma}(\epsilon') \cong [\frac{\epsilon' L(\epsilon')}{\epsilon' m_e c^2} \cdot \frac{\Delta R'}{c} \cdot \frac{1}{4\pi R^2 \Delta R'}] \cdot \sigma_{\gamma\gamma}(\epsilon') \cdot \Delta R' \ . \tag{2}$$

The blast wave shell, with comoving width $\Delta R'$, is at distance R from the explosion site when it radiates the photons that are measured with gamma-ray detectors at energies $\epsilon \cong \Gamma \epsilon'/(1+z)$. The total power gets boosted and redshifted by two factors of energy and time for a spherically expanding blast wave; thus $\epsilon L(\epsilon) = \Gamma^2 \epsilon' L(\epsilon')/(1+z)^2$ and $\epsilon \cong \Gamma \epsilon'/(1+z)$. By definition, $4\pi d_L^2 \epsilon S(\epsilon) = \epsilon L(\epsilon)$, where d_L is the luminosity distance and $S(\epsilon)$ is the spectral flux (ergs cm^{-2} s^{-1}ϵ^{-1}).

The $\gamma\gamma$ cross section peaks near threshold and reaches a value of $\sigma_{\gamma\gamma}(\epsilon' \sim 1) \approx \sigma_T$. An estimation of merely the $\gamma\gamma$ optical depth of near-threshold-energy photons in the blast wave frame – which are detected with $\epsilon \cong \Gamma/(1+z)$ – gives

$$\tau_{\gamma\gamma}[\epsilon = \Gamma/(1+z)] \simeq (\frac{1+z}{\Gamma})^2 \frac{d_L^2 S_0 \epsilon^{1-\alpha} \Delta R' \sigma_T}{R^2 m_e c^3} \ . \tag{3}$$

Here we parameterize the observed high-energy photon spectrum $\nu F_\nu = \epsilon S(\epsilon) = S_0 \epsilon^{-\alpha}$. Requiring $\tau_{\gamma\gamma} < 1$ and invoking the relation $\Delta R' = fR/\Gamma$, where $f \sim 1$ for an adiabatic blast wave [32,33], we place limits on the Lorentz factor Γ and the location R of the site where high energy radiation is produced. Suppose that a power-law spectrum of γ rays extending to energy $\epsilon_{\max}$ is measured. Then either $\Gamma \gtrsim (1+z)\epsilon_{\max}$, or

$$R \gtrsim \frac{d_L^2}{(1+z)} \frac{S_0 \sigma_T f}{\epsilon_{\max}^{\alpha+2} m_e c^3} = 2.7 \times 10^{21} \frac{S_{-6} d_{28}^2 f}{(1+z)\epsilon_{\max}^{\alpha+2}} \ \text{cm}, \tag{4}$$

where $S_{-6} = S_0/10^{-6}$ ergs cm^{-2} s^{-1} and $d_{28} = d_L/10^{28}$ cm.

For the specific case of GRB 990123 shown in Fig. 1, $\epsilon S(\epsilon) = 6.7 \times 10^{-6} \epsilon^{-1.1}$ ergs cm^{-2} s^{-1}, so that $S_{-6} = 6.7$ and $\alpha = 2.1$. Furthermore, $d_{28} = 3.1$ (see Table 1). The BATSE and COMPTEL observations of 4-8 MeV photons already imply that either $\Gamma > 20\text{-}40$ or $R \gtrsim 5 \times 10^{18} f$ cm. If 100 MeV photons had been observed coincident with this GRB (unfortunately, EGRET's spark chamber did not observe this GRB as it was too far off axis), then we could draw the conclusion that either $\Gamma \gtrsim 500$ or $R \gtrsim 3 \times 10^{13} f$ cm. This can restrict some forms of the internal shock model [34], with implications for neutrino production by GRBs.

Application to GRB 940217 provides looser constraints on R and Γ because we do not know its redshift, again highlighting the importance of redshift measurements. The $\gamma\gamma$ transparency constraints can be strengthened when one considers pair-producing interactions between high energy γ rays and lower energy photons [35]. The use of gamma-ray astronomy to infer properties of the expanding outflow will be well utilized by future AGILE and GLAST observations in the 100 MeV - GeV range, and also potentially from $\sim 0.1\text{-}1$ TeV emission observed with ground-based air or water Čerenkov telescopes.

V EXTERNAL SHOCK MODEL FOR GRBS

A minimum of nine parameters enter into a blast-wave model calculation for GRBs in the ESM (for details and references on the next 2 sections, see [36,37]). These can be grouped according to whether they are (i) intrinsic parameters associated with the properties of the central engine, (ii) environmental parameters that characterize the surrounding medium, or (iii) microscopic parameters that define the reinjection of swept-up hadron power into the nonthermal leptons in the blast wave.

The three intrinsic parameters are the directional energy $\partial E_0/\partial\Omega \to E_0/(4\pi) = 10^{54} E_{54}$ ergs/$(4\pi$ sr$)$ released by the central engine, the initial Lorenz factor Γ_0 of the blast wave, and the opening half-angle ψ of the collimated outflow. We take $E_{54} = 1$ and consider either uncollimated or collimated outflows with $\psi = 10°$. This opening angle relaxes the energy requirements by a factor of ~ 130 for a one-sided jet. (Two additional complications, not dealt with here, are angular gradients in outflows and lateral spreading of the blast wave.)

The initial blast wave Lorentz factor Γ_0 is closely related to the baryon loading of the fireball, because the optically thick fireball expands until most of its initial energy E_0 has been transferred to the kinetic energy of the outflowing baryons $\Gamma_0 M_b c^2$, where M_b is the mass of the baryons. As the blast wave sweeps up and captures material from the surrounding environment, it decelerates and becomes energized by the addition of nonthermal particles with Lorentz factors Γ in the comoving blast-wave frame. The circumburster environment is likely to be highly structured in all cases, but especially if the progenitor of a GRB is a massive star located in a star forming region where stellar winds could introduce inhomogeneities. Nevertheless, the surrounding density distribution is usually parameterized by the function $n(x) = n_0 x_{\text{dec}}^{-\eta}$. We take $n_0 = 100$ cm^{-3} and $\eta = 0$ as standard values, though $\eta = 2$ would be more appropriate for a wind. In the ESM, the measured durations of GRBs are comparable with the deceleration time scale $t_{\text{dec}} = [3(\partial E_0/\partial\Omega)/m_p c^2 n_0]^{1/3}/(c\Gamma_0^{8/3})$.

The microscopic parameters include the fraction ϵ_e of nonthermal swept-up proton kinetic energy transferred to nonthermal electrons, the injection index p of the electrons, and the maximum electron energy parameter $\epsilon_{\max}$, given through the kinematic limit $\gamma_{\max} = 4 \times 10^7 \epsilon_{\max}/\sqrt{B(\text{G})}$. The comoving magnetic field strength B is set by an equipartition argument. The value of the magnetic equipartition parameter ϵ_B is defined by $B^2/8\pi = 4\epsilon_B(\Gamma^2 - \Gamma)m_p c^2 n(x)$. We let $\epsilon_e = 0.5$, $\epsilon_{\max} = 1$, $p = 2.5$, and $\epsilon_B = 10^{-4}$, and furthermore assume that the microscopic parameters are time-independent. The low value of ϵ_B is required [36] to reproduce the generic eq.(1) spectrum.

VI MODEL SPECTRA AND LIGHT CURVES

The numerical simulation model [37] treats synchrotron, synchrotron self-Compton (SSC), synchrotron self-absorption and adiabatic loss processes, and follows blast-wave evolution self-consistently. The photons are attenuated by $\gamma\gamma$ absorption, but pair reinjection is not followed. Fig. 4 shows temporally evolving spectra for the standard uncollimated parameter set. The $\gamma\gamma$ process degrades only $\gtrsim 0.1$ TeV photons for the results shown here. Thus the internal attenuation of high-energy gamma rays in the ESM is not too severe and the SSC component is bright enough that TeV radiation is produced at a comparable νF_ν level as the synchrotron radiation. Dirty fireballs produce a larger relative νF_ν flux in the SSC component than in the synchrotron component. We [37] propose that the TeV radiation detected by Milagro from GRB 970417a and reported at this meeting [38] is the SSC emission from a nearby $z \lesssim 0.1$ GRB.

Fits to Fig. 2 data [15], taking into account BATSE triggering properties and the strong biases against detecting dirty fireballs with $\Gamma_0 \ll 300$, imply a very large population of undiscovered optical and X-ray transients with well-characterized properties. The clean fireball population produces sub-second transients peaking at GeV-TeV energies [39] that GLAST, AGILE, or Ĉerenkov detectors could discover. The distribution of baryon-loading parameters Γ_0 toward clean fireballs falls, however, below a power-law parameterization of the Γ_0 distribution [15], indicating that the space density of clean fireballs is less than that for fireballs producing detectable GRBs ($\Gamma_0 \sim 300$). Very clean fireballs ($\Gamma_0 \gtrsim 3000$) could produce TeV bursts of radiation. These can be distinguished from Hawking radiation and annihilating dark matter particles by their spectrum and afterglow.

The MeV, GeV, and TeV νL_ν light curves shown in Fig. 5 [37] are multiplied by time to show where most counts will be detected in logarithmic intervals of time. Synchrotron radiation forms the early MeV and GeV peaks – this is the GRB itself.

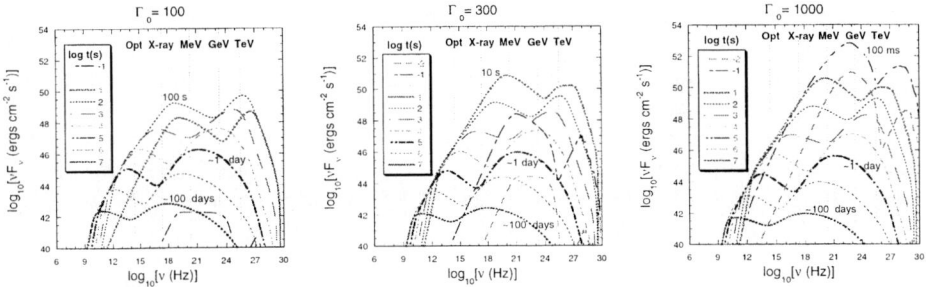

FIGURE 4. Calculations of SEDs from uncollimated GRB blast waves that are energized, decelerate and radiate by capturing material from a uniform surrounding medium with §V parameters. Only the initial Lorentz/baryon-loading factor Γ_0 differs between the three calculations. The duration decreases and the νF_ν flux and $E_{\rm pk}$ values increase with increasing Γ_0.

The SSC component forms the early TeV peak. The later peaks at MeV and GeV energies are due to the SSC component becoming increasingly dominant in these wavebands as the blast wave decelerates. The second maximum at GeV energies occurs at ≈ 5000 s, comparable to the duration of the extended emission observed from GRB 940217. Because the relative fluxes of prompt and delayed emission are greater at GeV energies than at MeV energies for these parameters, it is more probable that delayed GeV emission rather than MeV emission would be detected from a GRB, as in fact was observed with EGRET from GRB 940217. However, the particular calculation shown here corresponds to a uniform surrounding, whereas

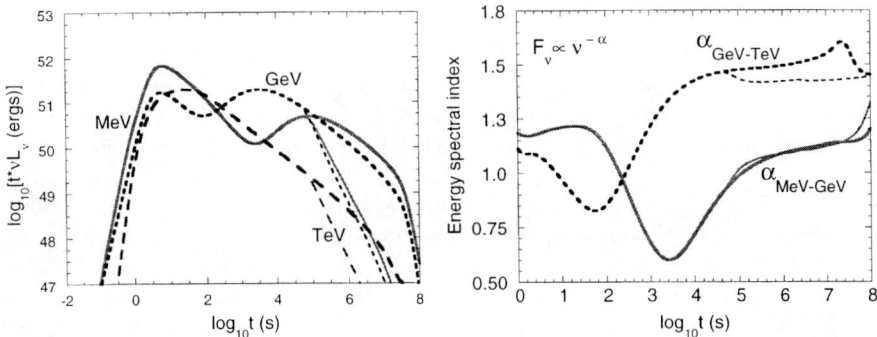

FIGURE 5. (left) Product of νL_ν flux and observing time t for the MeV, GeV, and TeV light curves using $\Gamma_0 = 300$ case in Fig. 4. (right) Temporal variation of the broadband MeV-GeV and GeV-TeV energy spectral indices. In both panels, thick curves are for uncollimated outflows and thin curves are for beamed outflows with $\psi = 10°$ and an observer along the symmetry axis of the jet.

the behavior of the light curve shown in Fig. 3 could be explained with the ESM only in terms of a highly structured medium. We emphasize that the full range of possible behaviors for spectral and temporal evolution, of which Fig. 5 represents only one possibility, has hardly been explored.

GLAST, with its larger effective area and field-of-view, should be able to monitor the evolution of the SSC spectral feature due to blast wave deceleration from many bright GRBs. Broadband MeV-GeV and GeV-TeV spectral indices due to blast-wave deceleration are plotted in the right panel in Fig. 5 for the standard parameter set studied here. During the early phase, the MeV-GeV photon spectral index corresponds to a soft cooled synchrotron spectrum, here with a value $\beta_{ph} \sim 2.25$ for the $p = 2.5$ injection electron spectrum, in accord with measurements of > 30 MeV EGRET spectra [24]. The GeV-TeV index is much harder because this waveband primarily samples the harder SSC component. After the prompt phase, the GeV-TeV index softens to a spectrum that is even softer than the cooled synchrotron spectrum due to effects of $\gamma\gamma$ attenuation, and the MeV-GeV index hardens as the SSC radiation sweeps into this waveband. The MeV-GeV index approaches the cooled synchrotron limit at later times. Spectral hardening in the MeV-GeV band in the early afterglow phase due to the deceleration of the blast wave as it interacts with a smooth external medium constitutes generic behavior of the ESM which can be tested with GLAST, and can be confronted by observations of GeV and TeV detectors of bright GRBs with smooth MeV light curves that signify a GRB source within a uniform surrounding.

VII HADRONS IN GRB BLAST WAVES

The nonthermal energy carried into GRB blast waves by hadrons is larger by a factor $\sim m_p/m_e$ than the energy carried by leptons, so hadronic effects can hardly be negligible. The physics of transferring energy from hadrons to leptons is just one of the many open questions in this field. An important related question is whether GRB blast waves accelerate ultra-high energy cosmic rays (UHECRs). The validity of this idea [40–42] was argued by comparing the energy densities of UHECRs with the globally averaged injection of energy by GRBs into a volume no greater than, for $\gtrsim 10^{20}$ eV UHECRs, the Zatsepin-Kuzmin-Greisen radius outside which UHECRs are degraded by photomeson production on cosmic microwave background photons. Photomeson neutrino production at $\gtrsim 10^{14}$ eV [34], and GeV γ-ray production from proton synchrotron radiation [43,44] are both potentially observable signatures of UHECR acceleration by GRBs.

The UHECRs are claimed to be accelerated either through a first-order shock [40] or second-order [41] stochastic Fermi process. The shock Fermi mechanism fails for collapsar models of GRBs [45] because only the first shock cycle produces a Γ^2 energy gain. Subsequent cycles give energy increases of only factors-of-2, because the shock catches up to the particle before it can complete more than a small fraction of its cycle.

The simplest approach is to assume [46,47] that no acceleration follows the capture of particles into the blast wave; of course, no UHECRs are then produced. Such a process could produce a low-level flux of γ rays at $E \sim 0.1\Gamma/(1+z)$ GeV and radio/optical synchrotron radiation from the process $p + p \to \pi + X \to \gamma, e^{\pm} + X$, where the low-energy protons are the thermal baryon-load material. This approach appears too inefficient, however, to describe flaring events. Magnetic turbulence injected by charged dust during the capture and isotropization process could, though gyroresonant processes, accelerate protons to ultra-high energies [48], as could the turbulence generated when the blast wave encounters inhomogeneities in the circumburster medium. The shock front will likely be Rayleigh-Taylor unstable, and this will also generate turbulence in the blast wave.

The population of GRBs with redshifts now permits a more quantitative estimate of the rate density of GRB sources. Stecker [49] argues that if GRBs follow the star-formation rate history of the universe, then a much smaller energy injection rate of UHECRs into the local universe occurs, so that GRBs cannot be the source of the UHECRs. This argument does not take into account the predicted but so-far undetected dirty fireball population, which can introduce a 2-3 orders-of-magnitude increase in the source density of GRBs [15] and therefore UHECRs. It has also been argued [50] that if energy is transferred very inefficiently from hadrons to electrons, then the total hadron energy in GRBs is $\sim m_p/m_e$ greater than implied by the gamma-ray measurements. Hence $\partial E/\partial \Omega \to 10^{56}$ ergs sr^{-1}. We resist this proposal because it multiplies difficulties in understanding the energetics of GRB sources.

Slow decay of GeV-TeV radiation from proton synchrotron radiation provides evidence in favor of hadrons in GRBs. Protons are much less radiative than leptons unless they are far more energetic; thus hadrons are more likely to be weakly cooled. When protons are injected with number index $s = 2$, the uncooled GeV proton synchrotron flux decays in the adiabatic regime with temporal index $\chi = 3/4$ (flux $\phi \propto t^{-\chi}$). In comparison, the optical and X-ray synchrotron radiation decays as $\chi = 1$ for strongly cooling electrons that are likewise injected with $s = 2$. The temporal decay from slowly cooling hadrons is thus slower than for strongly cooling leptons. Consequently GeV proton synchrotron radiation should decay more slowly than lepton synchrotron and SSC radiations. Before more concrete conclusions can be drawn, however, further studies are needed to distinguish between the behavior of the hadronic and SSC emissions, and to treat diffusive acceleration, cascade processes and UHECR escape from the GRB blast wave.

VIII SUMMARY

Gamma-ray transparency arguments push one irresistibly toward a relativistic blast-wave model of GRBs. The standard fireball/blast wave model implies strong GeV/TeV radiation from the SSC process [37]. Using parameters optimized to fit prompt hard X-ray and soft γ-ray emission from GRBs, our calculations show nearly coincident MeV/TeV light curves and extended GeV light curves due to the dominance of the SSC component at GeV energies in the early afterglow phase. In the framework of the ESM, GeV and TeV observations chart the evolution of the SSC component and hence the evolution and changes of the blast wave. Calculations of the MeV, GeV and TeV light curves were made for a standard parameter set, showing that the GeV band displays a soft-to-hard-to-soft evolution as the SSC component sweeps through this waveband. This spectral prediction applies to blast waves which decelerate in a uniform medium as evidenced by smooth GRB light curves; circumburster medium structure introduces many possible variations to the light curves and spectral behaviors not yet explored (compare Figure 3).

The possible existence of a class of very clean fireballs that produce $\lesssim 100$ ms flashes of GeV and TeV radiation is a straightforward prediction of the blast wave model. The related prediction of a large class of dirty fireballs finds good company with the hypothesis that GRB blast waves accelerate energetic hadrons, because the dirty fireballs provide a much more numerous source population with which to provide the energy of the UHECRs observed locally. Hadronic acceleration might reveal itself through the slow decay of GeV-TeV proton synchrotron radiation, but better studies are needed for quantitative predictions.

ACKNOWLEDGMENTS

CD thanks Anthony Crider for help in preparing Fig. 3 and for discussions about the BATSE data. He also thanks conference organizers Brenda Dingus and Mike Salamon for an exciting meeting. We acknowledge collaboration and useful discussions with M. Böttcher.

REFERENCES

1. Meegan, C. A., et al., *Nature* **345**, 143 (1992).
2. Costa, E., et al., *Nature* **387**, 783 (1997).
3. van Paradijs, J., et al., *Nature* **386**, 686 (1997).
4. Frail, D. A., Kulkarni, S. R., Nicastro, L., Feroci, M., and Taylor, G. B., *Nature* **389**, 261 (1998).
5. Kulkarni, S. R., et al., *Nature* **398**, 389 (1999).
6. Briggs, M. S., et al., *Astrophys. J.* **524**, 82 (1999).
7. Vietri, M., *Astrophys. J.* **478**, L9 (1997).
8. Waxman, E., *Astrophys. J.* **485**, L5 (1997).
9. Wijers, R. A. M. J., Mészáros, P., and Rees, M. J., *MNRAS* **288**, L51 (1997).
10. Rees, M. J., and Mészáros, P., *MNRAS* **258**, 41P (1992).
11. Mészáros, P., and Rees, M. J., *Astrophys. J.* **405**, 405 (1993).
12. Kobayashi, S., Piran, T., and Sari, R., *Astrophys. J.* **513**, 679 (1997).
13. Daigne, F., and Mochkovitz, R., *MNRAS* **296**, 275 (1998).
14. Dermer, C. D., Böttcher, M., and Chiang, J., *Astrophys. J.* **513**, L5 (1999).

15. Böttcher, M., and Dermer, C. D., *Astrophys. J.* **529**, 237 (2000).
16. Dermer, C. D., and Mitman, K. E., *Astrophys. J.* **513**, L5 (1999).
17. Meegan, C. A., et al., *Astrophys. J. Supp.* **106**, 65 (1996).
18. Meegan, C. A., et al., in *4th Huntsville Symposium on Gamma-Ray Bursts*, eds. C. A. Meegan, R. D. Preece, and T. M. Koshut, New York: AIP, 1998, p. 3.
19. Mallozzi, R. S. et al., *ibid.*, p. 273.
20. Kouveliotou, C., et al., *Astrophys. J.* **413**, L101 (1993).
21. Matz, S. M., Forrest, D. J., Vestrand, W. T., Chupp, E. L., Share, G. H., and Rieger, E., *Astrophys. J.* **288**, L37 (1985).
22. Connors, A., et al., in *4th Huntsville Symposium on Gamma-Ray Bursts, op. cit.*, p. 344.
23. Dingus, B. L., Catelli, J. R., and Schneid, E. J. 1998, in *4th Huntsville Symposium on Gamma-Ray Bursts, op. cit.*, p. 349.
24. Catelli, J. R., Dingus, B. L., and Schneid, E. J. 1998, in *4th Huntsville Symposium on Gamma-Ray Bursts, op. cit.*, p. 309.
25. Hurley, K. C., et al., *Nature* **372**, 652 (1994).
26. Janka, H.-T., Eberl, T., Ruffert, M., and Fryer, C. L. *Astrophys. J.* submitted (1999).
27. Popham, R., Woosley, S. E., and Fryer, C. *Astrophys. J.* **518**, 356 (1999).
28. Rees, M. J., and Mészáros, P., *Astrophys. J.* **430**, L93 (1994).
29. Krolik, J. H., and Pier, E. A., *Astrophys. J.* **373**, 277 (1991).
30. Fenimore, E. E., Epstein, R. I., and Ho, C., *Gamma-Ray Bursts*, eds. W. S. Paciesas and G. J. Fishman, New York: AIP, 1992, p. 158.
31. Baring, M. G., and Harding, A. K., *Astrophys. J.* **491**, 663 (1997).
32. Mészáros, P., Laguna, P., and Rees, M. J., *Astrophys. J.* **415**, 181 (1993).
33. Blandford, R. D., and McKee, C. F., *Phys. Fluids*, **19**, 1130 (1976).
34. Waxman, E., and Bahcall, J. *Phys. Rev. Lett.*, **78**, 2292 (1997).
35. Baring, M. G., these proceedings.
36. Chiang, J., and Dermer, C. D., *Astrophys. J.* **512**, 699 (1999).
37. Dermer, C. D., Chiang, J., and Mitman, K. E., *Astrophys. J.* submitted (1999), astro-ph/9910240.
38. McEnery, J., et al. 1999, these proceedings.
39. Dermer, C. D., Chiang, J., and Böttcher, M. *Astrophys. J.* **513**, 656 (1999).
40. Vietri, M., *Astrophys. J.* **453**, 883 (1995).
41. Waxman, E., *Phys. Rev. Lett.* **75**, 386 (1995).
42. Milgrom, M., and Usov, V., *Astropar. Ph.* **4**, 365 (1996).
43. Vietri, M., *Phys. Rev. Lett.* **78**, 4328 (1997).
44. Böttcher, M., and Dermer, C. D., *Astrophys. J.* **499**, L131 (1998).
45. Gallant, Y., and Achterberg, A., *MNRAS* **305**, L6 (1999).
46. Pohl, M., and Schlickeiser, R., *Astron. and Astrophys.* in press (1999), astro-ph/9911452.
47. Pohl, M., these proceedings.
48. Schlickeiser, R., and Dermer, C. D., *Astron. and Astrophys.* submitted (2000).
49. Stecker, F. W., *Astrophys. J. Lett.* submitted (1999), astro-ph/9911269.
50. Totani, T., *Astrophys. J.* **502**, L13 (1998).

High-Energy Spectral Signatures in Gamma-Ray Bursts

Matthew G. Baring[†]

Laboratory for High Energy Astrophysics, Code 661
NASA Goddard Space Flight Center, Greenbelt, MD 20771
baring@lheavx.gsfc.nasa.gov
[†] *Universities Space Research Association*

Abstract. One of the principal results obtained by the EGRET experiment aboard the Compton Gamma-Ray Observatory (CGRO) was the detection of several γ-ray bursts (GRBs) above 100 MeV. The broad-band spectra obtained for these bursts gave no indication of any high energy spectral attenuation that might preclude detection of bursts by ground-based Čerenkov telescopes (ACTs), thus motivating several TeV observational programs. This paper explores the expectations for the spectral properties in the TeV and sub-TeV bands for bursts, in particular how attenuation of photons by pair creation internal to the source modifies the spectrum to produce distinctive spectral signatures. The energy of spectral breaks and the associated spectral indices provide valuable information that can constrain the bulk Lorentz factor of the GRB outflow at a given time. These characteristics define palpable observational goals for ACT programs, and strongly impact the observability of bursts in the TeV band.

INTRODUCTION

High energy gamma-rays have been observed for six gamma-ray bursts by the EGRET experiment on CGRO. Most conspicuous among these observations is the emission of an 18 GeV photon by the GRB940217 burst [1]. Taking into account the instrumental field of view, these detections indicate that emission in the 1 MeV–10 GeV range is probably common among bursts, if not universal. One implication of GRB observability at energies around or above 1 MeV is that, at these energies, spectral attenuation by two-photon pair production ($\gamma\gamma \to e^+e^-$) is absent in the source. From this fact, early on Schmidt [2] concluded that if a typical burst produced quasi-isotropic radiation, it had to be less distant than a few kpc, since the optical depth $\tau_{\gamma\gamma}$ scales as the square of the distance to the burst.

This result conflicted with BATSE's determination of the spatial isotropy and inhomogeneity of bursts [3], which suggested that they are either in an extended halo or at cosmological distances (where $\tau_{\gamma\gamma} \sim 10^{12}$ for isotropic photons). Hence Fenimore et al. [4] proposed that GRB photon angular distributions are highly beamed and produced by a relativistically moving plasma, a suggestion that has become very popular. This can dramatically reduce $\tau_{\gamma\gamma}$ and blueshift spectral attenuation turnovers out of the observed spectral range. Various determinations of the bulk Lorentz factor Γ of the GRB medium have been made in recent years, mostly concentrating [5,6] on cases where the angular extent of the source was of the order of $1/\Gamma$. These calculations generally assume an infinite power-law burst spectrum, and deduce [7] that gamma-ray transparency up to the maximum energy detected by EGRET requires $\Gamma \gtrsim 100-10^3$ for cosmological bursts.

While power-law source spectrum assumption is expedient, the spectral curvature seen in most GRBs by BATSE [8] is expected to play an important role in reducing the opacity for potential TeV emission from these sources (Baring & Harding [9]). Such curvature is patently evident in 200 keV–2 MeV spectra of some EGRET-detected bursts, and its prevalence in bursts is indicated by the generally steep EGRET spectra for bursts [1,10,11]. In this paper, the principal effects introduced into pair production opacity calculations by spectral breaks in the BATSE energy range are considered, focusing the work of [9] to identify the properties of cosmological bursts in the 1 GeV–1 TeV range. These signatures are clearly distinguishable from absorption by background radiation fields, thereby defining diagnostics that future ground-based initiatives such as Veritas, MILAGRO, HESS and MAGIC, and space missions such as GLAST can provide for GRB studies.

SPECTRAL CURVATURE AND γ-γ ATTENUATION

The simplest picture [5,6] of relativistic beaming has "blobs" of material moving with a bulk Lorentz factor Γ more-or-less toward the observer, and having an angular "extent" $\sim 1/\Gamma$. For an infinite power-law spectrum $n(\varepsilon) = n_\gamma \varepsilon^{-\alpha}$, where ε is the photon energy in units of $m_e c^2$ (a dimensionless convention used throughout), for which the optical depth to pair creation assumes the form $\tau_{\gamma\gamma}(\varepsilon) \propto \varepsilon^{\alpha-1}\Gamma^{-(1+2\alpha)}$ for $\Gamma \gg 1$. As noted above, the input source spectrum needs to be modified, to explore the effects of a relative depletion of low energy photons in the BATSE range. The simplest approximation to spectral curvature is a power-law broken at a dimensionless energy $\varepsilon_{\rm B}$ ($= E_{\rm B}/0.511\,{\rm MeV}$):

$$n(\varepsilon) = n_\gamma \varepsilon_{\rm B}^{-\alpha_h} \begin{cases} \varepsilon_{\rm B}^{\alpha_l}\varepsilon^{-\alpha_l}, & \text{if } \varepsilon \leq \varepsilon_{\rm B}, \\ \varepsilon_{\rm B}^{\alpha_h}\varepsilon^{-\alpha_h}, & \varepsilon > \varepsilon_{\rm B}. \end{cases} \quad (1)$$

The optical depth determination for such a distribution utilizes results obtained in [12] for truncated power-laws. The resulting forms are presented in [9], and the optical depth $\tau_{\gamma\gamma}(\varepsilon)$ for attenuation of a broken power-law photon distribution has the basic form

$$\frac{\tau_{\gamma\gamma}(\varepsilon)}{n_\gamma \sigma_{\rm T} R} \propto \begin{cases} \dfrac{\varepsilon^{\alpha_h-1}}{\Gamma^{2\alpha_h}}, & \text{if } \varepsilon \lesssim \Gamma^2/\varepsilon_{\rm B}, \\ \dfrac{\varepsilon^{\alpha_l-1}}{\Gamma^{2\alpha_l}}, & \text{if } \varepsilon \gtrsim \Gamma^2/\varepsilon_{\rm B}, \end{cases} \quad (2)$$

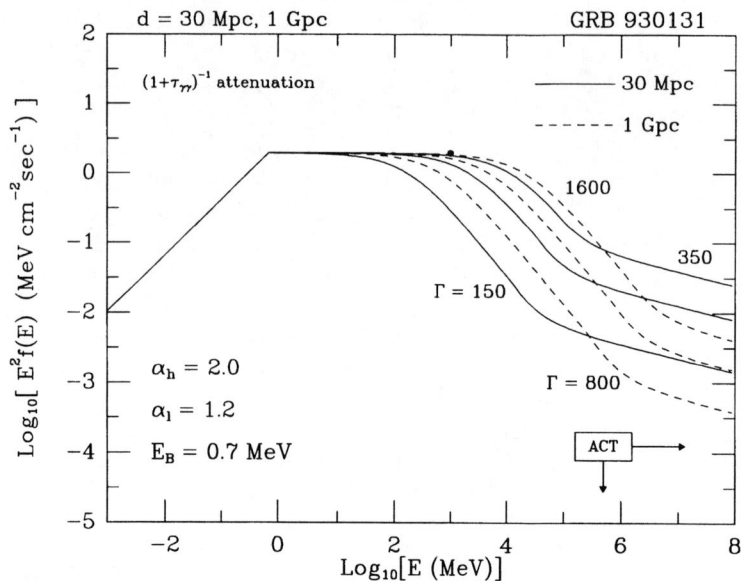

FIGURE 1. The γ-γ attenuation, internal to the source, for GRB 930131 at distances typical of nearby (solid curves, $\Gamma = 150, 250, 350$) and distant (short dashed curves, $\Gamma = 800, 1200, 1600$) cosmological origin, for bulk Lorentz factors Γ of the emitting region. The source spectrum (νF_ν format) was a power-law broken at $E_B = 0.7$ MeV, with spectral indices $\alpha_l = 1.2$ and $\alpha_h = 2.0$. The filled circle denotes the highest energy EGRET photon at 1000 MeV [7]. The threshold and sensitivity for ACT observations of later bursts is indicated by the "ACT" box.

that implies breaks in the absorbed portion of the hard gamma-ray spectrum that "image" the BATSE band break in the seed photons. More gradual spectral curvature can be treated by fitting the GRB continuum with piecewise continuous power-laws. A variability "size" $R_v = 3 \times 10^7$ cm ($= R/\Gamma$) is chosen here following [7,9], and the observed flux at 1 MeV normalizes the source density coefficient n_γ.

The results of the attenuation of the spectra in Eq. (1) are depicted in Fig. 1 using an attenuation factor $1/(1+\tau_{\gamma\gamma})$ that is appropriate for opacity skin depth effects. The source spectrum parameters are chosen to approximate the observed values for the "Superbowl burst" GRB930131, for two different extragalactic distance scales: the nearer one, 30 Mpc, is appropriate to scenarios where GRBs generate ultra-high energy cosmic rays. Clearly the attenuation is marked in the GeV–TeV band for the Lorentz factors Γ chosen, and could be reduced by increasing Γ. The onset of attenuation couples to Γ and the EGRET band spectral index α_h, and above this turnover the immediate spectral index is $1 - 2\alpha_h$. Precise knowledge of the GRB distance, such as through redshifts of accompanying optical afterglows, would facilitate the determination of tight constraints on Γ. For either distance scale in Fig. 1, there is a flattening (to index $1 - \alpha_h - \alpha_l$) in the TeV/sub-TeV band that is a consequence of the spectral break in the BATSE band: it arises at $\varepsilon \sim \Gamma^2/\varepsilon_B$.

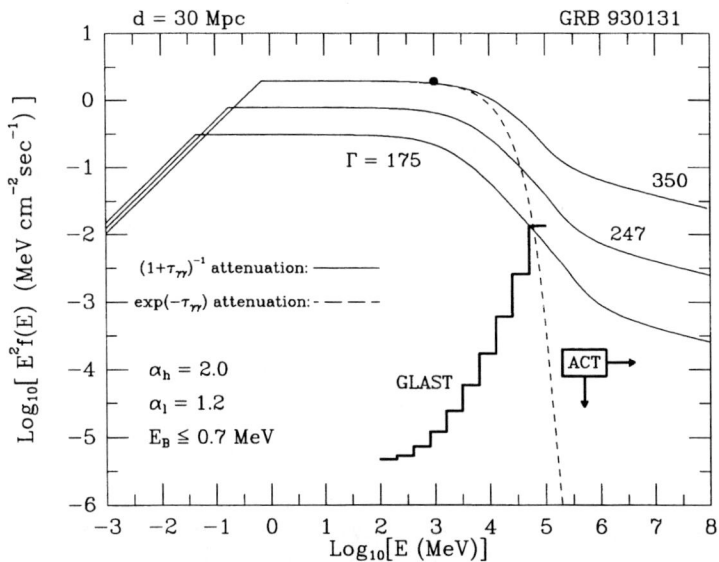

FIGURE 2. An evolutionary sequence for γ-γ attenuation, starting with a GRB930131 data fit, appropriate to an adiabatically-decelerating blast wave. The GLAST steady-source differential ($\Delta E/E = 2$ step-function) and approximate Whipple integral sensitivities (ACT box; derived from upper limits to later bursts) are depicted. A case of exponential attenuation (for the $\Gamma = 350$, dashed line) is illustrated; it would inhibit detections by ACTs.

The potential for observational diagnostics is immediately apparent. First, the extant EGRET data already provides a lower bound to Γ: the dot on Fig. 1 represents the highest energy photon from GRB930131, and clearly suggests that $\Gamma \gtrsim 250$ for $d = 30\,\mathrm{Mpc}$ or $\Gamma \gtrsim 800$ for $d = 1\,\mathrm{Gpc}$. Second, the sensitivity of ACTs is easily sufficient to detect bursts even with significant attenuation, so that they could well probe the spectral issues raised here. While the Whipple rapid search [13] postdated the EGRET detections, and produced merely upper limits as indicated in Fig. 1, an intriguing possible detection of a BATSE burst by the MILAGRITO forerunner to MILAGRO was announced at this meeting by McEnery et al. (these proceedings), foreshadowing advances to come.

Perhaps the greatest strides in understanding will be precipitated by broad-band spectral coverage afforded by simultaneous detection of bursts by GLAST and TeV experiments like MILAGRO. Fig. 2 displays a time-evolutionary sequence of GRB spectra, including the effects of γ-γ attenuation, and compares this with the potentially-constraining current Whipple integral sensitivity threshold (deduced from the results of [13]), and the projected GLAST *steady-source differential* sensitivity. The GLAST sensitivity is obtained from simulations (Digel, private communication) of the spectral capability for high latitude, steady sources in a one-year survey, i.e. roughly 8 weeks on source. The real GLAST sensitivity for transient GRBs of duration t_{dur} can be estimated to be roughly $[(8\,\mathrm{weeks})/t_{\mathrm{dur}}]^{1/2}$ times that depicted. Note that the differential sensitivity is the most appropriate measure for spectral diagnostic capabilities. Evidently, ACTs and GLAST working in concert will be able to determine the spectral shape and evolution of bright, flat-spectrum bursts like GRB930131 if the attenuation is no more dramatic than $1/(1 + \tau_{\gamma\gamma})$. The particular evolutionary scenario depicted in Fig. 2 is an adiabatic one for blast wave deceleration during the sweep-up phase, where the dependences on time t are $\Gamma \propto t^{-3/8}$, $\varepsilon_B \propto \Gamma^4 \propto t^{-3/2}$, and $\varepsilon_B^2 f(\varepsilon_B) \propto \Gamma^{8/3} \propto t^{-1}$ for the flux at the peak [14]. Shifts in the turnover energy and sub-TeV break energy, and correlations with BATSE flux and break energy should be discernible in bright sources.

It must be emphasized that these internal absorption characteristics are easily distinguishable from those of external absorption due to the cosmological infrared background along the line of sight [15,16]. Attenuation by such background fields couples to the redshift, not parameters internal to the source nor the shape of the spectrum in the BATSE and EGRET bands. Furthermore, it is always exponential in nature (i.e. of severity equivalent to the dashed curve in Fig. 2) since the emission region is distinct from the location of the soft target photons, and is patently independent of time. The possibility of confusing such with the internal attenuation that forms the focus of this paper seems minimal. Hence, the prospects for powerful spectral diagnostics in bright bursts with atmospheric Čerenkov telescopes and the GLAST mission promise an exciting time ahead for the field of high energy gamma-ray astronomy.

Acknowledgments: I thank Alice Harding and Brenda Dingus for helpful discussions, and Seth Digel for simulating GLAST spectral sensitivities.

REFERENCES

1. Hurley, K. et al. *Nature* **372**, 652 (1994).
2. Schmidt, W. K. H. *Nature* **271**, 525 (1978).
3. Meegan, C., et al. *Ap. J. Supp.* **106**, 65 (1996).
4. Fenimore, E. E., Epstein, R. I. & Ho, C. in *Gamma-Ray Bursts,* eds. Paciesas, W. S. and Fishman, G. J., (AIP, New York) p. 158 (1992).
5. Krolik, J. H. & Pier, E. A. *Ap. J.* **373**, 277 (1991).
6. Baring, M. G. *Ap. J.* **418**, 391 (1993)
7. Baring, M. G. & Harding, A. K. *Ap. J.* **491**, 663 (1997b).
8. Band, D., et al. *Ap. J.* **413**, 281 (1993).
9. Baring, M. G. & Harding, A. K. *Ap. J. Lett.* **481**, L85 (1997a).
10. Schneid, E. J., et al. *Astron. Astr. (Lett.)* **255**, L13 (1992).
11. Sommer, M., et al. *Ap. J. Lett.* **422**, L63 (1994).
12. Gould, R. J. & Schreder, G. P. *Phys. Rev.* **155**, 1404 (1967).
13. Connaughton, V. et al. *Ap. J.* **479**, 859 (1997).
14. Dermer, C. D., Chiang, J. & Böttcher, M. *Ap. J.* **513**, 656 (1999).
15. Stecker, F. W. & De Jager, O. C. *Space Sci. Rev.* **75**, 401 (1996).
16. Mannheim, K., Hartmann, D. & Funk, B. *Ap. J.* **467**, 532 (1996).

First Results of a Study of TeV Emission from GRBs in Milagrito

J. E. McEnery[1], R. Atkins[1], W. Benbow[2], D. Berley[3,10],
M.L. Chen[3,11], D.G. Coyne[2], B.L. Dingus[1], D.E. Dorfan[2],
R.W. Ellsworth[5], D. Evans[3], A. Falcone[6], L. Fleysher[7], R. Fleysher[7],
G. Gisler[8], J.A. Goodman[3], T.J. Haines[8], C.M. Hoffman[8],
S. Hugenberger[4], L.A. Kelley[2], I. Leonor[4], M. McConnell[6],
J.F. McCullough[2], R.S. Miller[8,6], A.I. Mincer[7], M.F. Morales[2],
P. Nemethy[7], J.M. Ryan[6], B. Shen[9], A. Shoup[4], C. Sinnis[8],
A.J. Smith[9], G.W. Sullivan[3], T. Tumer[9], K. Wang[9], M.O. Wascko[9],
S. Westerhoff[2], D.A. Williams[2], T. Yang[2], G.B. Yodh[4]

(1) University of Utah, Salt Lake City, UT 84112, USA
(2) University of California, Santa Cruz, CA 95064, USA
(3) University of Maryland, College Park, MD 20742, USA
(4) University of California, Irvine, CA 92697, USA
(5) George Mason University, Fairfax, VA 22030, USA
(6) University of New Hampshire, Durham, NH 03824, USA
(7) New York University, New York, NY 10003, USA
(8) Los Alamos National Laboratory, Los Alamos, NM 87545, USA
(9) University of California, Riverside, CA 92521, USA
(10) Permanent Address: National Science Foundation, Arlington, VA ,22230, USA
(11) Now at Brookhaven National Laboratory, Upton, NY 11973, USA

Abstract. Milagrito, a detector sensitive to γ-rays at TeV energies, monitored the northern sky during the period February 1997 through May 1998. With a large field of view and high duty cycle, this instrument was used to perform a search for TeV counterparts to γ-ray bursts. Within the Milagrito field of view 54 γ-ray bursts at keV energies were observed by the Burst And Transient Satellite Experiment (BATSE) aboard the Compton Gamma-Ray Observatory. This paper describes the results of a preliminary analysis to search for TeV emission correlated with BATSE detected bursts. Milagrito detected an excess of events coincident both spatially and temporally with GRB 970417a, with chance probability 2.8×10^{-5} within the BATSE error radius. No other significant correlations were detected. Since 54 bursts were examined the chance probability of observing an excess with this significance in any of these bursts is 1.5×10^{-3}. The statistical aspects and physical implications of this result are discussed.

I OBSERVATIONS AND ANALYSIS

Milagro, a new type of TeV γ-ray observatory sensitive at energies above 100 GeV, with a field of view of over one steradian and a high duty cycle, began operation in February 1999, near Los Alamos, NM. A predecessor of Milagro, Milagrito [5], operated from February 1997 to May 1998. During this time interval, 54 γ-ray bursts (GRBs) detected by BATSE [1] were within Milagrito's field of view (less than 45° zenith angle).

A search was conducted in the Milagrito data for an excess of events above the cosmic-ray background coincident with each of these γ-ray bursts. For each burst, a circular search region was defined by the BATSE 90% confidence interval, which incorporates both the statistical and systematic position errors [2]. The size of this 90% confidence interval ranged from 4° to 26° for the 54 GRBs in the sample. The search region was tiled with an array of overlapping 1.6° radius bins centered on a 0.2° × 0.2°grid. This radius was derived from the measured angular resolution of Milagrito and was selected prior to the search. The number of events falling within each of the 1.6° bins was tallied for the duration of the burst reported by BATSE. This duration is defined as the time required for BATSE to accumulate 90% of the γ-rays(T90). T90 ranged from 0.1 seconds to 195 seconds for the 54 bursts examined.

The angular distribution of background events on the sky was characterized using two hours of data surrounding each burst. This distribution was normalized to the number of events detected by Milagrito over the entire sky during the T90 interval (N_{T90}). The resulting background data were also binned in 1.6° bins spaced 0.2° apart. The Poisson probability that the excess of events in each 1.6° bin was due to a background fluctuation was calculated and the bin with lowest probability was taken as the candidate position of a TeV γ-ray counterpart to the BATSE burst. The background and signal counts in this bin were used to calculate a fluence or fluence upper limit for each burst.

II RESULTS

The flux sensitivity of Milagrito to γ-ray bursts depends on the zenith angle and duration of the burst, and on the instrument conditions at the time. During the lifetime of the Milagrito detector, data were taken with three different water depths (0.9 m, 1.5 m and 2.0 m). In addition, for the period February 1997 through the end of March 1997 a considerable amount of snow collected on the cover of the pond. Detector simulations were used to obtain effective area as a function of zenith angle for an assumed $E^{-2.0}$ spectrum for each of these configurations. These were then used to calculate flux upper limits for each burst. Flux upper limits in the range $10^{-6} - 10^{-8}$ $\gamma/cm^2/s$ were obtained for 53 of the 54 bursts in the sample.

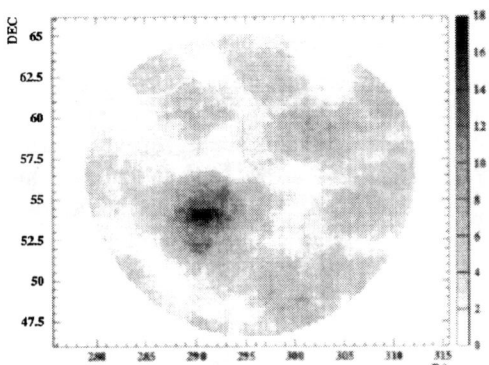

FIGURE 1. Number of events recorded by Milagrito during T90 in the BATSE error radius for GRB 970417a, each bin contains the number of events detected by Milagrito within a 1.6 degree radius.

Of the 54 bursts one, GRB 970417a, shows a substantial excess above background in the Milagrito data. The BATSE detection of this burst is a weak burst with a fluence in the 50–300 keV energy range of 1.5×10^{-7} ergs/cm^2 and T90 of 7.9 seconds. BATSE determined the burst position to be RA = 295.66°, DEC = 55.77°. The 90% positional uncertainty was 9.4°. The 1.6° radius bin with the largest excess in the Milagrito data is centered at RA = 289.89° and DEC = 54.0°, corresponding to a zenith angle of 21°. This position is 3.8° away from the position reported by BATSE; well within the BATSE 1-sigma position error 6.2°. The uncertainty in the position of the TeV candidate was determined by Monte-Carlo simulations to be approximately 0.5°. Figure 1 shows the number of counts in this search region for the array of 1.6° bins. The bin with the largest excess has 18 events with an expected background of 3.46±0.11. The Poisson probability for observing a signal at least this large due to a background fluctuation is 2.89×10^{-8}.

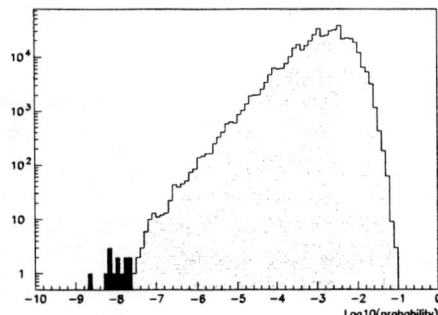

FIGURE 2. The distribution of minimum probabilities for the ensemble of simulated data-sets for GRB 970417a.

To obtain the significance of this result one must account for the size of the search region. The probability of obtaining the observed significance anywhere within the entire search region was determined by Monte Carlo simulations. A set of simulated signal maps was made by randomly drawing N_{T90} events from the background distribution. Each map was searched, as before, for a significant excess within the search region defined by BATSE. The probability of the observation in the actual data being due to a fluctuation in the background, after accounting for the size of the search region, is given by the ratio of the number of simulated data sets with probability less than that observed for the actual data to the total number of simulated data sets. The distribution of the probabilities for 4.65×10^6 simulated data sets is shown in figure 2; thirteen of which had Poisson probability less than 2.89×10^{-8}. We therefore find that the chance probability of such a detection within the entire 9.4° search region for GRB 970417a to be 2.8×10^{-5}. The probabilities for each of the other 53 bursts in the sample were obtained using the same method, the distribution of these probabilities, after correcting for the size of the search region, is shown in figure 3. The histogram on the left, plotted on a log-linear scale, illustrates the significance of the excess for GRB 970417a relative to the rest of the sample. The histogram on the right of this figure, plotted on a linear scale is flat, as expected. 54 bursts were examined. Therefore the chance probability of observing such a significant excess due to fluctuations in the background for any of these bursts is 1.5×10^{-3}.

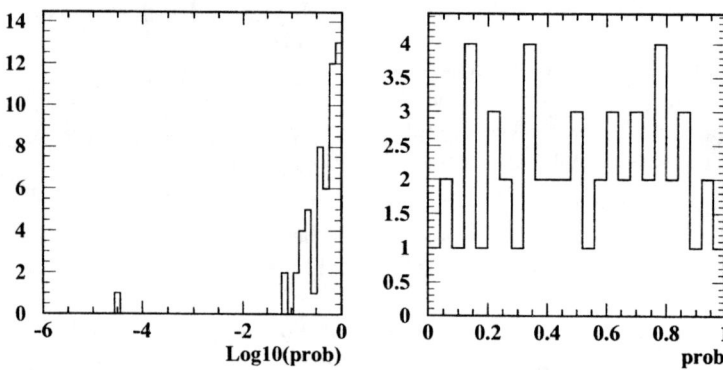

FIGURE 3. The distribution of probabilities, corrected for the size of the search region for the 54 GRBs in the sample, both plots show the same data with a linear and logarithmic scale for the x-axis

Although the initial search was limited to T90, for GRB 970417a longer time intervals were also examined. To allow for the positional uncertainty of the excess observed by Milagrito, the radius of the search bin was increased to 2.2° for this search. A search for TeV γ–rays integrated over long time intervals of one hour, two hours and a day after the GRB start time did not show any significant excess. Lightcurves where the data are binned in intervals of one second and of T90 (7.9 s) are shown in figure 4. A preliminary analysis reveals no statistically compelling evidence for TeV afterflares.

III DISCUSSION

If the observed excess of events in Milagrito is indeed associated with GRB 970417a then it represents the highest energy photons yet detected from a GRB in coincidence with the sub-MeV emission. The following discussion assumes that the excess observed by Milagrito was due to TeV γ-rays from GRB 970417a. The TeV spectrum and maximum energy of emission is difficult to determine from Milagrito data [5]. Monte Carlo simulations of γ-ray initiated air showers show that the

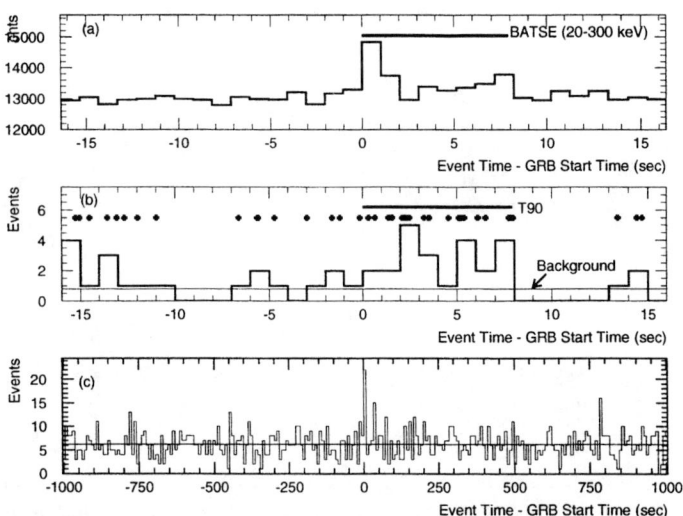

FIGURE 4. GRB 970417a: (a) The BATSE lightcurve, (b) Milagrito data within a 2.2° radius of GRB 970417a integrated in 1 second bins, the crosses indicate the arrival times of the events and (c) integrated in bins of 7.9 seconds (T90) for 2000 seconds

effective area increases smoothly with energy, making the definition of an energy threshold ambiguous. Figure 5 shows the implied fluence of this observation as a function of upper cutoff energy for a range of power-law input spectra.

Some information about the energies of the events observed for GRB 970417a can be obtained by considering the response of the summed untriggered counting rate of the individual PMTs in Milagrito. Detector simulations of the effect on PMT counting rates of γ-ray induced air-showers indicate that these rates are more sensitive than the standard shower data at energies below a few hundred GeV, but are only sensitive to very large fluxes [5]. No excess was observed in these rates, which implies that the air-showers detected by Milagrito were probably due to γ-rays at energies above several hundred GeV.

High energy γ-rays from sources at cosmological distances will be absorbed via electron-positron pair production with infrared photons in the intergalactic medium. Several studies find that the opacity due to pair production for above 200 GeV γ-rays exceeds one for redshifts larger than 0.3 [3,4]. Thus, if Milagrito has indeed detected high energy photons from GRB 990417a, it must be from a relatively nearby object.

IV CONCLUSION

An excess of events with chance probability 2.8×10^{-5} coincident both spatially and temporally with the BATSE emission for GRB 970417a was observed by

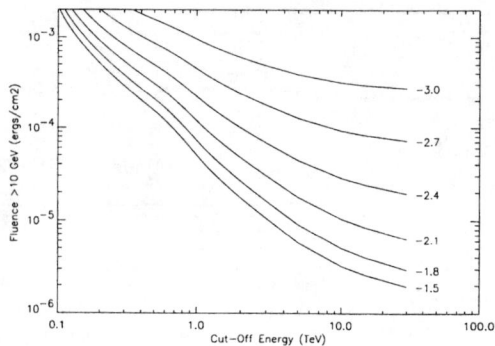

FIGURE 5. Implied fluences of this candidate for a range of assumed power-law spectra and high energy cutoffs (preliminary)

Milagrito. The chance probability that an excess of this significance would be observed from the entire sample of 54 bursts is 1.5×10^{-3}. The spectrum must extend with no cutoff to at least a few hundred GeV. The inferred TeV fluence from this result at least an order of magnitude greater than the sub-MeV fluence and the emission extends to at least several hundred GeV.

If the observed excess from GRB 970417a is not a fluctuation of the background, then a new class of γ-ray bursts bright at TeV energies may have been observed. A search for other coincidences with BATSE, to verify this result, will be continued with the current instrument, Milagro, which has increased sensitivity to TeV γ-ray bursts.

ACKNOWLEDGMENTS

This research was supported in part by the National Science Foundation, the U.S. Department of Energy Office of High Energy Physics, the U.S. Department of Energy Office of Nuclear Physics, Los Alamos National Laboratory, the University of California, the Institute of Geophysics and Planetary Physics, The Research Corporation, and the California Space Institute.

REFERENCES

1. W. S. Paciesas et al., (Astro-Ph-9903205) (1999)
2. M. S. Briggs et al., *Astrophys. J. Supp.* **122(2)**, 503 (1999)
3. M.H. Salamon and F. W. Stecker, *Astrophys. J.* **493**, 547 (1998).
4. J. R. Primack et al, *Astroparticle Physics* **11**, 93 (1999).
5. R. Atkins et al., *Nucl. Inst. and Methods* (1999) (submitted).

On the conversion of blast wave energy into radiation in AGN and GRBs

Martin Pohl and Reinhard Schlickeiser

Institut für theoretische Physik IV, Ruhr-Universität Bochum
44780 Bochum, Germany

Abstract. We address the important issue how the kinetic energy of collimated blast waves in AGN and GRBs is converted into radiation. It is shown that swept-up ambient matter is quickly isotropised in the blast wave frame by a relativistic two-stream instability, which provides relativistic particles in the jet without invoking any acceleration process. The fate of the blast wave and the spectral evolution of the emission of the energetic particles is therefore solely determined by the initial condition. We compare our model with existing multiwavelength data of AGN and find remarkable agreement.

INTRODUCTION

Existing models of the high energy γ-ray emission of AGN concentrate on the radiation process and on the temporal evolution of the spectrum of radiating particles, but generally neglect the problem of particle injection and acceleration, on which the observed behaviour of the sources places strong constraints. As an example, the TeV lightcurves of γ-ray emitting AGN often show secular variability on time scales of several months, on which rapid fluctuations are superposed [1]. This is difficult to account for, because the radiating particles need to be resupplied for every short time flare and it is unclear what kind of externally controlled energy reservoir may be used to provide the required source power. Therefore, if the particle acceleration problem is so difficult to deal with, it may be useful to construct a model in which particle acceleration is not required. In this paper we consider a strong electron-proton beam that sweeps up ambient matter and thus becomes enriched with relativistic particles without acceleration.

Viewed from the coordinate system comoving with the blast wave, the interstellar protons and electrons represent a proton-electron beam propagating with relativistic speed antiparallel to the x-axis. We demonstrate that very quickly the beam excites low-frequency magnetohydrodynamic plasma waves via a two-stream instability which isotropise the incoming interstellar protons and electrons in the blast wave plasma. Inelastic collisions between primary protons and the blast wave protons generate neutral and charged pions which decay into γ-rays, secondary electrons, positrons and neutrinos. Both, the radiation products from these interactions, and the resulting cooling of the primary particles in the blast wave plasma, are calculated. By transforming to the laboratory frame we calculate the time evolution of the emitted multiwavelength spectrum for an outside observer under different viewing angles. The deceleration of the blast wave is taken into account self-consistently. Since we do not consider any re-acceleration of particles in the blast wave, the evolution of particles and the blast wave is completely determined by the initial conditions. A more detailed description of our model will be published elsewhere [2].

THE TWO-STREAM INSTABILITY OF A PROTON-ELECTRON BEAM

All physical quantities are given in the blast wave frame, except those indexed with $*$, which are measured in the laboratory frame. In the blast wave frame the interstellar medium forms a weak, relativistic beam, which can excite plasma waves. The time-dependent behaviour of the intensities $I(k,t)$ of the excited waves is given by $\frac{\partial I_\pm}{\partial t} = \pm \psi I_\pm$ where the growth rate ψ is

$$\psi(k) \simeq \pi^2 c^3 [\frac{\partial J(\omega_R)}{\partial \omega_R}]^{-1} \text{sgn}(k) \sum_i \omega_{p,i}^2 (m_i c)^3$$
$$\times \int_{E_i}^{\infty} dE \, \frac{E^2 - 1 - (\frac{E}{N} - x_i)^2}{\sqrt{E^2 - 1}} \frac{\partial f_i}{\partial \mu} \delta(\mu + \frac{x_i}{\sqrt{E^2 - 1}}) \,, \tag{1}$$

and $N = ck/\omega_R$ is the index of refraction, and the limit of integration $E_i = \sqrt{1 + x_i^2}$ with $x_i = \Omega_{i,0}/kc$. To describe the influence of these excited waves on the beam particles we use the quasilinear Fokker-Planck equation [3] for the resonant wave-particle interaction. For Alfvén waves the index of refraction is large compared to unity, so that the Lorentz force associated with the magnetic field of the waves is much larger than the force associated with the electric field, and therefore on the shortest time scale these waves scatter the particles in pitch angle μ but conserve their energy, i.e. the isotropise the beam particles. The Fokker-Planck equation for the phase space density then displays only a term for pitch angle scattering with

$$D_{\mu\mu} \simeq \sum_n \frac{\pi \Omega_i^2 (1 - \mu^2)}{2 B^2} \int_{-\infty}^{\infty} dk \, I_n(k) \delta(\omega_R - kv\mu - \Omega_i) \tag{2}$$

We can estimate the isotropisation length by using the fully-developed turbulence spectra for calculating the pitch angle Fokker-Planck coefficient. For ease of exposition we assume that the initial turbulence spectrum has the form

$I(k,0) = I_0 k^{-2}$. Inserting our typical parameter values we obtain for the scattering length and the isotropisation time scale in the blast wave plasma

$$\lambda \simeq 10^{11} \frac{n_{b,8}^{1/2}}{\Gamma_{100}\, n_i^*} \quad \text{cm} \quad \text{and} \quad t_R = \lambda/c = 3.5\, \frac{n_{b,8}^{1/2}}{\Gamma_{100}\, n_i^*} \quad \text{s} \qquad (3)$$

If the thickness d of the blast wave region is larger than the scattering length, an isotropic distribution of the inflowing interstellar protons and electrons with Lorentzfactor $<\Gamma> = \Gamma(1-\beta_A\beta) \simeq \Gamma$ in the blast wave frame is indeed generated. In the following section we investigate the radiation products resulting from the inelastic interactions of these primary particles with the cold blast wave plasma.

RADIATION MODELLING OF THE BLAST WAVE

In the blast wave frame the external density $n_i = \Gamma n_i^*$, the blast wave surface area A is constant, and sweep-up occurs at a rate

$$\dot{N}(\gamma) = \pi R^2 c n_i^* \sqrt{\Gamma^2 - 1}\, \delta(\gamma - \Gamma)\,. \qquad (4)$$

The sweep-up is a source of isotropic, quasi-monoenergetic protons and electrons with Lorentz factor Γ in the blast wave frame. The isotropisation also provides a momentum transfer from the ambient medium to the blast wave. Therefore, the blast wave decelerates at a rate

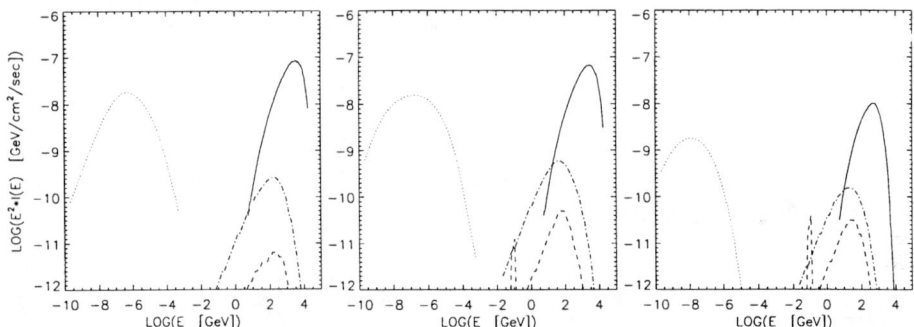

FIGURE 1. Spectral evolution of a relativistic blast wave having traversed a gas cloud of density $n_i^*=0.1$ cm^{-3} and thickness $5\cdot 10^{16}$ cm. The other parameters are: $\Gamma_0=200$, d=10^{13}cm, R=10^{14}cm, B= 1 G, $n_b=5\cdot 10^8$ cm^{-3}, for an AGN at z=0.1 viewed at an angle $\theta=0.3°$. Displayed are the individual spectra of the radiation processes π^0-decay (solid line), synchrotron emission (dotted), bremsstrahlung (dot-dashed), and annihilation (dashed). The panels show from left to right the νF_ν spectra after 360 seconds, 3600 seconds, and 36000 seconds observed time. The Lorentz factor of the blast wave did virtually not change, hence repeated cloud crossings would produce multiple outbursts with the same spectral evolution.

$$\dot{\Gamma} \simeq -\frac{A\, m_p\, c\, n_i^*\, (\Gamma^2-1)^{3/2}}{M_{BW}} \tag{5}$$

where the blast wave mass M_{BW} includes the relativistic particles.

Equation (4) states the differential injection of relativistic particles in the blast wave. Here we concentrate on the protons because they receive a factor of m_p/m_e more power than electrons, which also have a low radiation efficiency for $\gamma \ll 1000$. Electrons (and positrons) are supplied much more efficiently as secondary particles following inelastic collisions of the relativistic protons. Since no reacceleration is assumed, the continuity equations for particle i reads

$$\frac{\partial N_i(\gamma)}{\partial t} + \frac{\partial}{\partial \gamma}\left(\dot{\gamma}\, N_i(\gamma)\right) + \frac{N_i(\gamma)}{T_{ann}} = \dot{N}_i(\gamma) \tag{6}$$

where for positrons catastrophic annihilation losses on a timescale T_{ann} are taken into account. The injection rate of secondary electron is related to the rate of inelastic collisions. In Fig. 1 we show the spectral evolution of high energy emission from a collimated blast wave interacting with an isolated gas cloud.

CONCLUSIONS

We have shown that a relativistic blast wave can sweep-up ambient matter via a two-stream instability which provides relativistic particles in the blast wave without requiring any acceleration process. The high energy emission thereby produced has characteristics typical of BL Lacertae objects. In particular,
– The high energy spectra are very hard, in accord with the unspectacular appearance of TeV-bright sources at GeV energies.
– Variability at TeV energies can be produced on sub-hour timescales, in accord with the observed variability time scales of Mkn 421 and Mkn 501.
– For multiple outbursts the intensity follows the variation of the ambient gas density with little spectral variation, which is the observed behaviour of Mkn 501.
– X-ray synchrotron emission is produced in parallel to the γ-rays.

ACKNOWLEDGEMENTS

MP acknowledges support by a travel grant, Verbundforschung DESY 05 AG9PCA.

REFERENCES

1. Quinn, J., Bond, I.H., Boyle, P.J., et al., *ApJ* **518**, 693 (1999).
2. Pohl, M., Schlickeiser, R., *A&A*, submitted (1999).
3. Schlickeiser, R., *ApJ* **336**, 243 (1989).

Detection Techniques of μs Gamma-Ray Bursts using Ground-Based Telescopes

Frank Krennrich[1], Stephane Le Bohec[1] & Trevor Weekes[2]

[1]*Physics & Astronomy Department, Iowa State University*
Ames, Iowa 50011-3160, Osborne Drive
[2]*Harvard-Smithsonian Center for Astrophysics, Whipple Observatory*
670 Mount Hopkins Road, Amado, AZ, 85645

Abstract.
We propose to use imaging atmospheric Cherenkov telescopes (IACTs) to search for cosmic bursts of E > 200 MeV γ-ray emission. This energy regime has been the domain of the space based EGRET detector on the Compton Gamma Ray Observatory (CGRO), providing a good sensitivity for bursts lasting for more than 200 ms. Theoretical predictions of high-energy γ-ray bursts produced by quantum-mechanical decay of primordial black holes (Hawking 1971) suggest the emission of bursts on shorter time scales, lasting for a tenth of a microsecond or longer depending on particle physics. Upcoming next generation IACTs have the potential to detect microsecond scale γ-ray bursts, an observational window which is virtually unexplored. We discuss a technique, which is based on the detection of multi-photon-initiated air showers providing a sensitivity for bursts lasting nanoseconds to several microseconds.

INTRODUCTION

One of the most astounding predictions is that black holes should evaporate. Furthermore, this evaporation should proceed by emission of particles of increasing energy giving (Hawking 1974) a final tremendous explosion with duration determined by particle physics at energies well beyond the capabilities of accelerators. Primordial black holes (PBHs) may have formed as early as 10^{-23} s after the Big Bang, and those of initial mass $\approx 4 \times 10^{14}$ grams, would at present be near the end of their lifetime. The two most extreme particle physics scenarios determining the final evaporation time scale and average photon energy are given by the standard model and the Hagedorn model (Hagedorn 1970). In the latter, a burst of 250 MeV γ-rays as short as 10^{-7} s with a total energy of 10^{34} ergs would be the signature of such an event. In contrast, the standard model predicts a burst of 1 second duration (Halzen 1991) and an average photon energy of 10 TeV.

A search for sub-microsecond scale bursts using EGRET (dead time limited to time scales of > 200 ms) has been made by looking for multiple-γ-ray events arriving almost simultaneously within a single chamber gate (i.e., 600 ns); it produced an upper limit of $5 \times 10^{-2}/\text{yr}/\text{pc}^3$ (Fichtel et al. 1994). In an early experiment, first generation non-imaging ground-based atmospheric Cherenkov detectors were used to search on the shortest time-scales predicted (10^{-7} s), setting an upper limit of $4 \times 10^{-2}/\text{yr}/\text{pc}^3$ (Porter & Weekes 1978).

In this paper we discuss the potential of using IACTs to detect γ-ray flare phenomena on time scales of 0.1 - 10 microseconds. A short burst can be approximated as a thin plane wavefront of γ-rays traveling through space (wavefront event, hereafter), starting a multi-photon-initiated cascade when entering the earth's atmosphere. Measuring the angular distribution of Cherenkov light and the pulse profile from a short burst using IACTs, provides a means of distinguishing short bursts from background by cosmic rays. Previous efforts (Porter & Weekes 1978) used non-imaging Cherenkov detectors, and the suppression of cosmic rays was achieved by simultaneous recording by two telescopes separated at a distance of 400 km.

PHENOMENOLOGY OF WAVEFRONT EVENTS

The technique proposed here builds upon the atmospheric Cherenkov imaging technique that has been pivotal in establishing the field of TeV γ-ray astrophysics (for review see Ong et al. 1998; Weekes 1999).

Cherenkov light from a plane wavefront of multiple $E > 200$ MeV γ-rays can be detected with ground-based optical telescopes. A low energy multi-photon-initiated cascade differs significantly from a single-particle-initiated cascade e.g., a TeV photon or proton induced shower. Individual low energy γ-rays, when reaching the upper atmosphere, will typically generate one or a few (depending on energy) generations of electrons and positrons (collectively called electrons hereafter) by pair production and subsequent bremsstrahlung. Electrons, before falling below the critical energy, radiate Cherenkov light (6000 photons/electron/radiation length) which can be collected by an optical reflector at ground level. Although, a Cherenkov flash of a single sub-GeV γ-ray cascade is too faint to be detectable, a large number of γ-rays arriving within a short time can produce a Cherenkov signal strong enough to be recorded by an IACT. Three unique characteristics of the Cherenkov light image from a wavefront event should be noted.

a) The first one is the very large extent of the wavefront, which means it can be detected simultaneously by telescopes over vast distances. The images in all telescopes in an IACT array should be identical, independent from the distance between telescopes. This is different from single-particle-initiated shower images which are detectable over a limited area on the ground and, if detected, show a parallactic displacement between telescopes.

b) The second characteristic is the time profile of the Cherenkov pulse given by the burst profile (0.1 - 10 microseconds) - and thus is quite different from Cherenkov flashes of conventional air showers showing durations of 5-30 nanosecond.

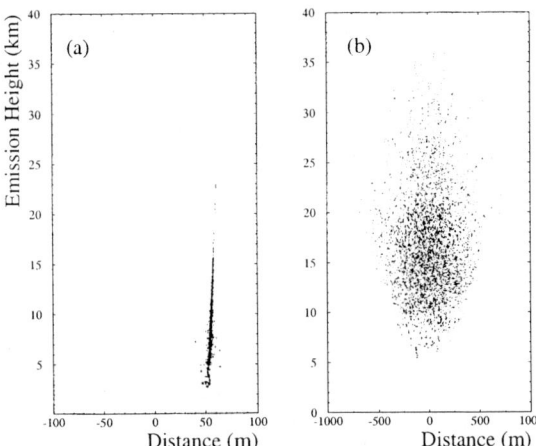

c) The images in the camera plane from a wavefront event are circular. They also will provide information about the arrival direction of the wavefront: the displacement of the image centroid from the optic axis of the telescope measures the arrival direction of the burst. Monte Carlo simulations were used to study the signatures of multi-photon-initiated cascades. The simulation code ISUSIM (Mohanty et al. 1998) includes the detector model of the Whipple Observatory 10 m telescope equipped with a 4.8° field of view camera. Fig. 2a shows the Cherenkov light image from a simulated wavefront event of 300 MeV γ-rays (fluence = 2.4 × 10^{-8}ergs/cm^2).

FIGURE 1. The lateral and longitudinal development of an air shower showing the origin of Cherenkov photons emitted: a) for a 1 TeV γ-ray shower, b) for a multi-photon-initiated shower of 300 MeV γ rays

The area and the gray-scale of the filled circles indicates the number of photoelectrons (p.e.) detected in each pixel[1]. Fig. 2b shows the image of a wavefront event (fluence = 4.8 × 10^{-8}erg/cm^2) arriving 1.13° off-axis, and it can be seen that the image is off-set by $\approx 1.1°$ from the center of the camera, with its centroid indicating the arrival direction of the burst. In the case of the image in Fig. 2b, a smoothly decreasing "halo" can be seen. The light beyond the central plateau (0.3° in radius) is caused mainly by the multiple-scattering of relatively low energy electrons. This halo is not easily recognizable in Fig. 2a (lower fluence), because of the smaller amount of Cherenkov light contaminated with night-sky background. The structure of the image can be described by its circular shape and its characteristic radius. The radial extend of the images is described by $Radius^2$. Fig. 3a shows the average radial light density profile of wavefront events exhibiting a F.W.H.M. of 0.8°. Fig. 3b shows the estimated $Radius$ distribution of wavefront events compared with detected cosmic-ray background events. Even though wavefront events have a characteristic shape, image analysis might not remove background completely.

---

[1] The noise from night sky fluctuations for a 100 ns gate width has been included, image processing as described in Reynolds et al. (1993) has been applied.
[2] The image shape is described here using a combination of the parameters, $Width$ and $Length$ (Hillas 1985), defined by: $Radius = (Width + Length)/2$.

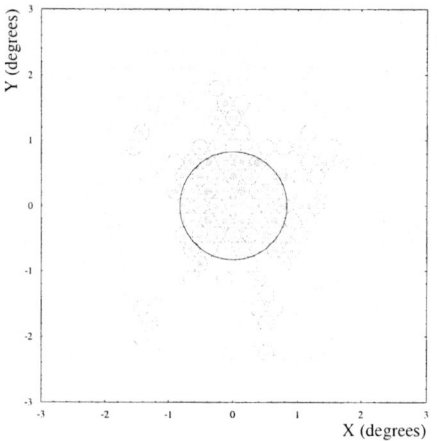
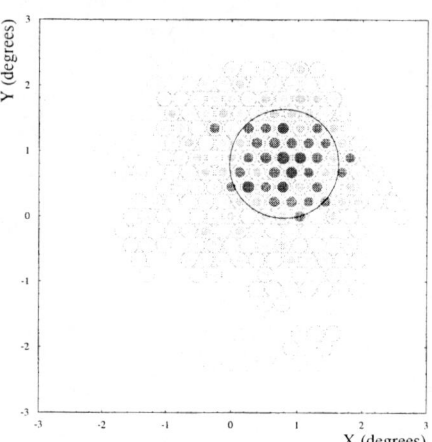

FIGURE 2.a The simulated image of 100 nanosecond burst of 300 MeV γ-rays. The circle indicates the angular extension and the shape of the image, whereas its center coincides with the burst arrival direction.

FIGURE 2.b The image of a 100 ns burst of 300 MeV γ-rays. The arrival direction was offset by 1.13° from the optic axis of the telescope. The circle indicates the plateau and the drop in the light density of the image.

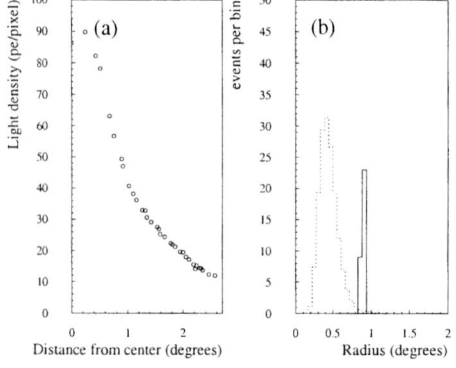
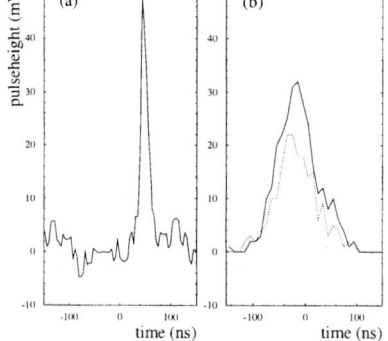

FIGURE 3.a The average radial light profile (light density vs. radial distance from image center) of wavefront events is shown. **b)** The *Radius* distribution of simulated 500 MeV wavefront events (solid line) and detected cosmic-ray background events (dashed line) is shown.

FIGURE 4.a The pulse shape of a cosmic-ray event is shown. The noise is due to fluctuations from the night sky background. **b)** The pulse profile of a simulated multi-photon-initiated cascade for 2 pixels one in the center of the image (solid line) and one by 1° off-center (dotted line) are shown (100 ns burst).

The width of the Cherenkov light pulse provides an additional signature to distinguish multi-photon-initiated cascades from single-particle-initiated air showers. Pulse shapes from cosmic-ray air showers are typically a few nanoseconds wide.

Fig. 4a shows the pulse shape of a typical Cherenkov light flash recorded with the Whipple Observatory 10 m telescope. In comparison we show (Fig. 4b) the simulated pulse profiles from a multi-photon-initiated cascade from a 100 ns burst of 500 MeV γ-rays of two different pixels: in the center of the image (solid line) and a pixel 1° off-center (dashed line). The pulse profiles of the multi-photon-initiated cascade are broad and only slightly shifted with respect to each other.

The trigger threshold for the detection of short bursts is a function of the fluence of the burst and is usually expressed in ergs/cm^2. The fluence is the product of the energy of the incoming particles and the number of particles per unit area impinging on the upper atmosphere. In order to trigger on a wavefront event we require 40 pixels to exceed the night-sky background fluctuations by 3σ, preventing accidentals and also ensures a good image reconstruction.

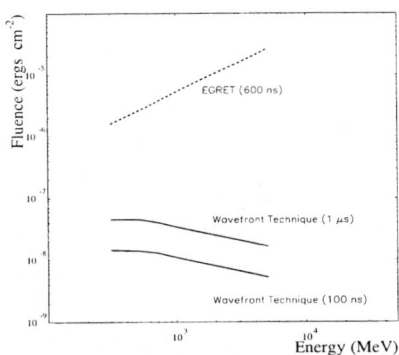

FIGURE 5. Fluence sensitivity for the wavefront technique.

Fig. 5 shows the fluence sensitivity as a function of energy for 100 ns and 1 μs burst time scale. For comparison to previous efforts we also show the sensitivity of the EGRET detector, based on the detection of multiple events within one readout cycle. We have assumed here a collection area of 0.15 m^2 and a minimum of 5 γ-rays to be detected. Over the energy range of 300 MeV to 1 GeV the sensitivity of the wavefront technique could exceed EGRET's sensitivity by a factor of 100 to 500 for 100 ns bursts. In the view of future upcoming highly sensitive IACT arrays (HESS and VERITAS), the microsecond burst detection technique could be explored in parallel to standard γ-ray observations.

REFERENCES

1. Fichtel C., et al., *ApJ* **434**, 557 (1994).
2. Hagedorn, R., , *A & A* **5**, 184 (1970).
3. Halzen, F., Zas, E., MacGibbon, J.H., Weekes, T.C., *Nature* **353**, 807 (1991).
4. Hawking, S.W., *M.N.R.A.S* **152**, 75 (1971).
5. Hawking, S.W., *Nature* **248**, 31 (1974).
6. Hillas, A.M., *Proc. of 19th ICRC (La Jolla)*, **3**, 445 (1985).
7. Krennrich F., Le Bohec, S., & Weekes, T.C., *ApJ* , **529**, in press (2000).
8. Mohanty, G., et al., *Astropart. Phys.* **9**, 15 (1998).
9. Porter, N. A., & Weekes, T. C., *M.N.R.A.S* **183** , 205 (1978).
10. Reynolds, P.T., et al., *ApJ* , **404**, 206 (1993).
11. Ong R.A., *Physics Reports* **305**, 93 (1998).
12. Weekes T.C., *these proceedings* (1999).

OTHER SOURCES

EGRET Unidentified Sources

Isabelle A. Grenier

Université Paris 7 & Service d'Astrophysique CEA Saclay
91191 Gif/Yvette, France

Abstract. Nearly 60% of the EGRET sources are still defeating our attempts at identifying them. The present strategy to elucidate their nature is three-fold : search for counterparts at other wavelengths; search for temporal signatures characteristic of pulsars, binaries or flaring sources; correlation studies to separate populations with distinct spectral/spatial/temporal behaviours. Although no clear picture has emerged yet as to the nature or variety of the objects involved, interesting clues have been recently gleaned from these three lines of work that may shed a new light on possible candidates besides the traditional pulsars and shell supernova remnants. Observational facts are briefly reviewed hereinafter and sprinkled with personal comments.

SEARCH FOR GALACTIC COUNTERPARTS AND PERIODICITIES

Source identification at other wavelengths is hindered by the lack of counterparts or by counterpart confusion in the large error boxes near the Galactic plane. Smaller GeV error boxes, when available, reduce chance probabilities. The latest 3rd EGRET catalogue positions (1), derived from integral photon fluxes > 100 MeV over 4 years of data, are almost as accurate, yet sometimes are 1 or 2σ away from the GeV positions. These discrepancies may reflect systematic errors due to the highly structured, underlying interstellar emission which is sampled at different angular scales above 100 MeV and 1 GeV. One should keep these limitations in mind when dealing with sources at low latitudes near steep background gradients or clouds.

Pulsars and their nebulae

Indirect pulsar activity, i.e. γ-ray production from the synchrotron nebula powered by the pulsar wind or from relic high-energy electrons may play an important role in the unidentified sources. Extreme cases of such activity, with emission extending to TeV energies, have been found in the Crab, Vela, and PSR1706-44 environs (2). MeV to TeV emission from the Crab nebula is successfully interpreted as synchrotron-self-Compton emission in the nebular magnetic field from electrons energized at the pulsar wind terminal shock. For the compact synchrotron nebula of PSR1706-44 and the "bubble" of relic electrons trailing off the Vela pulsar, inverse-Compton (IC) scattering of the cosmic microwave background (CMB) radiation, or of the local IR field, is favoured. In these small nebulae, the necessary magnetic field falls 3 to 6 times below the equipartition value and raises confinement problems over long periods. In this context, it may prove meaningful to note that ~20% of the low-latitude unidentified sources coincide (among other things) with synchrotron nebulae.

A non-thermal nebula is seen at keV energies towards 3EG 1809-2328 (3). It is possibly interacting with a 10^4 M☉ CO cloud. Electrons accelerated up to at least 20 TeV may account for the very hard X-ray synchrotron radiation in a 20 µG field while GeV electrons emitting bremsstrahlung (BREM) radiation in the surrounding cloud may explain the EGRET source. The predicted BREM spectrum should extend to TeV energies, well above the current instrumental limits in sensitivity.

3EG 1420-6038 is another noteworthy example. The Rabbit, a non-thermal, 10% polarized, center-filled radio and X-ray nebula, interpreted as a plerion (4), was first considered as an attractive counterpart, but it has "moved" out of the 3EG 95% confidence contour. The present error box overlaps much of the Kookaburra nebula, a bird-like structure with a thermal body and two non-thermal wings. The γ-ray source coincides with its north-east wing and with a radio+keV hot spot inside it, where a high-$\dot{E}$ radio pulsar has recently been discovered (5). The hot spot may well correspond to a pulsar wind nebula.

For the composite (shell+plerion) supernova remnant W44, γ rays can be produced by the hard shell electrons pervading a nearby cloud or up-scattering the remnant IR dust emission and ambient Galactic and CMB radiation fields (6). BREM radiation dominates in the cloud and has been advocated to account for the 2EG source (6). The latest 3EG and GeV positions for 3EG 1856+0114 have moved 1.5σ away into the remnant. While still overlapping a large part of the shell near the cloud (see Fig. 1), the 3EG error box is reasonably centered on the pulsar B1853+01 location and on the radio+keV plerion surrounding it (7). X-ray and radio emitting relic electrons are also present, confined in a 2-pc-long cometary tail. The $E^{-1.93 \pm 0.10}$ source

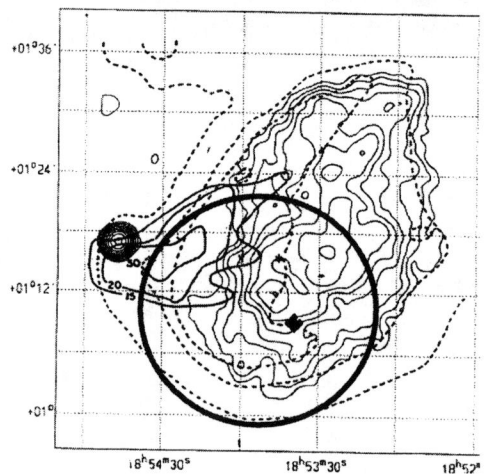

FIGURE 1. Composite 408 MHz (dotted lines), ROSAT (thin lines) contour map of W44 and of the ^{13}CO cloud (thick lines), after (7). The circle gives the 95% confidence region for 3EG 1856+0114. The cross marks the location of PSR 1853+01 and of its 1'× 2.5' synchrotron nebula.

spectrum is typical of a pulsar, unlike the mild flux variability. No γ-ray pulsation has been detected at the radio pulsar frequency (8). Away from the cloud, IC shell radiation dominates and could match the EGRET flux (6), however, half of the shell lies outside the 99% 3EG contour and the emission barycenter is clearly offset from the shell center. In this context, both contributions to the EGRET flux from the ~TeV electrons powering the compact wind nebula and from the lower-energy relic electrons up-scattering the remnant radiation fields should be revisited. The three EGRET sources discussed above exhibit E^{-2} spectra and a slightly variable flux on a month time scale which certainly need to be explored in any nebular modelling.

The position of 3EG 0617+2238, toward IC443, "moved" away from the hard X-ray region where the shell overtakes a dense cloud and where BeppoSAX recently detected a very hard (14 to >30 keV) component possibly powered by 10^{14} eV electrons (9). The 3EG position now points to the center of the shell. Could a pulsar be hiding there? The $E^{-1.96\pm0.10}$ spectrum and stable flux weakly support this possibility.

Another composite remnant, CTA1, coincides with 3EG 0010+7309. The source's steady flux, its hard $E^{-1.6}$ spectrum and cut-off energy of 2 GeV, and the presence of a ROSAT source with no optical counterpart at the heart of the X-ray center-filled nebula are strongly suggestive of a radio-silent γ-ray pulsar (10). Yet, there is a central synchrotron, keV, nebula which is driven by a 20 kyr, $\dot{E} = 1.7 \ 10^{36}$ erg/s, pulsar (11). So, plerionic emission should also be carefully investigated in CTA1.

A radio-silent γ-ray pulsar has also been suggested (12) as a counterpart to the stable, $E^{-2.08\pm0.04}$, 3EG 2020+4017 source toward γ Cygni. But, the association with the ROSAT source remains unclear: there is a 15th magnitude K0V star in the X-ray box and no compact keV nebula to support the pulsar hypothesis.

3EG 1048-5840 has been identified with the pulsar B1046-58 by detecting its γ-ray pulsation (13). It brings to 8 the number of known γ-ray pulsars. Seven are listed among the nine highest $\dot{E}/D^2$ ranked radio pulsars, thus indicating a high probability of intercepting the γ-ray beam when the narrow radio beam sweeps across the Earth, and a close relationship between the onset of high-energy showers and coherent bunching of radio emitting electrons. PSR B1046-58 is a 24 kyr, $\dot{E} = 2 \ 10^{36}$ erg/s, 3 kpc distant, pulsar. As for Vela, its lightcurve exhibits two γ-ray peaks with a familiar phase separation of ~0.4 and ~0.15 offset from the radio pulse. It radiates ~1 % of its spindown power into γ rays with a typical E^{-2} spectrum and steady flux. Another important property of this source is the large fraction (40%) of unpulsed emission detected above 400 MeV which may be associated with the compact (∅ <3') synchrotron nebula seen around the pulsar with ASCA (14). So, in addition to the Crab, PSR 1046-58 brings a second example of hard DC emission associated with a pulsar. No such emission has been reported from the other γ-ray pulsars.

The error box of 3EG 1837-0606 includes a bright ASCA point-source with emission trailing off to the east (5). The X-ray source coincides with a radio pulsar recently discovered at Parkes (M. Roberts, private communication). The steady γ-ray flux and $E^{-1.82\pm0.14}$ spectrum are quite consistent with a pulsar identification. Faint extended ASCA emission is also noted toward the steady 3EG 1734-3232 source and toward GeV J1907+0556 (≠ 3EG 1903+0550) (5).

Of the ten positional associations noted between 2EG unpulsed sources and radio pulsars (15, 16), one is clearly identified as PSR B1046-58 and only two remain valid using the refined 3EG error boxes (PSRB1853+01/W44 and PSR B1900+05).

In summary, at least 7 synchrotron nebulae show positional agreement with EGRET sources. Some are probably chance coincidences, however, their number suggests that we should carefully model the ~GeV emission from pulsar wind nebulae in each of these different circumstances. Relic radio electrons can indeed effectively scatter the optical-UV remnant radiation to produce 0.1-1 GeV photons. While the Crab nebula stands out as a super TeV plerion, we conjecture that less energetic cases may be seen as EGRET sources. The maximum energy of the wind particles scales with the voltage across the open magnetosphere ($\propto B.\Omega^2$), therefore with the square root of the spindown power $\sqrt{I\Omega\dot{\Omega}}$, itself proportional to the γ-ray luminosity L_γ (17). It is important to investigate how bright nebular IC emission can be for a variety of luminosities and nebular fields in the 0.1 GeV to TeV band.

Of particular interest for future CELESTE and STACEE observations will be the search for periodicities typical of pulsars or binary systems. The EGRET signal is too smeared out over the long exposures required for source detection to allow blind searches. Phase coherence is also lost because of the large position error. Cherenkov telescopes have a much finer (~0.15°) angular resolution and high sensitivities of order 10^{-10} γ cm^{-1} s^{1} > 50 GeV. Using the analysis techniques tested on Geminga (18, 19), one can expect a successful periodicity detection in the case of a 7σ source detection above 50 GeV, achieved after a 65h exposure, i.e. ~10 nights, for a 0.1 Crab source. Reducing the exposure time by lowering the energy threshold will be quite valuable. The present γ-ray pulsar sample is strongly biased to young, therefore bright, pulsars because they are detected at low latitude above the intense Galactic background. The spectral hardening of pulsar emission with age and the increase in cut-off energy for weaker magnetic fields (17) favours the detection of older or weakly magnetized pulsars by CELESTE and STACEE. The presence of nearby, 1-2 Myr old pulsars in the Gould Belt (20) may offer an excellent opportunity to study the evolution of pulsar emission with age and field strength and to verify the amount of TeV IC radiation produced in the outer gap (21).

Shell remnants

No γ-ray emission from freshly accelerated nuclei has been clearly identified yet. Associations between EGRET sources and a handful of remnants were proposed (22, 23, 24), but chance alignments are highly probable despite the refined 3EG or GeV localizations. As just discussed, the sources toward W44 and γ Cygni may well turn out to be due to direct or indirect pulsar activity. The case of IC443 is ambiguous. Toward W28 and 3EG 1800-2338, the 95% GeV contour encloses the runaway pulsar PSR B1758-23, but the 99% 3EG contour does not. Besides the remnant thermal emission, there is no obvious X-ray source in the field (5), so shell emission remains an interesting possibility for this source. Clearly, precise γ-ray imaging is required to distinguish cosmic-ray and pulsar emission. The TeV telescopes provide fine angular resolutions, but are still limited in sensitivity for extended sources. At GeV energies, GLAST will be able to resolve a few nearby remnants of large angular size.

Diffusive shock acceleration is widely believed to be active in these remnants. However, nuclei emission can easily be overwhelmed by electron BREM and IC radiation. Recent calculations for Cas A (25) have stressed the importance of a global, multi-wavelength modelling of the remnant radiation to provide constraints on the various parameters necessary to interpret the MeV to TeV intensities, such as the nebular field, ambient interstellar density, e$^-$/p$^+$ ratio, and, obviously, the maximum energies achieved against age-limited and space-limited acceleration and against radiation losses. The many facets of these models and the complex interplay between these parameters and the resulting particle and photon spectra are reviewed in this volume by M. Baring (and references therein).

Away from the plane, searches have been undertaken for faint extended, non-thermal, radio structures by filtering out the Galactic diffuse continuum (26, 27, 28). This technique, combined with the search for perturbations in the surrounding HI gas, has proved successful at revealing shells blown by a supernova blast wave or by the strong wind of a massive star. Three promising associations have been found toward 3EG 0725-5140, 3EG 1659-6251, and the trio 3EG 1834-2803, 1847-3219, & 1850-2652. In each case, an enhanced density of cosmic rays irradiating local 10^{2-3} M☉ clouds may account for the γ-ray flux through π^0 decay. Enhancements by a factor of 30 to 40 are typically required, but should be considered as upper limits since electron BREM and IC emission were not included in the calculation. The shell distances range from 300 to 700 pc. Their proximity and angular size of several degrees imply fairly evolved objects, most probably in the radiative phase. Cosmic rays should, however, remain in the remnant over periods as long as 10^5 years, slowly diffusing away from it. γ-ray production along the rim or inside the shell is not detected. The EGRET sources correspond to "hot spots" such as irradiated clouds. So, these objects may set useful upper limits to the cosmic-ray density in evolved remnants.

X-ray binaries

A few EGRET sources have been tentatively associated with X-ray binary systems, such as 2CG 135+01 (alias 3EG 0241+6103) with the radio source GT 0236+610. In this system, a neutron star, in an eccentric orbit around a B0Ve star, accretes matter from the stellar wind. Radio outbursts occur with a 26.5 day period and a 4-year modulation during supercritical accretion near periastron, or, alternatively, from pairs accelerated at the shock front between the pulsar and stellar winds. In the latter case, IC scattering of stellar optical photons by the radio electrons can produce γ rays (29, 30). The slight variability of the EGRET source, however, does not correlate with the radio phase (30), nor does the stable COMPTEL flux. The COMPTEL spectrum reasonably bridges the ROSAT and EGRET spectra, suggesting continuous emission from 10^2 eV to 10 GeV (29). The spatial coincidence with the EGRET source is not clear: the radio source lies near the edge of the GeV 95% region, but outside the > 300 MeV 95% circle. Yet, there are no other attractive counterparts in the field. So, the identification of the γ-ray source with the binary system needs confirmation.

Another very eccentric binary system, comprising a B2e star and a 48 ms pulsar, PSR B1259-63, in a 3.4 yr orbit, has been extensively studied near apastron (31) and periastron (32). Unlike GT 0236+610, it is not an accretion-powered system. Pairs are accelerated at the shock front between the pulsar and stellar winds. They radiate synchrotron radiation up to 10 keV near apastron and up to 200 keV near periastron, but no MeV to GeV flux has been recorded. The lack of γ-ray detection can be explained theoretically (33). X-ray emitting TeV electrons indeed up-scatter the stellar photons at much higher energies, in the 0.01-1 TeV band. Of particular interest to the TeV observers are the theoretical predictions (34, 33) of GeV and TeV emission, respectively, a few days about periastron, with a νF_ν flux of comparable magnitude at 10 keV and 100 GeV if IC losses dominate.

Another association has recently been proposed between the Be/X-ray binary system SAX J0635+0533 and the highly variable 3EG 0634+0521 source (35). The X-ray source exhibits a power-law spectrum in the 1- 40 keV band.

POPULATION STUDIES

Statistical studies of the temporal, spatial, and spectral distributions of the unidentified sources have been performed to test the heterogeneity of the sample and correlations with other classes of objects. The main difficulty resides in modelling the severe observational biases such as the non-uniform detection sensitivity across the sky (which varies with exposure, the underlying Galactic emission, and the unknown source luminosity function); the sensitivity to variability (which depends on exposure, source latitude and spectrum); and the reliability of spectral measurements (which varies with the source strength and background spectrum). Too often biases have been partially left out for convenience.

At low latitudes, the concentration of sources with a sharp Gaussian profile in $|b|$ ($\sigma_b = 1.6° \pm 0.3°$) and their shallow concentration toward the inner Galaxy are consistent with a population of standard candles with luminosity from 0.6 to 4 10^{35} erg s^{-1}, at distances of 1 to 4 kpc in the Galactic plane (36). Further evidence exists for a correlation with radio supernova remnants (SNR) and/or OB associations (22, 23, 37, 16). The latest compilation (24) lists 22 coincidences between 3EG sources and SNRs, with a combined chance probability $P < 10^{-5}$, and 26 coincidences with OB associations, with $P < 10^{-3}$. Adopting the counterpart distance and assuming isotropic emission, luminosities from 0.1 to 2 10^{35} erg s^{-1} are derived for these sources, in close agreement with the values inferred from the global spatial distribution. In 10 cases, the source coincides with both an SNR and an OB association, thus reviving the idea of SNOBs (38) where cosmic-ray acceleration proceeds in two steps: particles energized by the supersonic wind of a massive star are further accelerated by the passing shock wave of a nearby supernova remnant. So, half of the 3EG sources at $|b| < 10°$ are strongly related to star-forming sites. Pulsars can also explain most of the unidentified sources near the plane. Using the outer gap model, young pulsars born in the spiral arms within 4 kpc and travelling at large velocities can account for most of the sources at $|b| < 5°$ and for their spatial distribution (16). For the polar cap model, a 3 to 4 times wider beam than the standard polar-cap size can also reasonably reproduce the observed distributions at $|b| < 10°$, as well as the radio data (39).

So, the present limitation in our understanding of the low-latitude sources comes frustratingly from the over-abundance of candidates! Spatially as well as spectrally resolving the numerous γ-ray emitters related to active star-forming regions are the top priorities for any future instrument to identify the sources.

The situation is dramatically different at medium latitudes. Half of the sources at $|b| <10°$ and nearly all of them at $10° < |b| < 30°$ have no candidate counterpart. There is an excess of sources at medium latitudes that cannot be attributed to increased exposure or enhanced sensitivity away from the plane (40, 41, 42). The average spectral index of the sources at $|b| < 5°$ and in the $5°<|b|<30°$ range significantly differ (43): the low-latitude sources are harder ($<\gamma> = -2.18\pm0.04$) than the mid-latitude ones

($<\gamma>$ = -2.49±0.04). Their logN-logS distributions are also at variance (43): many faint sources are seen off the plane that cannot be so numerous at low latitudes. They would obviously not be resolved above the bright Galactic emission, but their integrated flux would largely exceed the upper limit to any unresolved point-source contribution to the interstellar emission. So, brighter, harder sources dominate at low latitude and weaker, softer sources at medium latitude.

Confronting the spatial distribution of the unidentified sources off the plane (|b|>2.5°) with various extragalactic and Galactic distributions (20), two populations have emerged with distinct spatial, spectral, as well as temporal properties that are briefly summarized here. The two sets have been named 'persistent' (P) and 'non-persistent' ($\bar{p}$) according to their behaviour in the 3rd EGRET catalogue: 67 P sources were significantly detected using the cumulative data from 1991 and 1995 (P1234) while 59 $\bar{p}$ sources were detected on one or several shorter periods only. So, persistence is not equivalent to constant flux. Variability measurements (44), however, show that the persistent sources are mostly stable or only moderately

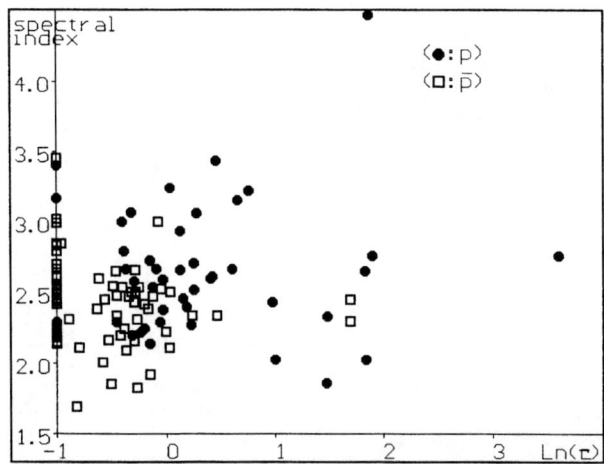

FIGURE 2. Distribution of spectral indices and fractional variabilities τ measured for the persistent ( ) and non-persistent (•) unidentified EGRET sources.

variable. In fact, P sources are on average less variable as well as harder than the $\bar{p}$ sources (Fig. 2). The fractional variability, τ, measures the ratio of the standard deviation to the average of a source flux: $\tau = \sigma_f/<f>$. It was obtained for all 3EG sources using a likelihood analysis of each viewing period (44). These results were combined independently for the two source sets. Max-likelihood averages of $<\tau>_p = 0.38 \pm 0.06$ and $<\tau>_{\bar{p}} = 0.95 \pm 0.18$ were obtained for the P and $\bar{p}$ sources, respectively. They are distinctly different, with a chance probability of 1.3×10^{-4} that the two means be from the same parent population. An average fractional variability τ of 0.14 ± 0.04 was obtained for the five pulsars of the 3EG catalogue. It agrees with the ~10% residual variability left after correction for the long-term drift of the instrument performance, or for the occasional flaring of a nearby source. The two samples also significantly differ in spectral index above 100 MeV, with max-likelihood averages for the P and $\bar{p}$ sources of $<\gamma>_p = 2.25 \pm 0.03$ and $<\gamma>_{\bar{p}} = 2.52 \pm 0.06$, respectively, and a chance probability of 2×10^{-7} of equal index. The τ and γ averages presented above do include the uncertainty in each τ or spectral measurement. No systematic bias due to a correlation of spectral index with interstellar background has been detected and its impact on index variations should be limited to 0.1 (43). We have investigated the possibility of a bias producing the softness/variability correlation, but found none of relevant magnitude. To the contrary, the softer/variable sources were detected with a lower significance implying a reduced sensitivity to flux variations.

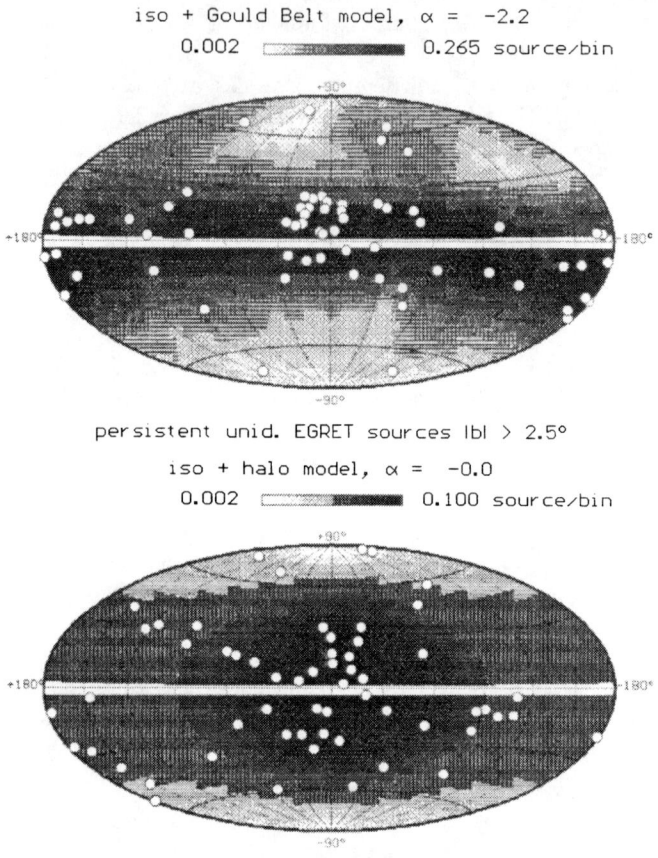

FIGURE 3. (l,b) maps of unidentified EGRET sources (p sources in (a), $\bar{p}$ sources in (b)) and source counts predicted from a combination of an isotropic distribution and (a) a Gould Belt population as traced by its young massive stars, (b) a homogeneous Galactic halo population, 20 kpc in radius. Predicted counts take into account the non-uniform sensitivity of the survey.

Furthermore, the P and $\bar{p}$ sources exhibit distinct spatial distributions, as shown in Figure 3. Spatial correlations with various combinations of Galactic populations, plus an isotropic component of extragalactic origin, were performed at $|b| > 2.5°$. The analyses take into account the non-uniform detection sen-sitivity across the sky. The unknown index, α, of the source luminosity function is left as a free parameter. Various distributions were chosen to cover sites that are likely to harbour γ-ray sources at medium latitudes: in a homogeneous *Galactic halo* (radius of 20 kpc); in a *thick Galaxy* (radial scale length of 9.3 kpc and scale height of 0.4 kpc, typical of radio pulsars); and *locally*, in the plane (any scale height z, no contrast in longitude) or in the inclined *Gould Belt*, as traced by the HI+2H$_2$ column-densities or by the column-density of young nearby OB stars of spectral type < B4. The method was validated using the 3EG AGN sample for which the best fit was always found to be the isotropic one.

In short summary, the goodness of fit to the P sources dramatically increases in two steps, first by allowing the sources to be Galactic ('bad luck' probability of a random effect $P = 10^{-8}$), and even more when allowing them to follow the inclined geometry of the clouds or stars in the Belt ($P \sim 10^{-12}$). The Belt rim extends to ~450 pc from the

Sun. So, a correlation with the Belt suggests a distance of a few hundred parsecs for a large majority of the P sources. No contrast in longitude is detected between the local plane and thick Galaxy fits that would require a typical distance of a few kpc. All four Galactic models yield a total of 50 ± 8 P sources in our vicinity. Moreover, the significant improvement from the Galactic to the Belt fit, with a 10^{-5} probability of a random effect, strongly suggests an origin in the inclined Belt for many P sources, as illustrated in Figure 3a. The Belt is swarmed with young massive stars. We estimated that, over the last few million years, this starburst expanding structure has produced 20 to 27 supernovae per Myr, i.e. at a rate that is 3 to 5 times higher than the Galactic average. This value is based on the present stellar content of the Belt, standard stellar lifetimes, a range of initial-mass-function indices from –2.0 to –1.1, and a constant birth rate. It is thus argued in (20) and in a forthcoming paper that γ-ray pulsars born in the Belt over the last 1-2 Myr, plus a few born in the Galactic plane, may plausibly explain most of the persistent sources. A pulsar origin is also supported by their relative stability and hardness. They would constitute a sample of older, closer, therefore intrinsically fainter pulsars than those already identified. Detecting their γ-ray beam at large, grazing angles would also be possible at such close distance. This is why, despite their age, spectral indices γ < -2.0 may be detected.

The $\bar{p}$ sources are apparently more dispersed, though still gathered at mid-latitudes (Figure 3b). The best fits are indeed obtained for a halo population or a thick Galactic distribution. The 'bad luck' probability of a random fluctuation from a truly isotropic parent population is $7\ 10^{-5}$. Local, flatter, Galactic distributions yield poorer fits. For instance, the fit improves from the Gould Belt to the halo distribution with a $2\ 10^{-3}$ probability of a random effect. From the results, 20 to 40 $\bar{p}$ sources can be attributed to a Galactic population with a large scale height. Using both P and $\bar{p}$ sources, a three-component correlation analysis significantly separates 55 ± 14 sources associated with the Gould Belt, 48 ± 12 sources in a halo, and a score of extragalactic sources.

Any trend between source properties remains subject to subtle observational biases that may not be understood yet. To first order, however, their reliability has been assessed. So, the distinct average temporal, spatial, and spectral properties of the p and $\bar{p}$ sources, together with the distinct logN-logS distributions observed near and away from the Galactic plane, strongly suggests the presence of at least three populations among the unidentified sources, besides the score of extragalactic sources present in the sample: **1)** bright objects a few kpc away in the Galactic spiral arms, obviously related to star-forming regions and plausibly powered by supernova remnants, OB stars, pulsars, or pulsar wind nebulae; **2)** fainter, relatively stable and hard sources, plausibly 1-2 Myr old pulsars, born in the inclined starburst Gould Belt; **3)** fainter, rather variable and soft Galactic sources with a large scale height or spread in the halo. Their distribution evokes that of the globular clusters, but none lies within the source error boxes. Old ms pulsars spawn γ rays in the 1-300 MeV band (45). Variable γ radiation is also expected from isolated, 35 M☉ black holes accreting from the interstellar medium (46). But, for the present, the nature of this new Galactic population of γ-ray sources remains a puzzle.

Acknowledgements: Many thanks to M. Baring, A. Harding, O. de Jager, P. Michelson, D. Thompson, M. Roberts, C. Dermer, and J. Paul for lively and informative discussions and congratulations to Brenda Dingus and Mike Salamon for a great meeting at Snowbird !

REFERENCES

1. Hartman R. C. et al., 1999, ApJS, 123, 79
2. Harding A. K., & de Jager O. C., 1997, "Towards a major atmospheric Cherenkov detector V", 64
3. Oka T., et al., 2000, ApJ, in press, astro-ph 9907261.
4. Roberts M. S. E., et al., 1999, ApJ, 515, 712.
5. Roberts M. S. E., Romani R. W., & Kawai N., 2000, submitted to ApJ
6. de Jager O. C., & Mastichiadis A., 1997, ApJ, 482, 874.
7. Harrus I. M., Hughes J. P., & Helfand D. J., 1996, ApJ, 464, L161.
8. Thompson D. J., et al., 1994, ApJ, 436, 229.
9. Preite-Martinez A., et al., 2000, Proc. 5th Compton Symp., in press.
10. Brazier K. T. S., et al., 1998, MNRAS, 295, 819.
11. Slane P. et al., 1997, ApJ, 485, 221.
12. Carraminana A., et al., 1997, Proc. 4th Compton Symp., 1267.
13. Kaspi V. M., et al., 1999, astro-ph 9906373.
14. Pivovaroff M. J., Kaspi V. M., & Gotthelf E. V., 1999, astro-ph 9906374.
15. Nel H. I., et al., 1996, ApJ, 465, 898.
16. Yadigaroglu I., & Romani R. W., 1997, ApJ, 476, 347.
17. Thompson D. J. et al., 1997, Proc. 4th Compton Symp., AIP 410, 39.
18. Mattox J. R., et al., 1996, A&AS, 120, 95.
19. Brazier K. T. S., & Kanbach G., 1996, A&AS, 120, 85.
20. Grenier I. A., & Perrot C., 1999, Proc. XXVIth Int. Cosmic Ray Conf., Salt Lake City, 3, 476.
21. Romani R. W., 1996, ApJ, 470, 469.
22. Sturner S. J., & Dermer C. D., 1995, A&A, 293, L17.
23. Esposito J. A., et al., 1996, ApJ, 461, 820.
24. Romero G. E., Benaglia P., & Torres D. F., 1999, A&A, 348, 868.
25. Ellison D., et al., 1999, Proc. XXVIth Int. Cosmic Ray Conf., Salt Lake City, 3, 468.
26. Combi J. A., Romero G. E., & Benaglia P., 1999, ApJ, 118, 659.
27. Combi J. A., Romero G. E., & Benaglia P., 1998, A&A, 333, L91.
28. Combi J. A., et al., MNRAS, 2000, in press.
29. van Dijk R., et al., 1996, A&A, 315, 485.
30. Kniffen D. A., et al., 1997, ApJ, 486, 126.
31. Hirayama M., et al., 1999, ApJ, 521, 718.
32. Tavani M., et al., 1996, A&AS, 120, 221.
33. Kirk J. G., Ball L., Skjaeraasen O., 1999, Aph, 10, 31.
34. Tavani M; & Arons J., 1997, ApJ, 477, 439.
35. Kaaret P., et al., 1999, ApJ, 523, 197.
36. Mukherjee R., Grenier I. A., & Thompson D. J., 1997, Proc. 4th Compton Symp., AIP 410, 394.
37. Kaaret P., & Cottam J., 1996, ApJ, 462, L1.
38. Montmerle T., 1979, ApJ, 231, 95.
39. Sturner S. J., & Dermer C. D., 1996, ApJ, 461, 872.
40. Grenier I. A., 1995, Adv. Space Res., vol. 15, (5)73.
41. Ozel M. E., & Thompson D. J., 1996, ApJ, 463, 105.
42. McLaughlin M. A., et al., 1996, ApJ, 473, 763.
43. Gehrels N., et al., 2000, Nature, in press.
44. Tompkins W., 1999, PhD thesis, Stanford University.
45. Kuiper L., et al., 1999, Proc. "Pulsar astronomy-2000 and beyond", astro-ph 9911268.
46. Dermer C. D., , 1997, Proc. 4th Compton Symp., AIP 410, 1275.

TeV Observations of X-Ray Binaries

P. M. Chadwick, K. Lyons, T. J. L. McComb, K. J. Orford,
J. L. Osborne, S. M. Rayner, S. E. Shaw, and K. E. Turver

*Department of Physics, Rochester Building, Science Laboratories, University of Durham,
Durham, DH1 3LE, U.K.*

Abstract. Further observations have been made of the X-ray binary Cen X-3 using the University of Durham Mark 6 imaging telescope. The results of these observations are reported and show that VHE gamma rays are emitted in all epochs when we have observed this object. No modulation of the VHE emission is seen, at either the pulsar or orbital period.

INTRODUCTION

The accreting X-ray binary Cen X-3 is an accurately measured system containing a 4.8 s pulsar in a 2.1 d orbit around an O-type supergiant. Results of observations of Cen X-3 using ground based gamma ray telescopes have been reported which have included evidence for sporadic outbursts of strong pulsed emission in the > 1 TeV band [1,2] and a constant but weaker unpulsed emission at > 400 GeV [3]. These results, together with the *CGRO* EGRET measurement of an outburst of pulsed GeV emission [4], indicate that Cen X-3 is a sporadic source of high energy gamma rays.

The discovery of very high energy (VHE) gamma rays from X-ray selected BL Lacs has been a highlight of high energy astrophysics in recent years [5]. They exhibit extremely short term time variability in TeV emission and a correlation between the emission of X-rays and TeV gamma rays [6], which provide constraints on the possible production models for TeV gamma rays. Many models involve the jets which are a feature of such objects. There have been suggestions that some galactic objects may share the jet properties more usually associated with AGNs [7] and jets have been suggested as sources of TeV emission from X-ray binaries [8–10].

We report the results of analysis of data taken during 1998 March and April and 1999 February. We present the results of a search for a possible correlation between > 400 GeV gamma rays recorded by the University of Durham Mark 6 telescope and X-ray emission according to measurements made with the *RXTE* and

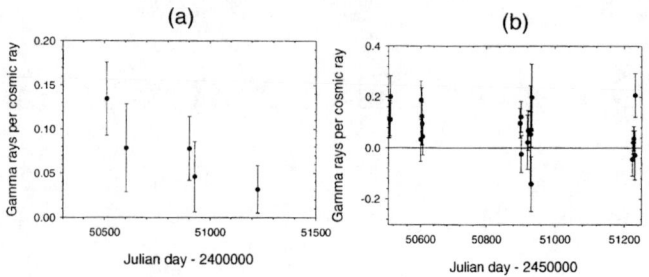

FIGURE 1. (a) The VHE gamma ray flux from Cen X-3 averaged over observing periods. (b) The VHE gamma ray flux from Cen X-3 plotted on a day-by-day basis.

$CGRO$/BATSE experiments. We also present the results of searches for variation of the emission at both the orbital and spin periods.

RECENT OBSERVATIONS OF VHE GAMMA RAYS

We report observations made with the University of Durham Mark 6 imaging gamma ray telescope operating at Narrabri NSW, Australia. The telescope has been described [11] and the results of initial observations of Cen X-3 have been reported previously [3]. Our Cen X-3 dataset now comprises data from 31 hrs of observation during 23 exposures in 1997 March and June (JD 2450508 – JD 2450606), 1998 March and April (JD 2450899 – JD 2450932) and 1999 February (JD 2451220 – JD 2451230).

Our earlier report [3] was based on data recorded in 1997 March and June (JD 2450508 – JD 2450606) only. Assuming a collection area of 10^9 cm^2 and that our selection procedure retained $\sim 50\%$ of the original gamma ray events, the time averaged flux was estimated to be $(2.0 \pm 0.3) \times 10^{-11}$ cm^{-2} s^{-1} for > 400 GeV. Ongoing simulations suggest that our current selection procedure retains 20% of the gamma rays. On this basis, the flux for the 1997 March and June (JD2450508 – JD2450606) data would be $(5.0 \pm 0.9) \times 10^{-11}$ cm^{-2} s^{-1}. The additional data taken in 1998 and 1999 provide fewer gamma ray candidates suggesting weaker TeV emission. An analysis of the total data yields a time averaged flux of $(2.8 \pm 1.4_{sys} \pm 0.6_{stat}) \times 10^{-11}$ cm^{-2} s^{-1}; the significance of the detection based on the total dataset is 4.7 σ.

THE TEV GAMMA RAY SIGNAL STRENGTH

Our recent work on PKS 2155–304 [12] has demonstrated a method of assessing the signal strength of gamma rays recorded by Cerenkov telescopes. It is suited to measurements made at different epochs and at different zenith angles when the telescope has different sensitivities and consequently a varying background cosmic ray detection rate. We have estimated the strength of TeV gamma ray emission by expressing it as a fraction of the cosmic ray background remaining after image shape and orientation selection [13]. In so doing we make allowance for variations in sensitivity, in first order, due to changes in efficiency of the telescope and variations in telescope performance with zenith angle. It is also assumed that the slopes of the gamma ray and cosmic ray spectra are similar.

In the present study, the average gamma ray signal strength from Cen X-3, expressed as a percentage of the cosmic ray background remaining after shape and orientation selection is $(7.0\pm1.5)\%$. The most straightforward, but not most powerful, test for constancy of emission is to repeat this process for the data recorded in each of the 5 dark periods as shown in Figure 1(a) and then test for consistency between the values obtained using a χ^2 test. On the basis of this test we find no internal evidence for monthly variability of the VHE signal; the data treated this way are consistent with a constant signal strength ($\chi^2 = 4.5$, 4 df).

X-RAY AND TEV GAMMA RAY CORRELATIONS

Cen X-3 is a strong but variable X-ray emitter. For example, the average daily rates for X-rays detected with the *RXTE*/ASM during 1997 and 1998 range from 0 to 32 counts s^{-1}; the data are variable on a time scale of days. The daily average for the *RXTE*/ASM count rates are available for 22 of the 23 days when TeV gamma ray observations were made[1].

The strength of pulsed X-ray emission was also available as a daily average from the BATSE archive for 1997[2]; during the 1998 and some of the 1999 VHE observations, the X-ray flux was below the threshold for BATSE detection. The BATSE data provide a series of independent X-ray measurements, including a measurement on the single day of the TeV gamma ray observations for which there is no corresponding *RXTE*/ASM measurement.

The VHE gamma ray signal plotted on a day by day basis is shown in Figure 1(b). There is no evidence for outbursts of TeV gamma ray emission on a timescale of days and the data are consistent with a constant TeV gamma ray flux ($\chi^2 = 22.1$, 22 df).

In Figure 2(a) we show the relation between the count rate of the *RXTE*/ASM data and our gamma ray signals. In Figure 2(b) we show a similar plot relating the individual BATSE pulsed X-ray fluxes and our gamma ray signals. We have no formal evidence for a correlation, although it is interesting to note that the day of highest detected gamma ray flux coincides with the day of most X-ray activity in the dataset (1997 Mar 4).

---

[1] Available on the web at http://space.mit.edu/XTE/asmlc/srcs/cenx3.html
[2] Original data obtained from the web at http://www.batse.msfc.nasa.gov/data/pulsar

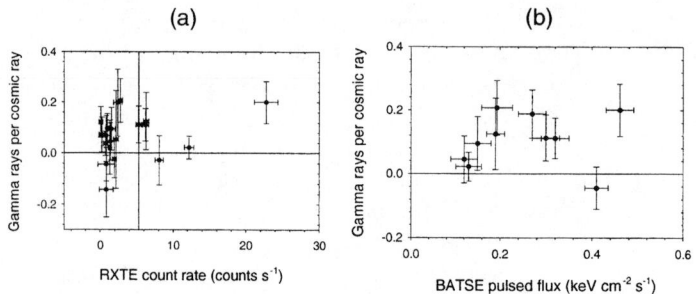

FIGURE 2. The relation between the daily VHE gamma ray flux from Cen X-3 and (a) the X-ray flux detected by ASM/*RXTE* and (b) the X-ray pulsed flux detected by BATSE.

MODULATION

We have searched for modulation of the gamma ray signal at the orbital period of the binary system. The orbital phase of each of our observations has been calculated using the ephemeris of Kelley et al. [14]. The results are shown in Figure 3. We conclude that there is no modulation of the VHE gamma ray emission at the orbital period.

The data have been subjected to a Rayleigh test for periodicity at a small range of periods around the BATSE period. Phase coherence between observations was not assumed. No significant periodicity was detected, leading to a 3σ upper limit to the pulsed flux of 2.0×10^{-12} cm^{-2} s^{-1} for an energy threshold of 400 GeV in the total dataset.

DISCUSSION

We have detected VHE gamma ray emission from Cen X-3 during each dark moon period that we have observed this object. The data are consistent with a weak but persistent emission, both when the VHE data is averaged over dark moon periods and when considered observation by observation. Although the observation that yields the strongest gamma ray flux occurs on the day when the daily averaged *RXTE* X-ray flux was the highest of any day on which we observed Cen X-3, there is no evidence for a formal correlation between the VHE gamma-ray and X-ray fluxes.

We have also tested for modulation of the VHE gamma ray flux at the orbital period of the binary system and at the pulsar period. We have no evidence for modulation of the VHE gamma ray emission at either period.

The processes considered for the production of TeV gamma rays in X-ray binaries have included beam dump models with both electrons and protons as the

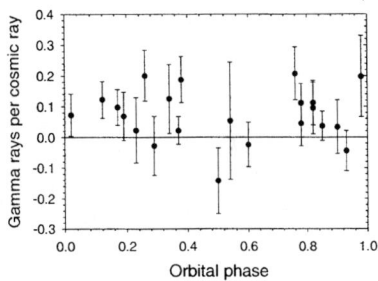

FIGURE 3. The measured VHE gamma ray rate during each observation of Cen X-3 plotted as a function of orbital phase.

accelerated particles (Vestrand and Eichler [8] for protons and Cheng et al. [15] for electrons). In addition, a co-rotating jet model has been suggested by Kiraly and Meszaros [10].

We are grateful to the UK Particle Physics and Astronomy Research Council for support of the project. The Mark 6 telescope was designed and constructed with the assistance of the staff of the Physics Department, University of Durham. This paper uses quick look results provided by the ASM/*RXTE* and BATSE teams.

REFERENCES

1. Carraminana, A., et al., *Timing Neutron Stars* ed. H. Ögelman & E. P. J. van den Heuvel (Dordrecht: Kluwer Academic Press), p 369 (1989).
2. Raubenheimer, B. C., et al., *Ap. J.*, **336**, 349 (1989).
3. Chadwick, P. M., et al., *Ap. J.*, **503**, 391 (1998).
4. Vestrand, W. T., Sreekumar, P. & Mori, M., *Ap. J.*, **483**, L49 (1997).
5. Weekes, T. C., et al., *Proc. 4th. Compton Symposium*, ed. C. D. Dermer, M. S. Strickman, & J. D. Kurfess (New York: AIP), **1**, 361 (1997).
6. Schubnell, M., *Proc. 4th. Compton Symposium*, ed. C. D. Dermer, M. S. Strickman, & J. D. Kurfess (New York: AIP), **2**, 1386 (1997).
7. Hjellming, R. M. & Han, X., in *X-ray Binaries*, ed. W. H. G. Lewin, J. van Paradijs & E. P. J. van den Heuvel (Cambridge: Cambridge University Press), p. 308 (1995).
8. Vestrand, W. T., & Eichler, D., *Ap. J.*, **261**, 251 (1979).
9. Hillas, A. M., *Nature*, **312**, 50 (1984).
10. Kiraly, P., and Meszaros, P., *Ap. J.*, **333**, 719 (1988).
11. Armstrong, P. et al., *Experimental Astron.*, **9**, 51 (1999).
12. Chadwick, P. M., et al., *Ap. J.*, **513**, 161 (1999)
13. Fegan, D. J., *J. Phys. G. Nucl. Part. Phys.*, **23**, 1013 (1997).
14. Kelley, R. L., et al., *Ap. J.*, **268**, 790 (1983).
15. Cheng, K. S., Ho, C., & Ruderman, M., *Ap. J.*, **300**, 522 (1985).

Flux Limits for TeV Emission from Pulsars

P. M. Chadwick, K. Lyons, T. J. L. McComb, K. J. Orford,
J. L. Osborne, S. M. Rayner, S. E. Shaw, and K. E. Turver

Department of Physics, Rochester Building, Science Laboratories, University of Durham, Durham, DH1 3LE, U.K.

Abstract. Observations have been made with the University of Durham Mark 6 telescope of a number of Southern hemisphere supernova remnants and young pulsars (Vela pulsar, PSR B1055-52, PSR J1105-6107, PSR J0537-6910 and PSR B0540-69). No VHE gamma ray emission, either steady or pulsed, has been detected from these objects. The implications of these results for theories of high energy gamma ray production in plerions and young pulsars are discussed.

INTRODUCTION

The *Compton Gamma Ray Observatory* (*CGRO*) telescopes have detected pulsed gamma radiation from 9 pulsars: the Crab, Vela, Geminga, PSR B1509-58, PSR B1706-44, PSR B1951+32, PSR B1055-52, PSR B0656+14 and PSR B1046-58. Of these medium and high energy gamma ray pulsars, the Crab [1] and PSR B1706-44 [2,3] are confirmed very high energy (VHE) gamma ray emitters, and the Vela remnant has been detected by the CANGAROO group [4]. Although the gamma ray emission from each of the pulsars detected with the EGRET telescope has a pulsed component, thus far no imaging VHE gamma ray telescope has detected pulsed radiation at TeV energies from any of the EGRET pulsars. Amongst limits to pulsed VHE emission are those appropriate to the Crab [5,6], Vela [4], and PSR B1706-44 [2,3].

We have previously reported limits on pulsed VHE gamma ray emission from a number of Southern hemisphere pulsars using the University of Durham Mark 3 non-imaging telescope [7-9]. We present here the results of VHE gamma ray observations of five plerions made using the Mark 6 imaging telescope; two EGRET sources (Vela and PSR B1055-52) and three X-ray emitting pulsars, PSR J1105-6107, PSR J0537-6910 and PSR B0540-69. We have searched for both steady and pulsed emission from these objects.

TABLE 1. Observing log for observations of pulsars made with the University of Durham Mark 6 telescope. The number of 15 minute ON-source scans obtained is shown, except for the LMC pulsars where the total exposure time is shown (see text).

Object	Date	No. of ON source scans	Object	Date	No. of ON source scans
Vela pulsar	1996 Apr 14	1	PSR J1105–6107	1997 Mar 31	5
Vela pulsar	1996 Apr 18	4	PSR J1105–6107	1997 Apr 1	6
Vela pulsar	1996 Apr 19	4	PSR J1105–6107	1997 Apr 3	9
Vela pulsar	1996 Apr 20	5	PSR J1105–6107	1997 Apr 4	9
Vela pulsar	1996 Apr 21	5	PSR J1105–6107	1997 Apr 5	7
Vela pulsar	1996 Apr 22	3	PSR J1105–6107	1997 Apr 6	9
Vela pulsar	1997 Feb 6	7	PSR J1105–6107	1997 Apr 7	4
Vela pulsar	1996 Feb 7	6	PSR J1105–6107	1997 Apr 8	2
LMC pulsars	1998 Mar 21	110 mins	PSR J1105–6107	1997 Apr 9	9
LMC pulsars	1998 Mar 22	130 mins	PSR J1105–6107	1997 Apr 10	7
LMC pulsars	1998 Mar 24	100 mins	PSR B1055–52	1996 Mar 19	5
LMC pulsars	1998 Mar 27	90 mins	PSR B1055–52	1996 Mar 20	5
LMC pulsars	1998 Mar 28	80 mins			

OBSERVATIONS

The Durham University Mark 6 telescope is described in detail elsewhere [10]. It consists of three 7 m diameter parabolic flux collectors mounted on a single alt-azimuth platform. A 109-element imaging camera with 0.25° pixels is mounted at the focus of the central mirror, with low-resolution cameras each consisting of 19 pixels (0.5°) mounted at the foci of the outer (left and right) flux collectors.

Data from all objects except the PSR J0537–6910/PSR B0540–69 field were taken in 15-minute segments. Off-source control observations were taken by alternately observing regions of sky which differ by ±15 minutes in RA from the position of the object. The use of alternate off-source segments which preceed and follow the on-source segment allow for any small residual secular effects. Data were accepted for analysis only if the sky was clear and stable and the gross counting rates in each on-off segments were consistent at the 2.5σ level. In the case of PSR J0537–6910 and PSR B0540–69, the field containing the two objects was tracked and kept in the field of view at all times during the observations.

A total of 36.5 hours of on-source observations under clear skies of the 5 objects was completed, and an observing log is shown in Table 1.

DATA ANALYSIS

Data reduction and analysis followed our standard procedure, which has been described in detail previously [11]. The selections were developed from our observations of PKS 2155–304, and allow for the variation of image parameters with

TABLE 2. Flux limits for observations of pulsars made with the University of Durham Mark 6 Telescope.

Object	Estimated Threshold (GeV)	Flux Limit (DC) ($\times 10^{-11}$ cm^{-2} s^{-1})	Flux Limit (pulsed) ($\times 10^{-11}$ cm^{-2} s^{-1})	Ephemeris Reference
Vela pulsar	300	5.0	1.3	[21]
PSR B1055−52	300	13	6.8	[22]
PSR J1105−6107	400	2.2	0.53	[23]
PSR J0537−6910	400	6.1	1.0	[24]
PSR B0540−69	400	6.1	1.1	[25]

image size.

The threshold energy for the observations has been estimated on the basis of preliminary simulations [12], and is in the range 300 to 400 GeV for these objects, depending on the elevation at which observations were made. The collecting areas which have been assumed are 5.5×10^8 cm^2 at an energy threshold of 300 GeV and 1×10^9 cm^2 at 400 GeV. These are subject to systematic errors estimated to be $\sim 50\%$. We have assumed that our current selection procedures retain $\sim 20\%$ of the γ-ray signal. All steady flux limits are 3 σ limits, based on the maximum likelihood ratio test [13].

The timing analysis proceded as follows. Every event time is recorded to 1 μs precision using a 1 MHz signal derived from a rubidium oscillator. The drift rate of this oscillator is routinely monitored by comparison with a GPS signal. This long term drift rate is found to be $< 10^{-11}$ s s^{-1}, and is stable. The recorded event times are corrected by the measured drift rate, leading to an absolute accuracy in recorded event times of better than 0.1 ms. The event times are corrected to the solar system barycentre using the JPL DE200 planetary ephemeris.

To check for the presence of a pulsed signal, the phase of each event was evaluated using the ephemeris nearest the observation date from the Princeton database [21] or other published sources. For data from the Vela pulsar, PSR B1055−52 and PSR J1105−6107 the events were binned in 20 phase bins. Rayleigh and χ^2 tests were performed on the binned data. The pulsed flux limits for these objects are based on the pulsed flux that would be required to yield a 3 σ excess in a single bin of a 20 bin lightcurve.

The data from PSR J0537−6910 and PSR B0540−69, for which no sufficiently accurate ephemerides were available, were subjected to a Rayleigh test over a small range of periods about the most likely period. Pulsed flux limits for PSR J0537−6910 and PSR B0540−69 are based on the percentage pulsed flux which would be required to produce a 3 σ pulsed detection using the Rayleigh test.

DISCUSSION

The flux limits obtained are shown in Table 2.

The only object considered in this study which has been detected at TeV energies is the Vela pulsar/nebula. An extrapolation of the flux detected with the CANGAROO telescope at 2.5 ± 1.5 TeV to our threshold energy of about 300 GeV suggests that we might expect to have detected the offset source described by Yoshikoshi et al. [4]. However, taking into account the errors on our flux and threshold energy estimates, CANGAROO's flux and energy threshold estimates and the errors on the measured spectral index, it is possible these results may be compatible.

The unpulsed TeV flux that might be expected from a number of pulsars (including Vela and PSR B1055–52) on the basis of the model of de Jager et al. [14] has been predicted [15]. Extrapolating these predictions to our energy threshold using their value of the spectral index leads to expected integral fluxes of 6.4×10^{-11} cm^{-2} s^{-1} for Vela and 2.9×10^{-11} cm^{-2} s^{-1} for PSR B1055–52. Our measured flux limit is not in conflict with the prediction for PSR B1055–302. Taking into account the errors in the flux and energy threshold measurements, our upper limit for steady emission from Vela is also not in conflict with these predictions.

Both PSR J0537–6910 and PSR B0540–69 are good candidates for steady TeV emission on the basis of their radio and X-ray characteristics. However, their distance from the earth (they are both situated in the LMC) means that an extended exposure will be necessary to make a significant detection.

The multiwavelength spectral energy distributions for several of the known gamma ray pulsars have been given by Thompson et al. [16] to which can be added pulsed flux limits for the Vela pulsar and PSR B1055–52 (present work), PSR B1706–44 [3] and PSR B1509–58 [9]. The addition of limits at $\sim 10^{26}$ Hz emphasises the appearance of a turnover in the spectrum in the 10 – 300 GeV region, following the pattern established in the extensive observations of the Crab pulsar.

The establishment of a pronounced spectral steepening above 10 GeV or so in the majority of these young gamma ray pulsars may help to constrain models of gamma ray production in pulsars. Early versions of the outer gap model [26] predicted large fluxes of pulsed TeV gamma rays from Vela-type pulsars, produced via the inverse Compton scattering of infra-red photons by primary electrons. The observations reported here emphasise that this class of model is unable to reproduce the TeV observations.

Detailed predictions of the expected spectrum of pulsed high-energy photons from several pulsars have been made for a number of models. The polar cap model predicts a very sharp cut-off in the pulsed high energy gamma ray spectrum of the Vela pulsar, with no emission occurring above 10 GeV [17]. The polar cap model of Sturner et al. [18], where the high energy gamma rays are produced via inverse-Compton scattering rather than curvature radiation, also predicts that no pulsed TeV emission should be seen from the Vela pulsar.

Modern versions of the outer-gap model (e.g. [19]) predict a cut-off in the pulsed spectrum of the Vela pulsar at around 10 GeV, due to the cut-off in the curvature radiation spectrum. However, this model predicts another component in the pulsed high energy spectrum due to inverse Compton scattering of the primary electrons on soft photons from the pulsar gap. This additional component would peak at an energy of a few TeV. The results reported here present no support for such an additional component; however, Burdett et al. [20] have pointed out that the absence of a TeV peak does not rule out the Romani model since its appearance depends on the density of local soft photons, which may not be correctly estimated.

We are grateful to the UK Particle Physics and Astronomy Research Council for support of the project. Pulsar ephemerides were extracted from the Princeton GRO/Radio Timing Database.

REFERENCES

1. Weekes, T. C., et al., *Astrophys. J.*, **342**, 379 (1989).
2. Kifune, T., et al., *Astrophys. J.*, **438**, L91 (1995).
3. Chadwick, P. M., et al., *Astropart. Phys.*, **9**, 131 (1998).
4. Yoshikoshi, T., et al., *Astrophys. J.*, **487**, L65 (1997).
5. Vacanti, G., et al., *Astrophys. J.*, **377**, 467 (1992).
6. Aharonian, F., et al., *Astron. Astrophys.* **346**, 913 (1999).
7. Brazier, K. T. S., et al., *Proc. 21st ICRC (Adelaide)*, **2**, 304 (1990).
8. Bowden, C. C. G., et al., *Proc. 22nd ICRC (Dublin)*, **1**, 424 (1991).
9. Bowden, C. C. G., et al., *Proc. 23rd ICRC (Calgary)*, **1**, 294 (1993).
10. Armstrong, P., et al., *Exp. Astron.*, **9**, 51 (1999).
11. Chadwick, P. M., et al., *Astrophys. J.*, **513**, 161 (1999).
12. Chadwick, P. M., et al., *Proc. 26th ICRC (Salt Lake City)*, **5**, 227 (1999).
13. Gibson, A. I., et al., *Proc. Intl. Workshop on Very High Energy Gamma Ray Astro.*, Bombay: Tata Institute, ed. P. V. Ramana Murthy & T. C. Weekes, p. 97 (1982).
14. de Jager, O. C., et al., *Proc.24th ICRC (Rome)*, **2**, 528 (1995).
15. Harding, A. K., & de Jager, O. C., *Towards a Major Atmospheric Čerenkov Detector – V*, Potchefstroom: Potschefstroom University for CHE, ed. O. C. de Jager, p. 64 (1997).
16. Thompson, D. J., et al., *Proc. Fourth Compton Symposium*, ed. C. D. Dermer, M. S. Strickman & J. D. Kurfess, AIP Conf Proc. 410, p. 39 (1997).
17. Daugherty, J. K., & Harding, A. K., *Astrophys. J.*, **458**, 278 (1996).
18. Sturner, S. J., Dermer, C. D., & Michel, F. C., *Astrophys. J.*, **445**, 736 (1995).
19. Romani, R. W., *Astrophys. J.*, **470**, 469 (1996).
20. Burdett, A. M., et al., *astro-ph/9906318*, (1999).
21. Arzoumanian, Z., Nice, D., & Taylor, J. H., *GRO/Radio Timing Database* Princeton: Princeton Univ., 1992.
22. Kaspi, V. M., et al., unpublished (1996).
23. Kaspi, V. M., et al., *Astrophys. J.*, **485**, 820 (1997).
24. Wang, Q. D. & Gotthelf, E. V., *Astrophys. J.*, **509**, L109 (1999).
25. Deeter, J. E., Nagase, F., & Boynton, P. E., *Astrophys. J.*, **512**, 300 (1999).
26. Cheng, K. S., Ho, C., & Ruderman, M. A., *Astrophys. J.*, **300**, 500 (1986).

The Diffusive Galactic GeV/TeV Gamma-Ray Background: Sources vs. Transport

Heinrich J. Völk*

*Max-Planck-Institut für Kernphysik
P.O. Box 103980, 69029 Heidelberg, Germany*

Abstract. The diffuse Galactic γ-ray background, as observed with EGRET on CGRO, exceeds the model predictions significantly above 1 GeV. This is particularly true for the inner Galaxy. We shall discuss here the contribution of the Galactic Cosmic Ray (GCR) sources, considered as unresolved, and in addition the possibility that the transport of the GCRs out of the Galaxy is not uniform over the Galactic disk. In both cases the spectrum of the diffuse gamma rays is harder than the GCR spectrum in the neighborhood of the Solar system, as observed in situ. The source contribution is a necessary and, as it turns out, significant part of the diffuse background, whereas the transport effect is one of several conceivable additional causes for the hard diffuse γ-ray spectrum observed.

INTRODUCTION

The observations of the diffuse Galactic γ-ray emission can be described rather well by a suitable model for the diffuse interstellar gas, GCR, and photon distributions (e.g. Hunter et al. 1997a). However, above 1 GeV the observed average diffuse γ-ray intensity, foremost in the inner Galaxy, $300° < l < 60°$, $|b| \leq 10°$, exceeds the model prediction significantly. As far as the energetic particles are concerned, there are at least two possible explanations for this discrepancy (e.g. Weekes et al. 1997; Hunter et al. 1997b, and references therein). The high-energy γ-ray excess may indicate that the GCR spectrum observed in the local neighborhood is not representative of the diffuse CR population in the Galactic disk. An unresolved distribution of CR sources is the other possibility. Since the γ-ray emission is the product of the energetic particle intensity on the one hand, and of the gas density or the photon density, on the other, it is of course possible that deviations from the above model assumptions for these latter densities across the Galaxy can also lead to changes in the observed energy spectrum of the diffuse gamma rays. We shall not discuss such deviations here. We shall rather evaluate the contribution of the sources, assumed to be the ensemble of Supernova Remnant (SNR) shells, following a recent calculation by Berezhko & Völk (1999). We shall also consider the transport of the particles from the same sources out of the Galaxy to naturally increase with decreasing Galactic radius (Breitschwerdt et al., 1991). We shall leave aside the possibility of new sources of particles, not known in the neighborhood of the Solar system.

GAMMA RAYS FROM THE ENSEMBLE OF SNRS

Since at best a handful of shell SNRs could be argued to have been detected up to now in gamma rays, we shall ignore their discrete contributions and consider the CR sources to be spatially averaged over the volume $V_g = 2.5 \times 10^{66}$ cm^3 of the Galactic gas disk, with a radius of 10 kpc and a thickness of 240 pc. The corresponding gas mass is $M_g = 4 \times 10^9 M_\odot$ (Dickey & Lockman, 1990). The source input rate in the form of energetic particle energy equals $\nu_{SN} \delta E_{SN}$, where we take $\nu_{SN} = 1/30$ yr, $E_{SN} = 10^{51}$ erg. The efficiency per SNR is $\delta < 1$. The total number of localized SNRs which still contain their shock accelerated CRs, called here the source CRs (SCRs), is given by $N_{SN} = \nu_{SN} T_{SN}$, where T_{SN} is their assumed life time, i.e. the time until which they can confine the accelerated particles in their interior. Thus N_{SN} is dominated by the population of old SNRs. We estimate $T_{SN} \simeq 10^5$ yr. After the time T_{SN} the SCRs rather quickly become part of the ordinary GCRs that presumably occupy a large Galactic residence volume uniformly.

Acceleration model

We assume the overall SCR number inside a single SNR to be given by a power law spectrum $N_{SCR} dE \propto \epsilon^{-\gamma_{SCR}} dE$ in energy ϵ in the relativistic range.

Averaged over the disk volume, the spatial density $n_{SCR}(\epsilon)$ of SCRs is given $n_{SCR}(\epsilon) = N_{SCR}(\epsilon) N_{SN}/V_g$, with energy density $e_{SCR} = N_{SN} \delta E_{SN}/V_g$. In terms of e_{SCR}, we have

$$n_{SCR}(\epsilon) = \frac{n_0^{SCR}(\gamma_{SCR} - 1)}{mc^2} \left(\frac{\epsilon}{mc^2}\right)^{-\gamma_{SCR}} \quad (1)$$

and

$$n_0^{SCR} = \frac{(\gamma_{SCR} - 2) e_{SCR}}{(\gamma_{SCR} - 1) mc^2}, \quad (2)$$

for $\gamma_{SCR} > 2$. The same expressions hold for the GCRs, given e_{GCR} and γ_{GCR}.

For the SCR we may quite possibly have $\gamma_{SCR} = 2$, and then

$$n_0^{SCR} = \frac{e_{SCR}}{mc^2 \ln(\epsilon_{max}/mc^2)}, \quad (3)$$

where $\epsilon_{max} \simeq 10^5\ mc^2$ is the maximum SCR energy.

The π^0-decay prodution rate is given by

$$Q_\gamma(\epsilon) = Z_\gamma \sigma_{pp} c N_g n(\epsilon), \qquad (4)$$

(Drury et al. 1994), which leads to the ratio $R = Q_\gamma^{SCR}/Q_\gamma^{GCR}$ of the γ-ray production rates due to SCRs and GCRs, given by

$$R(\epsilon_\gamma) = \frac{Z_\gamma^{SCR} N_{SN} \delta E_{SN}}{Z_\gamma^{GCR}(\gamma_{GCR} - 2) \ln(\epsilon_{max}/mc^2) V_g e_{GCR}}$$
$$\times \zeta \left(\frac{\epsilon_\gamma}{mc^2}\right)^{\gamma_{GCR}-2}, \qquad (5)$$

where ζ is the ratio N_g^{SCR}/N_g^{GCR}, N_g^{SCR} is the mean source gas density, and N_g^{GCR} denotes the average gas density in the disk.

With $\delta = 0.2$, $e_{GCR} \simeq 2 \times 10^{-12}$ erg/cm^3 for the relativistic part of the GCRs, and $\gamma_{GCR} = 2.75$ which results in $Z_\gamma^{SCR}/Z_\gamma^{GCR} = 10$ (Drury et al. 1994), we obtain

$$R(\epsilon_\gamma) = 0.16\zeta \left(\frac{T_{SN}}{10^5 \text{ yr}}\right) \left(\frac{\epsilon_\gamma}{1 \text{ GeV}}\right)^{0.75}, \qquad (6)$$

for $\gamma_{SCR} = 2$.

The total γ-ray spectrum measured from an arbitrary Galactic disk volume is then expected to be

$$\frac{dN^\gamma}{d\epsilon_\gamma} = \frac{dN_{GCR}^\gamma}{d\epsilon_\gamma}[1.4 + R(\epsilon_\gamma)], \qquad (7)$$

where the additional factor 0.4 is introduced to approximately take into account the contribution of GCR electron component to the diffuse γ-ray emission at GeV energies, and where $\frac{dN_{GCR}^\gamma}{d\epsilon_\gamma}$ is taken from the paper by (e.g. Hunter et al. 1997b).

"Leaky Box"-type model

We can derive very similar results from a leaky box-type balance equation

$$\frac{n_{GCR}(\epsilon)}{\tau_c} = \frac{N_{SCR}(\epsilon)}{V_c(\epsilon)}\nu_{SN}, \qquad (8)$$

where $V_c(\epsilon)$ is the energy-dependent residence volume occupied by GCRs that reach the gas disk during their constant mean residence time $\tau_c \simeq 3 \times 10^7$ yrs in $V_c(\epsilon)$. In the case of an extended Galactic Halo, $V_c(\epsilon \gg 1\text{GeV}) \gg V_g$ (Ptuskin et al. 1997). Using eq. (4) we can write

$$\frac{n_{SCR}}{n_{GCR}} = \frac{V_c T_{SN}}{V_g \tau_c} = \frac{T_{SN}}{\tau_g}. \qquad (9)$$

The GCR residence time in the disk volume

$$\tau_g = \tau_c V_g/V_c = \frac{xV_g}{vM_g} \qquad (10)$$

can be derived from the measured grammage $x = 14\,v/c\,(\epsilon/4.4\text{ GeV})^{-0.60}$ g/cm^2, for $\epsilon > 4.4$ GeV, and $x = 14v/c$ g/cm^2, for $\epsilon < 4.4$ GeV (Engelman et al. 1990).

At relativistic energies $\epsilon > mc^2$, the GCR spectrum and the overall SCR spectrum $N_{SCR} \propto \epsilon^{-\gamma'_{SCR}}$ are connected by the relation

$$\gamma'_{SCR} = \gamma_{GCR} - 0.6 = 2.15. \qquad (11)$$

Taking $\gamma_{SCR} = 2.15$, which leads to $Z_\gamma^{SCR}/Z_\gamma^{GCR} = 7.5$ (Drury et al. 1994), we obtain for $\epsilon_\gamma \geq 4.4$ GeV:

$$R(\epsilon_\gamma) = 0.06\zeta \left(\frac{T_{SN}}{10^5 \text{ yr}}\right) \left(\frac{\epsilon_\gamma}{1 \text{ GeV}}\right)^{0.6} \qquad (12)$$

(Berezhko & Völk, 1999).

The question is, of course, whether the SN confinement time T_{SN} is time dependent. Probably this dependence is $T_{SN}(\epsilon) = t_0(\epsilon/\epsilon_{max})^{-5}$, where t_0 is the sweep-up time when the SNR enters the Sedov phase and the shock speed begins to decrease with time. For average ISM parameters $t_0 \sim 10^3$ yr.

RESULTS INCLUDING THE SCRS

In Fig. 1 we show the measurements by Hunter et al. (1997a) and our two estimates for the total γ-ray emission, from GCRs plus SCRs. They demonstrate that the SCR contribution for the acceleration model exceeds the leaky box values for all energies. The reason is that for our empirical model the acceleration efficiency for the relativistic part of the spectrum is only $\delta \simeq 0.08$. This is probably due to the fact that the mean injection efficiency at the SNR shock is lower than the values typically assumed for a parallel shock by a factor of a few. We take the lower value for the γ-ray emission in Fig.1 as the most reliable estimate for the expected diffuse γ-ray emission, including the SCRs. Nevertheless the SCR distribution, which is about 10 percent at GeV energies, becomes dominant beyond 100 GeV, and exceeds the GCR emission at 1 TeV by almost a factor of 10. It would be very interesting to detect the diffuse Galactic γ-ray emission at 1 TeV in order to test this prediction.

Until now we have only discussed the γ-ray emission from hadronic SCRs. In fact, there are many reasons to assume that electrons are equally well accelerated in SNRs, even if their injection into the shock acceleration process is much less well understood. The inverse Compton emission by SCR electrons can be comparable with the hadronic emission, even though it does not contribute at TeV energies. For a more detailed discussion we refer to the paper of Berezhko & Völk (1999).

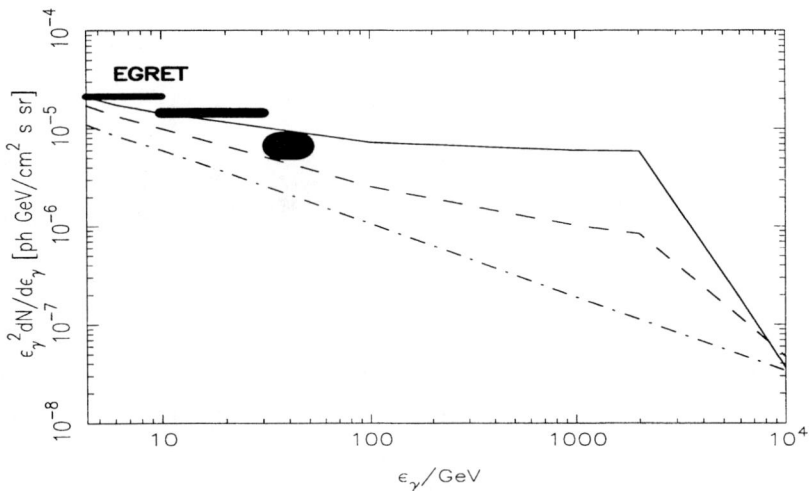

FIGURE 1. The differential diffuse γ-ray energy flux vs γ-ray energy above 4.4 GeV (cf. Berezhko & Völk, 1999). The heavy symbols are the EGRET measurements, and the dash-dot line is the model prediction of Hunter et al. (1997a). The full curve corresponds to our acceleration model with $\gamma_{SCR} = 2$, whereas the dashed curve corresponds to the Leaky Box model. Both theoretical curves incorporate energy-dependent loss from the acceleration region.

TRANSPORT EFFECTS

The models used to fit the γ-ray data from, say, EGRET assume a GCR energy spectrum that is uniform throughout the Galaxy. This tacitly assumes that the GCR transport properties leading to the escape from the Galaxy are everywhere the same. However that needs not be the case, and in fact is almost certainly not true. The dynamical processes leading to GCR escape depend on the strength of the regular magnetic field and on its fluctuation characteristics, as well as on the CR pressure, and the gravitational field. An example is the formation of Parker bubbles which remove the enclosed CRs through their boyant rise into the Halo and ultimately into the Intergalactic Medium. Another example which we wish to discuss here in some more detail, involves the Galactic Wind which is partly driven by the GCRs themselves (e.g. Breitschwerdt et al., 1991, 1993; Zirakashvili et al., 1996). In fact, the wind velocity perpendicular to the disk - in z-direction - is much larger in the central regions of the Galaxy than at larger radii, through the radial variation of the Galactic gravitational field alone (see Fig. 2). This implies that for a given particle energy the boundary seperating the dominantly diffusive transport perpendicular to the Galactic disk near the disk from the dominantly convective transport at greater halo heights *moves down* in direction to the Galactic midplane in the inner Galaxy. Since the GCR diffusion coefficient increases with energy, the position of this boundary will depend on energy. As shown by

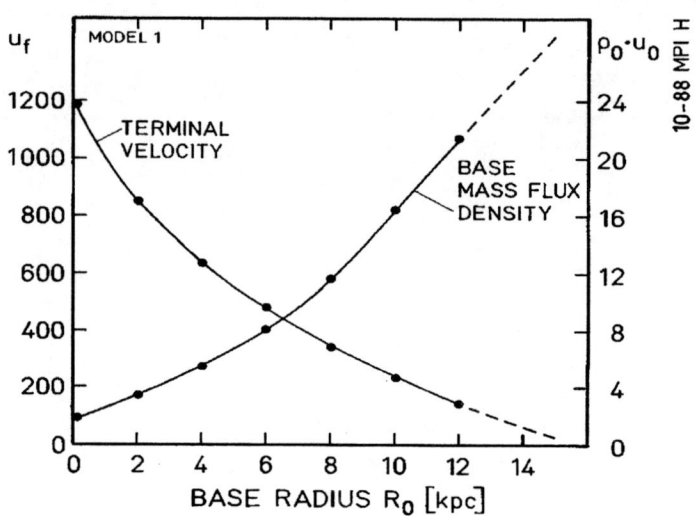

FIGURE 2. The terminal Galactic Wind velocity, and the base mass flux density, as functions of radius in the Galactic disk, cf. Breitschwerdt et al. (1991). All ISM parameters at the base of the wind were considered uniform. The radial variation of the Galactic gravitational field alone is sufficient to produce this radial gradient

Ptuskin et al. (1997), the energy spectrum of the GCRs is typically $\propto E^{-1.9}$ in the convection region compared to the standard spectrum $\propto E^{-2.7}$ in the diffusive confinement region of volume V_c discussed in subsecion 2.1. A line of site that intersects this boundary will therefore receive gamma rays from two regions of very different GCR energy spectra, emitting correspondingly harder spectra than does the diffusive confinement region alone. Qualitatively this implies a hardening of the truly diffuse γ-ray spectrum with Galactic longitude towards the inner Galaxy, for given latitude. However, the effect will disappear for high enough energies when the convective zone does no more extend into regions of significant gas density.

Thus, in contrast to the contribution of the sources, this transport effect looses importance at high energies.

It remains to work out this effect quantitatively. But its very existence illustrates the interest we should attach to the measurements of the diffuse Galactic γ-ray emission over an as wide as possible range of energies.

CONCLUSIONS

The foregoing discussion shows that there are at least two mechanisms of basic physical interest that contribute to a deviation of the diffuse Galactic γ-ray emission spectrum from what would be expected from CR observations in the Solar vicinity. The contribution from the SCRs is an inevitable one and is essentially sufficient to explain the data at least for the inner Galaxy; it should be part of the γ-ray emission model to begin with. Clearly this does not rule out effects from potentially existing new populations of CRs, especially electrons, or the influence of an increased strength, for instance, of the Interstellar radiation field. This is particularly true for high Galactic latitudes.

Acknowledgements The work on the role of the CR sources, summarized in the first part of the paper, has been done jointly with E.G. Berezhhko. I am also indebted to V.S. Ptuskin for a discussion on the effects of the Galactic Wind on the diffuse γ-ray emission.

REFERENCES

1. Hunter, S.D., Bertsch, D.L., Catelli, J.R. et al., *Astrophys. J.* **481**, 205 (1997a).
2. Weekes, T.C., Aharonian, F.A., Fegan, D.J., et al., *Proc. 4th. Compton Symposium, Williamsburg* **1**, 361 (1997).
3. Hunter, S.D., Kinzer, R.L., Strong, A.W., *Proc. 4th. Compton Symposium, Williamsburg* **1**, 192 (1997b).
4. Berezhko, E.G., Völk, H.J., *ApJ*, submitted (1999).
5. Breitschwerdt, D., McKenzie, J.F., Völk, H.J., *A&A* **245**, 79 (1991).
6. Breitschwerdt, D., McKenzie, J.F., Völk, H.J., *A&A* **269**, 54 (1993).
7. Dickey, J.M., Lockman, F.J., *ARA&A* **28**, 215 (1990).
8. Drury, L.O'C., Aharonian, F.A., Völk, H.J., *A&A* **287**, 959 (1994).
9. Ptuskin, V.S., Völk, H.J., Zirakashvili, V.N., Breitschwerdt, D., *A&A* **321**, 434 (1997).
10. Engelmann, J.J., Ferrando, P., Soutoul, A., et al., *A&A* **233**, 96 (1990).
11. Zirakashvili, V.N., Breitschwerdt, D., Ptuskin, V.S., Völk, H.J., *A&A* **311**, 113 (1996).

GeV Gamma-Ray Sources

R.C. Lamb[1] & D.J. Macomb[2,3]

[1] *Space Radiation Laboratory, California Institute of Technology, Pasadena, CA 91125, USA*
[2] *Astronomy Programs, Universities Space Research Association, USA*
[3] *Laboratory for High Energy Astrophysics, NASA/GSFC, Greenbelt, MD 20771, USA*

Abstract: We report on the preliminary extension of our work on cataloging the GeV sky to approximately 7 years of CGRO/EGRET observations with special emphasis on a search for transient sources. The search method and significance levels are presented. Our initial results on 13 possible transients indicate that 3 may be new gamma-ray sources. Sixteen new steady GeV sources are also detected, 3 of which have never been reported as gamma-ray sources.

INTRODUCTION

The sources detected by EGRET above 100 MeV are described in a series of papers culminating in the 3rd EGRET catalog (Hartman et al. 1999). There are several TeV sources which are not listed in the EGRET catalogs such as MRK 501 [1], 1ES 2344+514 [2], and SN 1006 [3]. Both blazars and supernova remnants can therefore have spectral shapes which make them preferentially detected in certain parts of a very broad gamma-ray energy range (the high-energy portion of which extends from 100 MeV to at least 10 TeV). This leads us to fully investigate the EGRET database above 1 GeV, an order of magnitude above the threshold for the standard EGRET catalogs.

We divide our search into two separate parts; a search for new steady sources and a search for GeV transients. Note that steady in this context means detectable in the full all-sky data, not showing a lack of variability. Our data consists of 99000 photons with energies above 1 GeV spread over 217 publicly available EGRET data sets taken over 7 1/2 years. We begin by extending our list of steady GeV sources beyond the catalog of Lamb and Macomb ([4], hereafter LM97). These sources then form the basis for our search for transients.

NEWLY DISCOVERED STEADY SOURCES

Finding transient sources requires understanding steady sources. Although this GeV photon database is only 10% large than in LM97, there is evidence for new sources. In addition, the 3rd EGRET Catalog [5] provides insight into weak source candidates. We prepared a database of nearly 99000 photons in maps that represent the full EGRET database through the early CGRO cycle 7 (Viewing period 710). The data preparation and treatment is as described in LM97. Briefly, photons above 1 GeV are binned in 0.5x0.5 degree all-sky maps then analyzed using maximum likelihood [6] and the diffuse model of galactic emission [7]. Sources detected above a set threshold are added to the background model and residual likelihood maps are prepared which allow a search for successively weaker sources.

Simulations of twelve all-sky maps using actual exposure indicates that a threshold of 4.5σ ensures that sources found in the full EGRET database are real. The number of spurious sources at this significance level is expected to be less than one. Ten sources beyond those in LM97 exceed this threshold. In addition to sources detected above 4.5σ, we include sources with significance above 4σ that have a 100 MeV counterpart in the 3rd EGRET catalog. This counterpart is defined by having the 95% GeV and 100 MeV error circles overlapping. Four new GeV sources are in this category. A third category of source are those whose significance is above 4σ, and are detected at a significance of at least 3σ in the 0.3-1.0 GeV energy band at exactly the position of the GeV excess (supported by simulations). Two such cases exist. Following these rules, Table 1 lists the steady GeV gamma-ray sources beyond those of LM97.

Table 1. New sources found in the full EGRET database, ordered by significance. For sources below 4.5σ, the qualifying characteristic is noted (3EG counterpart or probable 0.3-1.0 GeV detection).

Name	lii	bii	Sigma	Flux[A]	Rad[B]	Notes
GEV J2159-3024	17.45	-52.32	6.3	1.8±0.5	33	3EG J2158-3023, PKS 2155-304
GEV J2257-2755	24.45	-64.68	5.4	1.8±0.6	36	PKS 2255-282
GEV J1745-3014	358.86	-0.63	5.2	6.1±1.3	23[C]	3EG J1744-3011
GEV J1228+0159	289.89	64.30	5.2	0.9±0.3	33[C]	3EG J1229+0210, 3C 273
GEV J1824-1511	16.36	-1.03	5.1	5.6±1.3	32[C]	3EG J1824-1514
GEV J1017-5845	284.14	-1.60	4.9	4.0±1.0	33[C]	3EG J1013-5915
GEV J2204+4225	92.89	-10.48	4.9	2.2±0.7	41	3EG J2202+4217, BL Lac
GEV J1306-5920	304.81	3.47	4.6	2.9±0.8	29	---
GEV J1230-4839	299.39	14.07	4.5	1.6±0.5	44	5.1σ in VP 208.0
GEV J0526-6515	275.12	-33.32	4.5	1.2±0.4	41	LMC extended?
GEV J2057-4702	352.86	-40.58	4.4	2.1±0.8	52	3EG J2055-4716, QSO 2052-474?
GEV J0911+6548	148.38	38.74	4.3	1.1±0.4	59	3EG J0910+6556
GEV J2055+2548	70.69	-12.30	4.3	1.6±0.5	41	3σ (0.3-1 GeV)
GEV J1742-2039	6.70	4.91	4.2	2.0±0.6	24	3EG J1741-2050
GEV J1952+3251	68.67	2.94	4.1	2.6±0.8	28	3σ (0.3-1 GeV), PSR 1951+32
GEV J0614-3331	240.56	-21.74	4.1	2.0±0.8	57	3EG J0616-3310

[A] Units of 10^{-8} cm^{-2} s^{-1}, [B] Units of arcminutes, [C] larger of the radii for a fit to an ellipse is quoted

Most of these new steady sources are in the 3rd EGRET catalog. Of the six sources without a 3EG counterpart, GEV J2257-2755 has been detected above 100 MeV [8] but fell outside the time frame of the 3rd catalog. The source GEV J0526-6515 is very close to the LMC, which is in LM97, and this new source may be more diffuse emission. Above 1 GeV, the EGRET point spread may be small enough that the LMC is resolvable. This is being studied further. The source GEV J1952+3251 is consistent with PSR 1951+32, which was previously detected only through pulsar analysis [9] but is not in the 3EG catalog. Thus, three of our new sources have no true 100 MeV counterparts. The sources GEV J1306-5920, GEV J1230-4839 and GEV J2055+2548 are all new sources, two of the three being high latitude (lbiil > 10 degrees).

SEARCH FOR TRANSIENT SOURCES

Combining the Table 1 sources with those of LM97 gives us a basis for searching for transients. Our approach is to find week time-scale emission by analyzing all 217 public EGRET data sets. For each observation, the sources in LM97 and Table 1 are modeled as part of the background. A map of the residual maximum likelihood is then calculated. This gives the likelihood of a new source being present at each point. There are typically 25000 points per map for a total of 5.6 million likelihood values. The search for transients proceeds by listing points in any map which exceed a 3σ significance level. This gives about 3400 points on the sky for all viewing periods. Finally, we correlate all of these features to find instances of 3σ excesses within 1 degree of each other (a typical weak point source location radius) in separate but overlapping viewing periods. A new source is indicated by either a high significance in a single viewing period or by repeated outbursts in multiple viewing periods. Finally, we require that the source locations be within the inner 30 degrees of the EGRET field-of-view (16 degrees for EGRET reduced field) to avoid problems with systematic calibration uncertainties in the outer region of the EGRET field [10].

Table 2. Sources of transient GeV gamma rays ordered by galactic longitude.

Name	LII	BII	Sigma	Flux[A]	Rad[B]	V.P.s[D]	Notes
GEV J1653+3945	63.58	38.97	5.0	2.8±1.0	52	201 & 519	Likelihood sum=27.0; MRK 501
GEV J0426+1558	179.87	-22.40	3.8	4.9±2.5	57	1 & 616.1	3EG J0423+1707
GEV J0612+2910	182.68	5.16	4.9	12.9±4.5	56	2.1 & 213	Likelihood sum = 29.1
GEV J0448+1054	187.76	-21.09	4.3	8.0±3.4	57	36.5 & 39	3EG J0450+1105; PKS 0446+11
GEV J0339-0144	187.93	-42.20	5.6	12.0±4.3	51	419.1 & 420	5.1σ in VP 420; 3EG, B0336-019
GEVJ0424-0112	195.36	-32.81	5.3	5.7±2.0	27	21	5.3σ in VP 21; 3EG, PKS 0420-01
GEV J0502-0118	200.93	-24.70	5.1	8.8±4.1	56	413	5.1σ in VP 413;PKS 0458-02
GEV J0638+0446	207.19	-0.74	4.2	9.2±3.1	43	1 & 41	3EG J0634+0521
GEV J1223+2121	253.72	81.43	5.1	6.5±2.7	26	311.6 & 313	3EG J1224+2118; PKS 1222+21
GEV J1305-8232	303.42	-19.68	4.1	5.8±2.5	56[C]	17 & 224	3EG J1249-8330
GEV J1409-6126	312.13	0.02	5.1	15.6±4.0	26	314 & 424	3EG J1410-6147
GEV J1715-4044	346.77	-1.36	4.3	19.3±6.3	37	334 & 423	Likelihood sum = 27.9
GEV J1323+2206	359.91	81.20	5.8	10.1±3.8	22	308 & 313	5.3σ in VP 8.0; 3EG; B1324+2226

[A] Units of 10^{-8} cm^{-2} s^{-1}, [B] Units of arcminutes, [C] larger of the radii for a fit to an ellipse is quoted; [D] CGRO viewing period numbers

To date, two sets of simulations corresponding to the same exposures and sky pointings as the 217 actual data sets have been calculated. These simulations indicate that for a source detected in a single viewing period, a maximum likelihood statistic of 22 (4.7σ) is adequate to ensure less than one chance detection. For repeated outbursts, the cleanest separation of signal from chance coincidences comes from summing the likelihood values for the coincident pairs. For now, we concentrate on sources detected in two viewing periods, for which simulations indicate that a sum of the two likelihoods above 26 yields at most a single spurious detection. We also include sources that are detected at 3σ in at least two viewing periods and correspond to a 100 MeV catalog source regardless of the summed value.

The thirteen sources listed in Table 2 satisfy at least one of the three selection criteria. Four of the sources exceed 4.7σ in a single viewing period while three are detected only by summing likelihoods. Eight of these 13 transient sources have counterparts in the 3rd EGRET catalog, for which 6 are included solely on the basis of this association. The values listed in Table 2 are for the most significant detection, which may be the sum of two viewing periods even for those sources that qualify by

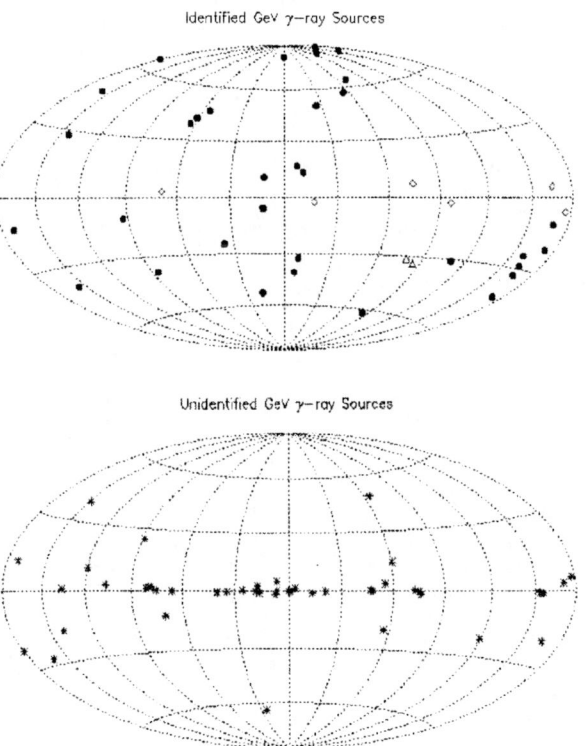

Figure 1. The positions of identified and unidentified GeV gamma-ray sources are plotted in galactic coordinates. These plots contain sources from LM97 and from this paper and the distributions are strikingly different.

being significantly detected in a single viewing period. The viewing periods used to calculate the source parameters are listed in the "VP" column. Two of the most interesting detections are of GEV J0502-0118 which is associated with a 1 Jy radio source and the TeV detected MRK 501 which has only recently been discovered to emit in the EGRET data [11].

DISCUSSION

A preliminary analysis of data above 1 GeV finds 29 new sources of GeV gamma-ray emission, 13 of which are found in a search for transient emission. Six of these sources are previously unreported. Of the 13 sources detected in the transient search, four are low latitude. These could be high-energy analogs to the 100 MeV transient GRO J1838-04 [12]. Many of the new sources have bright radio sources in their error circles, although most of the radio sources have 4.8 GHz fluxes below 1 Jy. Future work will emphasize other candidates, class studies, and comparing GeV and 100 MeV sources. Special attention will be paid to the unidentified sources, which as Figure 1 shows, tend to be a galactic population.

REFERENCES

1. Bradbury, S.M., et al. 1997, A&A, 320, L5
2. Catanese, M., et al. 1998, ApJ, 501, 616
3. Tanimori, T., et al. 1998, ApJ, 497, L25
4. Lamb, R.C. & Macomb, D.J. 1997, ApJ 490, 493 (LM97)
5. Hartman, R.C., et al. 1999, ApJS, 123, 79
6. Mattox, J.R. et al. 1996, ApJ, 461, 396
7. Bertsch, D.L., et al. 1993, ApJ, 416, 587
8. Macomb, D.J., Gehrels, N. & Shrader, C.R. 1999, ApJ, 513, 652
9. Ramanamurthy, P.V., et al. 1995, ApJ, 450, 791
10. Esposito, J.A., et al. 1999, ApJS, 123, 207
11. Sreekumar, P. et al. 1999, American Astronomical Society, HEAD meeting #31, #03.03
12. Tavani, M., et al. 1997, ApJ, 479, L109

THE NEW METAGALACTIC SOURCE OF GAMMA-QUANTA WITH ENERGY >10^{12}eV.

V.G. Sinitsyna

P.N. Lebedev Physical Institute, Russian Academy of Science,
Leninsky pr. 53, Moscow 117924, Russia.

Abstract. Observatory SHALON ALATOO has just announced the TeV discovered of NGC 1275. The TeV catalogue of extragalactic sources contains three objects investigated by the SHALON-ALATOO Observatory at energy above 0.8 TeV: Markarian 421, Markarian 501 and NGC 1275, which was detected in TeV energy range by the observation on the telescope SHALON-1. The flux appears to be about that of the Crab Nebula (the intensity of Active Galactic Nucleus NGC 1275 at energy above 0.8 TeV with the flux $(1.10\pm0.40)\cdot10^{-12}cm^{-2}s^{-1}$). Results obtained with the Mirror Cherenkov telescope SHALON-1 (mirror area more than 11 m^2) point out that the Active Galactic Nuclei are main sources of gamma-quanta and probably cosmic rays of the high energy.

More than forty years ago, when Cherenkov light from cascades of cosmic rays in the atmosphere on the background of a night sky had been detected, attention was paid to the possibility of observation gamma-astronomy in the energy range 10^{12}-10^{15} eV. In 1961-1963 A. Chudakov and his collaborators were the first who made the attempts to search gamma rays from the Galaxy center by means of an optical device with a resolution 1°-2°. From that time and till 1990 various sources detected in many experiments were mainly with variable sources that reasoned that sources detected by some observers weren't detected by others. From 1981 more precise methods were being developed in order to

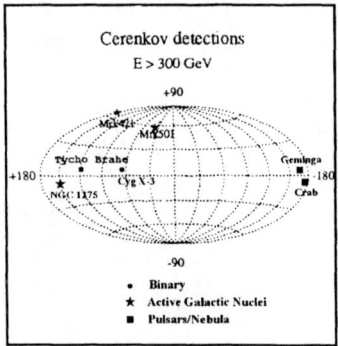

Fig. 1. *Sky map of gamma-quanta point sources in galactic coordinates that observed by gamma-telescope SHALON*

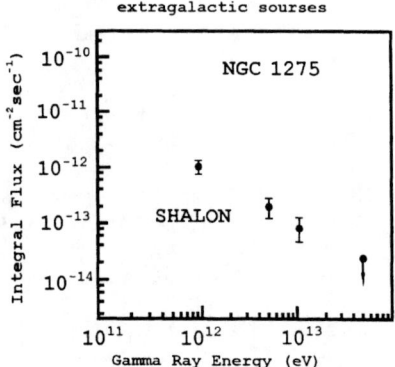

FIGURE 2. *The spectrum of the gamma- radiation of extra-high energies from active galactic nuclei NGC-1275*

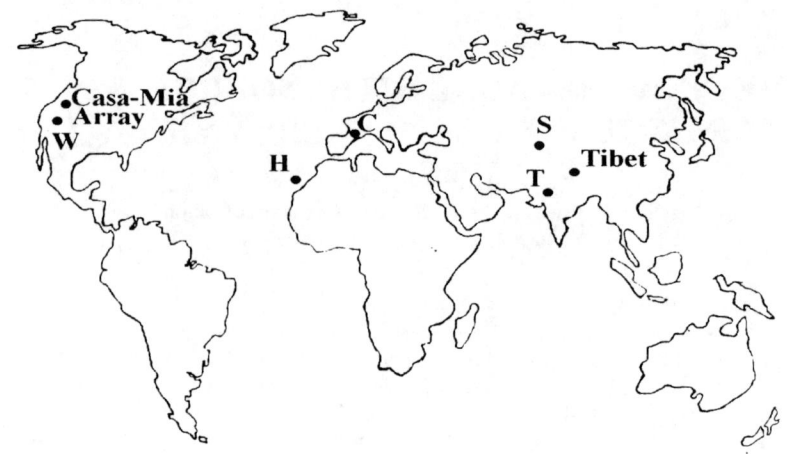

TABLE 1. Experiments reported to have detected Markarian 421 and Markarian 501 (FIGURE 3,4)

Experiment, Altitude	Site, Country	Area	Range of measurement	Full angle image. Pixel res.(°)•N	
Whipple, 2300 m	Arisona, U.S.A., 31°41 N110°53 W°	70 m²	300 GeV→ 12 TeV 5σ	3°	0.25°•150
CAT, 1650 m	French Pyrennees, 42° N 2° E	17,7 m²	220 GeV → 10 TeV	3.1°	0.12°•546
SHALON, 3338 m	ALATOO, Russia 42° N 75° E	11.2 m²	800 GeV → 50 TeV	8°	0.6°•144
TACTIC, 1300 m	Mt.Abu, India, 24° 39' N 72°47' E	9.5 m²	700 GeV →10 TeV	2.6°	0.31°•81
HEGRA, 2240 m	Canary Islands 28.75° N 17.89° W	8.5 m² 5 m²	700 GeV →>10 TeV 1.5 TeV → 15 TeV	4.3° 3.25°	0.25°•271 0.25°•127
Telescope Array, 1600 m	Mt. Cedar, Utah, U.S.A., 40.33° N 113.02° W	6 m²	600 GeV → 10 TeV 3σ	4°	0.25°•256
Tibet II EAS array, 4300 m	Yangbajing,Tibet, China, 30.1 ° N 90.53° E	3,7*10⁴m²	>3σ above 10 TeV	Angel resolution: 1° at E_γ =7 TeV	
CASA-MIA, 1450 m	Dugway, U.S.A., 40.2° N 112.8°W	23*10⁴m²	>45 TeV	Angel resolution: 0.15° at E_γ =70 TeV	

select electron-photon cascades in the atmosphere on the background of 10^3 more intensive flux of extensive air showers produced by protons and cosmic ray nuclei.

At the present time in north hemisphere a few telescopic devices for Cherenkov radiation with angular resolution ~0.1° (observatory Whipple in Arizona, CAT in French Pyrennees, Russian observatory SHALON in ALATOO et al., Table 1) allow to have stable results concerning to several galactic and extragalactic the sources of gamma-quanta in energy interval 10^{12}-$5 \bullet 10^{13}$ eV at fluxes less 10^{-12} cm^{-2} sec^{-1}. In the field energies above 10^{12} eV is shown, that observations of extensive air showers (EAS) generated by primary gamma-quanta is possible by other methods: as that EAS without muon, EAS without hadron, and even as to profile ionization light of atmosphere from the satellite. However at the primary energies below 10^{12} eV the observations of directed Cherenkov radiation from atmosphere for the present time is the most effective. One singles out the electron-photon showers in the EAS, which generated by protons and nuclei, you may improve by means of the increasing of the angular accuracy of direction, in which develops shower, that correspondences of the direction of primary gamma-quanta. Such improving of the gamma-astronomical observations is possible at the large area of mirror telescopes, like telescopes CAT and WHIPPLE. That is possible for the telescope SHALON by means of the increasing of the number of telescopes, placed into one group with the parallel axes of observation (it is possible so this will be done into Indian observatory Abu). At ALATOO observatory SHALON high-mountain station of Lebedev Physical Institute of Russian Academy of Science in Kazakhstan the development of gamma-astronomical observations is connected with the assemblage of the second telescope SHLAON. SHALON-2 will be set up at a distance 260m from telescope SHALON-1, and the propagation of EAS between the both gamma-telescopes (~10^4m^2) will be the subject of stereoscopic analysis of the of cascade development, which is different for the gamma and proton-nuclei showers.

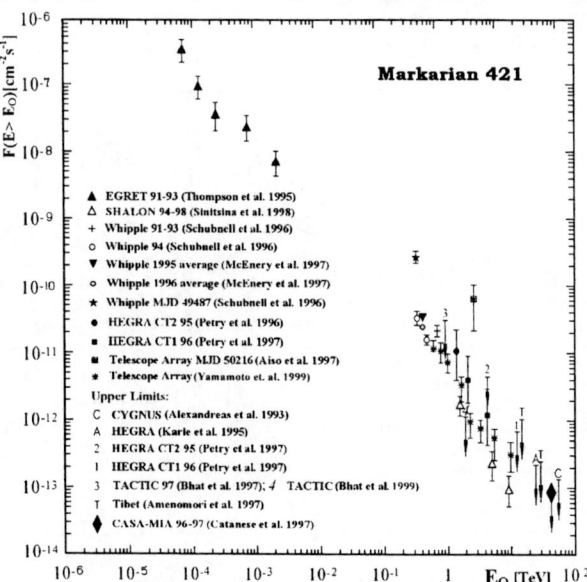

FIGURE 3. *The spectrum of the gamma- radiation of extra-high energies from active galactic nuclei Markarian 421*

TABLE 2. *The catalog of observing by SHALON telescope in TeV energies since 1994.*

Source	Type	Flux cm^{-2} s^{-1} E > 0.8 TeV SHALON	Distance
Galactic			kpc
Crab Nebula	Plerion	$(1.1\pm0.30)\bullet10^{-12}$	2.0
Cygnus X-3	Binary	$(4.2\pm0.80)\bullet10^{-13}$	11.0
Tycho Brahe	Supernova	$<2\bullet10^{-13}$ upper limit	2.0-5.1
Geminga	Supernova	$(5.7\pm4.0)\bullet10^{-13}$	0.25
Extragalactic			mpc
Mkn 421	AGN	$(1.09\pm0.41)\bullet10^{-12}$	124
Mkn 501	AGN	$(1.32\pm0.30)\bullet10^{-12}$	135
NGC 1275	AGN	$(1.10\pm0.40)\bullet10^{-12}$	71

For the present time it is known very finite number of the gamma-quanta sources with energy above 10^{12}eV. It is connected with the small intensity of observing fluxes and by large observation time that is greatly restricted by the weather and moon-less night periods. It is possible that this fact explain approximate equality of observing fluxes from different gamma-quanta sources both galactic and extragalactic. The last seems very wonderful, if the significant number of galactic objects with the less flux intensity will not be found and systematic number exceeding of extragalactic sources will not become essentially less then galactic sources number with the equal gamma-quanta flux intensity, so it is necessary to find the protons and nuclei sources of the cosmic rays with energy

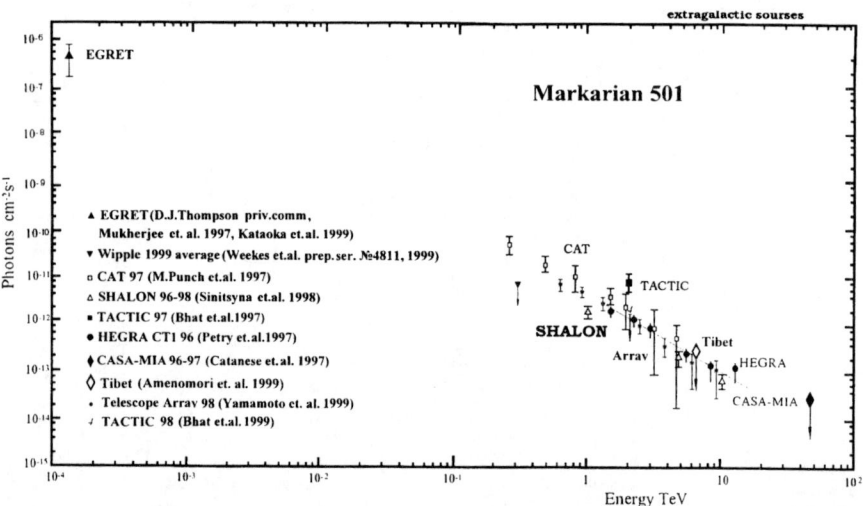

FIGURE 4. *The spectrum of the gamma- radiation of extra-high energies from active galactic nuclei Markarian 501*

10^{13}-10^{14} eV not in the our galaxy, but out of its area, it is because of the equal observational gamma-quanta flux intensity from the source near the observer and far off from observer means the difference of emitted flux in the square of the relation of distances to observing sources.

The estimations of diffuse gamma-quanta flux with energy >400 TeV, of the flux including in itself gamma-quanta from local sources, dictate to look about the intensity of gamma-quanta flux with equal or about 1 TeV of local sources <10^{-12}cm^{-2}s^{-1}.

TABLE 3. *Upper limit for diffusion flux of ultra high energy γ-rays in the cosmic rays*

Experiment	Level	Energy	Flux
EGRET		>1•10^9 eV	(1.45±0.05) •10^{-5}cm^{-2}s^{-1}sr^{-1}
Tien Shan - selection of muon and hadron - poor showers, 1984.	3338m	>4•10^{14} eV	<(3.4±1.2)•10^{-13}cm^{-2}s^{-1}sr^{-1}
EAS-TOP GRAD SASSO lab - selection of extensive Air Shower characterized by no muons recorder in 140 m^2 detector, 1996.	2005m	>8.7•10^{14} eV	<1.8•10^{-14}cm^{-2}s^{-1}sr^{-1}
CASA-MIA -poor in muons, 1997	1450m	>1•10^{14} eV	<6•10^{-13}cm^{-2}s^{-1}

CONCLUSION

It is more than thirty years ago by the researchers group under a A. Chudakov lead the first attempts of observation of very high energy gamma - quanta radiation of from Center of our Galaxy on directed Cherenkov radiation of air showers were undertaken. However, despite of large period of searches, it is possible to approve, that the search and observations of local gamma – quanta sources while that is carried on a bound of sensitivity of experimental installations. This statement well illustrates a sameness of extragalactic sources fluxes detected and published and comparison with intensity nearest on distance of a source in Crab Nebula. All of them have approximately identical observe intensity of a gamma - quanta fluxes with energy ≥10^{12} eV at difference of distances in kpc: 2 (Crab Nebula), 71·10^3 (NGC 1275), ~130·10^3 (Mkn 421 and Mkn 501).

REFERENCES

Nikolsky S.I., Sinitsyna V.G., 1987, VANT, Ser.TFE(1331), 30.
Nikolsky S.I., Sinitsyna V.G., 1989, Proc. Workshop VHE Gamma-ray Astr., 11-21.
Sinitsyna V.G., 1992, Proc. Workshop, Towards Major Atmospheric Cherenkov Detector-I, ed. P.Fleury, (Paris), 299-304; 1993, Detector -II, ed. R.Lamb, (Calgary), 91-101; 1995, Detector-IV, ed. M.Cresti, 133-140; 1997, Detector-V, ed. O.De Jager, Kruger Park, 136.
Djannati - Atai, A., CAT collab., 1997, Detector-V, ed. O.De Jager, Kruger Park,21-25.
Bhat, C et al. 1997, the same, p. 196-201; 1999 26th ICRC, 4, 104.
Petry, D., et al., 1997, Proc. 25th ICRC., 3, 241; 1996, A&A, 311, L13.
Karle, A.D., et al., 1995, Astropart. Phys., 4, 1.
Aiso, S., et al., 1997, 25th ICRC, 3, 261; Yamamoto.et.al 1999 26th ICRC, 3, 386.
McEnery, J.E.,et al.,1997, Proc. 25th ICRC, 89.
Thompson, D.J., et al.1995, ApJS, 101, 259.
Schubell, M.S., et al., 1996, ApJ, 460, 644.
Alexandereas, D.E.,et al., 1993, ApJ, 418, 832.
Amenomori, M.,et al., 1997, Proc. 25th ICRC, 3, 297; 1999 26th ICRC, 3, 418.
Catanese et.al. 1997 Proc. 26th ICRC,3, 301-304.

IMAGING ATMOSPHERE CHERENKOV OBSERVATORIES

The Status Of The Whipple 10m GRANITE III Upgrade Program

J.P. Finley[1,2], I.H. Bond[3], S.M. Bradbury[3], A.C. Breslin[4], J.H. Buckley[5], A.M. Burdett[2,3], M. Carson[4], D.A. Carter-Lewis[6], M. Catanese[2,6], M.F. Cawley[7], S. Dunlea[7], M. D'Vali[3], D.J. Fegan[4], S.J. Fegan[1,8], J.A. Gaidos[1], T.A. Hall[1], A.M. Hillas[3], D. Horan[4], J. Knapp[3], F. Krennrich[6], S. LeBohec[6], R.W. Lessard[1], C. Masterson[4], B. McKernan[4], J. Quinn[4], H.J. Rose[3], F.W. Samuelson[6], G.H. Sembroski[1], V.V. Vassiliev[2], T.C. Weekes[2]

[1]Department of Physics, Purdue University, West Lafayette, IN 47907, USA
[2]Fred Lawrence Whipple Observatory, Harvard-Smithsonian CfA, Amado, AZ 85645, USA
[3]Department of Physics, Leeds University, Leeds, LS2 9JT, UK
[4]Experimental Physics Department, University College, Belfield, Dublin 4, Ireland
[5]Department of Physics, Washington University, St. Louis, MO 63130, USA
[6]Department of Physics and Astronomy, Iowa State University, Ames, IA 50011, USA
[7]Physics Department, St. Patrick's College, Maynooth, County Kildare, Ireland
[8]Physics Department, University of Arizona, Tucson, AZ 85721, USA

Abstract. The Whipple Collaboration has been carrying out an upgrade program, GRANITE III, on the 10m telescope atop Mt. Hopkins. The 3 year program, 1996 through 1999, has involved a large field-of-view camera, a hardware pattern recognition trigger, and finally a small pixel camera. Results of the first and second stage of the program indicate that a large field-of-view yield an increased collection area for cosmic-ray triggers and the hardware pattern trigger reduces the energy threshold of the telescope by greatly suppressing night sky accidental triggers over a simple multiplicity coincidence trigger. The final phase of the program, being completed during the summer and fall of 1999, will be the installation of a small pixel camera. The pixel size of $\sim 0°.12$ will yield the lowest threshold energy at which the 10m Whipple telescope has ever operated.

INTRODUCTION

The field of Very High Energy (VHE) gamma-ray astronomy is the newest branch of high energy astrophysics and a great deal of progress has been made in this decade with the discovery of several classes of VHE sources[1] (for a recent review of the field see Catanese & Weekes 1999). The recent progress was in large part due to the development of the atmospheric Cherenkov imaging technique that was pioneered by the Whipple collaboration at the 10m telescope located at the Fred Lawrence Whipple Observatory in southern Arizona. Over the current decade the focal plane cameras in use at the existing VHE observatories have progressively been expanded in both field-of-view and pixelization with the aim of increasing the sensitivity of the telescopes to gamma-rays and, simultaneously, reducing the threshold energy of the instruments.

The Whipple collaboration commenced an upgrade program to the focal plane detector of the 10m telescope, dubbed the GRANITE III upgrade, in 1996 that was to last 3 years. The upgrade was carried out in 3 distinct phases: (a) commissioning of a large field-of-view camera (~ 5° diameter) with $0°.25$ pixels, (b) commissioning of a hardware pattern recognition trigger, and (c) the installation and commissioning of a small pixel camera (~ $0°.12$ pixel spacing). Phase (a) was carried out during the summer and fall of 1997 while phase (b) was completed during the winter of 1998. The final phase of the upgrade, the installation and commissioning of the small pixel camera, is being carried out during the summer and fall of 1999. In this article we will describe the results of the first 2 phases of the upgrade and describe the plans for the final phase.

UPGRADE PHASE A: THE LARGE FIELD-OF-VIEW CAMERA

During the summer and fall of 1997 the Whipple collaboration installed a 331 pixel large field-of-view camera to investigate the gain that can be realized with the large collection area that such a camera provides. The field of view of this camera was ~ 5° diameter with a pixel spacing in the focal plane of ~ $0°.25$ center-to-center. A schematic layout of the focal plane camera is given in Figure 1 below. In addition to

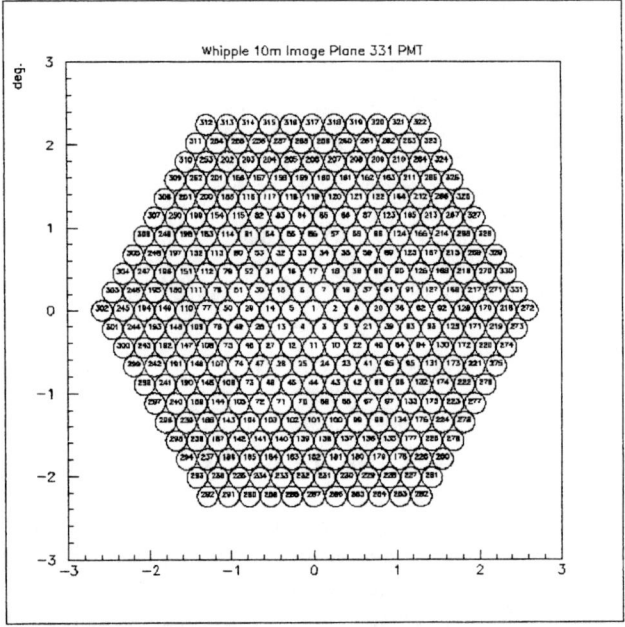

FIGURE 1. The layout of the $0°.25$ pixels in the image plane of the phase (a) large field-of-view camera. The horizontal and vertical axes are degrees in the focal plane of the 10m telescope.

the increased field-of-view a set of light cones were installed to help decrease losses due to the dead space between the pixels. These light cones were simply overlapping simple truncated cones and were fabricated out of an aluminum substrate with aluminized mylar bonded to the reflective surfaces. To assess the gain that this camera yielded a series of off-axis Crab nebula observations were carried out that were compared to similar observations made with the previous 3°.5 field-of-view camera. In the case of each camera the Crab nebula was positioned on-axis and then 1° and 1°.5 off-axis and the gamma-ray rate compared. The results of that comparison are given in Figure 2 for both the 3°.5 and 5°.0 cameras below. Inspection of Figure 2 shows that the larger field-of-view yields ~ a factor of 3 in increased sensitive area for the 5°.0 camera over the 3°.5 camera. This increase is larger than the simple geometric factor of 2 that one might expect since it is possible for showers outside of the geometric field-of-view to trigger the telescope. Therefore showers having larger impact parameters can then produce enough light to trigger the telescope.

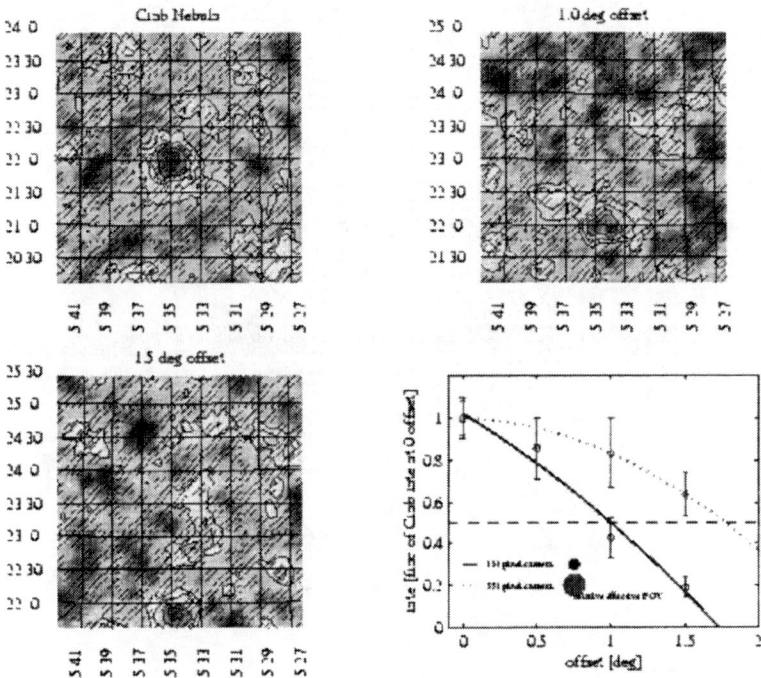

FIGURE 2. Offset Crab nebula observations carried out with the 331 pixel 5°.0 field-of-view camera. The images are on-axis (upper left), offset by 1° in declination (upper right), and offset by 1°.5 in declination (lower left). The diagram in the lower right shows the relative Crab nebula rate as a function of off-axis angle for the 3°.5 camera (solid line) and the 5°.0 camera (dashed line).

UPGRADE PHASE B: THE HARDWARE PATTERN SELECTION TRIGGER

During the summer and fall of 1998 a hardware pattern selection trigger[3] (PST), developed at the University of Leeds in partnership with Hytec Electronics, was installed and commissioned on the 10m telescope. The hardware PST is essentially a logic device that examines overlapping sectors of 19 pixels in the camera and ascertains if there are adjacent pixels within a sector that are above a pre-determined hardware trigger level. Figure 3 shows the overlapping sectors on the 331 pixel camera plane. The PST is designed to eliminate accidental triggers due to the night sky brightness, an example of which is pictured in Figure 4, and preferentially select triggers which are topologically consistent with the expectation for a gamma-ray shower, an example of which is pictured in Figure 4. The suppression of the accidental night-sky triggers allows for a lower energy threshold to be achieved over a simple multiplicity coincidence trigger that has been the operational mode at the 10m telescope. To ascertain the gain in energy threshold attributable to the PST a telescope

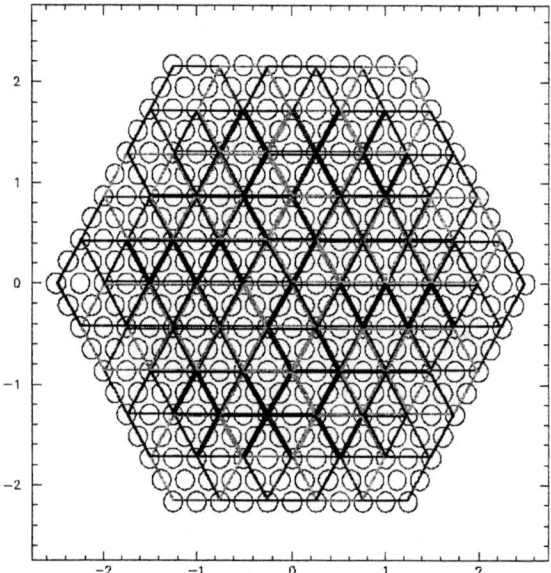

FIGURE 3. The overlapping sectors of 19 pixels on the 331 pixel camera plane. A trigger condition is satisfied when an adjacency condition is satisfied in any one of the sectors.

bias curve can be acquired with and without the PST and compared. Figure 5 displays such a bias curve with telescope triggering rates for a simple 2-fold multiplicity trigger as well as a 2-fold PST trigger and a 3-fold PST trigger as a function of the threshold setting of the constant fraction discriminators in mV. Figure 5 clearly shows that the PST reduces the night sky accidental rates by at least a factor of 10 over the simple multiplicity trigger and by a factor of more than 100 when a 3-fold adjacency is required. Therefore the triggering level of the telescope can be set closer to the noise region (the rapidly rising curve on the left in Figure 5) which allows for an overall lower energy threshold.

To test the reduction in the energy threshold observations of the Crab nebula using the 2-fold PST were acquired and compared to the gamma-ray rate with the simple multiplicity trigger. In Figure 6 the Crab nebula rate as a function of time is displayed for the pre-PST 331 pixel camera and the post-PST 331 pixel camera. It is clearly evident that the use of the PST resulted in a factor of ~ 2 increase in the overall Crab nebula rate which corresponds to an energy threshold reduction of ~ 40% for an $E^{-1.5}$ integral spectrum which describes the nebular emission[2].

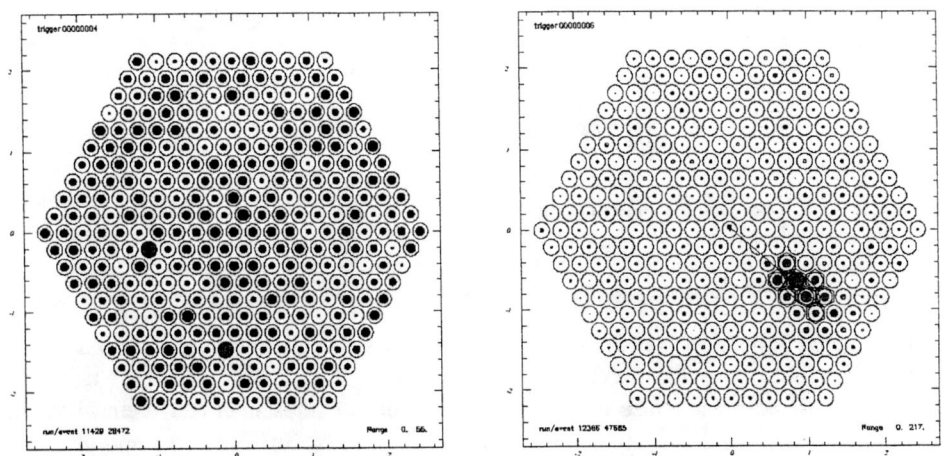

FIGURE 4. A 2 pixel accidental trigger on the 331 pixel camera image plane (left) and a gamma-ray like image (right). The horizontal and vertical scales are in degrees and the size of the circles in each pixel is proportional to the total signal in that pixel.

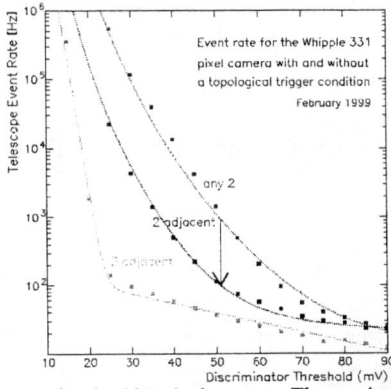

FIGURE 5. A telescope bias curve for the 331 pixel camera. The vertical axis is the telescope trigger rate and the horizontal axis is the discriminator setting of each of the individual PMTs in the camera (in mV). The uppermost curve is a simple multiplicity coincidence trigger while the 2 lower curves are for 2-fold and 3-fold adjacent triggers respectively. The night sky accidentals cause the rapidly rising part of the curve on the left while the cosmic-ray triggers are the slowly changing curves on the right

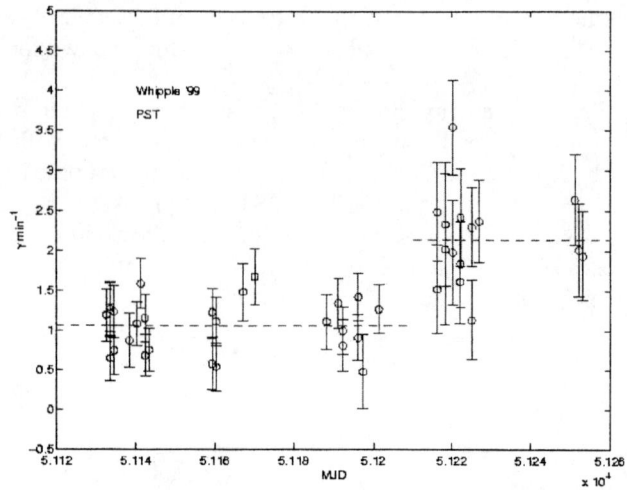

FIGURE 6. The Crab nebula gamma-ray rate as a function of time showing the effect of the PST. The PST was installed and commissioned on ~ 51210 MJD. The average gamma-ray rate prior to the PST was ~ 1.1 γ's/min while the post-PST rate was ~ 2.2 γ's/min (the horizontal dashed lines). The data were from the '98-'99 observing season.

A reasonable indication that the increased rate of the Crab nebula was due to a lower energy threshold was the increase in the number of triggers that are attributable to local μ rings and arcs. These lower energy μ induced events manifest themselves in the length over size distribution for the triggering events. Since μ rings and arcs are long events in the image plane yet fairly faint (i.e. small size) compared with the more compact and brighter cosmic-ray images they populate a region of the length/size distribution which is distinct from the cosmic-rays (i.e. larger values). Pre-PST and post-PST length/size distributions indicate an increase in the μ induced events that are triggering the telescope and therefore a lower energy threshold.

UPGRADE PHASE C: THE SMALL PIXEL CAMERA

The final phase of the upgrade is the installation of a small pixel camera. The camera consists of 379 13mm PMTs constituting a 2°.6 diameter inner camera surrounded by 111 28mm PMTs which fills out the field-of-view to 4°.0 diameter. The triggering pixels in this camera are the inner 331 pixels. A schematic layout of the camera is given in Figure 7. The center-to-center spacing of the inner pixels is 0°.12 in the 10m

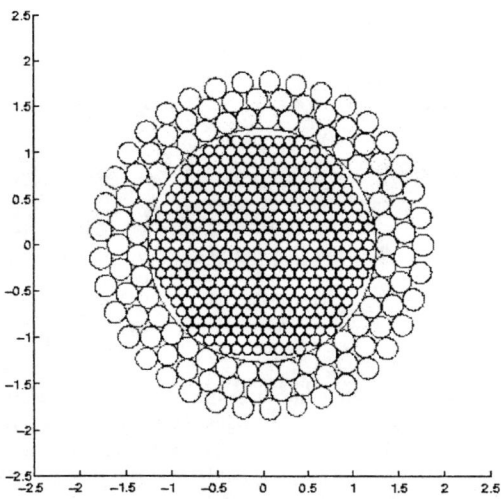

FIGURE 7. A schematic layout of the 490 pixel camera to be installed at the 10m in 1999.

focal plane while the outer pixels are spaced at ~ $0°.24$ but in a circular rather than close packed arrangement. The outer rings of 28mm PMTs will be utilizing analog fiber optics[4] to transmit their signals from the telescope to the electronics area as a test of the durability and reliability of these fibers in an observatory setting. The fibers were developed at the University of Leeds and are scheduled for installation and evaluations during the late fall of 1999. The installation of the camera was carried out during the summer of 1999 and first light was achieved in the early fall of 1999. The combination of the large aperture of the 10m telescope and the fine pixelation of the focal plane camera in consort with the PST will yield the lowest energy threshold (< 200 GeV) at which the Whipple 10m telescope has ever operated. We anticipate some new detections during the 1999-2000 observing season and, with any luck, a few surprises as well.

ACKNOWLEDGEMENTS

The authors would like to thank the U.S. Department of Energy, PPARC, and Enterprise Ireland for their generous support. The authors would also like to thank K. Harris, E. Roache and all the support staff at the Whipple Observatory.

REFERENCES

1. Catanese, Michael, and Weekes, Trevor C., PASP 111, 1193-1222 (1999).
2. Hillas, A. M., et al., ApJ 503, 744-756 (1998).
3. Bradbury, S. M., et al., A&A 320, L5-L9 (1997).
4. Rose, H.J. et al., Leeds University *preprint* (1999).

Calibration of the CAT Telescope

Frédéric Piron, for the CAT collaboration

Laboratoire de Physique Nucléaire des Hautes Energies
Ecole Polytechnique, route de Saclay, 91128 Palaiseau Cedex, France

Abstract. Due to the lack of test-beams in ground-based γ-ray astronomy, detector calibration has been a major challenge in this field. However, with the use of Cherenkov ring-images due to cosmic-ray muons and of strong γ-ray signals, the CAT telescope could be rather well monitored and understood. Here we present a few outstanding aspects of this work.

INTRODUCTION: THE CAT DETECTOR

The CAT (Cherenkov Array at Thémis) telescope records Cherenkov flashes due to VHE atmospheric showers through its 17.8m^2 mirror. Its camera is located 6m from the mirror and has a 4.8° full field of view, consisting of a central region of 546 0.12° angular diameter phototubes arranged in a hexagonal matrix and of 54 surrounding tubes in two "guard rings" (Fig.1a). Fast electronics allows a relatively low γ-ray detection threshold energy of 250 GeV (at Zenith), and the fine grain of the camera permits an accurate image analysis. The experiment and the analysis method are fully described elsewhere [1,5]. Briefly, after selecting the most significant triggers (total charge $Q_{tot}>30$ photo-electrons), good discrimination between γ and hadron-induced showers is achieved by looking at the shape and the orientation of the images (see the events on Fig. 1a): since γ-ray images are rather thin and ellipsoidal while hadronic images are more irregular, a first cut is applied which selects images with a "γ-like" shape; it is based on a χ^2 fit to a mean light distribution predicted from electromagnetic showers, and a probability $P(\chi^2)>0.35$ is required. Then, since γ-ray images are expected to point towards the source position in the focal plane whereas cosmic-ray directions are isotropic, a second cut $\alpha<6°$ is used in the case of a point-like source, where the pointing angle α is defined as the angle at the image barycentre between the actual source position in the focal plane and that of the image which is reconstructed by the fit [1]. As a result, this procedure rejects 99.5% of hadronic events while selecting 40% of γ-ray events. Fig 1b is an example of the pointing angle distribution obtained on the Crab nebula: the signal is clearly seen in the first bins, while a second signal can be

---

[1] The resolution per event is of the order of the pixel size, i.e. ~0.1°.

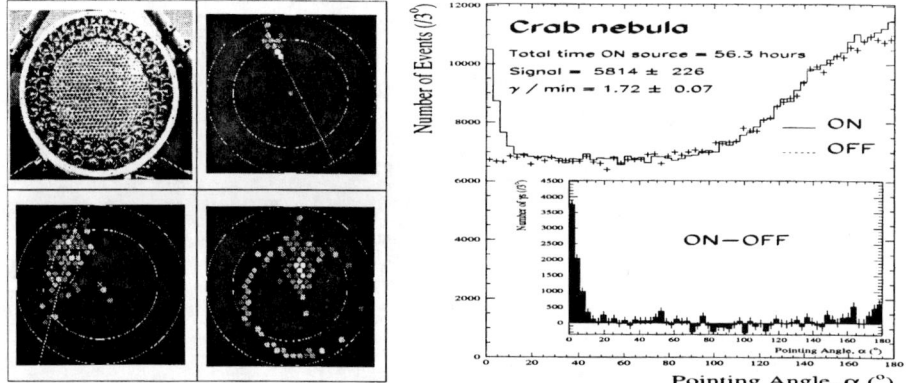

FIGURE 1. *(a)* The CAT camera and three typical events: the first image is presumably due to a γ shower because of its fine and cometary shape, and because it is pointing towards the source position at the center of the camera; the second image is certainly due to a hadron because it is more spread out and pointing elsewhere; finally, the third image is clearly hadronic and would be easily rejected, because it is signed by the presence of a muon ring. *(b)* Pointing angle distribution from a sample of data taken on the Crab nebula between 1996 and 1999, for zenith angles ranging from 21° to 35°. Cuts on total charge and shape have been applied (see text). The inset shows the ON−OFF distribution (OFF source runs are used to estimate the hadronic background): within 6°, the total significance is 25.6σ for a ratio of durations $T_{ON}/T_{OFF} = 2.4$ (the OFF distribution has been renormalized). The corresponding significance for an equal amount of ON and OFF runs would be 32.2σ, i.e. 4.3σ in one hour.

seen at $\alpha \sim 180°$, due to γ-ray images whose direction has been mis-reconstructed by the fit [2].

CALIBRATION OF THE DETECTOR

Hardware monitoring

Mrk 501 exhibited a remarkable series of flares during the whole year 1997 [3], which have been very useful for detector calibration. As an example, Fig. 2a illustrates the very good quality of the mechanical monitoring for data taken with a zenith angle between 0° and 44°: the signal appears right at the actual position of the source, thus validating the angular correction which is applied on each event to compensate for the unavoidable and zenith-dependent slight mis-alignment of the optical axis of the camera [3].

---

[2] The global rise of the background distribution for large values of α corresponds to large hadronic images which were cut by the edge of the camera. This effect becomes fainter for larger zenith angles, since images form closer to the center of the camera; see [6] for illustration.
[3] This is due to the weight of the camera, which bends the arms of the telescope, and to the telescope azimuth and altitude axis mis-alignment, especially at large zenith angles.

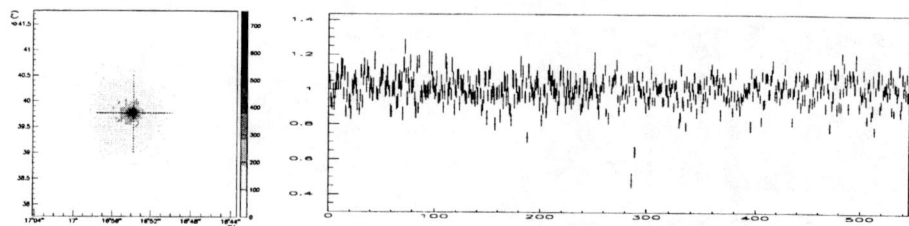

FIGURE 2. *(a)* Projected distribution (bin size=0.05°) of the reconstructed shower angular origins for all data of Mrk 501 in 1997: no background substraction has been performed, and events are selected by the shape cut only. The cross marks the actual position of Mrk 501. *(b)* Ratio of the charge integrated in each of the 546 small inner phototubes to that predicted by a model of muon rings. ∼100 muon images have been used for these statistics. The dispersion of the ratio around unity is mainly due to the uncertainty on the gain values.

Cosmic-ray muons falling onto the mirror yield ring-like images in which the light distribution can be easily predicted. The fine grain of the CAT camera allows a fine analysis of these images [4]. In this way, the overall conversion factor between ADC counts and incident Cherenkov photon number can be directly checked. This factor involves optical efficiencies, as well as phototube pedestals and gains. As an example, Fig 2b shows that the camera is correctly calibrated, except for a few channels which are not used in the analysis. In this study, particular attention has been paid to the wavelength-dependent aspect, by taking special runs using different UV-filters placed in front of the camera.

Validation of the simulations

On April 16th, 1997, Mrk 501 reached ∼8 times the level of the Crab nebula. The signal-to-noise ratio for this night is 2.7, corresponding to a γ-ray beam with only 30% contamination. The good atmospheric quality of this night allows the use of this beam for calibration through comparison with simulations. Fig. 3a shows the perfect agreement on the distributions of Hillas parameters [7]: it is shown for events selected both with the single orientation cut and with the complete selection including the χ^2 fit. The distribution of the final number of pixels N^{pix} retained by the fit (see [5]) is also well reproduced. Furthermore, the good agreement observed on the fourth-brightest-pixel's charge Q^{4th} validates the simulations at the trigger level, since the trigger condition requires four pixels above threshold. Finally, Fig. 3b shows the χ^2/ndf and $P(\chi^2)$ distributions [4], as well as the previously discussed parameters, expressed as functions of the total image charge Q_{tot}: here again the agreement is very good. This allows the simulation to be used to calculate

---

[4] The $P(\chi^2)$ distribution is not flat, as would be expected if the χ^2 were performed using variables with Gaussian errors. The description of Cherenkov light fluctuations in showers development is a very difficult task (see [5] for discussion).

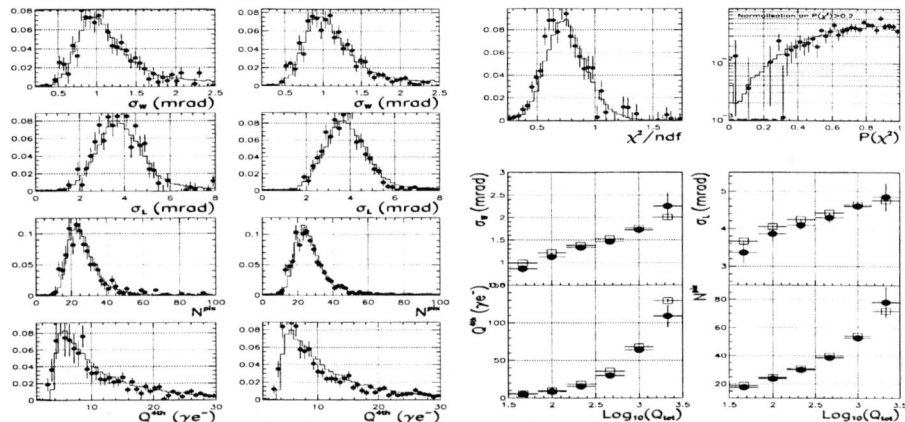

FIGURE 3. Comparison of γ-ray shower simulations (full lines or open squares) with the data (ON−OFF) from the flare of Mrk 501 on April 16th, 1997 (filled points). Statistical errors are negligible for simulations. *(a)* Distribution of several variables, within the orientation cut $\alpha < 6°$ (left column) and with the cut on shape $P(\chi^2) > 0.35$ also (right column): Hillas parameters *width* (σ_W) and *length* (σ_L), number of pixels in the image N^{pix}, and fourth-brightest-pixel's charge $Q^{4\text{th}}$; *(b)* χ^2/ndf and $P(\chi^2)$ distributions, and previous parameters as functions of $\text{Log}_{10}(Q_{\text{tot}})$.

the γ-ray effective detection area within the selection cuts, as well as the energy-resolution function Ψ. This is shown in Fig. 4. In particular, a clear positive bias in the energy reconstruction is visible for low values of the injected energy (Fig. 4b):

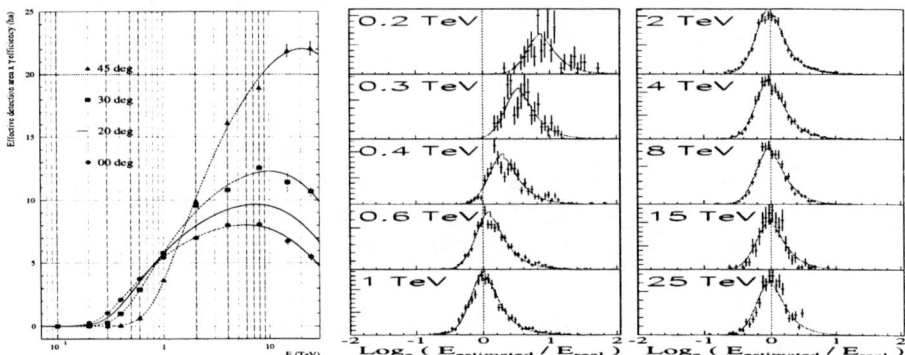

FIGURE 4. *(a)* γ-ray effective detection area (in 10^4m^2), including the effect of event-selection efficiency, as a function of the energy. Each point corresponds to simulations, while full lines come from an analytical 2D-interpolation over energy and zenith angle; *(b)* Energy-resolution functions $\Psi(E \to \widetilde{E}, \cos\theta)$ vs $\log(\widetilde{E}/E)$, for a fixed zenith angle $\theta = 30°$ (y-axis units are arbitrary). Each plot is defined for a fixed value of the *real* energy E, ranging from 200 GeV to 25 TeV, and runs on the *estimated* energy $\widetilde{E}$. Full lines come from an analytical 3D-interpolation over the three variables E, $\widetilde{E}$ and θ.

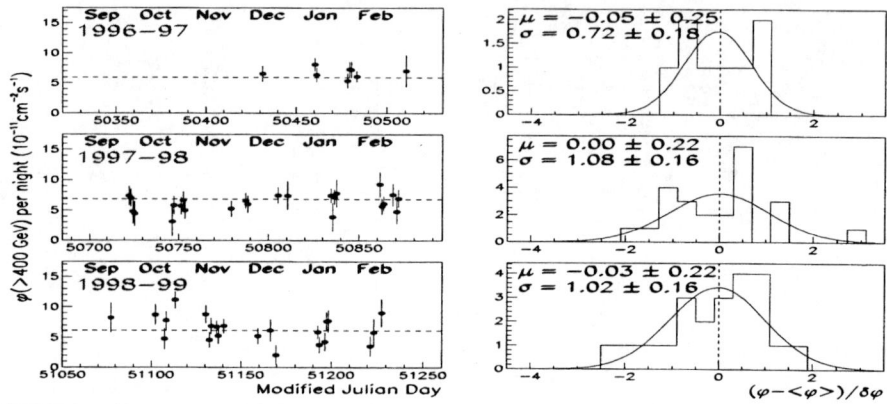

FIGURE 5. 1996 to 1999 nightly integral flux of the Crab nebula above 400 GeV, from a sample of data with zenith angles up to 35°. The dotted lines represents the mean flux for each period, and the corresponding residuals are shown on the right panel.

it is due to a trigger effect which selects those events which benefited from a positive fluctuation of Cherenkov light during the shower development. This energy over-estimation disappears when going towards higher energies, where Ψ becomes more Gaussian, with a zero mean value and a width $\sigma \sim 20\%$.

CONCLUSION

More details concerning calibration and spectrum measurement will be given in a forthcoming paper. The good stability obtained to date on the signal from the Crab nebula (Fig. 5), using the acceptances and energy-resolution function discussed above, illustrates the good quality of running conditions and the good understanding of the detector. Future effort will be devoted to the analysis of large zenith angle data (see [6] for a first study) and to the cross-calibration with the CELESTE experiment, operating on the same site with an energy threshold of $\sim$50 GeV; the recent observation of the first common events between both detectors [2] is an encouraging result in this direction.

REFERENCES

1. Barrau, A., *et al*, *Nucl. Instr. Meth.* A **416**, 278 (1998).
2. De Naurois, M., *et al*, *Proc. XXVI ICRC* **5**, 211 (Salt-Lake City, 1999).
3. Djannati-Ataï, A., *et al*, *A&A* **350**, 17 (1999).
4. Iacoucci, L., *Ph.D. thesis*, Ecole Polytechnique, Palaiseau, France (1998).
5. Le Bohec, S., *et al*, *Nucl. Instr. Meth.* A **416**, 425 (1998).
6. Mohanty, G., *et al*, *Proc. XXVI ICRC* **3**, 452 (Salt-Lake City, 1999).
7. Weekes, T.C., *et al*, *ApJ* **342**, 379 (1989).

Initial Performance of CANGAROO-II 7m telescope

Hidetoshi Kubo[1], S.A. Dazeley[2], P.G. Edwards[3], S. Gunji[4],
S. Hara[1], T. Hara[5], J. Jinbo[6], A. Kawachi[7], T. Kifune[7], J. Kushida[1],
Y. Matsubara[8], Y. Mizumoto[9], M. Mori[7], M. Moriya[1],
H. Muraishi[10], Y. Muraki[8], T. Naito[5], K. Nishijima[6],
J.R. Patterson[2], M.D. Roberts[7], G.P. Rowell[7], T. Sako[11],
K. Sakurazawa[1], Y. Sato[7], R. Susukita[12], T. Tamura[13],
T. Tanimori[1], S. Yanagita[10], T. Yoshida[10], T. Yoshikoshi[7], and
A. Yuki[8]

[1] Department of Physics, Tokyo Institute of Technology, Meguro, Tokyo 152-8551, Japan
[2] Department of Physics and Mathematical Physics, University of Adelaide, South Australia 5005, Australia
[3] Institute of Space and Astronautical Science, Sagamihara, Kanagawa 229-8510, Japan
[4] Department of Physics, Yamagata University, Yamagata 990-8560, Japan
[5] Faculty of Management Information, Yamanashi Gakuin University, Kofu, Yamanashi 400-8575, Japan
[6] Department of Physics, Tokai University, Hiratsuka, Kanagawa 259-1292, Japan
[7] Institute for Cosmic Ray Research, University of Tokyo, Tanashi, Tokyo 188-8502, Japan
[8] STE Laboratory, Nagoya University, Nagoya, Aichi 464-860, Japan
[9] National Astronomical Observatory, Tokyo 181-8588, Japan
[10] Faculty of Science, Ibaraki University, Mito, Ibaraki 310-8521, Japan
[11] LPNHE, Ecole Polytechnique. Palaiseau CEDEX 91128, France
[12] Computational Science Laboratory, Institute of Physical and Chemical Research, Wako, Saitama 351-0198, Japan
[13] Faculty of Engineering, Kanagawa University, Yokohama, Kanagawa 221-8686, Japan

Abstract. CANGAROO group constructed an imaging air Cherenkov telescope (CANGAROO-II) in March 1999 at Woomera, South Australia to observe celestial gamma-rays in hundreds GeV region. It has a 7m parabolic mirror consisting of 60 small plastic spherical mirrors, and the prime focus is equipped with a multi-pixel camera of 512 PMTs covering the field of view of 3 degrees. We report initial performance of the telescope.

INTRODUCTION

CANGAROO group planned to construct an array of four 10m telescopes for stereoscopic observations of air Cherenkov lights [2]. In 1995 the construction of one imaging telescope was approved (CANGAROO-II project), however, due to the limit of the fund we decided to construct the telescope with a 7m mirror, of which frame and base can sustain a 10m mirror because we plan to expand the mirror from 7m to 10m in early 2000. In December 1998 all components of the telescope and the imaging camera were shipped to Australia. The construction of CANGAROO-II at Woomera, South Australia started in January 1999, and finished in the middle of March(Fig.1) [4]. In May the tuning of both electronics and trigger condition were carried out, and we started normal observations from June. Here we present the electronics system and the brief report of the performance as an imaging Cherenkov telescope. Performance of mirrors is described in detail by Kawachi et al. [1].

FIGURE 1. CANGAROO-II 7m telescope at Woomera, South Australia

PERFORMANCE OF TELESCOPE

Figure 2 shows the front view of the camera attached in the focal plane, where that of 3.8m telescope is also shown for comparison. This camera consists of 512 pixels to cover a field of view (FOV) of diameter $\sim 3°$. Each pixel covers $0°.115 \times 0°.115$ (16mm × 16mm), and 13 mm ϕ photomultiplier (PMT: Hamamatsu R4124UV) was used as a pixel detector. The photocathode of this PMT has an area of 10 mmϕ and covers about 35% of the FOV. The array of light collector were attached in front of the PMTs in order to increase the collection area of the camera by twice. Sixteen PMTs are housed in one module unit with a common bleeder circuit. PMTs are operated with a low gain of $\sim 10^5$ to avoid the gain drop for more than a few ten minutes after the passage of bright stars. Buffer amplifiers (LeCroy TRA402) are installed in the module unit to obtain the total gain of $\sim 10^7$ after amplification. The whole camera consists of these 32 module units.

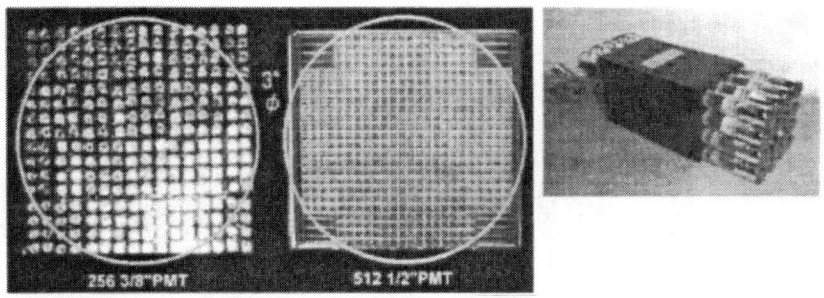

FIGURE 2. Front view of the cameras of the 3.8m telescope (left) and the new 7m telescope (middle). PMT module unit with 16 PMTs (right)

The signal from each PMT is transmitted through twisted cables of 36m in length to electronics circuits in the hut beside the base of the telescope. The arrival timing of lights detected with each PMT is recorded by multihit TDCs, and pulse heights are measured by charge ADCs in unit of a PMT base. The detail is described by Mori et al. [3]. At present triggers are generated when there are more than 4 PMTs whose pulse heights are larger than 4 photoelectrons, and the linear sum of any PMT module unit exceeds the night sky background fluctuation. Inner 16 of 32 PMT module units are concerned with the trigger, which covers a $\sim 1°.8$ diameter of the FOV. In this condition triggers were generated at ~ 10 Hz. Images taken with CANGAROO-II are shown in Fig. 3. Propagation of a shower triggered by hadron, and a ring triggered by muon are clearly seen. In 3.8m telescope, muon events were rarely detected since its detectable energy was relatively high (~ 1.5 TeV). The timing distribution of these events is shown in left panel of Fig. 4. The arrival timing of the muon event concentrates within 15 ns, on the other hand, that of the hadron event is distributed in 30 ns. The arrival timing accumulated for all events is also shown in right panel of Fig. 4. In case that the correction for the time jitter is not applied, the timing is distributed in 50 ns. After correction, the timing almost concentrates within 30 ns. The image parameter of off-source observation is shown in Fig. 5. The comparison with simulation is ongoing.

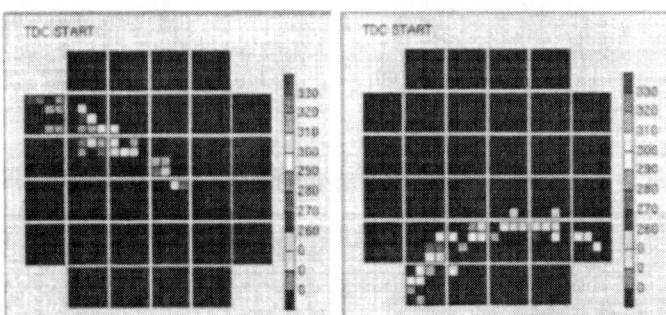

FIGURE 3. Images observed with CANGAROO-II, triggered by hadron(left) and muon(right).

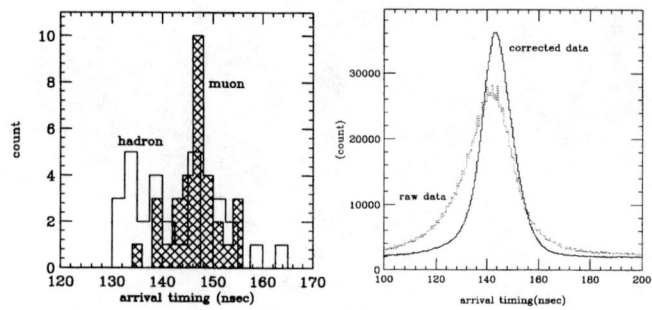

FIGURE 4. Arrival timing of hit PMTs for one event shown in Fig. 3(left), and for all events(right).

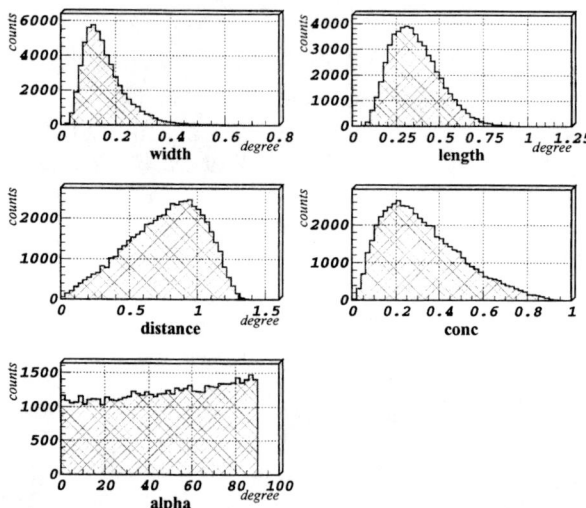

FIGURE 5. Image parameter of an off-source observation for 5 hours.

Figure 6 shows the distribution of the linear sum of all PMTs for triggered events. The slope becomes flatter as energy decreases. This might be due to the decrease of the detection efficiency at the trigger level for hadrons below ∼1 TeV, which was expected for the large telescope having a fine imaging pixels from the simulation study as shown in Fig. 7. From comparison between the simulation and the observed spectrum, the energy threshold of CANGAROO-II 7m telescope is estimated to be ∼300 GeV.

SUMMARY

The construction of new CANGAROO-II 7m telescope has been completed in March 1999, and the telescope has detected many shower and muon events. The energy threshold is estimated to be ∼ 300 GeV. The detailed study of the trigger condition and analysis of both on-source and off-source observations are ongoing.

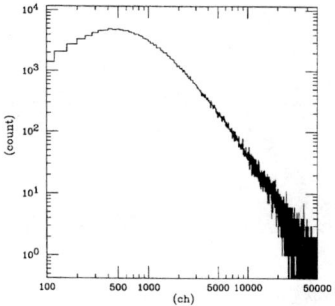

FIGURE 6. Observed distribution of the linear sum of all PMTs for triggered events

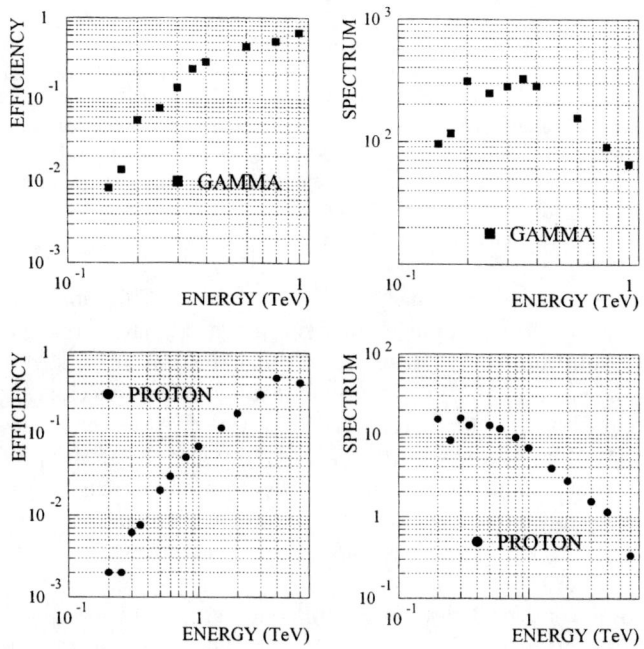

FIGURE 7. Simulated spectrum and efficiency for gamma-rays (upper) and protons (lower).

REFERENCES

1. Kawachi, A. et al., in these proceedings.
2. Mori, M. et al., in these proceedings.
3. Mori, M. et al., *Proc. 26th ICRC*, OG.4.3.31 (1999).
4. Tanimori, T. et al., *Proc. 26th ICRC*, OG.4.3.04 (1999).

Data analysis techniques for stereo IACT systems

Werner Hofmann

Max-Planck-Institut für Kernphysik, P.O. Box 103980, D-69029 Heidelberg, Germany

Abstract. Based on data and Monte-Carlo simulations of the HEGRA IACT system, improved analysis techniques were developed for the determination of the shower geometry and shower energy from the multiple Cherenkov images. These techniques allow, e.g., to select subsamples of events with better than 3' angular resolution, which are used to limit the rms radius of the VHE emission region of the Crab Nebula to less than 1.5'. For gamma-rays of the Mrk 501 data sample, the energy can be determined to typically 10% and the core location to 2-3 m.

Systems of imaging atmospheric Cherenkov telescopes (IACTs) for TeV gamma-ray astronomy allow the stereoscopic reconstruction of air showers, and provide improved angular resolution, energy resolution, and rejection of backgrounds such as showers induced by cosmic rays, local muons, or random triggers caused by night-sky background light. For systems with more than two telescopes, the shower parameters are overdetermined, allowing important cross-checks of the performance of the telescope system and of the reconstruction algorithms. In particular, the event-by-event determination of the position of the core permits to directly measure effective detection areas, and to estimate the systematic errors in the flux measurement [1,2].

In this talk, I will cover recent developments concerning improved algorithms to reconstruct shower direction and shower energy, and their tests using data from the HEGRA IACT system [3,2]. Detailed information as well as a more complete list of references can be found in [4–6].

Reconstruction of the shower geometry [4]. The traditional reconstruction algorithm used in HEGRA determines the shower direction by intersecting the axes of all Cherenkov images, regardless of the quality of individual images (Fig. 1(a)). In particular in events combining some bright images with dim images, the latter, with their poorly determined image parameters, can spoil the angular resolution. The angular resolution can be improved by estimating, for each image, the errors on the image parameters and by properly propagating these errors (Fig. 1(b)). In addition, one can use the shape of the image, in particular the *width/length* ratio, to estimate the *distance d* between the image centroid and the source, and use this information to derive, for each telescope, an error ellipse for the source location (actually, two ellipses, because of the head-tail ambiguity). The ellipses from different telescopes are then combined to locate the source (Fig. 1(c)). Finally, another approach (d) is to fit the intensity distribution of the images using a set of image templates, rather than parameterizing images by their Hillas parameters. Fig. 1(e) shows the angular resolution for achieved for different event classes. As expected, the techniques (b)-(d) outperform the simplest algorithm (a). The fit (d) is generally best, but the improvement compared to the much simpler and faster algorithm (c) is not dramatic. In addition to an improvement in the angular resolution, algorithms (b)-(d) provide, for each event, an estimate of the angular resolution (Fig. 2(a)), which can be used, e.g., to reject poorly reconstructed events.

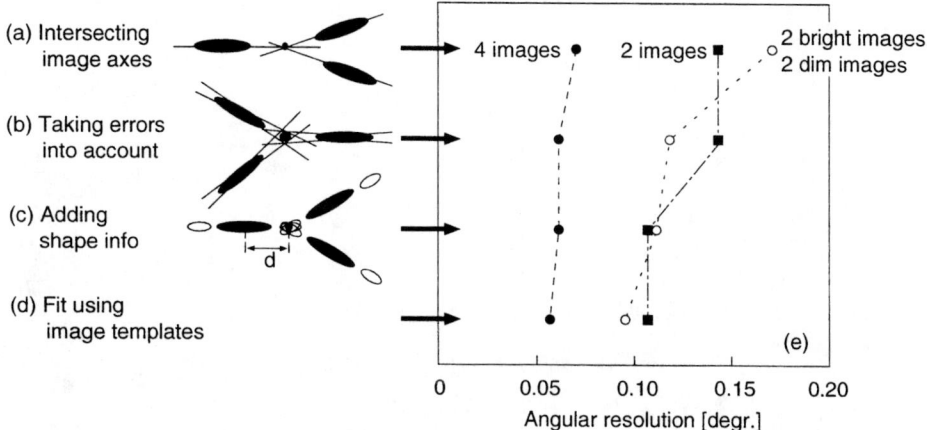

FIGURE 1. (left) Illustration of different algorithms to reconstruct the shower direction from the multiple Cherenkov images. (right) Resulting mean angular resolution for different data sets.

Size of VHE emission region of the Crab Nebula [6]. Well-reconstructed events reach an angular resolution on the same scale as the characteristic size of the Crab Nebula. One can use such events to search for evidence for an extended VHE emission region. Fig. 2(b) shows the angular distribution of events with an estimated angular error of less than 3' in each projection, relative to the direction to the Crab. The width of the distribution is, within statistical errors, identical with the width expected for a point source on the basis of simulations (Fig. 2(c)) and with the width of the gamma-ray distribution observed for Mrk 501. Therefore, we can only give an upper limit on the size of the emission region. Including systematic effects, e.g. due to pointing errors, we find a 99% upper limit on the rms radius $<r^2>$ of the TeV emission region of 1.5'. This value is comparable to the radius at radio wavelengths, but significantly larger than the size at x-ray energies. Standard models for the VHE gamma-ray emission of the Crab Nebula assume that the same electron population is responsible for x-rays via synchrotron emission, and for TeV gamma-rays via the IC process, and predict a small TeV emission region, well below the experimental limit. Possible hadronic production of gamma-rays, on the other hand, could take place at significantly larger distances from the pulsar.

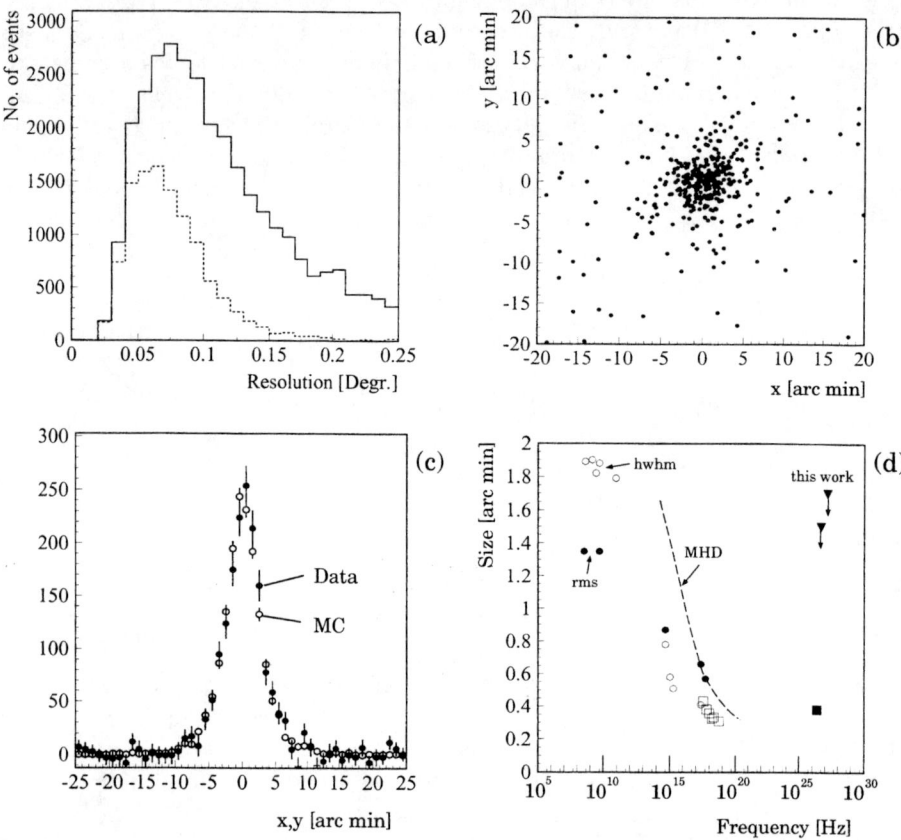

FIGURE 2. (a) Distribution of the estimated angular error for gamma-ray events. The full line includes all events, the dashed line only those where all four telescopes triggered. (b) Angular distribution relative to the direction to the Crab pulsar, for gamma-rays with estimated reconstruction errors of less than 3' in each direction. (c) Distribution of gamma-ray directions, compared to Monte-Carlo simulations assuming a point source. (d) Rms radius of the photon emission region in the Crab nebula, as a function of frequency, including the upper limits obtained at TeV energies.

Core determination [5]. The shower core is usually located by intersecting the image axes, starting from the telescope locations. The precision of the core determination is therefore given by the precision with which the image axes can be determined, typically $O(5°)$. If the source location is known, as is the case, e.g., for the Mrk 501 data sample, one can alternatively determine the image axis as the line connecting the image of the source and the image centroid. With a typical distance between the source and the image centroid of 1° and a measurement of the centroid to $O(0.02°)$, the image axis is then known to $O(1°)$. Using this technique, Monte Carlo simulations predict that the precision for the shower core improves from about 6 m to 10 m for the normal method, to about 2 m to 3 m, depending on the core distance (in each case, properly taking into account the errors on the measured image parameters). The exact knowledge of the core position is particularly important for the energy determination, when the observed light yield is translated into an energy estimate.

Energy determination [5]. Earlier studies comparing event-by-event the light yield observed in different telescopes indicated correlated fluctuations in the light yield of individual showers [1]. Monte-Carlo studies point to the fluctuation in the height of the shower maximum as the primary source for these correlated fluctuations. Fig. 3(a) illustrates that for distances up to about 100 m from the shower axis, the light yield varies significantly with the height of the shower; only beyond the Cherenkov radius of about 120 m is the light yield stable. An obvious approach to improve the energy resolution is therefore to measure the height h_{max} of the shower maximum, and to include it as an additional parameter, writing $E_i = f(size_i, r_i, h_{max})$, where $size_i$ is the image *size* measured in telescope i at a distance r_i from the shower axis. With an IACT system, the height of the shower maximum, or, more precisely, the height of maximum Cherenkov emission, can be determined essentially by triangulation, using the relation between the *distance* d_i from the image to the source, r_i, and h_{max}: $d_i \approx r_i/h_{max}$. The actual algorithm [5] uses a slightly more complicated relation, reflecting the fact that light arriving at small r_i is generated by the tail of the shower rather than by particles near the shower maximum. The algorithm reaches a resolution in shower height of 530 to 600 m rms.

Fig. 3(b) illustrates the effect of the various possible improvements to the energy resolution. Whereas the conventional algorithm provides a resolution of 18% to 22% for the 1 TeV to 30 TeV range, the shower-height correction provides a resolution of about 12% to 14%, and the combination of the shower height correction with the improved core determination assuming a known source yields 9% to 12% resolution.

Before applying this technique to the actual data to obtain improved energy spectra, one needs to make sure that systematic effects are under control at a level consistent with the improved resolution. While the redundant data from the IACT system provide sufficient information to check this, the analysis is not yet finished. A first test of the new method with Mrk 501 data results in a spectrum consistent with earlier analyses, possibly with a slightly steeper spectrum in the cutoff region beyond 6 TeV.

Summary. The analysis algorithms discussed here represent clear improvements over the first-generation algorithms used in the reconstruction of data from the HEGRA IACT system; it is also clear that further improvements are possible and that at this point we do not fully use all the information provided by multiple IACT images of an air shower. The algorithms do not only improve the angular resolution and the energy resolution; they also help to boost the significance of faint signals. For example, instead of simply counting all events reconstructed within a

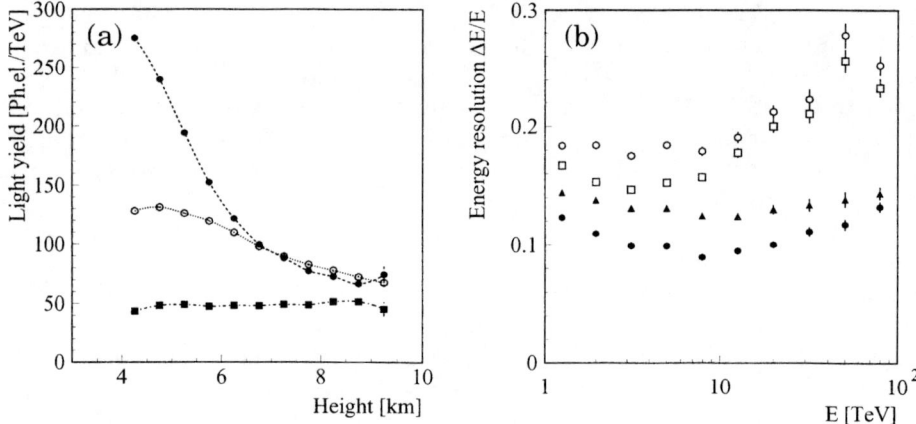

FIGURE 3. (a) Light yield (in Photoelectrons/TeV) as a function of the height of maximum Cherenkov emission, at core distances around 40-50 m (full circles), 90-100 m (open circles) and 140-150 m (full squares). (b) Energy resolution as a function of energy, for the conventional energy reconstruction (open circles), with improved core determination (open squares), with shower height correction (full triangles), and with shower height correction and improved core determination (full circles).

certain angular distance form a source, one can form a weighted sum, weighting events according to their expected signal-to-background ratio, as determined event-by-event from the estimated angular error and misidentification probability. First tests of such methods indicate in an increase in the significance for the detection of weak sources by up to 80%.

Acknowledgments. Many of the members of the MPIK CT group have contributed in one way or another to the development and tests of the advanced analysis techniques discussed here; in particular, I. Jung, A. Konopelko, H. Lampeitl, H. Krawczynski and G. Pühlhofer should be mentioned.

REFERENCES

1. W. Hofmann, Proceedings of the Int. Workshop "Towards a Major Atmospheric Cherenkov Detector V", Kruger Park, (1997), p. 284, and astro-ph/9710297 (1997).
2. F.A. Aharonian et al., Astron. Astrophys. 342 (1999) 69, 349 (1999) 11.
3. A. Daum et al., Astroparticle Phys. 8 (1997) 1.
4. W. Hofmann et al., Astroparticle Phys., in press, and astro-ph/9904234.
5. W. Hofmann et al., Astroparticle Phys., in press, and astro-ph/9908092.
6. F.A. Aharonian et al., in preparation.

Optimum spacing between imaging atmospheric Čerenkov telescopes in the future 50 GeV multi-telescope arrays

Alexander K. Konopelko

Max-Planck-Institut für Kernphysik,
Heidelberg D-69029, Postfach 10 39 80, Germany
e-mail: alexander.konopelko@mpi-hd.mpg.de

Abstract. Regarding the optimum design of future low energy (50–100 GeV) arrays of imaging air Čerenkov telescopes an important issue is a choice of optimum arrangement (layout) of telescopes within the array. Here we present the basic results of such studies addressed by use of detailed Monte Carlo simulations.

INTRODUCTION

Currently, the TeV Very High Energy (VHE) γ-ray astronomy experiences an unprecedented upgrade of its experimental instrumentation (for review see [1]). Three major forthcoming detectors, CANGAROO III, H.E.S.S., and VERITAS, will exploit a concept of an array of several imaging air Čerenkov telescopes (IACT) operating simultaneously in the so-called *stereoscopic mode* [2]. Recent studies based, in the main, on simulations (e.g., discussion in [3]) assist to infer the optimum basic parameters of a single IACT: *(i)* effective area of optical reflector $\sim 10\,\mathrm{m}^2$; *(ii)* camera field of view [1] – 4.5°; *(iii)* angular size of photo multiplier in the camera focal plane – 0.15°. These parameters have been derived for general assumptions respecting the TeV γ-ray source (point γ-ray source with the Crab-like energy spectrum). Note that the optimum telescope design might have very different parameters, if it is constrained for the most part, by the specific observational goals.

In addition to the design of a single telescope there is a number of important aspects of the array performance. For instance the choice of telescope layout may significantly influence resulting sensitivity of array. Recent observations, made by the HEGRA system of 5 IACTs with relatively small reflectors ($\sim 8.5\,\mathrm{m}^2$), as well as relevant simulations have shown that four Čerenkov light images of air shower allow

---

[1] Observations of diffuse γ-ray sources will need camera of a larger field of view ($\geq 5°$) in order to map the emission effectively over a broad angular regions.

good reconstruction of shower parameters (i.e., determination of shower arrival direction, primary energy, shower maximum *etc*) and provide a strong rejection of the background cosmic rays (see [4]). Use of more than four IACTs, simultaneously, does not noticeably improve the accuracy of the shower reconstruction. Thus, future multi-telescope arrays could be gathered from the subsystems (cells) of 4 IACTs operating in stereoscopic mode.

The accuracy of shower reconstruction depends on the impact distance of a shower core from the system of telescopes. A 4 IACTs cell with certain separation between the telescopes samples the final distribution of triggered air showers over the impact distance and consequently define the resulting accuracy of shower reconstruction, effective collection area and finally, array sensitivity. Apparently the relevant simulations could be useful to make a choice of the optimum telescope separation. Here we discuss the results of such simulations.

MONTE CARLO SIMULATIONS

The ALTAI computational code was used for generating γ-ray- and proton- induced air showers in the energy range 10 GeV - 50 TeV for two zenith angles of 20° and 60°. In present simulations for each individual shower we saved the response of an extended array of IACTs, arranged in a rectangular lattice of 1000x600 m² with 33 m step (589 nodes in total). This method is very effective and robust in performing such extensive simulations. In alternative approach one can save all photons hitting the observation level for each individual shower, and afterwards sample the telescope responses for a certain array layout. However such method is exclusively memory consuming and results in very complicated and rather slow data handling procedure.

Our calculations have been done for 10 m ($S \simeq 82$ m²) telescopes equipped by the camera in two designs: (1) 271 pixels of 0.25°; (2) 721 pixel of 0.15°. The mean contribution of night sky light for 0.15° pixel, trigger time gate of 10 ns, and photon-to-photoelectrons conversion efficiency of 0.1 was taken as 1 ph.e. In calculations we used the detector response functions (mirror reflectivity, point spread function, quantum efficiency of photo multiplier *etc*) which correspond to currently operating HEGRA IACTs system.

OPTIMIZATION CRITERION

Sensitivity of IACT array is determined by a few basic characteristic functions: effective detection areas of γ-rays and cosmic rays (A_γ, A_{CR}), acceptance of γ-rays and cosmic rays after applying the cuts on orientation (d) and shape (s) of the Čerenkov light images ($\kappa_\gamma^{(d,s)}$, $\kappa_{cr}^{(d,s)}$). The resulting so-called signal-to-noise ratio, calculated for 1 hr observations, is given by

$$S/N = \eta^{(d)}\eta^{(s)}R_\gamma J_\gamma (R_{cr}J_{cr})^{-1/2} t^{1/2} \qquad (1)$$

where

$$R_\gamma = \int A_\gamma(E) F_\gamma(E) dE, \quad R_{cr} = \int d\Omega \int dE F_{CR}(\Omega, E) A_{CR}(\Omega, E) \qquad (2)$$

are the detection rates for γ-rays and cosmic rays, respectively, assuming certain fluxes – F_γ, F_{CR}. $\eta^{(d,s)} = \kappa_\gamma^{(d,s)} / (\kappa_{cr}^{(d,s)})^{1/2}$ are the enhancement factors (Q-factors) after application of orientation and shape analysis cuts (γ-ray selection criteria). In general, the optimum design of the IACT array should give the maximum S/N ratio.

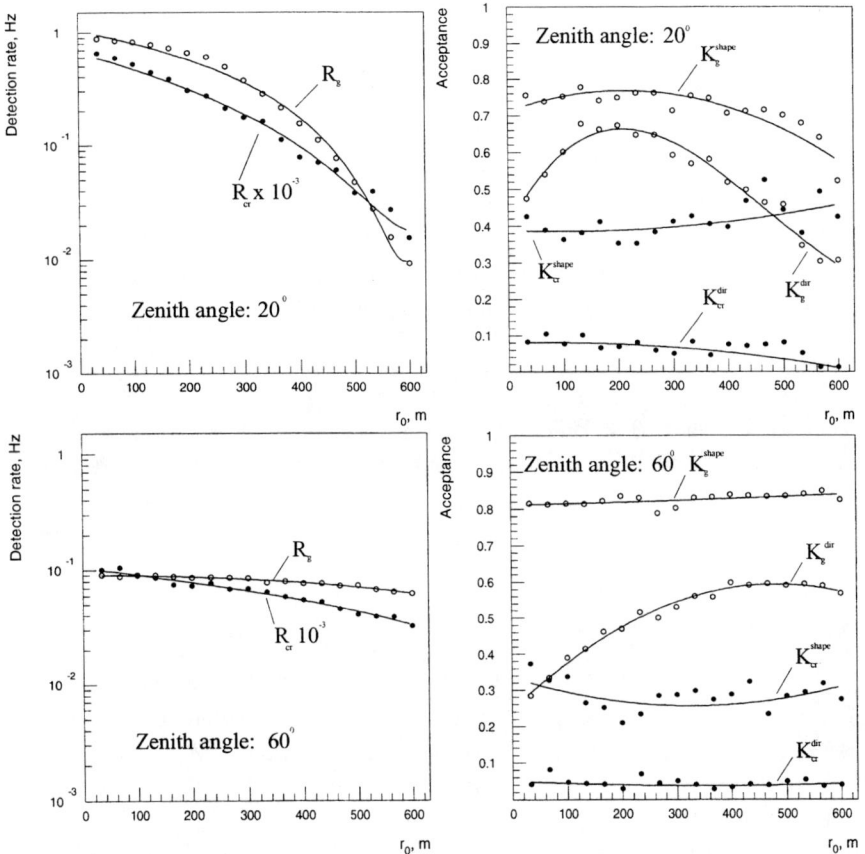

FIGURE 1. Detection rates and acceptances after application of orientation and shape cuts, $K_{\gamma,cr}^{dir}$, $K_{\gamma,cr}^{shape}$, for γ-ray and cosmic ray air showers simulated for a 4 IACTs cell with separation between the telescopes – r_0. Data corresponds to the simulations at zenith angles of 20° and 60°.

The functions $R_\gamma, R_{cr}, \eta^{(o,s)}$ depend on *(i)* single telescope trigger (local trigger), as well as on the system trigger (global trigger), *(ii)* analysis methods used for the

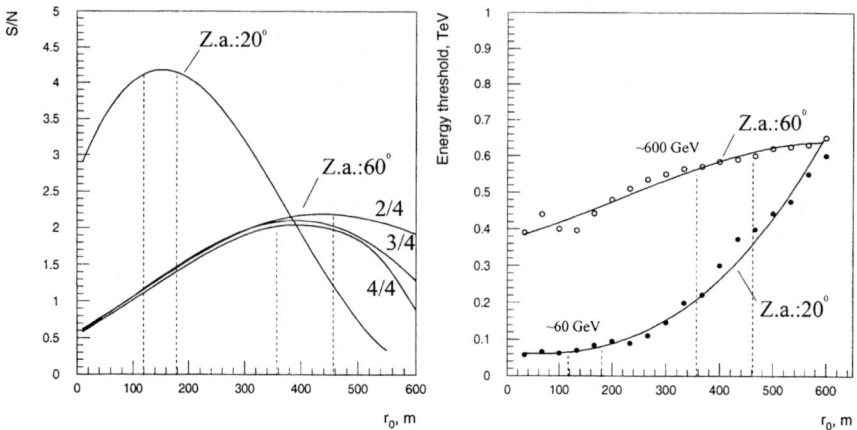

FIGURE 2. Signal-to-noise ratio (left panel) and corresponding energy threshold (right panel) as a function of the separation between the telescopes, r_o.

data reduction; *(iii)* separation between the telescopes. Here we used the trigger logics for array which allows to get the minimum possible trigger threshold in order to reduce as low as possible the energy threshold of array ($\simeq 50$ GeV). Note that the results on optimum telescope separation might be very different in case of much larger energy threshold of the array ($\gg 100$ GeV), the larger distances may be preferable in such a case. For the shower reconstruction and rejection of cosmic rays we used here the standard tools developed for the HEGRA IACTs system [4].

RESULTS

For the Crab-like energy spectrum of the γ-rays and standard cosmic ray spectra we calculated the functions $R_\gamma, R_{cr}, \eta^{(d,s)}$ for the 4 IACTs cell of a different spacing (see Figure 1). One can see in Figure 1 that γ-ray detection rate, R_γ, drops down with increase of r_o-parameter because of reducing the effective collection area of γ-rays. At the same time for relatively large separation ($r_o \simeq 120 - 160$ m) the reconstruction accuracy increases and the same angular cut ($\theta^2 < 0.1°$) provides much higher acceptance of γ-rays, κ^d (see Figure 1). The efficiency of the shape cut (we analyzed the data by *mean scaled Width*) does not strongly depend on r_o. Note that the fall of the γ-ray detection rate for larger separations (> 100 m at $20°$ zenith angle) will be much faster for telescopes equipped with the camera of relatively narrow field of view ($\ll 4°$) because the images will be truncated by the camera edge.

Resulting signal-to-noise ratio is shown in Figure 2. For $20°$ zenith angle the calculated signal-to-noise ratio reaches maximum in the range of $r_o = 120 - 180$ m whereas at $60°$ zenith angle the "plateau" of S/N curve starts at about 320 m. One can see from Figure 2 that the S/N maximum is relatively broad (150 ± 30 m for 20°) and any choice of array spacing within this range could provide almost the same S/N ratio. However, the energy threshold of array rises up with increase of r_o (see Figure 2). In order to keep the energy threshold at $\simeq 50$ GeV for small zenith angles (20°), and correspondingly, at $\simeq 500$ GeV for the large zenith angles (60°) the spacing between telescopes in array has to be about 120 m and 360 m, respectively. Note that independent studies based on the data taken with the HEGRA system of 5 IACTs allow to make similar conclusion respecting the optimum separation between the telescopes in array [5].

Finally, the layout of an array of 16 IACTs can be chosen in form of rectangular grid with the separation between the telescopes of 120 m. In this case the separation between the peripheral telescopes in array will be of about 360 m. Such distance is optimum for the large zenith angle observations. The array can be used *(i)* as a single detector observing an object with all 16 telescopes; *(ii)* as four subsystems (cells) of 4 IACTs operating independently; and even *(iii)* as 16 stand alone telescopes. Evidently the operation mode will be chosen according to particular observational program (e.g., precise spectroscopic measurements, routine monitoring of a number of objects, *etc*).

These is an alternative layout of array in form of four cells with 4 IACTs at relatively large distance ($\gg 250$ m) in order to have four independent subsystems. Note that almost all low energy γ-ray events (≤ 100 GeV) are concentrated within the cell of 4 IACTs and the detection area is determined by the geometrical area of a cell. The array of 16 telescopes arranged in form of rectangular grid corresponds to 9 cells whereas other design has only 4 such cells and apparently provides lower rate of ≤ 100 GeV γ-rays but at the same time gives a rather high rate of ≥ 100 GeV γ-ray showers.

CONCLUSIONS

Base on Monte Carlo simulations we found that the arrangement of 10 m class imaging air Čerenkov telescopes in rectangular grid with the spacing between the telescopes of 120 m is close to the optimum one. Such array may effectively fulfill most of the physics goals relevant for the future γ-ray observations in the energy range above 50 GeV.

REFERENCES

1. Catanese, M., Weekes, T.C., *Publ. Astron. Soc. Pac.*, **111**, 1193, (1998).
2. Aharonian, F., et al., *Astroparticle Physics*, **6**, 343-368 (1997).
3. Konopelko, A., *Astroparticle Physics* **11**, 263-266 (1999).
4. Hofmann, W., this proceedings, (1999).
5. Hofmann, W., et al., *Astroparticle Physics*, in press (1999).

"Convergent observations" with stereoscopic HEGRA CT system

Hubert Lampeitl and Werner Hofmann
for the HEGRA Collaboration

Max-Planck-Institut für Kernphysik, P.O. Box 103980, D-69029 Heidelberg, Germany

Abstract. Observations of air showers with the stereoscopic HEGRA IACT system are usually carried out in a mode where all telescopes point in the same direction. Alternatively, one could take into account the finite distance to the shower maximum and orient the telescopes such that their optical axes intersect at the average height of the shower maximum. In this paper we show that this "convergent observation mode" is advantageous for the observation of extended sources and for surveys, based on a small data set taken with the HEGRA telescopes operated in this mode.

The HEGRA collaboration is operating a system of currently five imaging atmospheric Cherenkov telescopes (IACTs) for the stereoscopic observation of VHE cosmic γ-rays [1]. The telescope system is located on the Canary Island La Palma, at the Observatorio del Roque de los Muchachos (ORM), at 2.2 km asl. The system telescopes feature a mirror area of 8.5 m^2 and a focal length of 5 m and are equipped with 271-pixel photomultiplier cameras. Based on the multiple views obtained for each shower, the orientation of the shower axis in space as well as the location of the shower core can be determined. Compared to single IACTs, stereoscopic IACT systems provide superior angular resolution, energy resolution, and background rejection [1–3]. During typical observations with the HEGRA IACT system, the optical axes of all telescopes are parallel and either point directly to the source, or – in the so-called wobble mode – at a point displaced by $\pm 0.5°$ in declination relative to the source. In the latter case, the rate in a region displaced by the same distance from the optical axis, but in the opposite direction, is used to estimate off-source background rates.

With all telescopes pointing in exactly the same direction, both the operation of the system and the data analysis are simplified, but one may wonder if the detection characteristics could not be improved by canting the telescopes towards each other, such that their optical axes intersect roughly at the height of the shower maximum. Such an alignment of telescopes would guarantee that the most luminous region of an air shower is optimally viewed by all telescopes simultaneously.

The two alternatives are shown in Fig. 1, which also serves to illustrate the trigger

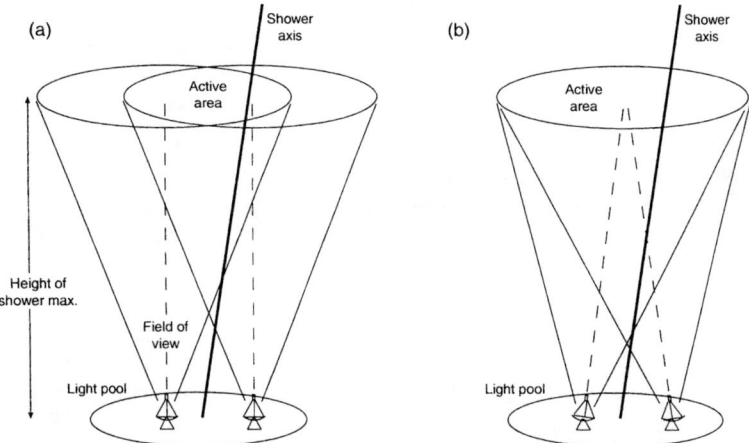

FIGURE 1. Illustration of the geometry with parallel (a) and canted (b) telescopes. The dashed lines show the optical axes of the telescopes. The "active area" indicates where the shower maximum can be observed in both telescopes.

characteristics of IACT arrays. To first approximation, an individual IACT will trigger on an air shower if two conditions are fulfilled [4]:

1. the telescope has to be located within the light pool of the shower, with its typical radius of about 120 m, and
2. the shower maximum has to be within the field of view of the camera.

For the HEGRA telescopes, with their 4.3° field of view, the latter condition implies that the shower maximum – at TeV energies typically located 6 km above the telescopes – should be within 225 m from the optical axis of the telescope. For showers propagating parallel to the optical axis, the second condition is automatically fulfilled, once a telescope is within the light pool of the air shower. The camera field of view adds an additional constraint only for showers at angles of more than 1° relative to the optical axis (at least as far as triggering is concerned – to avoid truncation of images, one may want to require in addition in the subsequent image analysis that the centroid of the image is at least 0.5° away from the edge of the field of view; this will still result in a field of view of about 175 m radius at the shower maximum). Therefore, canting of telescopes is most likely not an issue for studies of point sources near the center of the field of view; it may however be important for the observation of extended sources as well as for surveys of larger areas of the sky, where it is important to maximize sensitivity over a large solid angle. For two telescopes, the situation is relatively obvious from Fig. 1: for parallel pointing, the range of locations of the shower maximum, and hence the accessible solid angle, is much more restricted than for canted telescopes. For three or more telescopes, the conclusion depends on the locations of the telescopes and the trigger conditions; parallel pointing will reduce the solid angle for a coincidence of all N telescopes, but may increase the angle if only 2 out of N telescopes are required in the trigger and in the subsequent analysis.

Estimates of detection rates were carried out for the actual geometry of the HEGRA IACT system, with telescopes located at three of the four corners of a square of about 100 m side length, and another telescope located at the center of the square (the remaining corner telescope had at that time – summer 1998 – an older camera and was not yet included in the IACT system). The rate estimates were based on the simplified model discussed above, assuming a radius of the light pool of about 120 m and a usable field of view (without edge distortion of the images) of 3.6°. Two cases were compared: 1) parallel optical axes of all telescopes, and 2) telescopes canted such that the axes intersect in a height of 6 km, with the nominal pointing of the IACT system defined as the pointing of the central telescope. The results – event rates for a given source flux and observation time as a function of the distance of the source from the optical axis of the central telescope – are shown in Fig. 2(a)-(d). The simulations indicate almost identical total detection rates for the two pointing modes (Fig. 2(a)). However, for sources more than 0.5° from the optical axis of the system, the "convergent observation mode" provides a significantly larger fraction of four-telescope events (Fig. 2(d)). Since both the angular resolution and the cosmic-ray rejection improve with the number of triggered telescopes, the convergent mode is clearly favorable.

To verify these model predictions experimentally, three hours of observations of the Crab nebula were performed with canted telescopes. The cosmic-ray background provides a uniform flux of particles and allows to study detection rates as a function of the distance to the optical axis of the IACT system. At last qualitatively, these characteristics should be similar for hadronic showers and for the more interesting γ-ray induced showers. Limited observation time and low counting rate prevented us from scanning the field of view using the Crab nebula as a γ-ray source.

Fig. 3 illustrates the effect of the canting: the positions of the images in the different cameras coincide, whereas for the parallel pointing mode they are displaced by about $\delta \approx d/h \approx 1°$ along the direction connecting the locations of the telescopes. Here, d is the spacing of the telescopes and h the height of the shower maximum. As a consequence, in convergent mode it is unlikely that images in some of the telescopes are truncated; either all telescopes have well-contained images, or all images suffer from edge problems.

Due to slight differences in the weather conditions, the telescopes trigger rates varied somewhat between the observations taken in convergent tracking mode, and the reference data set. Since the pointing mode may influence the trigger rates, one cannot simply normalize the data sets on the basis of the raw trigger rates. To derive the correct normalization factor, the field of view in convergent mode was artificially truncated to 2.4° diameter, and in the reference data set – with parallel pointing – the convergent pointing was mocked up by selecting in software a 2.4°

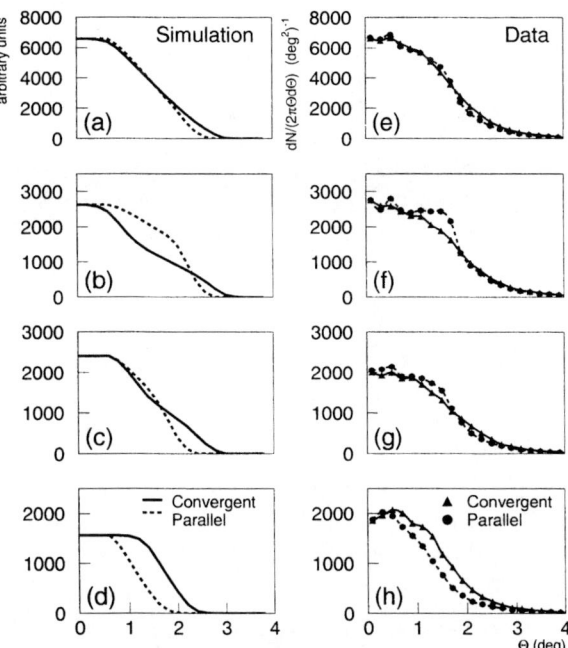

FIGURE 2. Left: simple geometrical model for the number of detected showers as a function of the angle between the shower axis and the axis of the telescope system, for parallel axes of the telescopes (dashed line) and for canted telescopes (full line). (a) Total rate, (b) 2-telescope events, (c) 3-telescope events, (d) 4-telescope events. Right: experimental detection rates for cosmic-ray showers, for all events (e), 2-telescope events (f), 3-telescope events (g) and 4-telescope events (h), for parallel (points) and canted alignment of axes (triangles).

field of view shifted by the canting angle. After these software trigger cuts, the detection rates can be compared, resulting in a $24 \pm 2\%$ correction.

The corrected cosmic-ray detection rates as a function of the angle relative to the system axis are shown in Fig. 2(e) for the total rate, and separately for 2-telescope (f), 3-telescope (g), and 4-telescope events (h). Here, only images within the central $3.6°$ of the field of view were accepted, to exclude truncated images. The observed pattern matches that predicted by the simple model: very similar total rates, but a clear enhancement of the 4-telescope rate at larger angles in case of the convergent pointing mode, at the expense of 2 and 3-telescope events. For the 4-telescope events, the diameter of the effective field of view is increased by about $0.8°$.

During the data taking in convergent mode, the Crab nebula was positioned $0.5°$ off the system optical axis. Under these conditions, one would not expect any

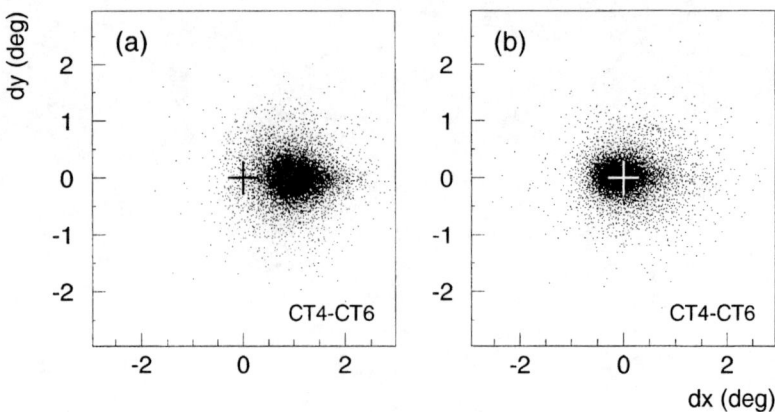

FIGURE 3. Difference in the centroid positions of images in two cameras for parallel (a) and convergent observations (b). The camera coordinate systems are rotated in a way that the x-axis points along the line connecting the telescopes.

difference in detection rates between the two modes, and indeed the Crab signals in the two modes are well consistent within the statistical errors of about 20%.

In summary, for stereoscopic IACT systems, the convergent tracking mode – canting the telescopes towards each other such that their optical axes intersect at the height of the shower maximum – improves the detection capabilities in particular for sources near the edge of the field of view, and is advised for observations of extended sources and for surveys.

Acknowledgments. The support of the German Ministry for Research and Technology BMBF and of the Spanish Research Council CYCIT is gratefully acknowledged. We thank the Instituto de Astrofisica de Canarias for the use of the site and for providing excellent working conditions. We gratefully acknowledge the technical support staff of Heidelberg, Kiel, Munich, and Yerevan.

REFERENCES

1. A. Daum et al., Astroparticle Phys. 8 (1997) 1.
2. N. Bulian et al., Astropart. Phys. 8 (1998) 223.
3. F.A. Aharonian et al., Astron. Astrophys. 342 (1999) 69.
4. C. Köhler et al., Astroparticles Physics 6 (1996) 77.

Experimental results on the optimum spacing of stereoscopic imaging atmospheric Cherenkov telescopes

Werner Hofmann, for the HEGRA Collaboration

Max-Planck-Institut für Kernphysik, P.O. Box 103980, D-69029 Heidelberg, Germany

Abstract. For stereoscopic systems of imaging atmospheric Cherenkov telescopes (IACTs), a key parameter to optimize the sensitivity for VHE γ-ray point sources is the intertelescope spacing. Using pairs of telescopes of the HEGRA IACT system, the sensitivity of two-telescope stereo IACT systems is studied as a function of the telescope spacing, ranging from 70 m to 140 m. Data taken during the 1997 outburst of Mrk 501 are used to evaluate both the detection rates before cuts, and the sensitivity for weak signals after cuts to optimize the significance of signals.

IACT stereoscopy – the simultaneous observation of air showers with multiple imaging atmospheric Cherenkov telescopes (IACTs) under different viewing angles – has become the technique of choice for most of the next-generation instruments for earth-bound gamma-ray astronomy in the VHE energy range, such as VERITAS [1] or HESS [2]. The stereoscopic observation of air showers allows improved determination of the direction of the primary and of its energy, as well as a better suppression of backgrounds, compared to single IACTs. A crucial question in the layout of stereoscopic systems of IACTs is the spacing of the identical telescopes. The coincidence rate between two telescopes will decrease with increasing spacing; on the other hand, the angle between the views and hence the quality of the stereoscopic reconstruction of the shower geometry will improve. For existing or planned IACT systems, the spacing between nearest telescopes is 70 m (HEGRA), 80-85 m (VERITAS [1]), 100-120 m (HESS [3,4]), 70 m and 120 m (Telescope Array [5]), and 140 m (WHIPPLE-GRANITE [6]).

Optimization of IACT system geometry is heavily based on Monte-Carlo simulations. Over the last years, the quality of these simulations has improved significantly, both concerning the reliability of the shower simulation and concerning the details of the simulation of the telescopes and their readout. Nevertheless, a direct experimental verification of the dependence of the performance of IACT systems on the intertelescope distance would be highly desirable. We report here on results of such a study, using the HEGRA IACT system.

The HEGRA IACT system at the Observatorio del Roque de los Muchachos on La Palma consists of five telescopes, four of them arranged roughly in the form of a square with 100 m side length, with the fifth telescope in the center. During 1997, when the data for this study were taken, only four of the five telescopes were included in the HEGRA IACT system; the fifth telescope (one of the corner telescopes) was still equipped with an older camera and was operated in stand-alone mode. The four telescopes can be used to emulate six different two-telescope stereo systems: three combinations with intertelescope distances around 70 m (from the central telescope to the three corner telescopes), two combinations with about 90 m and 110 m (the sides of the imperfect square), and one combination with 140 m (the diagonal). These combinations cover essentially the entire range of interest. To evaluate the sensitivity of different pairs of telescopes, the large sample of gamma-rays [7] acquired during the 1997 outburst of Mrk 501 is used. Data from pairs of (triggered) telescopes are analyzed, ignoring the information provided by the other telescopes. Since only two telescopes are required to trigger the system, there is no trigger bias or other influence from those other telescopes.

The HEGRA telescopes, the general data analysis and the performance of the IACT system are described in more detail in [8,7] and references given there.

In this analysis, a slight difficulty arises since one compares combinations of different telescopes, rather than varying the distance between two given telescopes. While the four HEGRA system telescopes used here are identical in their construction, they differ somewhat in their age and hence the degree of mirror deterioration, in the quality of the alignment of the mirror tiles, and in the properties of the PMTs in the cameras, which show systematic variations between production batches. These effects have to be determined from the data, and compensated.

As a first step in the analysis, the detection rates were studied as a function of intertelescope distance. Here and in the following, only data taken at zenith angles $< 20°$ were used, i.e., basically vertical showers.

To equalize the energy thresholds of all telescopes, pairs of telescopes were used to reconstruct showers and events with cores located at equal distance from both telescopes were selected. By comparing the mean *size* of the images in the two telescopes, and in particular by comparing the mean signal amplitude in the highest pixels (which determine if a telescopes triggers or not), and can derive correction factors which can be used to equalize the response of the telescopes. Three of the four telescopes were found to be identical within a few %, one had a sensitivity which was lower by about 25%. In the analysis, pixel amplitudes are corrected correspondingly, and only camera images with two pixels above 20 photoelectrons are accepted. This software threshold is high enough to eliminate any influence of the hardware trigger threshold (at about 8 to 10 photoelectrons in two pixels, depending on the data set), even after the worst-case 25% correction is applied. In a similar fashion, slight differences in the efficiencies of cuts on image shapes etc. are determined and corrected.

The resulting detection rates of cosmic rays and of gamma-rays – after subtraction of the cosmic-ray background – are shown in Fig. 1(a). The rates measured for

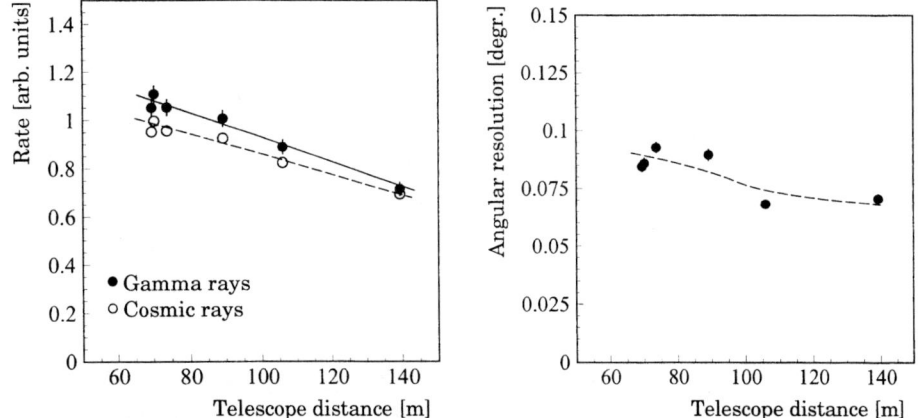

FIGURE 1. (a, left) Detection rates of cosmic rays and of gamma rays as a function of the spacing between stereoscopic pairs of telescopes, using a software trigger requirement of more than 20 photoelectrons in two pixels of both cameras. The normalization is arbitrary. Gamma-ray events were selected using very loose cuts on pointing relative to Mrk 501 ($< 0.45°$) and on the *mean scaled width* of the images (< 1.3), keeping essentially all gamma-rays. Cosmic-ray background in the on-source sample was subtracted on a statistical basis. The lines are drawn to guide the eye. (b, right) Angular resolution provided by pairs of telescopes as a function of telescope spacing. A software trigger threshold of 20 photoelectrons is used. The angular resolution is determined by fitting a Gaussian to the difference between shower direction and direction to the source, projected on two orthogonal axes.

the three different telescope combinations around 70 m spacing agree within 5%, indicating the precision in the adjustment of telescope thresholds. Detection rates decrease with increasing distance; between 70 m and 140 m, rates drop by about 1/3, with a marginally steeper dependence for gamma rays as compared to cosmic rays.

As a measure of the sensitivity for weak sources, the ratio $S/\sqrt{B}$ of the gamma-ray rate to the square root of the cosmic-ray rate was used. The significance of background-dominated signals scales with this ratio. For optimum sensitivity, $S/\sqrt{B}$ is optimized by tighter cuts on the pointing of showers, and on the image shapes. Fig. 1(b) shows the angular resolution provided by pairs of telescopes as a function of spacing. The three pairs at 70 m show differences at the level of 10%, indicating small differences in the quality of the mirror alignment and of the telescope alignment. Angular resolution improves slightly with increasing spacing; however, given the 10% systematic variations, this effect is of marginal significance.

To determine the sensitivity for the different combinations of telescopes, cuts on pointing and on the *mean scaled width* of the Cherenkov images [7] were optimized for each combination, resulting in pointing cuts around 0.1° and cuts on the *mean*

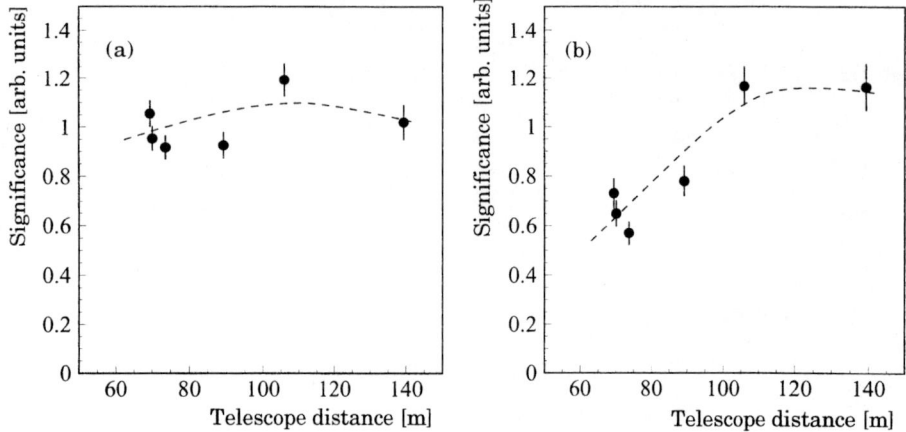

FIGURE 2. (a) Significance $S/\sqrt{B}$ of the Mrk 501 signal for different pairs of telescopes, with optimized cuts on shower pointing and image shapes in the two telescopes. The significance is normalized to ≈ 1 for the 70 m spacing. The dashed line is drawn to guide the eye. (b) As (a), but with an additional cut of a stereo angle of at least 45° between the two views. The normalization if the same as for (a), so that the two data sets can be compared directly. A weaker 20° cut on the stereo angle, as used in some HEGRA analyses, has little effect on the significance, and yields the almost same results as shown in (a).

scaled width around 1.05. Fig. 2(a) shows the resulting significance (defined as $S/\sqrt{B}$) of the Mrk 501 signal, after applying small corrections for differences in the sensitivity between telescopes. The errors shown are statistical, and are dominated by the low statistics in the cosmic-ray background sample after cuts. One finds that the significance is almost independent of telescope spacing, over the 70 m to 140 m range covered.

Concerning systematic uncertainties, we note that the results for the three combinations at 70 m agree reasonably well. Variations of parameters such as the software trigger threshold, or the exact values of the pointing cuts and angular cuts produce stable results, with systematic variations of at most 10%.

The absolute significance of a point source detected with two stereoscopic IACTs, and its variation with the intertelescope distance will, of course, depend on additional cuts which may be applied in the analysis. Additional requirements may e.g. concern the stereo angle, defined as the angle between the two views of the shower axis provided by the two telescopes, or, equivalently, the angle between the image axes in the two cameras. If the stereo angle is small – as it is the case for events with shower impact points on or near the line connecting the two telescopes, and for very distant showers – the two views coincide and the spatial reconstruction of the shower axis is difficult. In some analyses, one will therefore add the requirement of a minimal stereo angle between the two views, in order to increase the reliability of the stereoscopic reconstruction. For a minimum stereo angle of 20° – as used in some HEGRA work – the resulting significance remains virtually unchanged compared to the data shown in Fig. 2(a). With a large minimum stereo angle of 45° (Fig. 2(b)), the distance dependence becomes more pronounced; for small telescope distances, significance is reduced, while for large distances it remains or is even slightly enhanced. For small intertelescope distances, the stereo cut reduces the effective area, since it limits the maximum impact distance of showers. On the other hand, cameras with very limited field of view will result in a loss of sensitivity for large distances.

These results confirm Monte Carlo simulations (see e.g. [4]) which generally show that the exact choice of telescope spacing is not a very critical parameter in the design of IACT systems. The results shown apply, strictly speaking, only to two-telescope systems. In IACT arrays with a large number of telescopes, one may place telescopes at maximum spacing in order to maximize the effective area of the array, under the condition that most individual showers are observed by two or maybe three or four telescopes. Alternatively one may choose to place telescopes close to each other, such that an individual shower is observed simultaneously by almost all telescopes, improving the quality of the reconstruction and lowering the threshold, at the expense of detection area at energies well above threshold. In the first case, one would probably use a spacing between adjacent telescopes between 100 m and 150 m; in the second case, one would limit the maximum spacing between any pair of telescopes to this distance.

Acknowledgments. The support of the HEGRA experiment by the German Ministry for Research and Technology BMBF and by the Spanish Research Council CYCIT is acknowledged. We are grateful to the Instituto de Astrofisica de Canarias for the use of the site and for providing excellent working conditions. We gratefully acknowledge the technical support staff of Heidelberg, Kiel, Munich, and Yerevan.

REFERENCES

1. VERITAS Letter of Intent, T.C. Weekes et al. (1997); VERITAS Proposal (1999).
2. HESS Letter of Intent, F. Aharonian et al. (1997).
3. F.A. Aharonian et al., Astroparticle Phys. 6 (1997) 343.
4. A. Konopelko, Astropart. Phys., in press, and astro-ph/9901365.
5. N. Hayashida et al., Astroph. J. 504 (1998) L71; T. Yamamoto et al., Astropart. Phys. 11 (1999) 141.
6. R. Lamb, Proceedings of the Int. Workshop "Towards a Major Atmospheric Cherenkov Detector IV", Padua, (1995), M. Cresti (Ed.), p. 386; F. Krennrich et al., Astropart. Phys. 8 (1998) 213.
7. F.A. Aharonian et al., Astron. Astrophys. 342 (1999) 69.
8. A. Daum et al., Astroparticle Phys. 8 (1997) 1.

Kernel Analysis in TeV Gamma-Ray Selection

P.Moriarty* and F.W.Samuelson[†]

Galway-Mayo Institute of Technology, Galway, Ireland
[†]*Iowa State University, Ames, IA 50011*

Abstract. We discuss the use of kernel analysis as a technique for selecting gamma-ray candidates in Atmospheric Cherenkov astronomy. The method is applied to observations of the Crab Nebula and Markarian 501 recorded with the Whipple 10 m Atmospheric Cherenkov imaging system, and the results are compared with the standard Supercuts analysis. Since kernel analysis is computationally intensive, we examine approaches to reducing the computational load. Extension of the technique to estimate the energy of the gamma-ray primary is considered.

INTRODUCTION

In atmospheric Cherenkov imaging, a large optical reflector is used to focus the Cherenkov light from an extensive air shower on to an array of photomultiplier tubes. For a gamma-ray primary, the image thus formed is approximately elliptical and can be characterised by a set of parameters representing the shape, position, orientation and light content of the image (1). For a gamma-ray source at the centre of the field of view, the major axis of the ellipse will point towards the centre of the camera. Air showers generated by hadron primaries are generally less compact than those due to gamma rays and have no preferential pointing (2). Accordingly, selection of gamma-ray events in the presence of the far more numerous hadronic background events can be achieved by applying appropriate cuts on the image parameters, as exemplified by the Supercuts criteria developed by the Whipple Collaboration (3). The kernel analysis approach described here is one of a number of multivariate parameter-based selection techniques outlined by Samuelson (2).

BASIS OF TECHNIQUE

The image parameter set used (see for example ref.(4)) consists of: the *width* and *length* of the image ellipse; the *distance* from ellipse centre to camera centre; the natural logarithm of the *size* (a measure of total light content in the image); and *alpha* (the angle between the major axis of the ellipse and the line from ellipse centre to camera centre). An event is then represented as a vector $\mathbf{x} = (\alpha, w, l, d, \log(s))$. Each event is assigned a score $r = f_G/f_B$, where f_G is the likelihood that the event is gamma-initiated and f_B is the likelihood that it is a background event. f_G is estimated from a set of N_G gamma-ray simulations (5) with parameter vectors $\mathbf{g}_i = (\alpha_i, w_i, l_i, d_i, \log(s_i))$:

$$f_G = \frac{1}{N_G} \sum_{i=1}^{N_G} K(\mathbf{x} - \mathbf{g}_i) \qquad (1)$$

The *kernel function* $K(\mathbf{x} - \mathbf{g}_i)$ is effectively a point spread function describing the influence of $\mathbf{g}_i$ at $\mathbf{x}$. f_B is estimated in a similar way from a set of N_B real background events with parameter vectors $\mathbf{b}_i$. In principle, $K(\mathbf{z})$ can be any scalar function of its argument $\mathbf{z}$. In practice, the form used is a multivariate Gaussian :

$$f_G = \frac{1}{G} \sum_{i=1}^{N_G} \exp\{-(\mathbf{x} - \mathbf{g}_i)^T \xi_G^{-1} (\mathbf{x} - \mathbf{g}_i) / 2h_G^2\}$$
$$f_B = \frac{1}{B} \sum_{i=1}^{N_B} \exp\{-(\mathbf{x} - \mathbf{b}_i)^T \xi_B^{-1} (\mathbf{x} - \mathbf{b}_i) / 2h_B^2\} \qquad (2)$$

where ξ_G and ξ_B are the covariance matrices of $\mathbf{g}_i$ and $\mathbf{b}_i$, respectively, G and B are normalisation factors, and h_G and h_B are scale factors (6).

ANALYSIS OF CRAB NEBULA DATA

The analysis described in the previous section was applied to observations of the Crab Nebula recorded in early 1997 with the Whipple 10 m 151-pixel imaging system (7, 8). The database used comprises 11 ON/OFF pairs (during an ON run the telescope tracks the direction of the source for 28 minutes, while for an OFF run the telescope tracks a control position offset by 30 minutes in right ascension). For each event recorded, the image is parametrised to give the vector $\mathbf{x}$. f_G and f_B are calculated on the basis of Eq.(2) using a set of 5000 simulations for $\mathbf{g}_i$ and 5000 real background events for $\mathbf{b}_i$, and hence the score r is found. For any given score value r_c, the significance S is determined by finding N_{on}, the number of ON events with $r > r_c$, and N_{off}, the number of OFF events with $r > r_c$:

$$S = \frac{N_{on} - N_{off}}{\sqrt{N_{on} + N_{off}}} \qquad (3)$$

The plot of S versus $\log(r_c)$ in Fig.1 shows a peak for $\log(r_c) = 5.14$, with $S = 20.1\sigma$. The corresponding numbers of ON and OFF events are given in Table 1, along with the results obtained using Supercuts criteria optimised for the 151-pixel camera (8). It can be seen that the kernel analysis achieves a higher significance by more efficient rejection of background events. This is not strictly a fair comparison, however, since the cut on r_c is optimised on this particular data set. This point is dealt with later.

TABLE 1. Comparison of Kernel and Supercuts results for Crab Nebula data.

	N_{on}	N_{off}	excess	S
Kernel, $\log(r_c) > 5.14$	1084	328	756	20.1σ
Supercuts	1679	865	814	16.1σ

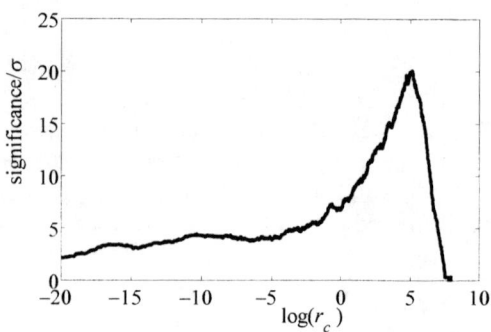

FIGURE 1. Kernel analysis applied to Crab Nebula data.

Fraction Of Gamma-Ray Events Retained

The cut on r_c required to accept a given fraction of gamma-ray events can be found by applying the analysis to the gamma-ray simulations. Using the cut values thus obtained with the Crab Nebula data, the kernel analysis achieves a higher significance than Supercuts while retaining roughly the same fraction of gamma-ray events (Fig.2). At 70% retention, the significance is as good the Supercuts value, and is still high even at 90% retention, an important consideration for spectral analysis. An examination of the fraction of gamma rays retained as a function of energy shows that the kernel technique performs particularly well at higher energy.

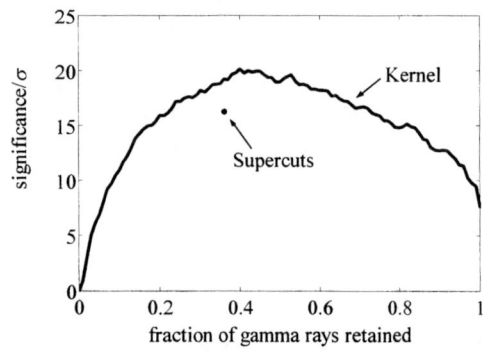

FIGURE 2.: Significance as a function of gamma-ray retention for Crab Nebula data.

REDUCTION OF ANALYSIS TIME

Kernel analysis is computationally intensive, since *every* event is compared with *every* simulation and with *every* background event. Two main possibilities for reducing the analysis time have been examined, pre-selection of the events to be analysed and lattice analysis.

Pre-Selection

Most events in the data set are not gamma-initiated, and many can be eliminated with loose cuts on individual parameters *before* applying the kernel analysis. The pre-selection cuts shown in the last line of Table 2, for example, reduce the analysis time by a factor ~12, with little effect on the results obtained.

TABLE 2. Effect of pre-selection cuts.

	N_{on}	N_{off}	excess	S
No cuts	1084	328	756	20.1σ
$\alpha < 30°$, $w < 0.30$, $l < 0.40$	1089	319	770	20.5σ
$\alpha < 20°$, $w < 0.25$, $l < 0.40$	1088	317	771	20.6σ

Lattice Analysis

The expressions for f_G and f_B represent convolution of the simulation and background points with a point spread function (the kernel function), giving smoothed approximations to the parameter space probability distributions for gamma-ray events and background events. The values of f_G and f_B can therefore be computed for a lattice of points in parameter space, and interpolated for each data event. In the present case, a five-dimensional lattice with a total of 1.4×10^6 points was used. While evaluation of f_G and f_B for the lattice is slow, this computation has to be carried out only once. Linear piecewise interpolation to obtain the scores for data events is then extremely fast. As can be seen from the comparison in Table 3, the lattice calculation gives virtually the same results as the full kernel calculation.

TABLE 3. Lattice calculation applied to Crab Nebula data.

	N_{on}	N_{off}	excess	S
Lattice calculation	1107	333	774	20.4σ
Full kernel calculation	1084	328	756	20.1σ

Lattice Analysis Of Markarian 501 Data

The lattice approach was used to analyse observations of the AGN Markarian 501. The database comprises 31 ON/OFF pairs, recorded in 1997 with the same camera configuration as for the Crab Nebula observations. The cut on $\log(r)$, optimised on the Crab data, was applied without modification to the Markarian 501 data. The results are shown in Table 4. In this completely independent test, the method described here is again seen to be much more efficient than Supercuts at rejecting background events.

TABLE 4. Lattice calculation applied to Markarian 501 data.

	N_{on}	N_{off}	excess	S
Lattice calculation	3222	428	2794	46.2σ
Supercuts	4079	1233	2846	39.0σ

ENERGY ESTIMATION

An estimate of the energy of a candidate gamma-ray event can be obtained as

$$\widetilde{E} = \frac{\sum_{i=1}^{N_G} E_i \exp\left\{-(\mathbf{x}-\mathbf{g}_i)^T \xi_G^{-1} (\mathbf{x}-\mathbf{g}_i)/2h_G^2\right\}}{\sum_{i=1}^{N_G} \exp\left\{-(\mathbf{x}-\mathbf{g}_i)^T \xi_G^{-1} (\mathbf{x}-\mathbf{g}_i)/2h_G^2\right\}} \qquad (4)$$

where E_i is the energy of the simulated gamma-ray event $\mathbf{g}_i$. Applying this to the simulations themselves, it is found that good energy reconstruction can be achieved in the range 0.3-10 TeV if the energy estimate is modified to $\widetilde{E}' = 0.96\widetilde{E}^{1.22}$. The energy resolution over this energy range is ~0.3-0.4.

CONCLUSIONS

The multivariate approach embodied in the kernel analysis makes maximal use of the inter-relationships between image parameters. Although the full kernel function calculation is rather slow, the lattice approximation permits rapid analysis of observations. The technique gives excellent gamma-ray selection, good spectral performance and good energy reconstruction in the energy range 0.3-10 TeV.

ACKNOWLEDGMENTS

We thank the Whipple Collaboration for the use of the Crab and Mrk 501 data sets. This work was supported by the U.S. Department of Energy and by Enterprise Ireland.

REFERENCES

1. Hillas, A. M., *Proc. 19th ICRC*, La Jolla, **3**, 445-448 (1985).

2. Samuelson, F., *Exp. Astron.*, submitted (1999).

3. Punch, M., *et al*, *Proc. 22nd ICRC*, Dublin, **1**, 464-467 (1991).

4. Moriarty, P., *et al*, *Astropart. Phys.* **7**, 315-327 (1997)

5. Kertzman, M. P., and Sembroski, G. H., *Nucl. Inst. Meth.* **A343**, 629-643 (1994).

6. Scott, D. W., *Multivariate Density Estimation*, New York: John Wiley & Sons Inc., 1992.

7. Cawley, M. F., *et al*, *Exp. Astron.* **1**, 173-193 (1990).

8. Quinn, J., *et al*, *Ap. J.* **518**, 693-698 (1999).

Observations at Large Zenith Angles

F. Schröder* for the HEGRA Collaboration and D. Heck[†]

Fachbereich Physik, Universität Wuppertal, Gaußstr. 20, D-42097 Wuppertal, Germany
[†]*Institut für Kernphysik, Forschungszentrum Karlsruhe, D-76021 Karlsruhe, Germany*

Abstract. Cherenkov telescope observations at zenith angles > 70° are capable of providing large collection areas for high energy γ-induced air showers. In order to provide a full Monte Carlo simulation of the large zenith angle observations the air shower simulation code CORSIKA was modified to treat particles in a curved geometry. First results of studies with the stand alone telescope HEGRA CT1 are presented.

INTRODUCTION

In the last few years Cherenkov telescopes have shown their capability to detect photon fluxes from discrete point sources on the level of 10^{-12} cm^{-2} s^{-1} above 300 GeV. Most of these observations have been made at zenith angles < 40° since at larger zenith angles the energy threshold is higher and less Monte Carlo simulations are existing. This constraint reduces the number of observable γ-sources. To overcome this limitation a study of the higher zenith angle region on the basis of appropriate simulations is necessary.

The observation of astrophysical sources at larger zenith angles offers another important feature. Because of the large increase of the geometrical collection area, it is possible to achieve a better acceptance for high energy γ-primaries [10]. By these means a Cherenkov telescope is able to expand its energy range for measuring the source spectrum [1,5,6,11,12].

Possibilities and problems of this observation mode are investigated using HEGRA CT1 [7,8] as example. The HEGRA experiment located on the Canary island La Palma operates six Cherenkov telescopes, five are combined in a system while the CT1 is a stand alone telescope. CT1 has a high resolution PMT camera consisting of 127 pixels in an 3.25° field of view (pixel diameter 0.25°) and, since 1998, a mirror area of 10 m^2.

γ-HADRON SEPARATION

In order to get a higher sensitivity for γ-induced air showers from a particular source, a separation of these showers from the dominant hadron background is required. In the case of modern Cherenkov telescopes such as CT1 this is possible due to their high resolution camera. Due to the telescope's imaging capabilities the whole air shower is visible in the camera and differences in the shower development are fixed in the measured image of the air shower's Cherenkov light. One method to apply a γ-hadron separation is to calculate appropriate parameters which characterize the image. The method of Hillas [4] is to calculate the second moments of the pixels intensity distribution in the camera frame. Thereby one achieves an observable for the longitudinal (LENGTH) and transverse (WIDTH) extension of the shower, its direction (ALPHA), energy (SIZE), the distance to the shower's maximum (DIST, MDIST) and for the impact parameter of the shower core (DIST).

At small zenith angles most of the γ-hadron separation is obtained by cuts on the transverse extension of the shower (WIDTH) and its direction (ALPHA).

GEOMETRICAL ASPECTS

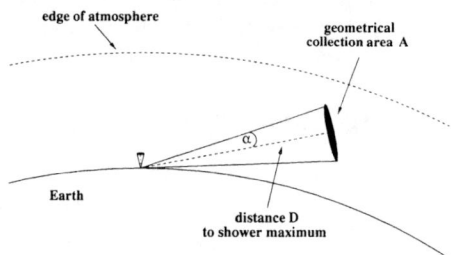

FIGURE 1. Increased collection area at large zenith angle observation.

In contrast with vertical air showers with a typical height of the shower maximum of 10 km, the maxima of horizontal showers are further away from the telecope due to the larger amount of slant depth the particles have to penetrate. Since telescopes are able to detect air showers with large impact parameters corresponding to large DIST angles in the camera plane, the geometrical collection area A is a function of the zenith angle θ and thus from the distance D between telescope and shower maximum. This function is given by Sommers & Elbert [10]:

$$A(\theta) = \pi \, (\, D(\theta) \tan \alpha \,)^2 \propto D^2 \, .$$

The angle α which is related to the position of the shower maximum in Figure 1 is found to be approximately equivalent to the MDIST parameter for showers parallel to the telescope axis (HEGRA CT1: $\alpha \approx$ MDIST $\approx 1° \approx$ Cherenkov angle at shower maximum). Since the signal-to-noise ratio is a function of the distance D and the quality factor Q of the γ-hadron separation (signal / noise $\propto \sqrt{A} \cdot Q \propto D \cdot Q$), it is necessary to study the shower characteristics at large zenith angles to find out whether the increase of collection area is offset by a decrease of the quality factor.

TABLE 1. Increase of slant depth and geometric path in a plane and a curved (spherical) atmosphere as seen at sea level.

Zenith Angle [°]	flat Atmosphere Length [km]	Path [g/cm^2]	curved Atmosphere Length [km]	Path [g/cm^2]
0	113	1037.2	113	1037.2
30	130	1197.6	130	1197.2
45	160	1466.8	158	1465.1
60	226	2074.4	220	2067.3
70	330	3032.5	311	3006.8
80	650	5972.9	529	5771.5
85	1294	11900.4	771	10583.0
89	6463	59429.3	1098	25942.7
90	∞	∞	1204	36479.9

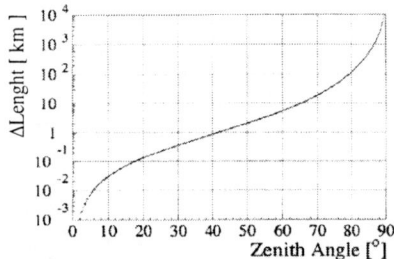 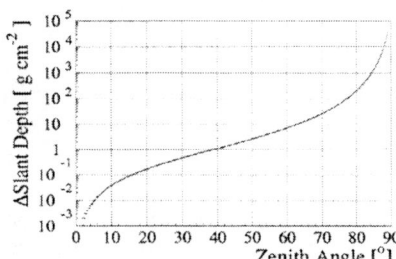

FIGURE 2. Difference of geometric path (left) and slant depth (right) between a plane and a curved (spherical) atmosphere as seen at sea level.

NEW MONTE CARLO SIMULATIONS

Since the standard version of the air shower simulation code CORSIKA [2] treats the shower development in a plane atmosphere (Figure 3 left), modifications have been performed for a complete simulation at large zenith angles. To avoid lengthy formulas for a treatment in a spherical coordinate system with corresponding long CPU times the description of particle transport is kept, but the horizontal step size is limited (see Figure 4). Longer transport distances are subdivided into appropriate segments to be treated in a local flat atmosphere. After each traversed segment the particle coordinates are transferred into the next local Cartesian coordinate system with its vertical axis pointing to the center of Earth. Thus the curved Earth's surface is approximated piece by piece by flat segments with limited horizontal extension (Figure 3 right).

When transferring the coordinates into the next local flat system, the atmospheric thickness makes a jump due to the treatment in a local flat atmosphere. To keep this jump negligible (Δthickness/thickness $< 1\%$ and Δthickness < 0.5 g/cm^2) a thickness

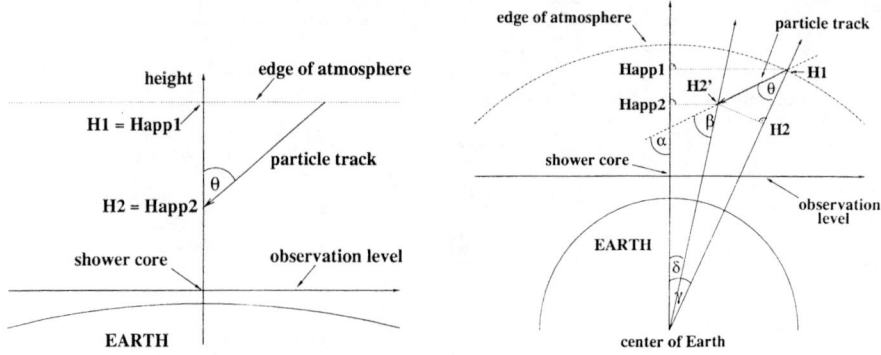

FIGURE 3. Left: CORSIKA geometry in the standard version. Right: Scheme of the new treatment of particles in the modified CORSIKA version. Heights and zenith angles: H1, H2, θ in the first local frame; H2', β in the next local frame; Happ1, Happ2, α in the 'detector frame'.

dependent horizontal step size limitation is introduced. Thus the step size limit varies exponentially between 20 km at the top of the atmosphere and 6 km at sea level. A similar treatment has been described by [9] (see also [3]). Additionally to the modified particle transport the Cherenkov photon tracks are corrected due to the atmospheric refraction using precalculated tables gained by numerical integration.

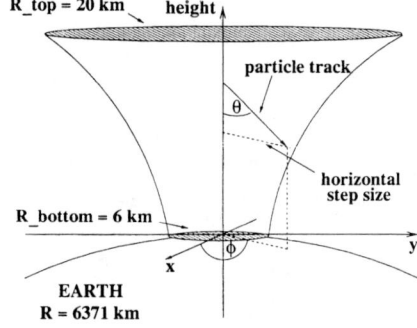

FIGURE 4. Condition for defining a local plane Cartesian coordinate system.

Comparing results from the modified CORSIKA version with those from the standard one differences e.g. for the height of shower maximum (Figure 5) appear at large zenith angles $> 70°$. Due to the larger distance to the shower maximum and thus the higher atmospheric extinction the images in the telescope's camera are smaller and more concentrated. As a consequence of the relatively large pixels of HEGRA CT1 (0.25°) the Hillas parameters [4] are not well determined. Especially the LENGTH and WIDTH parameter have nearly the same size and therefore the ALPHA parameter is hard to determine. Special cuts have to be developed. This work is still in progress.

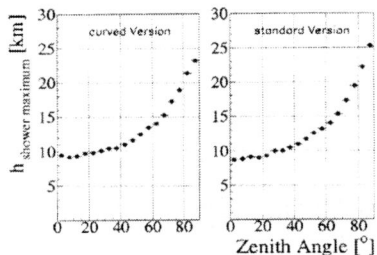
FIGURE 5. Shower maximum in a curved and a plane atmosphere; differences above 70° are visible.

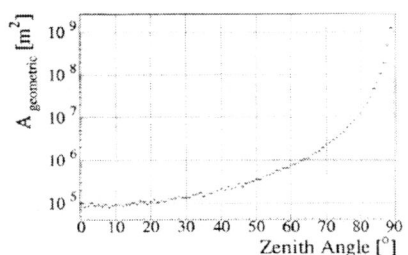
FIGURE 6. Simulation of the increase of geometrical collection area (curved atmosphere) with zenith angle.

RESULTS

As a first result of a Monte Carlo study the increase of geometrical collection area depending on the zenith angle is presented (Figure 6). The increase of more than one magnitude for zenith angles $> 70°$ agrees with the expectation. The calculation of the effective collection area, taking into account the atmospheric extinction as well as the telescope's trigger and γ-efficiency, is in progress.

REFERENCES

1. Chadwick, P. M. et al., preprint astro-ph/9904405 (1999)
2. Heck, D. et al., Report FZKA 6019, Forschungszentrum Karlsruhe (1998)
3. Heck, D. et al., Proc. 26th ICRC, Salt Lake City, **1**, 498 (1999)
4. Hillas, A. M., Proc. 19th ICRC, La Jolla, **3**, 445 (1985)
5. Konopelko, A. et al., preprint astro-ph/9906405 (1999)
6. Krennrich, F. et al., ApJ **511**, 149 (1999)
7. Mirzoyan, R. et al., Nucl. Instr. Meth. A, **351**, 513 (1994)
8. Rauterberg, G. et al., Proc. 24th ICRC, Rome, **3**, 460 (1995)
9. Sciutto, S., Auger technical note GAP-98-032 (1998)
10. Sommers, P. & Elbert, J. W., J. Phys. G: Nucl. Phys. **13**, 553 (1987)
11. Tanimori, T. et al., ApJ **429**, L61 (1994)
12. Tanimori, T. et al., ApJ **492**, L33 (1998)

Monte Carlo Simulations for High Zenith Angle

A. Ibarra[1], J.C. González[2], J. Cortina[2],
J. A. Barrio[1] and V. Fonseca[1]

[1] *Facultad de Ciencias Físicas. Universidad Complutense 28040 Madrid, Spain.*
[2] *Max-Planck Institut für Physik, Föhringer Ring 6, 80805 München, Germany.*

Abstract. With the advent of new generation Imaging Atmospheric Cherenkov Telescopes (IACTs), with improved sensitivity, it will be mandatory to extend observation to *High Zenith Angles* (HZA). This will allow us to extend by a large factor the observation time as well as the statistics for any gamma-ray source. Here we present some results obtained for several zenith angles and for different energies of the primary particle, obtained with a modification in the CORSIKA code, which will allow us to simulate and study the behavior of atmospheric Extensive Air Showers at HZA. In order to obtain more realistic results, the effect of the atmospheric attenuation, including Rayleigh scattering, Mie scattering and Ozone absorption, is included as well.

INTRODUCTION

Imaging Atmospheric Cherenkov Telescopes (IACTs) have played a key role over the last ten years in establishing Very High Energy (VHE) γ-ray astronomy as a mature discipline. These instruments suffer nevertheless from some drawbacks, among them a very small duty cycle ($\leq 18\%$) and a moderate dynamic range in energy, just somewhat higher than one order of magnitude (typically 300 GeV - 10 TeV).

These drawbacks can be partially overcome by the use of the so-called High Zenith Angle (HZA) technique. Initially, IACTs observed sources at Low Zenith Angles (LZA), below 45°, aiming to reach the lowest energy threshold achievable by the instrument. It was pointed out by Sommers and Elbert [1] that extending observations to zenith angles up to 80° would allow to enlarge the detectable energy by around one order of magnitude. Detection rate reduction due to the increase of energy threshold would be compensated by the sizeable enhancement of effective area coming from the development of the Extensive Air Showers (EAS) maximum further away from the observer.

Another obvious advantage of this technique is the possibility to enlarge the amount of sources to be observed and increase the duty cycle of the source. This technique has been applied in several γ-ray experiments with good results [2–5].

In this paper we perform a detailed Monte Carlo study on γ-ray induced EAS for different zenith angles up to 80°. The next sections, describe the simulation of the showers and how we checked it. Finally, we have analyzed the data produced by the simulation, and some conclusions are obtained on the impact of our results in the new generation of IACTs.

AIR SHOWER SIMULATIONS

The Monte Carlo program CORSIKA [7], modified to permit HZA showers, was used to simulate γ-showers. The modification allows CORSIKA to handle spherical earth atmosphere, by using a correct *air-mass factor*, $m(\theta, H_v)$

$$m(\theta, H_v) = \frac{\sqrt{R_e^2 cos^2(\theta) + 2 \cdot R_e \cdot H_v + H_v^2} - R_e cos(\theta)}{H_v} \quad (1)$$

where θ is the zenith angle, H_v is the vertical height and R_e is the earth radius. γ-showers were simulated at several energy points, from 10 GeV to 10 TeV, and for various zenith angles, ranging from 0° to 80°. Atmospheric absorption effects on Cherenkov radiation were taken into account by including Rayleigh, aerosol and ozone absorption as modeled by Elterman [6].

DEVELOPMENT OF γ-EAS

Observing γ-rays under HZA will lead to a dramatic growth in the amount of atmosphere traversed by the particles of the generated atmospheric showers before they arrive to the observer. Obviously a correct treatment of curved atmosphere leads to a finite thickness of atmosphere at very HZA, as opposite to the infinite thickness obtained by the flat earth model.

HZA showers develop further from the observer than low zenith angle ones. Sommers and Elbert introduce a rough formula to calculate the distance to the shower maximum. Table 1 compares the distance to the shower maximum as calculated using Sommers-Elbert formula with that obtained with our simulation for several zenith angles.

At 80 degrees, our simulation predicts a lower distance to the shower maximum than Sommers-Elbert's ones. The reason for this discrepancy is the *air-mass factor*. Sommers and Elbert [8] use the factor $sec(\theta)$, which is known to overestimate the distance to the shower maximum for zenith angles higher than 60°.

Figure 1 shows the average longitudinal development in slant thickness (i.e. thickness measured along the shower axis) obtained for γ-ray showers of 10, 100 and 10000 GeV primary energy and 0°, 40°, 60° and 80° zenith angle. As expected, the distributions depend only mildly on zenith angle.

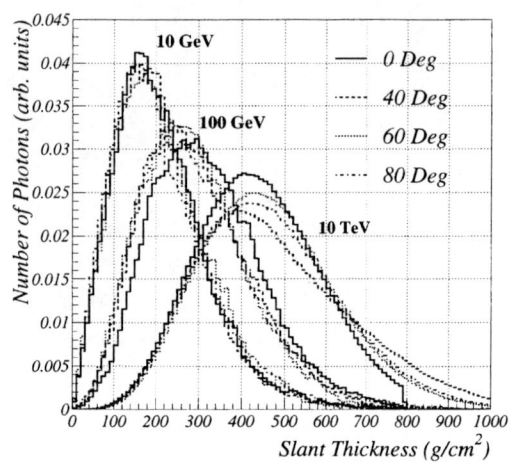

FIGURE 1. Longitudinal distributions of γ-shower Cherenkov photons as a function of slant thickness, for three primary energies (10 GeV, 100 GeV and 10 TeV) and four incident zenith angles (0°, 40°, 60° and 80°).

TABLE 1. Distance to the shower maximum for 100 GeV γ-showers

Zenith Angle	Sommers's formula	Our Simulation
0°	7 Km	9 Km
40°	14 Km	15 Km
60°	27 Km	29 Km
80°	129 Km	113 Km

LATERAL DISTRIBUTION

The above described effects have a decisive impact on the γ-shower Cherenkov photon signature at the ground level, as can be seen in Fig. 2. HZA γ-showers reach their maxima further away from the observer. As a consequence Cherenkov photons traverse a larger amount of atmosphere, and therefore suffer more attenuation. Consequently Cherenkov photon density at ground diminish with increasing zenith angles. At 80°, the Cherenkov photon density is three orders of magnitude lower than that at 0°, at core distances below 150 m.

A second important feature, also shown in Fig. 2, is how the *hump* fades away and moves towards larger impact parameters as the zenith angle increases. This is explained as a simple geometrical effect caused by the γ-showers developing at larger distances from observer. This geometrical effect also contributes to some extent to the decrease of the Cherenkov photon density near the shower core.

ATTENUATION VS. WAVELENGTH

The large amount of atmosphere traversed by Cherenkov photons produced by HZA showers has also dramatic effects on the wavelength distribution of those photons reaching the ground, as can be seen in Fig. 3.

Rayleigh scattering is the main attenuation process for all the wavelength (excluding the ozone absorption for wavelength between 290 and 340 nm). This effect reduces the amount of Cherenkov photons reaching the ground and shifts the maximum of the wavelength distribution towards longer wavelength for HZA showers, as we can see in the Fig. 3 (it follows a λ^{-4} power law). In the limiting case, for γ-rays arriving at zenith angles around 80°, the maximum of the wavelength distribution peaks at around 500 nm almost independently of primary energy. This leads to conclude that only red-extended photosensor would allow us to fully detect the Cherenkov light distribution at HZA.

CONCLUSIONS

We have studied the features of γ-ray showers reaching the ground under high zenith angles by means of detailed Monte Carlo simulations. Results exhibit the expected reduction of Cherenkov photons at ground. We have also described how the *hump* position moves towards larger radial distances from the shower core as a consequence of the larger distances of the shower maximum to the observer. Finally

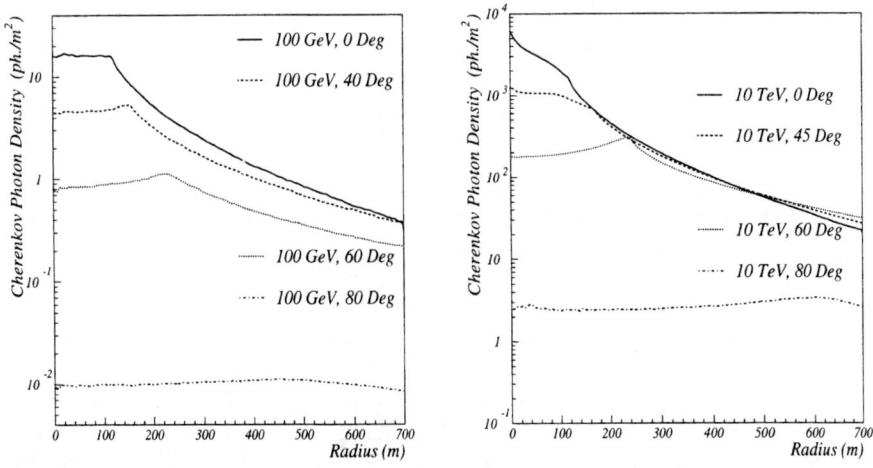

FIGURE 2. γ-shower Cherenkov photon density as a function of the distance to the shower core. Two primary energies are presented (100 GeV and 10 TeV) at four zenith angles (0°, 40°, 60° and 80°).

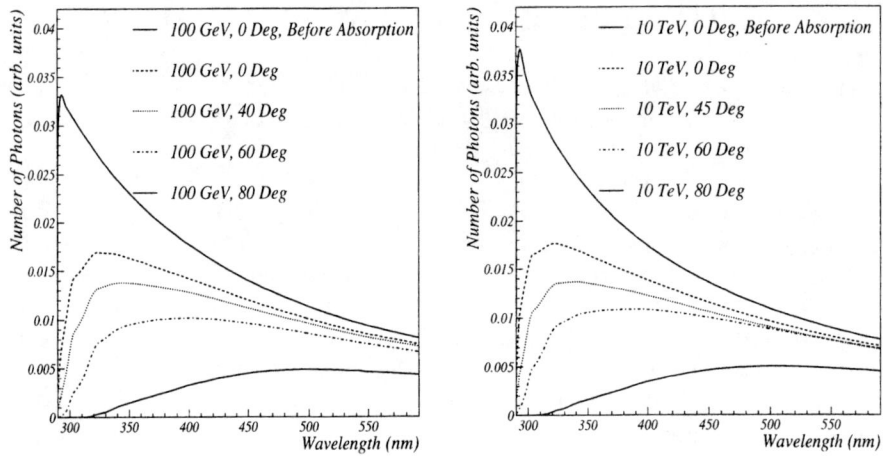

FIGURE 3. Absorption profiles for Cherenkov radiation generated in EAS. Two primary energies are presented (100 GeV and 10 TeV) at four zenith angles points (0°, 40°, 60° and 80°).

we have shown the displacement of the Cherenkov photon wavelength distribution towards longer wavelengths.

From the above results we conclude that, although current IACTs can apply the HZA technique, only the next generation of IACTs will take fully advantage of this technique. In order to do so, they will need to incorporate large mirror areas.

ACKNOWLEDGMENTS

This work is partially supported by the Spanish CICYT.

REFERENCES

1. Sommers, P. & Elbert, J. W. 1987, J. Phys. G: Nucl. Phys. 13, 553.
2. Krennrich, F., et al., 1997, *ApJ* **481**, 758.
3. Krennrich, F., et al., 1999, *ApJ* **511**, 149.
4. Tanimori, T., et al., 1998 *ApJ* **492** L33-L66.
5. Mohanty, P., 1999, OG.2.2.03 in Proc. 26th ICRC (Salt Lake City, 1999).
6. Elterman L. *Handbook of geophysics and space environments*, ch. 7, S.L. Valley editor, NY. 1965.
7. Heck, D., et al., *CORSIKA: A Monte Carlo Code to Simulate Extensive Air Showers*, Tech. Report Forschungszentrum Karlsruhe, FZKA 6019, 1998.
8. Kondratyev, K. *Radiation in the atmosphere*, vol. 12, Academic Press editor, NY. 1969.

Geomagnetic Effects on the Performance of Atmospheric Čerenkov Telescopes

P. M. Chadwick, K. Lyons, T. J. L. McComb, K. J. Orford,
J. L. Osborne, S. M. Rayner, S. E. Shaw, and K. E. Turver

Department of Physics, Rochester Building, Science Laboratories, University of Durham, Durham, DH1 3LE, U.K.

Abstract. We have reported the results of the interaction between the geomagnetic field and the development of atmospheric electron cascades as manifested by the distortion of the atmospheric Čerenkov signal. We will summarise these effects and describe our initial attempt to remove this distortion for those cases where the cascade develops perpendicular to a strong (> 0.35 G) field and so enhance the hadron rejection procedures of gamma ray telescopes.

INTRODUCTION

In a recent paper [1] we reported the results of both simulations and measurements which demonstrated the effect of the interaction between the geomagnetic field and the development of atmospheric electromagnetic cascades. We drew attention to the fact that the images of the Čerenkov radiation produced by the electron cascades can be distorted and rotated by the geomagnetic field.

The geomagnetic field can have an important effect on the operation of atmospheric Čerenkov telescopes (ACTs) when observing in some directions and that observations with our Mark 6 telescope [2] at the Narrabri site are particularly susceptible.

With imaging ACTs the identification of very high energy (VHE) gamma rays (covering the energy range ~ 100 GeV – ~ 30 TeV) in the presence of numerous hadronic cosmic rays depends on the shape and orientation of the Čerenkov light image. Of particular importance is the "pointing" of the images produced by gamma rays towards the camera centre (the source position).

The most important result demonstrated in [1] was that images of the Čerenkov radiation from electromagnetic cascades initiated by hadrons may be elongated or broadened and rotated by the field. The shape change of the image results in an increase in the threshold and a decrease in the count rate of gamma ray ACTs. The rotation of the image results in a decrease in the efficiency of the ACTs to suppress the problematic cosmic ray background on the basis of the characteristic "pointing" of the gamma ray Čerenkov image.

We expect that these rotation effects may be reduced using appropriate techniques since the effects of the geomagnetic field do not destroy the information contained in the image.

We report here the results of our first attempt to compensate for the geomagnetic effect on our observations of VHE gamma ray sources. We also summarise the factors governing the importance of the geomagnetic effect at various locations in the world and when telescopes of varying performance (mirror quality and camera resolution) are considered.

In figure 1 we depict images in the camera of an ACT in the simple case when the projection of the geomagnetic field is vertical in the plane of the camera. The stretching of the images will be in a side-to-side direction and we sketch the images expected in different regions of the camera. At the left and right sides of the camera the images are elongated. For images aligned vertically the image shapes are broadened. In both these cases the pointing of the image is substantially maintained and $ALPHA$ is unaffected. (The angle between of the long axis of the image and the direction defined by the radius vector to the source is denoted by $ALPHA$ [3].) The important cases occur when the image is pointing at an intermediate angle. In these cases the sideways extension of the gamma ray image results in a systematic rotation away from the source position (camera centre). This is the origin of a substantial part of the broadening of the distributions in $ALPHA$ noted in many of our observations.

On the basis of simulations of the Čerenkov radiation produced in atmospheric cascades initiated by gamma rays with a spectrum of energies, we have derived the expected rotations of the images. These depend upon the angle, $MAGANG$, between the developing cascade axis and the geomagnetic field direction (both projected onto the focal plane). The simulations are for field strengths in the range 0.0 – 0.5 G.

We define the quantity signed $ALPHA$ (denoted by $SALPHA$) by

$$ALPHA = |SALPHA| \qquad (1)$$

and the sign of $SALPHA$ is such that a clockwise rotation of the observed long axis from the source direction is positive. We show in figure 2(a) the correlation of the observed angle $SALPHA$ (including the rotational effects) with $MAGANG$ for a field strength of 0.45 G (which is typical of our observations). The rotation of the image is greatest at a $MAGANG$ of approximately 60°, 120°, 240° and 300°. The data are for simulated 500 GeV gamma rays at a distance of 0.65° from the camera centre. The peak-to-peak modulation of the mean $SALPHA$ against $MAGANG$ is a function of distance of the image from the camera centre and image brightness. Gamma ray images closer to the camera centre, which are characteristically less elongated, are more affected.

Demonstration of this behaviour in observational data is difficult because it is confined to gamma rays and is not shown by hadrons since they do not, in general,

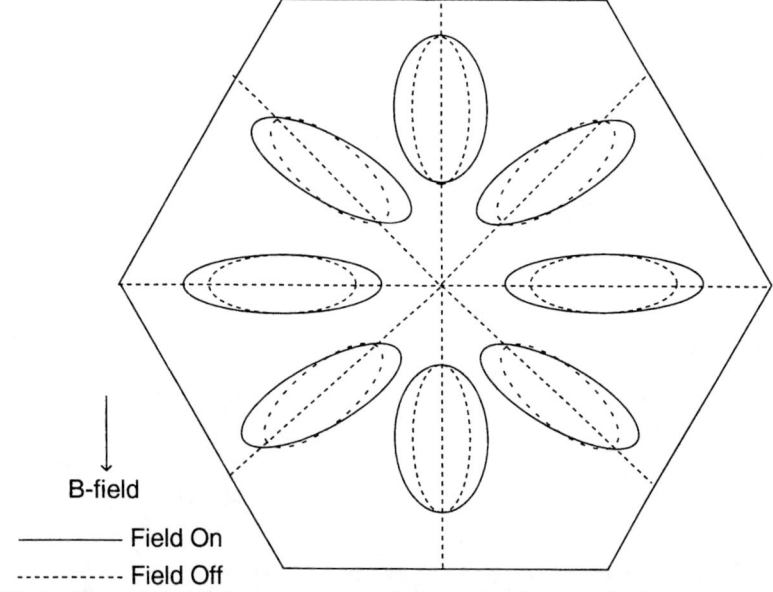

FIGURE 1. The effects of the geomagnetic field on the shape and orientation of images of VHE gamma rays. The projection of the geomagnetic field in the camera is vertically downwards. Undistorted gamma ray images are shown by dotted lines; solid lines show the shape and direction after distortion by the geomagnetic field.

point to the camera centre. However, we have some evidence that the distortion in $ALPHA$ for gamma ray candidates in our datasets show this effect.

On the basis of simulation data we have produced an initial algorithm which applies a correcting rotation which depends on the position of the image in the camera and the brightness of the image. This algorithm is peculiar to the effects in our telescope and for our camera characteristics, with a maximim correction to $ALPHA$ of typically 30° for the camera-centred events. The result of the application of the algorithm to the simulated data in figure 2(a) is shown in figure 2(b). Most of the rotation effects are removed.

The correcting algorithm has been applied to the PKS 2155–304 data from observations at zenith angle less than 45° [4]. Without the correction, this subset of the data has an excess on source with $\alpha < 22.5°$ which is significant at the 4.4 σ level. After the correction has been applied and using the same gamma ray selection criteria, this excess increases to 4.9 σ. this small increase is consistent with the simulation predictions and further improvements may be possible by re-optimising the selection criteria for the corrected data.

In figure 3 we show the variation of the strength of the geomagnetic field across the surface of the earth. Clearly, observations made in Australasia may be subject to strong effects with peak field values of > 0.55 G, especially for observations in

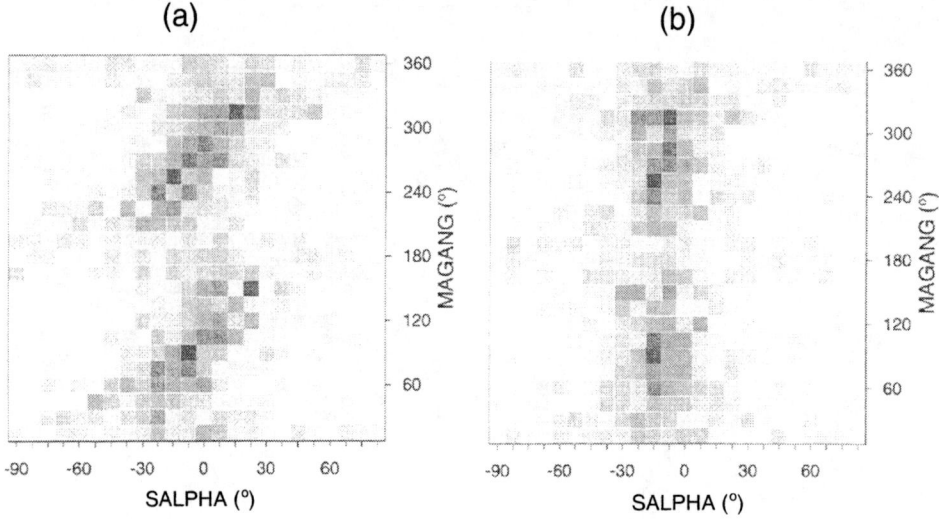

FIGURE 2. The correlation between the value of the pointing angle *SALPHA* and magnetic field angle *MAGANG* for simulated 500 GeV gamma rays in a magnetic field of 0.45 G which includes the effects of the optics, pixellation and noise. (a) is for uncorrected simulations while in (b) the simulated data have been rotation corrected for the effects of the geomagnetic field.

the directions of many Southern sources. On the contrary, observations in South America would be subject to a maximum value of the field of 0.2 G and telescopes with characteristics of the Mark 6 telescope would be free of observable geomagnetic effects.

Many of the sites of other ACTs, including those in the Northern hemisphere and India, are subject to values of the transverse field in certain directions of > 0.4 G. We have previously suggested that for fields up to 0.35 G, the distorting effect is of limited consequence for telescopes with the optical and imaging properties of first generation (eg our Mark 6) telescopes. However, the new telescopes now in operation or in development which have improved imaging performance with a typical mirror point spread function and pixel resolution of about 0.1° may suffer from a similar loss of efficiency to the Mark 6 telescope in Australia, but resulting from the influence of smaller fields (~ 0.2 G). This is because although the value of the rotation of the image is less in such smaller fields, the telescopes depend for improved efficiency on the measurement of gamma rays characterized by smaller values of recorded *ALPHA*.

We are grateful to the UK Particle Physics and Astronomy Research Council for support of the project and the University of Sydney for the lease of the Narrabri

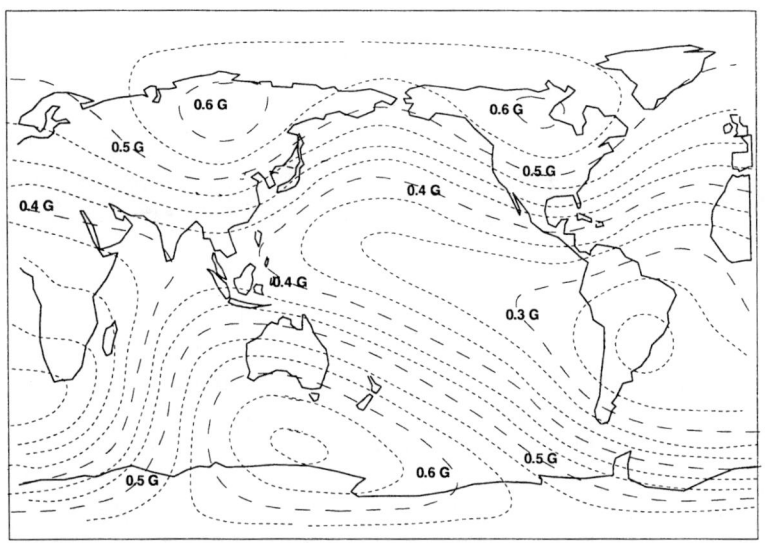

FIGURE 3. The variation of the strength of the geomagnetic field across the earth's surface (from http://ngdc.nooa.gov/seg/potfld/geomag.html).

site. The Mark 6 telescope was designed and constructed with the assistance of the staff of the Physics Department, University of Durham.

REFERENCES

1. Chadwick, P. M., et al., *J. Phys. G: Nucl. Part. Phys.*, **25**, 1223 (1999).
2. Armstrong, P. et al., *Experimental Astron.*, **9**, 51 (1999).
3. Fegan, D. J., *J. Phys. G: Nucl. Part. Phys.*, **23**, 1013 (1997).
4. Chadwick, P. M., et al., *Astrophys. J.*, **513**, 161 (1999).
5. Chadwick, P. M., et al., *Astropart. Phys.*, **9**, 131 (1998).
6. Chadwick, P. M., et al., *Astrophys. J.*, **503**, 391 (1998).

A Concept for the Readout of Multichannel Detectors by Using Analog Signal Transmission Via Optical Fibres Coupled to a Fast CCD

R. Mirzoyan*,[1], E. Lorenz*, and J. Rose[†]

*Max-Planck-Institute for Physics
Foehringer Ring 6, 80805 Munich
†Department of Physics and Astronomy, University of Leeds, Leeds, U.K.

Abstract. Recent developments in the field of electro-optical components allowed one to transform fast analog electrical signals into fast light pulses in a wide dynamic range and to send them via optical fibres over relatively long distances with very low time dispersion and amplitude losses. Here we propose to use the analog signal fibre transmission technique in combination with fast CCDs for the data acquisition of multichannel detectors as, for example, for the read out of imaging cameras of atmospheric Cherenkov telescopes. Hundreds of signal channels can be read out by using commercially available single CCD module in fast gated mode. The latter can provide high amplitude resolution and an acquisition rate of up to a few hundred Hz. Such a system can provide significantly lower costs compared to traditionally used amplitude digitising systems.

INTRODUCTION

Significant progress has been achieved in recent years in converting very fast, analog electrical signals into optical ones and sending them via optical fibres over relatively long distances. At the end of the fibre the light pulses are converted back into electrical signals. Wide bandwidth of such a system provides very low time dispersion and also low amplitude damping. Also, it should be noted that optical fibres are immune to electromagnetic pick-up and are substantially lighter and compacter compared to conventional coaxial cables. It is quite attractive to try to combine the technique of analog light transmission with commercially available multichannel light measuring sensors, such as digital CCDs or digital pixel image sensors, as a cheap read out system for multichannel detectors (such as, for example, for imaging cameras of atmospheric Cherenkov telescopes). Several years ago a

---

[1]) On leave from Yerevan Physics Institute, Yerevan, Armenia.

proposal for research and development on the central tracking detector has been published [1]. The very low light level from scintillating fibres had to be measured by an image intensified CCD. Recent achievements in the analog light transmission and availability of high frame rate CCDs allow one to modify the above mentioned idea. One can transform analog electrical signals into light pulses and then send them via optical fibres to multichannel light sensors where they can be measured. Because of usually rather long integration time of a single frame in a CCD, when measuring fast signals one has to introduce a very fast shutter between the optical fibres and the CCD in order to eliminate background light as much as possible. LCD shutters are relatively slow, providing response times in the range of $\sim$5 μsec. Other shutters such as Pockel's cells or gated image intesifiers (II) are $nsec$ fast.

Current Status of Analog Signal Transmission via Fibres

An analog optical transmission system in which a LED is used as a light source has been described in reference [2]. Ibidem it is shown that electrical pulses of $\sim$6 $nsec$ width and 1.2 V amplitude fed into an optical fibre of $2km$ length, disperse to 8 $nsec$ width and 0.45 V in amplitude at the end of the fibre. The system has a relatively small dynamic range. Recent report [3] describes an analog transmission scheme with improved parameters. This scheme was tested to transmit pulses of 1 $nsec$ rise time and 2 $nsec$ width without deterioration over fibre lengths of $\sim$120m. Another progress of the scheme is the large dynamic range reaching $\sim$2000. This substantial improvement stems from the use of a VCSEL (Vertical Cavity Surface Emitting Laser) diode instead of a LED. When used for transmitting pulses over distances of $\geq$ 100m the total cost of the latter scheme is matching that for the commonly used high bandwidth coaxial cables.

Fast Readout CCDs

There are two main reasons for the speed limitation in CCDs. The first one is due to the basic operational principle of the CCD, i.e. the sequential readout of pixels and the second one is because of the need of not too fast control pulses (sharp rise and fall times of pulses degrade the charge transfer). In spite of these limitations there exist a way to speed-up the overall readout by, for example, as much as 4 times. Typically one divides the CCD matrix into upper and lower read out halves and in addition splits each transferred to the given shift register line into left and right read out halves providing thus four parallel read out ports in total.

1kHz frame rate 8-bit digital CCD

To our knowledge the fastest commercially available, full matrix CCD, MD4256C from EG&G RETICON, can be read out with 1000 frames per second. The

MD4256C is an 8-bit digital, high speed, split-frame transfer, 4 port parallel output CCD camera. It has 65536 pixels, each of 16×16 μm^2 size. Pixel rate is 20 MHz per tap. The CCD has an imaging region of 256 columns each with 256 pixels and two protected from light storage regions, each with a size of 130×256 pixels, in opposite from the imaging region directions. An integrated image/frame is split into two parts and fast shifted into storage regions. While integrating the next frame one reads out the two storage regions each via two ports. Light integration time of a frame is $927 \mu sec$ and $60 \mu sec$ are used for the frame transfer.

Frame buffer & PC interface SB 4001

EG&G RETICON provides also a data acquisition PC board (SB4001) for the operation of MD4256C. The main features of the unit are:

- Data can be acquired at a sustained rate of 160 Mbytes/second into the buffer memory of up to 256 Mbytes (4000 full frames in 4 seconds)

- Four digital 8-bit input channels, each up to 40 MHz/channel

- Programmable timers, generating frame start signal

- Optional high-speed daughter board interface, allowing one, for example, to compress or to process an image

- Up to 8 SB4001 boards can be installed in a single computer

- Acquisition and buffering may be started or stopped from keyboard or by external inputs

Gated Image Intensifiers

A rather large variety of gated IIs exists. Focusing can be either electrostatic or so-called proximity type. One-stage, 2nd generation proximity type II consists of a photocathode, a single microchannel plate (MCP) and a phosphor screen. The electrons from the photocathode are accelerated by the applied electrical field and hit the MCP where they are multiplicated a few thousand times. These amplified electron "clouds" are further accelerated and hit the phosphor screen releasing optical photons. Such a type II can provide light gain of $\sim 10^3$. By altering the sign of the applied to the photocathode voltage one can close (positive level is applied) or open (negative level is applied) the II (the electronic shutter function). In the closed state the photon flux is attenuated by a factor of 10^7. Different types of photocathode materials and phosphor screens are available. Switching time can be as fast as $2 - 5 nsec$.

THE PROPOSED READOUT OF MULTICHANNEL DETECTORS

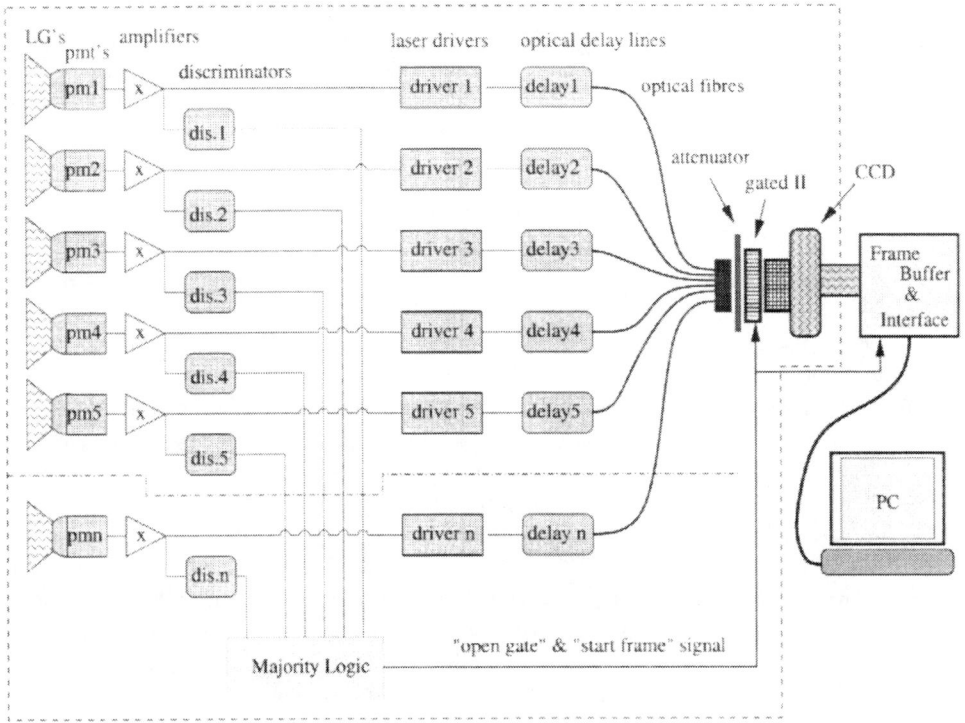

FIGURE 1. Schematic view of the proposed readout system.

The schematic view of the proposed readout system is shown on Fig.1. Pulses from each detector channel are amplified, discriminated and fed into the majority logic. The latter generates an output signal which opens the gate of the II and starts a frame. The amplified analog signals, after an appropriate delay in optical cables (they are cheap and do not deteriorate the pulse shape), are fed into the II which is coupled to the CCD. The gate signal sets an inhibit to the II for the time duration necessary to read out the already started frame. Note that for linearity reasons the VCSEL diode should be biased and a threshold current of $\sim 3.5 mA$ flows through it, providing a low level DC light emission. The closed shutter is not completely opaque. Therefore some part of light from the biased VCSEL diode will pass through and will be detected by the CCD in its $\sim 1 msec$ frame integration time. One can reduce this effect by inserting an appropriate attenuator of light between the glass fibre bundle and the II. During the open gate of, for example, $10-20 nsec$ the DC light from the laser diode will be integrated together with the genuine light pulse and will play the role of a pedestal. Each fibre will illuminate a relatively large group of neighbour pixels, say $\sim$25 pixels in total. When the given frame is fed into the SB4001 data acquisition board, providing the daughter board one can sum-up the contents of all 25-pixels in groups for corresponding channels as well as one can set to "0" all the pixels in the dead area between individual fibres. By assuming a real data taking rate of $\sim$100 Hz, ($\sim$10% dead time), providing the "0" suppression and the summ-up pixels in groups, one obtains a data stream of $\sim$0.1 Mbytes/second. This means that a single SB4001 board with its 256 Mbyte buffer memory can store images for 2560 seconds, i.e. for almost 3/4 hours. Meanwhile this data can easily be send to the data collecting PC and saved.

DISCUSSION

The merits of the proposed readout system are to a great extent due to commercial availability of fast-frame CCDs which are usually available as a complete set with all necessary power supplies, electronic boards and operational software (the cost of the above described system, including the gated II, is $\sim$20 kE). The analog signal transmission scheme may also become commercially available. Instead of using bundles of thick coaxial cables, crates for electronics and traditional amplitude digitizers for each channel of a multichannel system one can use the above described compact read out system. Currently, industry develops another concept for the read out of video cameras. These C-MOS pixel arrays can be addressed and read out individually. Thus in future even faster read out speed can be expected. The above mentioned concept is not restricted to the use in air Cherenkov telescopes. For example, it might also be considered for the read out of a system such as, say, Air Watch/OWL [4], where the shift function of the (for example, linear) CCD might be directly used as a relatively slow speed FADC.

CONCLUSIONS

The proposed readout system is a robust and cheap alternative compared to traditional amplitude digitizing systems. Depending on experimental needs one can also use light sensors with substantially higher amplitude resolution (10-16 bits) than the one described above.

REFERENCES

1. CERN/DRDC 90-29, 1990.
2. A. Karle, T. Mikolajski, S. Cichos, et al., NIM A 387 (1997) 274-277.
3. J. Rose, I. Bond, A. Karle, et al., Proc. 2nd Conf. on Photon Detection, Beaune, France, 21-25 June (1999); also, to appear in NIM A.
4. R.A. Chapman, Y. Takahashi, J.O. Dimmock, et al., late paper in 25th ICRC (Durban), (1997).

The FADC Readout of the MAGIC Telescope

J. Cortina, E. Lorenz and R. Mirzoyan
for the MAGICT collaboration

Max-Planck Institut für Physik, Föhringer Ring 6, 80805 München, Germany.

Abstract. The MAGIC Telescope will use FADCs to sample the camera pixel signals. This allows to analyse features of the Cherenkov time pulses with ~ 1 ns resolution and to minimize data taking deadtime and noise. Here we report on the first tests of a 300 MHz FADC 32-channel unit and on some special solutions for enlarging the signal's dynamical range. In the long run MAGIC will be provided with > 1 GHz FADCs in order to profit from the full speed capabilities of the PMs in use [1 ns FWHM].

INTRODUCTION

The MAGICT collaboration has taken the first steps to construct a 17 m diameter imaging air Cherenkov telescope for γ-ray observations above 10 GeV (details in [1,2]). A number of reasons necessitate a FADC-based data acquisition system:

(i) *Night sky (and moon light) noise minimization.* The Cherenkov light flashes, particularly those from γ-showers, are very "sharp" in time. FWHM values of < 1-2 ns are common. With ≤ 3.3 ns digitizer time-slices, a resolution below 1 ns can be reached for large pulses, whereas for small pulses, extending barely above noise, a resolution of ca. 2-3 ns can be achieved. Therefore a good correlation of the different pixel contents with the image and an efficient noise reduction should be possible by selecting optimal time slices around the signal.

(ii) *Event buffering* during second level trigger decision. Modern trigger concepts break up the traditional trigger into a multilevel decision system due to the steady increase in pixel number and faster signals. A 2 level trigger will call for FADC memories of 8 k depth.

(iii) *Improved γ/h separation* from measurement of the arrival time distribution of the Cherenkov light in different pixels.

(iv) *Correlated readout with other telescopes* in an array configuration. If many telescopes are combined for precision measurements, it is indispensable to use some intermediate storage device for the combination of the low-level pixel content.

DESCRIPTION OF THE FADC SYSTEM

A group from the university of Siegen has built a unit of 300 MSamples per second Flash ADCs. It accomodates 32 channels in four 8-channel modules inside a eurocrate, along with a fifth module for synchronization and I/O interface. The interface is made via a single PCI slot in a PC presently working under Linux OS (Real-Time Linux is also under test). An implementation of the full DAQ in a PC-based system allows to reduce costs with respect to a VME- or VXI-based one.

The FADCs have an 8-bit dynamical range. Input signals are continously sampled and stored in an 8 kB deep ring buffer. The FADC stops sampling when a trigger arrives and saves a predefined number of samples at 30 MHz to a 32 kB deep CMOS buffer. Once this memory is full, the content is in turn sent at 2 MHz to the PC. This system is too slow for a 1 kHz rate telescope like MAGIC so a redesign of the readout is currently under way. The FADC front section will not change in the final design.

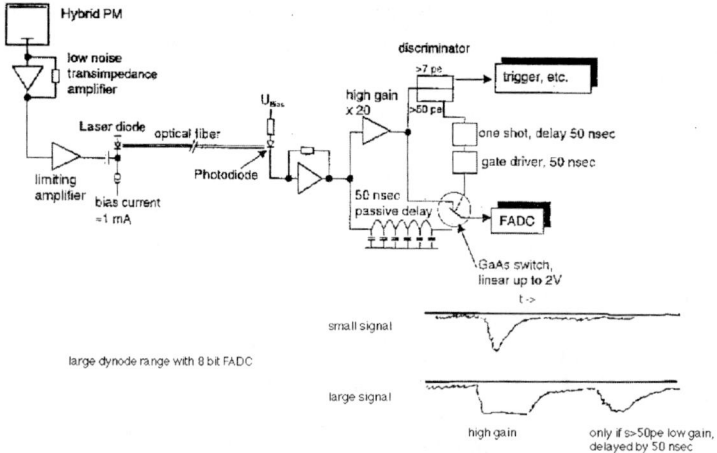

FIGURE 1. The data acquisition chain envisaged for MAGIC.

The data acquisition chain of the telescope is sketched in figure 1. After a first amplification the PM signal is sent through an optical fiber [3] to the central electronics room where it is split. Part of the signal (*high gain branch*) is amplified again and sent to the trigger system and the FADCs. Another part (*low gain branch*) is delayed in 50 ns and digitized right after the *high gain* signal. Using two branches allow to enlarge the system dynamical range. The GaAs switch enables to use only one FADC channel for each PM.

TESTS OF THE FADC SYSTEM

In its first phase MAGIC will use very fast classical 6 stage PMTs and amplifiers resulting in Cherenkov pulses as short as 2 ns FWHM. Such a short pulse is optimal for trigger requirements but is not long enough to permit an accurate pulse reconstruction using 300 MS/s FADCs. Therefore the signal entering the FADCs has to be stretched. On the other hand, stretching to very long pulses is not desirable either, since the longer the pulse the more noise caused by the night sky enters the digitized pulse, i.e., the lower the dynamical range of the system.

TABLE 1. Number of time bins, RMS of the noise, charge at saturation (both in arbitrary units) and dynamical range for different widths of the generator input pulse

Pulse width	N. bins	σ_{noise} (arb)	Q_{sat} (arb)	Dyn. range=$\frac{Q_{sat}}{3 \times \sigma_{noise}}$
3 ns	5	5.0	1300	87±10
4 ns	5	5.0	1450	97±10
5 ns	7	6.1	1650	90±8
7 ns	11	7.4	1900	86±8

We have used a fast pulse generator to feed the FADCs with input signals of different widths in order to estimate dynamical range and ability to reconstruct the pulse width, as well as linearity and accuracy in the charge reconstruction. Charge and width of the input pulse are determined very accurately with a fast digital scope. The width of the digitized pulse is then estimated by fitting it to a gaussian. This provides a rough estimate of our ability to deconvolve the original pulse shape from the digitized pulse. The charge is estimated by counting entries in the digitized pulse. We define the dynamic range of the FADC as $Q_{sat}/3 \times \sigma_{noise}$, where Q_{sat} is the charge for which the pulse hits the top of the ADC 8-bit scale and σ_{noise} is the RMS of the noise in the time bins used to reconstruct the pulse. The number of time bins required to digitized the pulse depends on its width. Table 1 lists this number along with the noise, Q_{sat} and the dynamical range achieved for different input pulse widths.

Figure 2a shows the reconstructed charge in the FADC as a function of input charge for 5 ns wide input pulses. The response is linear over the range under consideration (from $3 \times \sigma_{noise}$ up to Q_{sat}). Figure 2b shows the RMS of the reconstructed charge as a function of input charge. The resolution falls below 5% for large pulses.

Figure 3a displays again the dynamical range for different pulse widths. The arrows in figure 3b depicts the range of charge resolutions achieved for different input widths. The results range from 33% for the smallest signals (by definition, as we start considering signals only above $3 \times \sigma_{noise}$) down to below 5% for the largest ones. Obviously the charge resolution improves for longer pulses. Figure 3c shows the width resolution as a function of pulse width. As expected the resolution of the width improves with increasing width. In practice 3 ns wide pulses are very hard to reconstruct for any pulse charge and the arrow does not represent a real improvement with charge.

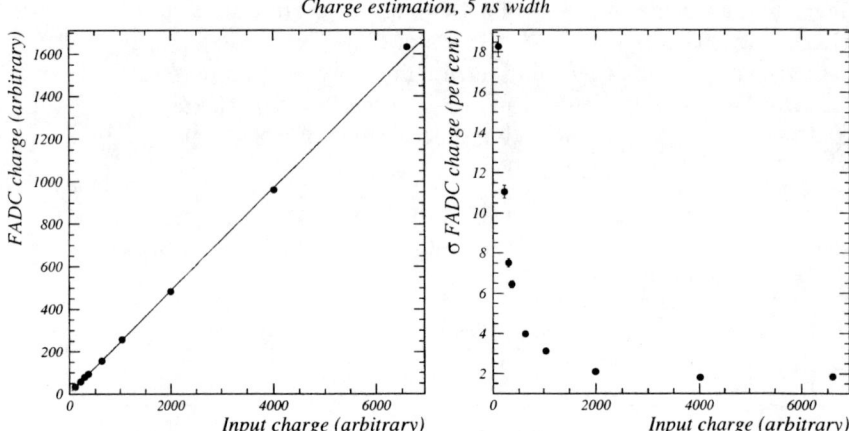

FIGURE 2. a) Reconstructed charge in the FADCs as a function of input charge, as measured in a fast scope, b) RMS of the charge reconstruction versus input charge. Both a) and b) were measured for 5 ns wide input pulses.

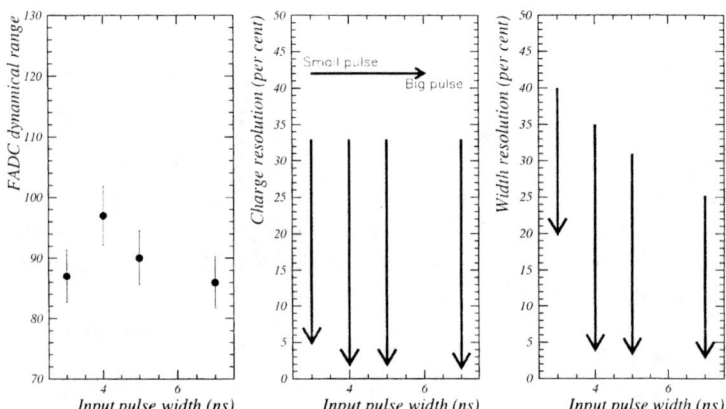

FIGURE 3. a) FADC dynamical range, b) charge resolution and c) width resolution as a function of input pulse width.

We have also made use of a more realistic setup simulating Cherenkov light pulses with a blue Nichia NLPB320A LED onto an ET 9116A PMT. This is the 6-dynode bialkali PMT which has been selected for use in MAGIC [4]. The signal was amplified by a 1 GHz HP 02170 amplifier. After amplification the signal has a width of 2.3 ns. We have developed a pulse shaper to stretch this signal to 5 or 7 ns FWHM.

The above mentioned tests were repeated for this second setup. Some results are shown in figure 4. The pulses stretched to 5 ns exhibit a dynamical range around 100 which is comparable to the one measured for generator pulses. The width resolution ranging from 40% to 3% is comparable to that obtained for generator pulses. Due to a bad design the stretcher to 7 ns amplified too much the PMT electronic noise. This resulted in too high noise and a reduced dynamical range.

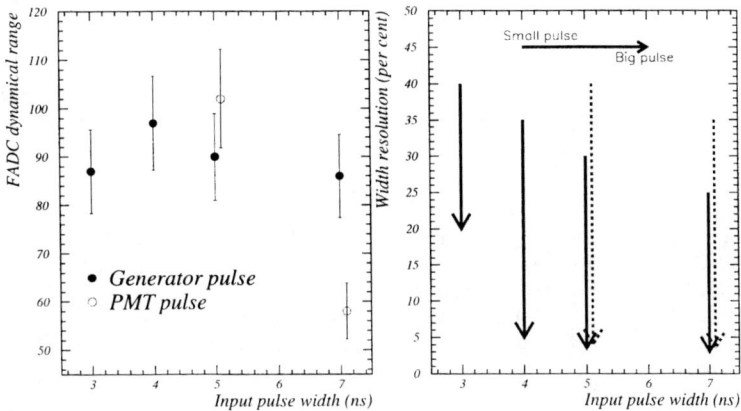

FIGURE 4. a) FADC dynamical range, and b) width resolution as a function of input pulse width for generator- (solid lines) and PMT- (hashed lines) produced input pulses.

We conclude that already pulses stretched to 5 ns width can be used to achieve a good shape reconstruction with an acceptable dynamical range around 90 in each branch. Charge resolution ranges from 33% for the smallest pulses to less than 5% for the largest ones. Further improvement in the dynamical range can be achieved by reducing the intrinsical noise of the FADCs. (In their present configuration this noise most probably stems from the power supplies switching noise.) These FADCs will be installed in the first phase of MAGIC with essentially the described features. In a second phase, the telescope will be provided with 1 GHz FADCs.

ACKNOWLEDGEMENTS

We kindly acknowledge the Siegen group for their assistance.

REFERENCES

1. Lorenz E., *these proceedings*.
2. Barrio J. A. et al, *The MAGIC Telescope Design Study* Munich (1998).
3. Lorenz E., Mirzoyan R. and Rose J., *these proceedings*.
4. Ostankov A. et al, *Proc. 2nd Conference on Photon Detection*, Beaune, 21-25 June 1999, to be published in NIM A.

The New Data Acquisition System of the First Telescope in HEGRA

J. Cortina[1], J.A. Barrio[2] and G. Rauterberg[3]
for the HEGRA collaboration

[1] *Max-Planck Institut für Physik, Föhringer Ring 6, 80805 München, Germany.*
[2] *Facultad de Ciencias Físicas. Universidad Complutense 28040 Madrid, Spain.*
[3] *Institut für Kernphysik, Universität Kiel, 24118 Kiel, Germany.*

Abstract. The first imaging Cherenkov telescope of the HEGRA collaboration (CT1) detects VHE gamma rays in stand-alone mode. Last year CT1 was equipped with new high-reflectivity mirrors and a next-neighbour trigger which have enabled reduction of the energy threshold to around 700 GeV. In addition completely new software for the data acquisition has been developed and installed in the past months. Here we report on the features of the new system and some of the tests which have been performed with it.

INTRODUCTION

CT1 is the first Cherenkov telescope of the HEGRA collaboration. It is located at the HEGRA site on the Canary island of La Palma and is equipped with a camera consisting of 127 0.25° diameter pixels. It has an equatorial mount and works in stand-alone mode with an energy threshold around 700 GeV. In the past two years CT1 has undergone a number of improvements in both its hardware and its software.

On the hardware side we have replaced the old mirrors with new ones provided with higher reflectivity and added a new ring of mirrors to the telescope disk. A new next-neighbour trigger has been installed which has allowed us to lower the accidental trigger rate and further decrease the energy threshold.

On the software side the old Macintosh-based data acquisition system has been upgraded to a state of the art PC-based DAQ running under Linux.

HARDWARE IMPROVEMENTS

CT1 was originally equipped with 18 round glass mirrors making up a total reflective area of 5 m^2. The resistance of the stand prevented any further increase in the mirror area using new glass mirrors. We found a way to overcome this limitation by using some newly developed diamond-turned light-weight aluminium mirrors [1,2].

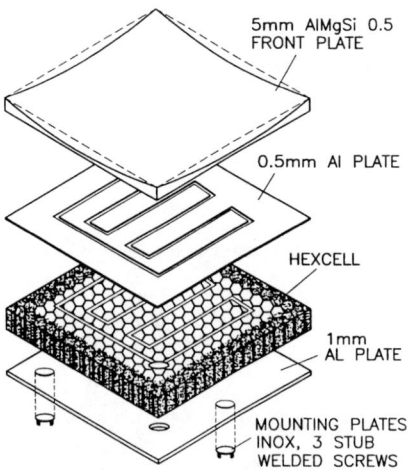

EXPLODED VIEW OF A MIRROR ELEMENT

FIGURE 1. An exploded view of the aluminium mirror inner structure. The actual mirrors in CT1 are of hexagonal shape.

Figure 1. shows an exploded view of one of the mirrors. It is are made up of a HEXCELL Al core, sandwiched between the reflecting Al front plate and an Al backplate. A resistive heating wire is integrated in order to prevent dew formation or ice deposit on the surface (the usefulness of the new heating was demonstrated already on the following winter on days of high humidity and intense icing which even managed to bend down some of the old mounts). The reflective surface is generated by diamond turning. A wet-formed anodic Al_2O_3 layer of 120 nm thickness protects the delicate aluminium surface. They were built with a hexagonal shape in order to optimally cover the disk. Each mirror weights 6 kg, has an area of 0.31 m^2, a mean reflectivity of 83-85% between 300 and 500 nm and focal point spread of 2 arc minutes. Figure 2 shows the spectral reflectivity after anodization. The light weight of the new mirror allowed us to add another ring of mirrors to the original disk. All in all the reflecting area doubled to a total of 10 m^2.

The old trigger was based on simple majority of any 2 out of 127 pixels within a 13 ns time window. Due to the night sky noise and large signals from ion feedback [4], the individual trigger threshold had to be set to the equivalent of 15 pe's in order to keep the accidental trigger low. We replaced this trigger for a next-neighbour logic one [5] which enables us to reduce the single pixel threshold to 12 pe's. This

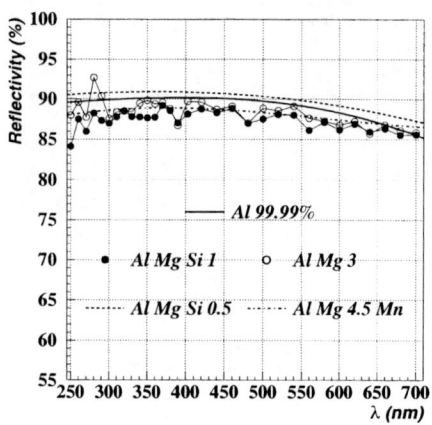

FIGURE 2. Spectral reflectivities for different alloys and surface protection materials. Al Mg 4.5 Mn is the alloy adopted for CT1.

in turn brings the trigger rate from 15 Hz (mostly accidental triggers) down to 3 Hz.

The enhanced mirror area along with the higher sensitivity results in a γ energy threshold reduction from 1.5 TeV to around 700 GeV. In winter 1998 we measured the Crab Nebula with a significance of 3.7 $\sigma/\sqrt{t}$ and a rate of 27 $\gamma/hour$. The time necessary for a 6 σ flux detection has been reduced from 76.1 hours to less than 4 hours [3].

SOFTWARE IMPROVEMENTS

The camera of CT1 consists of 127 EMI-9083A photomultipliers. The signals of these PMTs are directed to a central station where they are processed using CAMAC and NIM electronics: the charge is digitized in a 10 ns window using Le Croy 2249A ADCs; the next-neighbour trigger is implemented as a NIM module; and a number of other modules allow to monitor ADC currents, single trigger rates and general status of the telescope. The CAMAC electronic modules were in turn accessed by a Macintosh computer through two special crate controllers [6]. The system was slow, difficult to upgrade and had run into a good number of limitations. We decided to exchange it for a new concept based on a PC running under Linux.

Two new Wiener CC16 crate controllers were installed which are interfaced to a 300 MHz Pentium II using PC16-Turbo ISA cards. The PC runs under Linux 2.0.36. Linux is not a real-time operating system. Thus in order to minimize the deadtime imposed by the normal telescope operation we had to split the data acquisition system in a fast event builder and a slow user interface.

A new driver was developed for the CAMAC interface cards allowing very fast interrupt-driven acquisition of the pixel data. The dead time is well below 0.5%

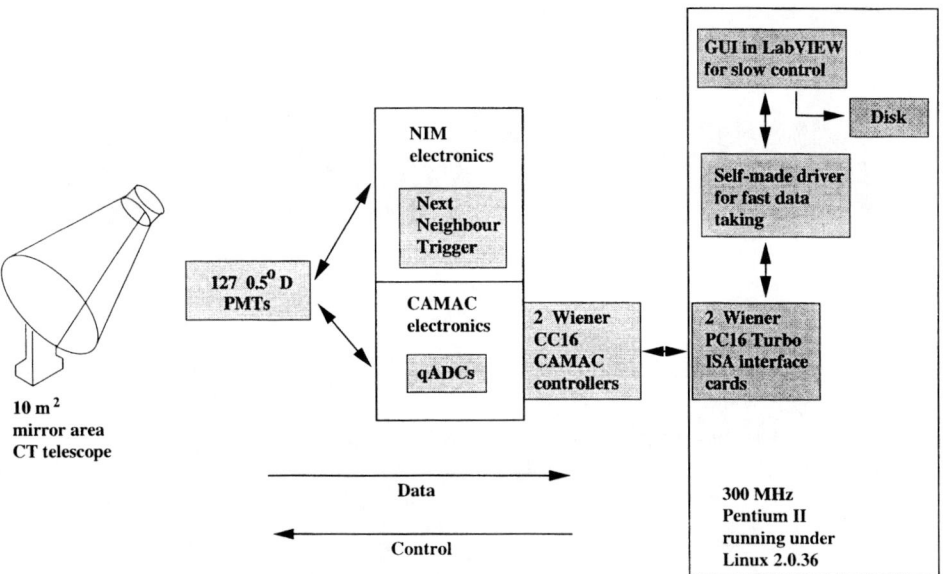

FIGURE 3. Diagram of the new data acquisition system.

and the system has proven to work in a stable way up to rates in the order of 50 Hz (more than 10 times our current data taking rate). The driver itself acts as an event builder acquiring events at interrupt time with less than 10 μs latency time and piping them to a slower control program through a memory FIFO. An overview of the system is displayed on figure 3.

The telescope control and interface to the user are realized in a program written in National Instruments LabVIEW programming language. LabVIEW is provided with multithreading capabilities which fit very well the requirements of our system and an easy to program graphical interface eviroment. The interface (see figure 4) enables the telescope operator to fully manage the different telescope functionalities and data taking. It saves telescope events and control data to hard disk. With respect to the old system the tracking has been optimized for an equatorial mount telescope. New features of the system include online run booking; full access to data, run books and program documentation via www; an interface to the xephem catalogues (90,000 sources) and, in the near future, online data analysis, an interface to a weather station and an autonomous GPS unit, a CCD-based sky transparency monitor and an autoguiding system.

ACKNOWLEDGEMENTS

This work is supported by the Spanish CICYT and the German BMBF. We gratefully acknowledge the IAC for the excellent working conditions at La Palma.

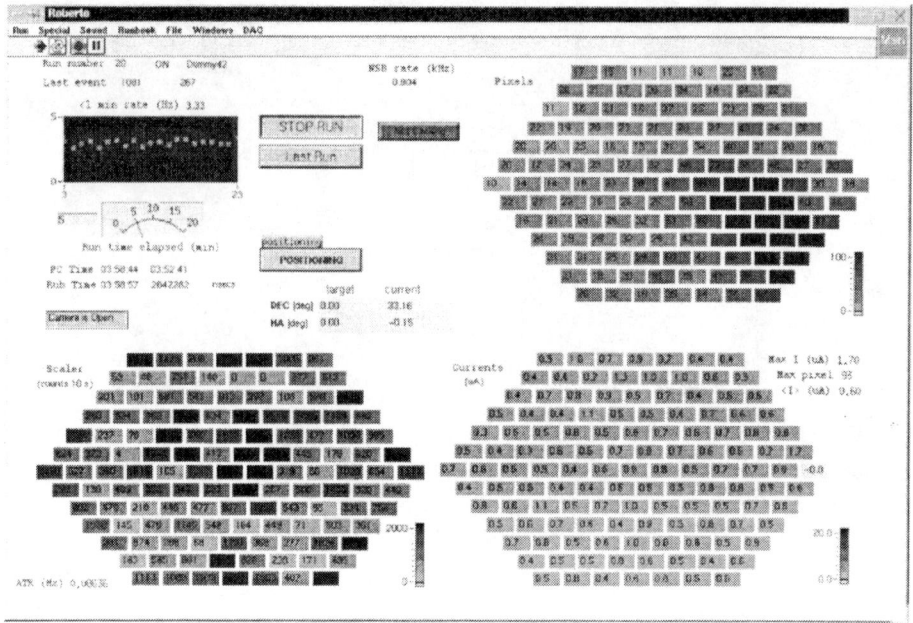

FIGURE 4. The Graphical User Interface allows easy monitoring of the telescope status and steering of its functionalities. This image is a snapshot of a real event, showing the values of the pixel ADCs for the event along with average currents, single pixel trigger rates and other telescope status information.

REFERENCES

1. Barrio J. A. et al, *Proc. Towards a Major Atmospheric Cherenkov Detector V*, 374 (1997).
2. Barrio J. A. et al, *The MAGIC Telescope Design Study* Munich (1998).
3. Kestel M. et al, *Proc. 16th. European Cosmic ray Symposium*, 515 (1998).
4. Mirzoyan R., Lorenz E., Petry D., Prosch C. *NIM A* **387** 74 (1997).
5. Bulian N. et al, *Astrop. Phys.*, **8** 223 (1998).
6. Rauterberg G. et al, *Proc. 24th ICRC* **3** 460 (1995).

GigaHertz Analogue Memories in Ground-based Gamma-Ray Astronomy

M. Punch[1], J.-P. Denance[2], P. Nayman[2], F. Toussenel[2], M. Rivoal[2],
J.-P. Tavernet[2], P. Goret[3], L.-M. Chounet[4], B. Degrange[4], P. Espigat[1],
P. Fleury[4], C. Renault[2], P. Vincent[2]

1. PCC & IN2P3/CNRS, Collège de France, 11 Place Marcelin Berthelot, 75231 Paris, France
2. LPNHE & IN2P3/CNRS, Universités Paris 6/7, Tour33 RdC, 4 Place Jussieu, 75252 Paris, France
3. Service d'Astrophysique, CEA-Saclay, Bât. 709, Orme des Merisiers, 91191 Gif-sur-Yvette, France
4. LPNHE & IN2P3/CNRS, Ecole Polytechnique, 91128 Palaiseau CEDEX, France

Abstract. The ARS0 ("Analogue Ring Sampler") GigaHertz Analogue Memory was developed for the ANTARES experiment by DAPNIA-SEI. Its application for an imaging Cherenkov camera is considered. This chip contains five analogue ring buffers with 128 cells each, with sampling on the nanosecond time-scale. After the arrival of a trigger signal, the analogue data may be read out at a slower rate, up to several hundred microseconds later. These characteristics allow a fully-integrated, compact imaging camera to be built based on this circuit. Results of tests on prototype circuit boards, for a possible application to the HESS project, are presented.

INTRODUCTION

The HESS project

The HESS experiment (1) is an international project for a system of stereoscopic Cherenkov telescopes, to be located in Namibia, in the Southern Hemisphere (23°S, 16°E, 1800 m a.s.l.). For the phase I, it will consist of four high-resolution imaging telescopes, each with a diameter of 12 m, surface area 80 m^2, a focal length of 15 m, and a camera consisting of ~850 photo-multipliers (gain 2×10^5), with 0.16° resolution and a field of view of 5°. Stereoscopic operation is foreseen for 2002.

The electronics requirements for such a "third-generation" atmospheric Cherenkov detector are quite strict. A large dynamic range is necessary since the γ–ray energies to be measured range over three orders of magnitude, in addition to the dynamic range of the images themselves. Good linearity is needed in order to characterize the images and to extract the γ–ray energy and impact parameter. The capability to see the single photoelectron peak is very useful for calibration of the electronics "downstream" of the photo-cathode. Since the system is auto-triggered, these characteristics must be maintained after a storage time of ~50 ns during which the trigger decision is being made. In order to preserve the signal quality and rapidity, the electronics should be situated in the camera, which necessitates that it be compact. Finally, given the large number of channels, a low cost per channel is desirable.

The ARS0 Analogue memory chip

The ARS0 was developed by DAPNIA-SEI for use in the ANTARES underwater neutrino detector (2). It is an analogue ring-sampling ASIC (application-specific integrated circuit), containing five channels, each of which acts as an analogue ring-buffer with 128 switched-capacitor cells. The sampling rate can be chosen from 300 MHz to 1.5 GHz; for our purposes clocking the ARS at 8 MHz chooses a 1 GHz rate. The dynamic range of each cell is 1000 and the linearity is ~9 bits, working on positive signals only (0–3 V). After the arrival of a trigger, sampling continues until the pointer is positioned a user-defined number of steps before the trigger arrival time. The required number of samples can then be read at a much lower rate (here 700 kHz).

APPLICATION OF THE ARS0 TO HESS

The overall schematic of the electronics (the "Analogue Memory Card") is shown in Figure 1. This card contains the analogue read-out channels coupled with a 12-bit ADC, together with the read-control system implemented using an ALTERA FPGA (field-programmable gate array), and some of the trigger logic.

Since the large dynamic range needed for the experiment cannot be obtained on a single channel of the ARS, it is necessary to use two channels per photo-multiplier (PM): high-gain (×-50) and low-gain (×-3) channels, with ranges of 1–100 γe and 16–1600 γe, respectively.

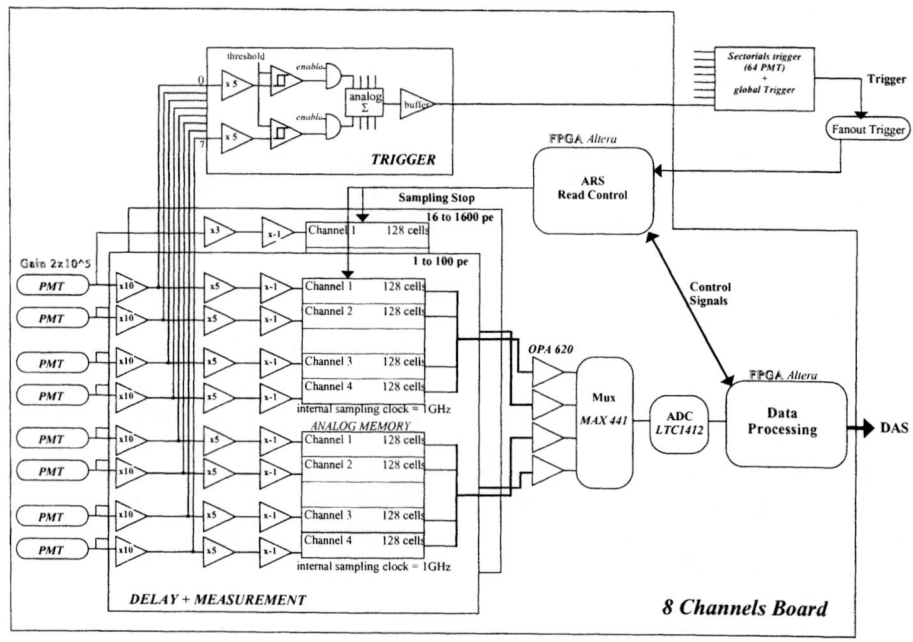

FIGURE 1. Schematic of the Electronics.

For reasons of modularity and implantation of the circuits, we have used four channels of each ARS (i.e., the fifth channel is unused) and four ARSs per Analogue Memory card, so each card can handle eight PMs. The four outputs of each ARS that are used are tied together, giving the equivalent of a 4 to 1 multiplexer. The outputs of the 4 ARSs are then multiplexed (with a MAX411 4 to 1 multiplexer) and input to the 12-bit ADC (LTC1412, 1.22 mV/d.c., conversion time 330 ns).

The signals in a channel can be summed in a "read-window" of 20 ns to give the integrated charge, the summation being performed in the FPGA. The read-out, conversion, and treatment time is 1 μs per cell, giving 320 μs per card, the cards being read-out in parallel. A test mode allowed in which all the channels in the read-window are sent.

Test Results

Tests were preformed using a prototype circuit, which was similar to that which would be used in the final detector, but for which the read-out is performed using a VME FIFO card read by a PC. For the tests, not only were all the ADC data read-out, but also the physical address in the ARS memory (7 bits) corresponding to the sample.

A LeCroy 9210 signal generator was used to simulate the pulse from a photo-multiplier (see Figure 2(a)). The pulse has a FWHM of 2.9 ns, an area of 8.59 nV.s, and a charge of 171.8 pC. For the PM gain that will be used, the single γe would give 32 fC. The pulse is passed through a wide-band Telonic attenuator which attenuates it in a range up to 110 dB, allowing pulses of varying numbers of γe to be simulated. Figure 2(b) shows the sampled signal in the high-gain channel for the read-window used. The form of the signal in the low-gain channel is practically identical.

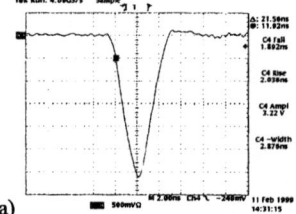
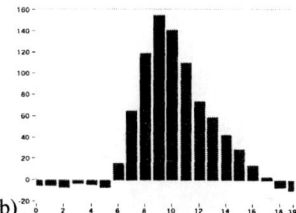

FIGURE 2. The form of a) the input test signal b) the output of the ARS for the high-gain channel.

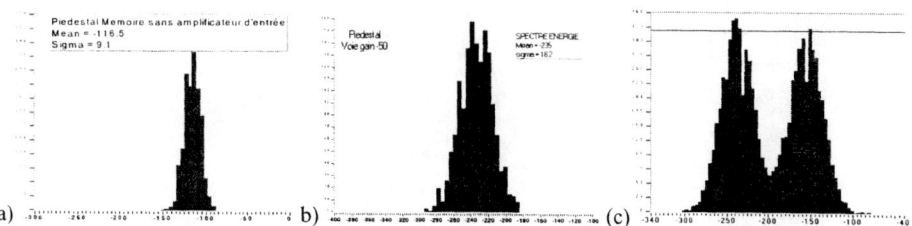

FIGURE 3. Pedestal measurements for a 20-sample read-window, (a) without and (b) with the input amplifier. (c) shows the single photoelectron spectrum along with the pedestal.

Figure 3 shows the pedestal measurement for the high-gain channel, both without and with the input-amplifier. Without the input-amplifier, we find $\sigma_{Ped} = 9.1$ d.c. for the 20-sample read-window. This is the noise of the memory itself. We can estimate the sigma for a cell: $\sigma_{Cell} = \sigma_{Ped}/\sqrt{20} \approx 2$ d.c. Given the ADC conversion factor, this gives the mean sigma per cell: $\sigma_{Cell} \approx 2.5$ mV, comparable to the value given in the ARS specifications. With the input amplifier, σ_{Ped} is measured to be 18.2 d.c. and 14.5 d.c. for the high and low-gain channels, respectively. It can be seen that the ARS and the input amplifiers introduce little noise. Figure 3(c) shows the spectrum of the signal equivalent to the single γe, with the pedestal, for the high-gain channel. With a Gaussian fit, we get a mean of 80 d.c. above the pedestal and a sigma of $\sigma_{\gamma e} = 20$ d.c., which gives a ratio of σ/Q of 25% in addition to that of the PM itself.

Figure 4 shows the linearity of the low and high-gain channels as a function of the number of γe. The useful ranges for the high and low-gain channels are up to 100 γe and 16–1600 γe, respectively. In these ranges, the non-linearities are low. The largest relative error is for the high-gain channel with a 5% error at 1 γe, or only 4 d.c. on 80.

Given the careful design of the printed circuit-board, the influence of one channel on another is dominated by the cross-talk within an analogue memory. To test the cross-talk, we injected a signal on channel 1 and measured the signals on the other three channels (whose amplifiers were not connected). The effect of the cross-talk is shown in Figure 5, in γe as a function of the number of γe input to channel 1. Note that the maximum of the cross-talk is 6 γe for an input of 1000 γe. The fall-off in the curves beyond this is due to saturation in the input channel. The existence of this

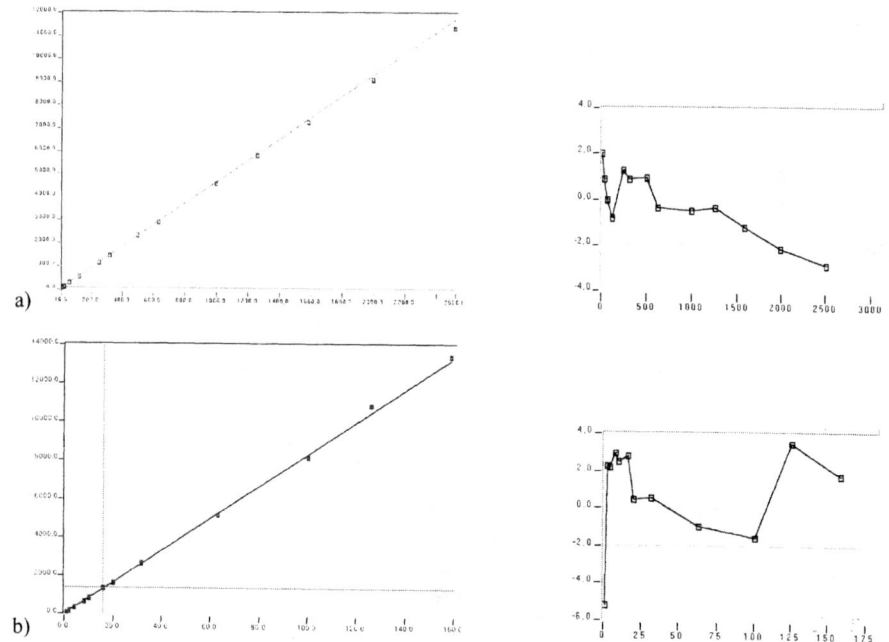

FIGURE 4. Linearity of (a) the low and (b) the high-gain channels, in γe, and their relative errors in %.

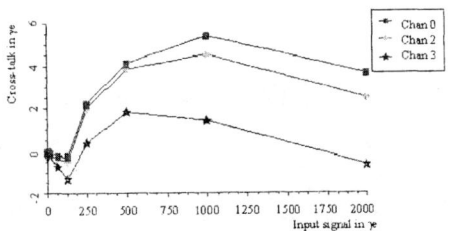

FIGURE 5. Cross-talk between the channels. The signal is input on channel 1.

cross-talk led us to put the high and low-gain channels on separate memory chips. The cross-talk between individual cells in adjacent channels is much greater, but since the pulse induced in the adjacent channel is bipolar, taking the sum over the cells in the read-window reduces it to the acceptable level shown here. However, it effectively precludes using the ARS as a multi-channel GigaHertz sampler.

The variation of the mean pedestal as a function of the temperature was measured to be 1 γe for a 5° temperature change. Therefore, the telescope's cameras must have good temperature control to within ~1°.

CONCLUSIONS

The Analogue Memory ARS0 appears to be well-adapted for use in ground-based γ-ray astronomy. It allows the construction of compact, integrated data-acquisition system, in which the analogue information is stored in camera with little degradation while the trigger is being formed, the digitization takes place in the camera, with only digital signals being sent out from camera. The dynamic range necessary for the analysis of Cherenkov events is achieved by using a dual-range system. The problem of the quantity of data to be sent and stored is solved with the use of an integrated adder. The signals can be sent from the camera using, for example, an ethernet connection, which avoids the problems associated with the transmission of analogue signals over long distances. It provides a relatively low-cost solution for the electronics of such a detector, since the majority of the electronics is custom-built and the number of electronics crates and cables is greatly reduced.

ACKNOWLEDGMENTS

We would like to thank François Darniaud & David Lachartre of DAPNIA-SEI, Saclay, for the use of a number of ARSs and for helpful discussions on their use, and Alain Castera for use of a test PC and programming of the LabView read-out system.

REFERENCES

1. Kohnle, A., et al., Proc. 26th ICRC, Utah, **5**, 239 (1999)
2. Feinstein, F., et al., Proc. 26th ICRC, Utah, **2**, 488 (1999)

A Cherenkov Camera with Integrated Electronics based on the "Smart Pixel" Concept

Norbert Bulian, Thomas Hirsch, Werner Hofmann, Thomas Kihm, Antje Kohnle, Michael Panter, and Michael Stein

Max-Planck-Institut für Kernphysik,
Heidelberg D-69029, Postfach 10 39 80, Germany

Abstract.
An option for the cameras of the HESS telescopes, the concept of a modular camera based on "Smart Pixels" was developed. A Smart Pixel contains the photomultiplier, the high voltage supply for the photomultiplier, a dual-gain sample-and-hold circuit with a 14 bit dynamic range, a time-to-voltage converter, a trigger discriminator, trigger logic to detect a coincidence of X=1...7 neighboring pixels, and an analog ratemeter. The Smart Pixels plug into a common backplane which provides power, communicates trigger signals between neighboring pixels, and holds a digital control bus as well as an analog bus for multiplexed readout of pixel signals. The performance of the Smart Pixels has been studied using a 19-pixel test camera.

INTRODUCTION

HESS is a next-generation array of imaging atmospheric Cherenkov telescopes (IACTs) to be built in the Southern Hemisphere in the Khomas Highland of Namibia. Details of the HESS project are given elsewhere in these proceedings (see Hofmann et al.). In its final stage, the HESS experiment should consist of 16 telescopes each with a $\sim$100 m^2 reflector and a high-resolution camera in the focal plane. The camera initially contains $\sim$700 photomultiplier pixels each viewing 0.16° of the sky (4.3° field-of-view), with a subsequent expansion to $\sim$850 photomultiplier pixels (5° field-of-view). The schedule foresees to start with scientific operation of the first 4-telescope subsystem in 2002 (Phase 1). The energy threshold of about 40 GeV, the angular resolution of less than a few arc minutes and the flux sensitivity of the full system above 100 GeV of a few 10^{-12} ph cm^{-2} s^{-1} should allow the observation of a significant number of objects and new source populations at GeV-TeV energies.

THE SMART PIXEL DESIGN

The Smart Pixel design was driven by the desire to

- Achieve a low energy threshold and good spectroscopic measurements by providing the shortest possible path for fast signals in order to allow short gate times, thereby minimizing night sky background noise and uncertainties due to timing jitter along the signal path.
- Use a highly modular design with multiple slow control and monitoring information, which allows the simple test and exchange of modules as well as a fast identification of faulty modules.
- Avoid a large number of long cables routing fast signals from the camera to an electronics unit at the base of the telescope.

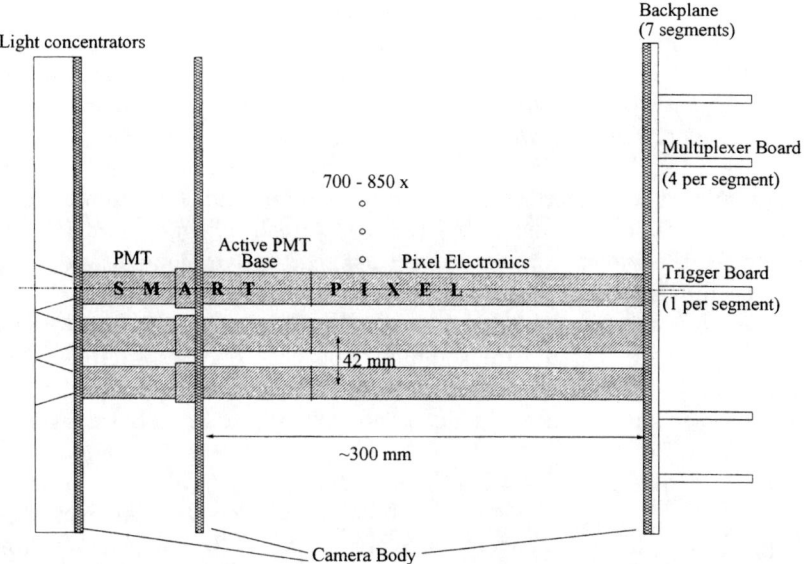

FIGURE 1. The layout of the Smart Pixel camera.

The Smart Pixel design incorporates analog signal processing electronics, the trigger, and slow control electronics in the camera (see Figure 1). A Smart Pixel contains the photomultiplier (PMT), the high voltage supply for the PMT, a dual-gain sample-and-hold circuit, a trigger discriminator, trigger logic to detect a coincidence of $X = 1\ldots 7$ neighboring pixels, a time-to-voltage converter for trigger timing, an I/V-converter for the PMT current, and an analog ratemeter. The Smart Pixels plug into a common backplane which provides power, communicates trigger signals between neighboring pixels, and holds a digital control bus as well as an analog bus for the multiplexed readout of the pixel signals. Multiplexer boards (1 per 32 pixels) for the multiplexing of the analog signals and trigger boards (1 per 127 pixels) for the dispatching of the trigger signal plug into the backside of the backplane.

The analog multiplexed outputs of each pixel include the PMT high- and low-gain outputs of the sample-and-hold, the PMT DC current, the mean trigger rate, two temperatures, supply and reference voltages, and the HV voltage and current signals. With the digital bus, it is possible in each pixel to set the multiplexer, the width and delay of the dual-gain sample-and-hold gate, the pedestal offsets, the X of the X/7 trigger, the range of the analog ratemeter, and the HV. Further functions include the enabling / disabling of the trigger and the HV as well as a global reset.

The camera is triggered if a $X/7 > q_0$ photoelectrons condition is met ($X=1\ldots 7$). The discriminator output of each pixel is fed into the microcontrollers of the pixel itself and the six neighboring pixels. The X/7 trigger fires if the discriminator output of the pixel itself and (X-1) of the six neighbor signals overlap. The entire camera is divided into seven segments with 100 to 127 pixels each, and for each segment the X/7 output signals of the pixels are ORed on a trigger board, dispatched to the trigger boards of the other segments, and then fanned out to all pixels. The time for the camera trigger formation is ~55 ns. The analog signal is delayed for this time with a thin coaxial cable bundled into each pixel, and then stored in the sample-and-hold. The trigger signal is also sent to the central station to form the global, multi-telescope coincidence trigger. If a camera trigger is not followed by a global trigger, the camera is reset after 2 μs, i.e., the sample-and-hold is cleared and the discriminator output is enabled.

Only about 50 cables are needed between the camera and a local electronics station. These include ~30 multiplexed analog outputs, one per group of 32 pixels, the digital bus lines, the camera trigger output, the global trigger input, an analog reset for the case that a camera trigger is not followed by a global trigger, a test input and a digital reset.

The camera readout is as follows: the pixel multiplexer of all pixels is set to the first value (the high-gain PMT signal). All multiplexers on the multiplexer boards (1 per 32 pixels) are then successively set to pixels 1,2,...,32. The pixel multiplexers of all pixels are then set to the second value, etc. With a readout frequency of 5 MHz, it takes 6.4 μs to read out a single analog value for all pixels. Typically, for each event one would read out the low- and high-gain PMT signal and the time-to-amplitude converter (TAC) giving the trigger timing. Thus, the readout time is ~20 μs. Assuming one ADC per 32 pixels, after readout the data are arranged in the ADC memory with words 0 to 31 as the high gain of 32 pixels, words 32 to 63 as the low gain of 32 pixels, words 64 to 95 as the TAC values of 32 pixels, and optional monitoring information following, if requested. Besides the event readout, the slow control and monitoring signals would be read out periodically.

The HV supply and PMT base are on a separate board behind each PMT built by industry according to our specifications. The board contains a DC-DC converter, allowing a DAC-controlled HV to be set for each individual pixel, and has voltage and current monitoring. The last three dynodes of the voltage divider network are actively stabilized, allowing a reduction in power consumption compared to conventional resistive dividers.

The total power consumption for a 850-pixel camera is 3.5 kW. The power consumption is 0.2 W/unit for the HV supply and 3.6 W/unit for the pixel electronics. Tests with a full-size model of a quadrant of the camera show that air cooling is sufficient to remove the heat and keep the pixels at constant temperature independant of ambient temperature. The Smart Pixels are inserted into tubes through which air is funneled via small holes in the backplane.

To test the performance of the Smart Pixels, a 19-pixel prototype camera was built up. A photo of the camera can be seen in Hofmann et al. in these proceedings. A custom-built VME-based sender module sends commands via the digital bus to the pixels. The readout frequency for the prototype is 2 MHz. A Hytec VTR2535T 12-bit 10 MHz transient recorder, synchronized with the sender clock, is used to digitize the signals. A pulse generator and a programmable 127 dB attenuator are used to simulate PMT pulses. The performance of the sample-and-hold (pulse response, linearity, noise, dynamic range, out-of-gate signal suppression, droop, fast clear), of the trigger electronics, (minimal discriminator threshold, X/7 trigger efficiency vs. overlap time, trigger timing), and of the slow control electronics (stability of the HV output, DC current linearity, trigger control, temperature readout) were investigated.

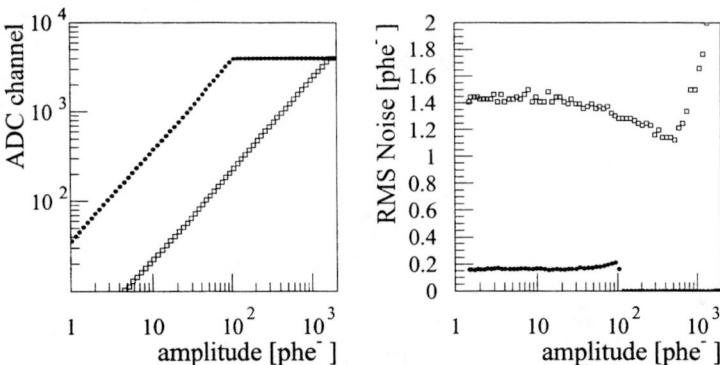

FIGURE 2. The left-hand plot shows the ADC channel as a function of the signal amplitude in photoelectrons (phe$^-$) for the high gain (filled circles) and the low gain (open squares) branch. The right-hand plot shows the rms noise as a function of signal amplitude.

A typical linearity curve is shown in Figure 2. The gate width of the sample-and-hold is 12 ns, the input pulse has a 4.5 ns width (FWHM). Typical pixel output signals for the high-gain branch are a gain of 19 mV/phe$^-$ and a range of 0 - 105 phe$^-$, and a gain of 1.2 mV/phe$^-$ and a range of 0 - 1700 phe$^-$ for the low-gain branch. The useful range is 0 - 2 V. As one can see from Figure 2, the electronics

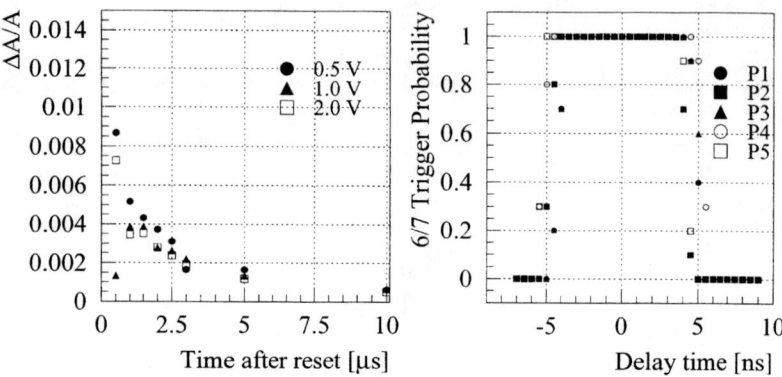

FIGURE 3. The left-hand plot shows the relative increase in amplitude as a function of the time between the reset and the next signal pulse for three output amplitudes. The right-hand plot shows the 6/7 trigger probability as a function of the relative delay time between signals for five neighbor pixels.

have single photoelectron resolution capability. The gate width of the sample-and-hold is adjustable from 6 - 18 ns in 3 ns steps. The time-to-amplitude converter measures the time between the discriminator firing and the camera trigger signal with a resolution of 25 mV/ns, and is used to verify that the gate is centered on the pulse.

The left-hand plot of Figure 3 shows the performance of the fast clear. Waiting an additional 0.5 μs after the reset before enabling the trigger for the next event implies a maximal amplitude increase due to the former signal of 1%. This implies an acceptable deadtime of 2.5 μs for local trigger events not followed by a global trigger. The right-hand plot of Figure 3 shows the performance of the X/7 trigger. The 6/7 trigger probability is measured while changing the relative timing of the input signals. The discriminator output is 8 ns long. There are slight ≤1 ns timing differences between the pixels. A 2.5 ns minimal overlap is necessary for the X/7 trigger to fire.

CONCLUSIONS

The performance of the prototype camera has shown the feasibility of the Smart Pixel design for the HESS project. Tests of the prototype camera with PMTs will be performed. The total cost of the camera electronics is estimated to DM 440,- per pixel, including the cost of components, PCB, and mounting of components, for the HV supply, the pixel electronics, the backplane, multiplexer and trigger boards. For a ~850 pixel camera, the total cost is 370 kDM, not including the photomultipliers, the camera mechanics, the power supplies, and the data acquisition electronics.

An Optical Reflector System for the CANGAROO-II Telescope

Akiko Kawachi for the CANGAROO Collaboration[1]

Institute for Cosmic Ray Research, University of Tokyo Tanashi, Tokyo 188-8502, Japan[2]

Abstract. We have developed light and durable mirrors made of CFRP (Carbon Fiber Reinforced Plastics) laminates for the reflector of the new CANGAROO-II 7 m telescope. The reflector has a parabolic shape (F/1.1) with a 30 m^2 effective area which consists of 60 small spherical mirrors. The attitude of each mirror can be remotely adjusted by stepping motors. After the first adjustment work, the reflector offers a point image of about 0°.14 (FWHM) on the optic axis. The telescope has been in operation since May 1999 with an energy threshold of ~300 GeV.

INTRODUCTION

The CANGAROO collaboration has started operation of a new imaging Cherenkov telescope of 7 m reflector. With an effective light collecting area of 30 m^2, it enables us to observe very high-energy gamma-rays of the southern sky with a 300 GeV energy threshold [1,2]. The telescope is as well a prototype of the project of an array of four 10 m telescopes, CANGAROO-III [3].

MECHANICAL DESIGN

The reflector is an F/1.1 paraboloid with a diameter of 7 m. In order to use timing information to reject night sky background, we chose a parabolic shape which provides isochronous collection of photons. Sixty spherical mirrors, each of which has an 80 cm diameter and a curvature radius between 16–17 m, were arranged according to their curvature radii from the inner to the outer sections of the reflector, with the shorter focal length mirrors innermost. In the prime focal plane, there is a multi-pixel camera with 0.°12 pitch covering about 3° of FOV.

The design of the supporting frame of the reflector was based on a commercially available communications antenna, and the frame is mounted by 9 honeycomb panels. Several mirrors (6–9) were installed onto a honeycomb panel and the alignment

---

[1] see [3] for a complete name list of the collaboration.
[2] E-mail: kawachi@icrr.u-tokyo.ac.jp

of the mirrors in each panel was roughly adjusted ($\lesssim 0.°3$) with a laser beam before the shipping to Australia. With these adjustments, we checked our remote adjustment system as well as it saved on-site labor. The total weight of the dish including the mirrors and adjustment system is very light; only 4.3 ton (6.3 ton when the camera stays are included), so that we could greatly reduce cost and assembling labor. The structure was designed to be operated at the average velocity 30 km/hr, operational up to 100 km/hr of the wind load. By tracking various stars at 12–85 degrees in the elevation angle and at all the azimuthal angles, gravitational deformations of the structure were measured to cause less than $1'$ of deviation in the pointing accuracy at the focal plane

The present support frame allows us to extend the reflector up to 10 m diameter with additional 54 mirrors, and the extension is to be completed by early 2000.

SMALL SPHERICAL MIRRORS

The small spherical mirror is of 80 cm in diameter, of about 2 cm thick, and weighs only 5.5 kg. Sheets of CFRP and adhesives were laid on a metal mold, sandwiching a core of low density, high shear strength foam to avoid twisting deformations, and a polymer sheet coated with laminated aluminum was applied on the top of the layers as a reflecting material. The laminates were vacuum bagged and cured to 120°C in an autoclave pressure vessel.

For protection against dust, rain, and sunshine, the mirror surface was coated with fluoride. The reflectivity is over 80 % at 340 nm–800 nm, falls off rather slowly down to 40 % at 250 nm. We found dusts on the surface deteriorate the reflectivity to about 75 % after several months, however, it was confirmed with a sample in a year-time-scale that the reflectivity repeatedly recovers easily by water washing. The mirror surface is free from dewing until the relative humidity exceeds 83 %.

The curvature radii of the mirrors show a flat distribution between 15.9–17.1 m, with an average of 16.45 m. The mirrors were arranged on the support according to their radii to make a smooth $f=8$ m paraboloid, and the individual facets were adjusted toward the focal point by the method described later. The image size of each mirror was measured with a light source 5.8 km away. A typical size is $0.°08$ (FWHM), and 50 % of the photons concentrates within $\sim 0.°1\ \phi$.

REMOTE ADJUSTMENT SYSTEM

Two watertight stepping motors are installed at the back surface of each mirror, and the attitude of a mirror can be remotely adjusted in two perpendicular directions. The minimum step size corresponds to about 1×10^{-4} degree at the focal plane, and adjustable up to ± 3-degree. The accuracy of 1×10^{-3} degree is retained when motors are switched off. All mirrors are adjusted one by one using two motor drivers with relay switches controlled by a computer.

Our adjustment work on-site used a distant light source at night. All small mirrors were covered but one with plastic lids, and its image on a screen at the focal plane was monitored by a CCD camera installed at the center of the reflector. The attitudes of the mirrors were adjusted by moving stepping motors using feedback information from CCD images so that the image center should lay at the focal point.

As a result of the first adjustment work, the deviation of the small mirror axis orientations is $\sim 0°.03$ on average, a larger value than expected. The deviation was mainly caused by temporal fluctuations of the CCD camera geometry over different nights we applied the adjustment. Removing this effect, it is estimated that the orientations can be adjusted within an error of $0°.01$.

PERFORMANCES OF THE REFLECTOR

The optical property of the reflector in total was measured using images of several stars tracked by the telescope. Images on the focal plane screen were taken by a CCD camera at the reflector center.

A On the Optical Axis

In Figure 1, an image of Sirius on the optic axis is shown in units of CCD pixels. A pixel corresponds to 6.7×10^{-3} degree. One pixel of the camera ($0°.12$ square)

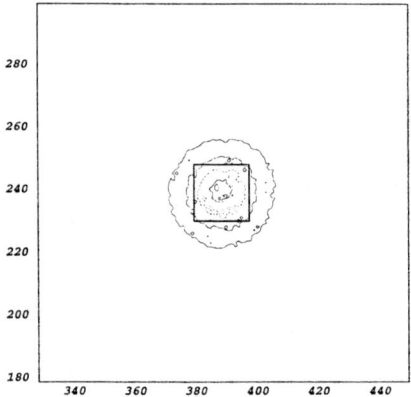

FIGURE 1. A CCD image of Sirius on the optic axis. The axes are in unit of CCD pixels, corresponding to a 6.7×10^{-3} degree pitch. z axis is in arbitrary unit. A square overdrawn is a scale of a camera pixel ($0°.12$ square).

is overdrawn for scaling. An image size of 0°.14±0°.01 (FWHM) is deduced, and 30±4 % of the photons concentrates in a single camera pixel. The characteristic difference in size between the images of gamma-rays and protons is on the scale of 0.°1–0.°2. Our optical quality meet the requirement to start with, though there is room for improvement. A preliminary analysis of the CANGAROO-II data shows thin muon rings whose average width is ∼0.°11 [4]. Contribution of mirror aberrations to broadening the images of rings can be estimated as comparable to that of multiple scatterings in the atmosphere [5].

B Off the Optical Axis

Parabolic mirrors have greater off-axis aberrations. The aberration is rather serious for a Cherenkov imaging telescope since a relatively wide field of view (∼3°)

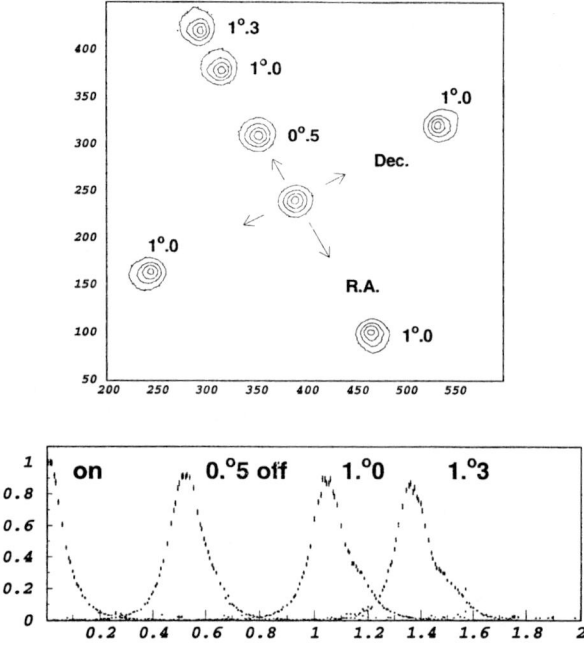

FIGURE 2. *top*) A synthesized figure of the CCD images obtained by displacing the pointing coordinates of a star. *bottom*) Radial point spread functions at the different pointing coordinates; (α, δ), $(\alpha-0.°5, \delta)$ $(\alpha-1.°0, \delta)$ and $(\alpha-1.°3, \delta)$, respectively. Vertical scales are normalized by the peak height of the on-axis psf.

is needed for image analyses of atmospheric showers. We compared off-axis images of Sirius displacing the pointing coordinates both in right ascension and in declination. As a result of symmetric configuration and alignments, displacement to all the directions caused deformations consistent with each other (Figure 2 *top*)). Figure 2 *bottom*) shows radial point spread functions of the Sirius pointed on and off in right ascension angles by $0.°5$, $1.°0$ and $1.°3$, respectively, We can estimate concentration decrease by about 18 % at the edge of the FOV.

C   Gravitational Deformation

Since the adjustment work of the small mirrors was performed pointing horizontally, possible deflections of the alignment system at other attitudes had to be calibrated. The effect of gravitational deformations on the reflector was measured by comparing the images of stars taken at various elevation angles of the telescope. For elevation angles between 15–70 degrees, the images show no dependence on the elevations either in shape or in size. Thus the deformations are confirmed to be negligible within an error of $0.°01$.

SUMMARY

The new CANGAROO-II 7 m telescope has been completed and operations has begun. The reflector, F/1.1 paraboloid, has a point spread function of $0.°14$ (FWHM) over 3 degrees of FOV, with $\sim$20 % loss of light concentration by aberration at the FOV edge.

ACKNOWLEDGMENTS

The small mirrors of CFRP laminates have been developed in collaboration with Mitsubishi Electric Corporation, Communication Systems Center. AK was supported for this work by a JSPS postdoctoral fellowship.

REFERENCES

1. Tanimori T. et al., *Proc. 26th ICRC* (Salt Lake City, Utah, USA), OG4.3.04 (1999).
2. Kubo H. et al., in these proceedings (1999).
3. Mori M. et al., in these proceedings (1999).
4. Okumura K. et al., in preparation.
5. Vacanti G. et al., *Astropart. Phys.*, **2**, 1-11 (1994).

A study of the effect of polarizing filters in Imaging Cherenkov Telescopes

I. de la Calle[1], J.L.Contreras[1], V.Fonseca[1]

[1] *Facultad de Ciencias Físicas. Universidad Complutense 28040 Madrid, Spain.*

Abstract. In this contribution we study the net polarization of the Cherenkov Light emitted in Extensive Air Showers of energy around 1 TeV, by means of an Air Shower Monte Carlo simulation. The study investigates the modification of the Hillas image parameters and telescope performance due to a particular arrangement of polarizing filters in front of the camera of a Cherenkov Telescope.

INTRODUCTION

Cherenkov light emitted in Extensive Air Showers shows a net linear polarization pointing towards the shower core, in what we will call *radial direction*. Its origin can be easily understood in the limiting case of a shower with no transverse dimensions. Effects that deviate from this ideal case (multiple scattering, short distances to the core, transverse momentum in hadronic interactions, etc) reduce the degree of polarization. The situation is illustrated in figure 1.

The optical properties of Imaging Atmospheric Cherenkov Telescopes , IACTs, imply that the *radial* polarization on the ground translates into a net *radial* polarization on the camera of an IACT, but only for showers coming from the source direction. This fact has led to the proposal of an scheme to take advantage from this situation (Contreras, Calle, Cortina 1998). It consists in placing polarizers on each photo-multiplier with their transmitting axis oriented towards the center of the camera.

With this setup we expect an *a priori* reduction of 50% in the Light of Night Sky background (LONS), an important reduction of the light yield of proton showers ($\approx$ 45% from figure 1) and a smaller one for γ showers ($\approx$ 30%). Light from the higher part of the shower, where the transverse dimensions of the shower is smaller and the directions of the emitting particles follow more closely the shower axis, should posses a higher degree of polarization. Finally, it can be argued that since light arriving from the shower axis corresponds to the peak of the light pulse (at radii smaller than the *hump*), it defines the shower front. This light also

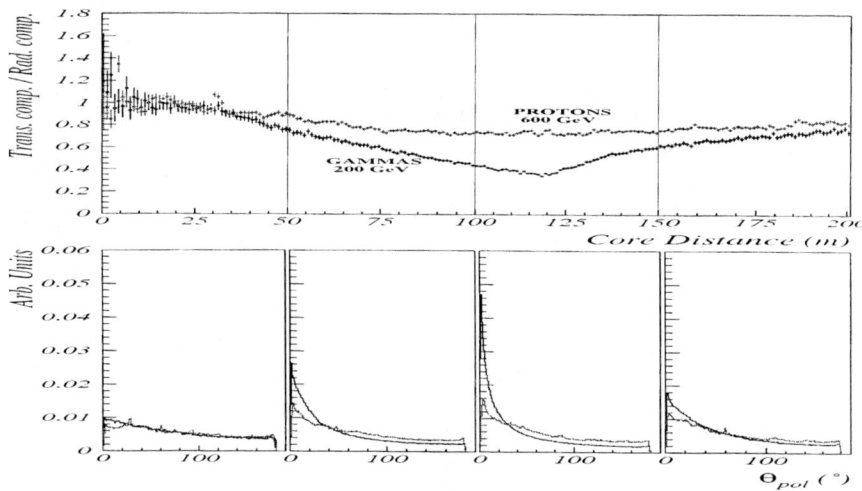

FIGURE 1. Upper plots : Ratio of the transverse to radial components squared of the polarization vector on ground. Bottom plots: Distribution of the angle between the polarization vector and the radial direction. The plots are the result of superimposing data corresponding to 10 showers at normal incidence for γs and 20 showers uniformly distributed in (0^0 - 3^0) for protons. The radial direction is defined as the line joining the impact point of the photon with the shower core.

shows the highest degree of polarization and therefore, selecting it with a polarizer should lead to narrower time pulses. The two last effects can be seen in the plots of figure 2.

SIMULATION CHARACTERISTICS

To study the proposed setup we have used a Monte Carlo simulation based in Corsika 5.6 (Capdevielle et al. 1992) with minor modifications to track the polarization of the Cherenkov photons. We have generated γ showers of energies in the range [200 GeV - 1.6 TeV], in discrete energy points spaced by 100 GeV, and proton showers with energies in the range [600 GeV - 4.8 TeV] in steps of 300 GeV. The difference in energy between both sets was meant to get similar amounts of Cherenkov light for both kinds of events at $E_{proton} = 3 \times E_\gamma$. Gammas were generated at vertical incidence while protons were generated uniformly in solid angle within 3^0 of the zenith

For each energy point and species a total of 100 showers was generated. The image of each showers was obtained on many IACTs distributed on the ground, inside a circle of 250 meters around the shower core. This procedure lead to many events highly correlated, whose results will be accurate with respect to sampling fluctuations, but not in regard to shower to shower fluctuations.

The characteristics of the IACTs simulated were taken from the HEGRA CT system [4], with drastic simplifications: the telescope mirror was approximated by a parabola, for the ray tracing simulation. The instrument efficiency was represented by a global factor, which takes into account several effects such as atmospheric absorption, mirror's reflectivity, quantum efficiency of the PMs, etc. Polarizer effect were taken into account with perfect polarizers i.e. no absorption in them. Each event was simulated with and without the polarizing filters in front of the camera for comparison.

The definition of the trigger threshold is important, since it affects the effective area. The condition used for the case without polarizing filters was to demand two neighbor pixels with more than 10 photoelectrons, out of the total 271 pixels of the camera [7]. For the polarizers case, the demand to keep the same (within 30%) probability of random triggers led to a the requirement of 8 phe/pixel. The reduction of the trigger condition is possible due to the cut of LONS induced by the polarizers.

Finally, images were analyzed using a standard second moment analysis and values of the *super-cuts* in agreement with those used in the HEGRA collaboration analysis. The same set of super-cuts was used in the *polarizers* and *no polarizers* cases.

RESULTS

In the study of the performance of the *polarizers* setup we have considered two different scenarios: observations during a *dark night*, with a low LONS, and high

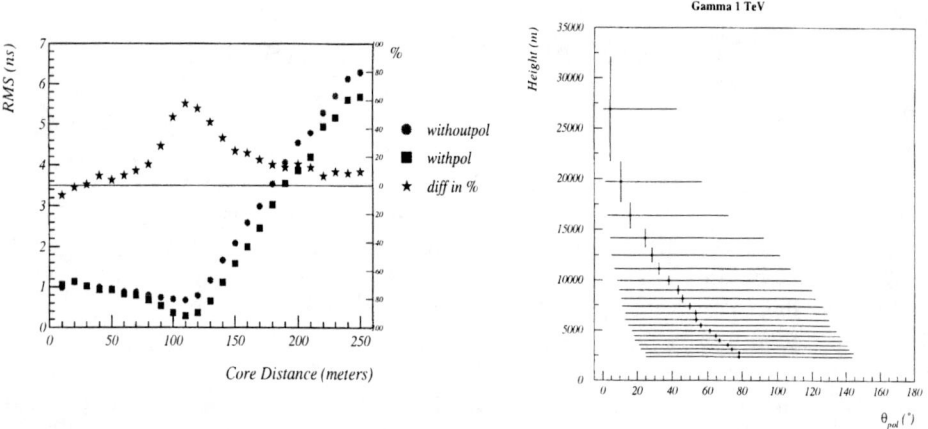

FIGURE 2. Left hand side: fall time of the Cherenkov light pulse (time interval containing 34% of the photons after the peak) for a 1 TeV γ event, as a function of the distance to the core. Right hand side: mean value of the polarization angle in bins of one radiation length, for 10 merged 1 TeV γ events with the first interaction point fixed at 35 Km a.s.l.

noise observations, as detailed afterwards.

For the low noise case we have used the value of the LONS measured at the HEGRA site at La Palma (Mirzoyan and Lorenz 1994), typical of good observation site and conditions. In this situation we can summarize our results by saying that the introduction of the polarizing filters leads to:

- Effective collection areas that are always smaller, both for γ and proton showers, even taking into account the reduced trigger threshold. Nevertheless the loss of effective area is bigger for proton showers (around 25%) than for γ showers (around 15%). So, the signal over noise ratio $\frac{S}{N}$ is improved while the significance, $\frac{S}{\sqrt{N}}$ decreases slightly.

- For images which survive the trigger condition the loss of light, as estimated by the parameter *size*, is slightly above 30% for γs and around 40% for protons.

- More compact shower images, reflecting the loss of light from the periphery of the image. The difference being around 10% - 20% depending on the energy.

- A small change, in the order of 10% - 20%, of the quality factor derived from a standard analysis. The *polarizers* setup tends to improve the quality factor for high energies and decrease it for lower energies. These results are affected by big uncertainties arising from the Monte Carlo statistics.

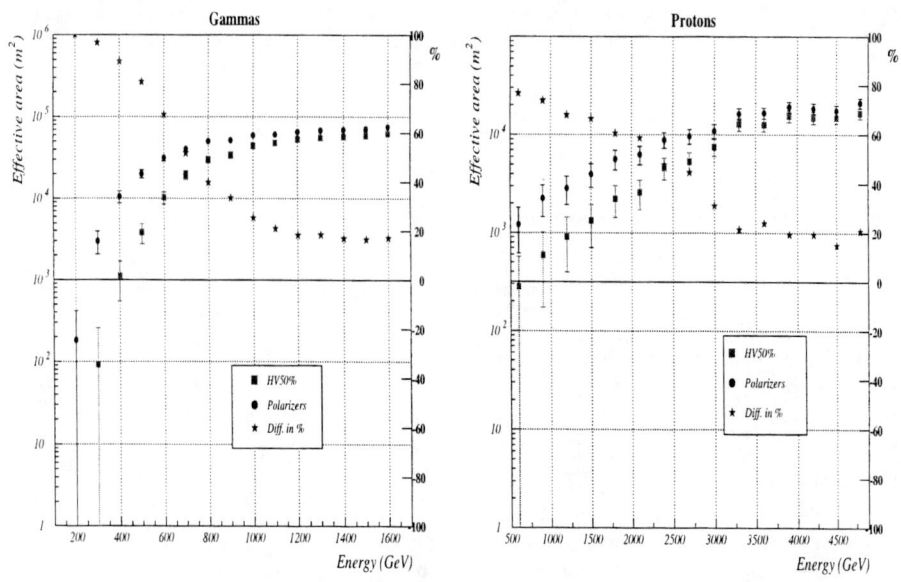

FIGURE 3. Effective areas for γs and protons for the case with high LONS and two different alternatives: reduced PM gain by 50% and use of polarizing filters

The results can be resumed by saying that in good observation conditions the gain (or loss) from the introduction of the polarizing filters is small or nonexistent. An optimization of the image analysis for this case and the use of narrower temporal gates (made possible by the reduction in the length of the light pulses) could improve this conclusion. On the other hand , the consideration of real polarizers (as opposed to our 100% efficiency approximation) would surely degrade it.

In the second situation studied the level of LONS was multiplied by a factor of 10, representing worse observing conditions, which could typically appear in observations with a not too high level of moon light, see for example (Kranich 1998). The usual procedure to deal with this level of noise is to reduce the gain of the Photo-Multipliers, PMs , of the camera by lowering the voltage. We have thus compared the effect of a reduction of 50% in the gain of the PMs with the introduction of polarizing filters, both setups having the same net effect on LONS. The trigger threshold condition was adjusted to get the same random probability as in the *normal* case. The results obtained can be seen in figure 3.

The outcome of comparing both alternatives is that the use of polarizers increases the effective area by a factor of (20% -100%) with respect to the plain reduction of PM gain, both for γs and for protons. The gain is more important for γs than for protons, increasing therefore both $\frac{S}{N}$ and $\frac{S}{\sqrt{N}}$.

With the caveat of the approximations used (mainly the polarizer transparency) we conclude that the use of polarizers can help to improve the performance of IACTs in high noise situations. It must be remembered that moon observations are not just an example, but a logical extension of the IACT technique, needed to improve their duty cycle.

ACKNOWLEDGEMENTS

This work is supported by the Spanish CICYT. We gratefully acknowledge many members of the HEGRA collaboration for useful comments and discussions

REFERENCES

1. A.M. Hillas *Space Science Reviews* **75** vol 1, pp 17-30 (1996)
2. J.L. Contreras, I. de la Calle, J. Cortina
   Proceedings16th European Cosmic Ray Symposium pp 467-470
3. R. Mirzoyan & E. Lorentz MPI-PhE/94-35 , December 1994
4. Daum et al. Astrop. Phys. 1997, 8 1-11 (1997).
5. D. Kranich et al. Astrop. Phys. 12 1,2 pp 65-74. Oct 99
6. J.N. Capdevielle et al., *KfK Report 4998* (1992).
7. N. Bulian et al., Astrop. Phys 8, 223-233 (1998).

Sensing Atmospheric Conditions using MIR Radiometers

P. M. Chadwick, K. Lyons, T. J. L. McComb, K. J. Orford,
J. L. Osborne, S. M. Rayner, S. E. Shaw, and K. E. Turver

Department of Physics, Rochester Building, Science Laboratories, University of Durham, Durham, DH1 3LE, U.K.

Abstract. MIR (8 − 14μm) radiometers have been used to sense the temperature of the sky as a measure of cloud and other obscurants in the field of an atmospheric Čerenkov telescope. Comparisons between the MIR signature and the performance of the Mark 6 gamma ray telescope are reported.

INTRODUCTION

We report a technique involving measuring the radiative temperature of the night sky in order to sense the presence of cloud and other obscurants against the cold clear sky. The method is quick, sensitive and independent of the performance of an atmospheric Čerenkov telescope. A more detailed account appears elsewhere [1].

Measurements of the sky temperature have been made with mid infra red (MIR) radiometers (Heimann model KT 17 and KT 19[1]) sensitive in the 8 − 14 μm waveband which coincides with an atmospheric window. The radiometers have a temperature range −75°C − +100°C, a temperature resolution of ±0.2°C, and an aperture of 2° defined by a germanium lens. The radiometers were calibrated using black body sources whose temperature and emission characteristics were known.

Measurements were made with radiometers mounted on the University of Durham Mark 6 gamma ray telescope [2] in Narrabri, NSW, Australia and on steerable alt-azimuth platforms and fixed mounts in Durham, England. Data from measurements made at various angles to the zenith, throughout the day and night during different seasons over a period of years are available. Cloud base measurements were made in 1998 using a LIDAR ceilometer for a subset of the data. Full meteorological data are available for all observations.

---

[1] Manufactured by Heitronics Infrarot Messtechnik Gmbh., Weisbaden, Germany.

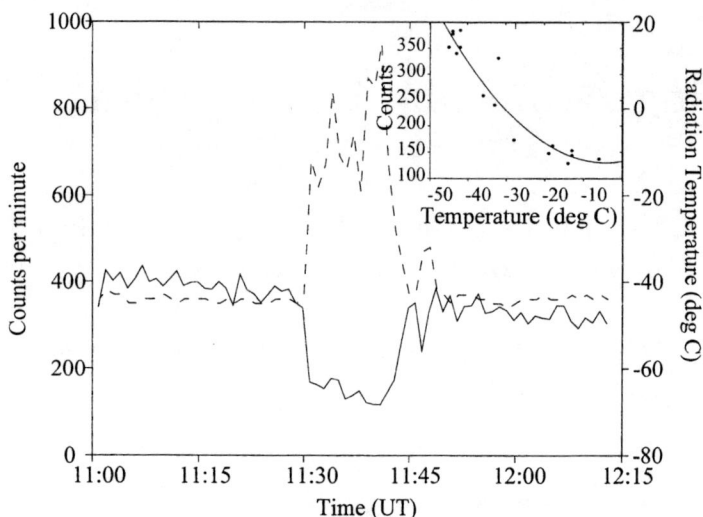

FIGURE 1. The correlation between the background counting rate of the Mark 6 gamma ray telescope (solid line) and the radiative temperature of the sky (broken line). The main figure shows the variation of the telescope counting rate and radiative temperature with time. The inset figure shows the variation of the count rate with measured sky radiative temperature.

RESULTS

We show in Figure 1 both the time variation of telescope cosmic ray counting rate and sky radiative temperature (main figure) and the correlation between the background count rate of the telescope and the radiative temperature of the sky (inset figure) during an observation using the Mark 6 telescope at Narrabri. Under clear sky conditions the short term variations in telescope counting rate are consistent with Poissonian counting statistics. Determinations of the sky radiometric temperature have a standard error of about 1° which is constant for all sky temperatures measured. The sky temperature is seen to be very sensitive to the presence of the warmer clouds and correlates inversely with the background count rate of the telescope. For example, the data in Figure 1 show that a count rate under clear skies of 400 ± 20 cpm reduces under cloud to 200 ± 14 cpm. The corresponding increase in sky temperature is from $-40° \pm 1°C$ to $0° \pm 1°C$. The MIR-measured temperature thus presents an instantaneous, sensitive, significant and independent monitor of the clarity of the sky in the beam of the gamma ray telescope.

We show in Figure 2 the measured temperature of a clear sky using the Narrabri radiometer as a function of zenith distance of observation in the range $0° - 70°$. The plot is based on data from observations during two nights when there was a small difference in air temperature. The radiative temperature of the clear sky increases as the zenith distance (θ) is increased and the measurement is made through an increasing slant thickness of warm atmosphere.

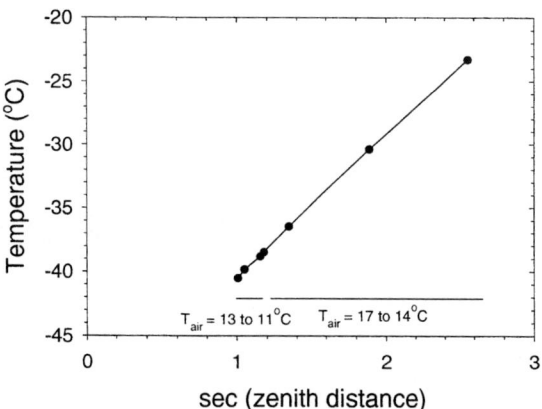

FIGURE 2. The variation in the clear sky radiometric temperature with zenith distance from 0° to 67°. The data were obtained on two nights and there is a small variation in the air temperature between the observations at zenith distance < 30° and > 30°, as shown in the figure.

The profile of the sky temperature (measured at the zenith near Durham, England) during a typical 24 hour period during which the sky was clear is shown in Figure 3. The air temperature at screen height is also plotted and it is noted that the vertical clear sky temperature tracks the air temperature at screen height. Such data taken over a long period are the basis for a simple phenomenological model linking the observed sky radiometric temperature with the air temperature and humidity at screen height.

Figure 4 shows measurements of the sky temperature at the zenith and 60° to the zenith during a 24 hour period when clear skies were replaced by total overcast clouds with a base at 600 m at about 1200 UTC. The temperature of the clear sky is seen to be about 40°C cooler than that of the low overcast cloud which is, as expected, very similar to the ground level air temperature.

The temperatures of clouds with bases at different heights were measured at the zenith using a radiometer aligned with a LIDAR ceilometer — see Figure 5. The radiometer has an aperture of about 2° and on occasion a part of the cold clear sky is viewed when the cloud base indicated by the narrow angle LIDAR is low. This is the origin of some spread of the data in Figure 5. An increase in cloud base height from 310 to 2170 m corresponds to a decrease in measured cloud temperature from +10°C to 0°C.

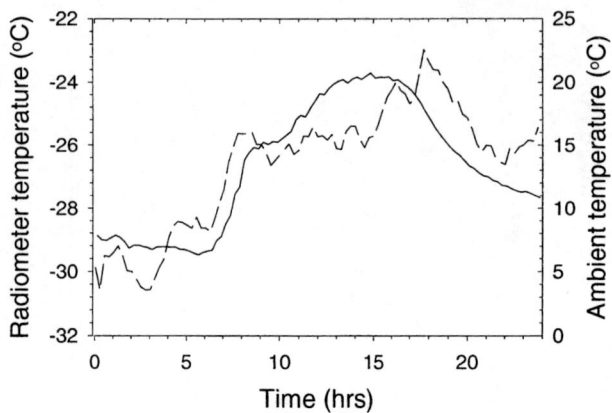

FIGURE 3. The variation in the clear sky radiometric temperature (solid line) over a 24 hr period. The profile of the air temperature at screen height (broken line) is also shown.

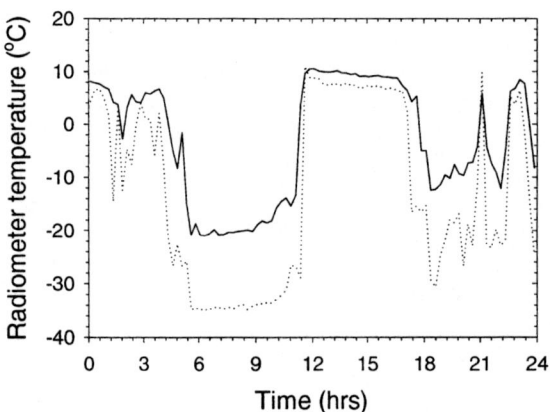

FIGURE 4. The sky radiative temperatures at the zenith (broken line) and at 60° to the zenith (solid line) in intervals of cloud and clear sky during a 24 hour period.

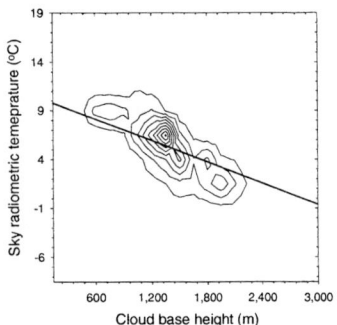

FIGURE 5. The measured cloud radiative temperature plotted against cloud base height. The contours (which have linear spacing) specify the frequency of observations of a given cloud radiative temperature and base height combination. The maximum contour corresponds to 180 observations with that combination of cloud base height and radiative temperature and the lowest to 20 observations. The line represents the model prediction — see Buckley et al. [1].

SUMMARY

The use of a MIR radiometer sensitive to the 8 – 14 μm waveband mounted on a ground based gamma ray telescope provides a simple and inexpensive method of reliably monitoring the atmospheric clarity in the telescope's field of view. The response of the radiometer is significant, prompt, strongly correlated with the background counting rate of air Čerenkov experiments, and independent of the performance of the telescope itself.

We are grateful to the UK Particle Physics and Astronomy Research Council for support of the gamma ray project and Muir Matheson Limited for the loan of the LIDAR system. Observations made in Northeast England were part of a program conducted by the Industrial Research Laboratories, University of Durham on behalf of the Science and Technology Division, Defence Clothing and Textiles Agency. These data are reproduced with the permission of the Director, Science and Technology Division, Defence Clothing and Textiles Agency.

REFERENCES

1. Buckley, D. J. et al., *Exp. Astron.*, in press.
2. Armstrong, P., et al., *Exp. Astron.*, **9**, 51 (1999).

OTHER TeV OBSERVATORIES

STACEE-32: Design, Performance, and Preliminary Results

Rene A. Ong*

*Enrico Fermi Institute, The University of Chicago, Chicago IL, 60637

Abstract. The Solar Tower Atmospheric Cherenkov Effect Experiment (STACEE) is designed to detect astrophysical sources of γ-rays at energies between 25 and 500 GeV. STACEE uses large solar mirrors (heliostats) to collect the atmospheric Cherenkov radiation produced in γ-ray air showers. The use of a large mirror collection area will allow STACEE to probe γ-ray sources at energies above the reach of the Compton Gamma Ray Observatory (CGRO), but below the reach of conventional Cherenkov telescopes. During the 1998-99 observing season, a portion of STACEE using 32 heliostats was installed at the National Solar Thermal Test Facility (NSTTF) of Sandia National Laboratories (Albuquerque, NM). This initial configuration (STACEE-32) observed a number of astronomical sources, including the Crab and several active galactic nuclei (AGN). Here we report on these observations. We highlight the experimental configuration and the preliminary results from the Crab data. The full STACEE experiment using 64 heliostats will be commissioned in 2000.

SCIENTIFIC MOTIVATION

A revolution in high energy gamma-ray astrophysics began ten years ago with the launch of the EGRET detector aboard the CGRO and with the high significance detection of sources at TeV energies by ground-based telescopes. EGRET has discovered over 150 sources of γ-radiation at energies up to 20 GeV [1]. At the same time, improvements in ground-based instruments using the atmospheric Cherenkov technique have resulted in the detection of at least seven point sources at energies above 250 GeV [2]. We now know that very high energy γ-rays are copiously produced by some of the most powerful astrophysical objects in the Universe, including pulsars, supernova remnants (SNRs) and AGN.

Of the many sources detected by EGRET, most are not yet seen by the ground experiments. For example, more than sixty AGN at a variety of redshifts are seen by EGRET, but only a few nearby sources have been detected from the ground. These results imply that the spectra of AGN cut off at energies between 20 and 250 GeV. The fact that only nearby AGN are seen at very high energies suggests that the γ-rays are absorbed or attenuated on their journey to Earth. High energy γ-rays interact with photons at infrared (IR) energies via the pair-production process.

The level of the intergalactic IR background is not well known, but measurements of absorption features of AGN should provide constraints on its flux and spectral shape [3]. In turn, these constraints could give us valuable information about the epoch of galaxy formation and about the composition of dark matter in the Universe [4]. Absorption features might also place important constraints on γ-ray emission mechanisms for AGN.

For these reasons, and others, the energy range between 20 and 250 GeV is expected to yield a wealth of scientific discovery. To date, however, it remains *terra incognita*. Exploring the energy window between 20 and 250 GeV is the primary scientific goal for STACEE. A similar experiment, CELESTE, is starting operations at a site in France [5].

STACEE-32 DESIGN

Concept

When high energy γ-rays enter the Earth's atmosphere, they interact and produce extensive air showers of highly relativistic charged particles. These charged particles emit Cherenkov radiation which is beamed to the ground. A pancake of Cherenkov photons approximately 250 m in diameter strikes the ground within a very short period of time (~ 10 ns).

The total amount of Cherenkov light generated in an extensive air shower is directly proportional to the energy of the primary γ-ray. As one goes down in energy, Cherenkov photon density decreases. Thus, the energy thresholds of atmospheric Cherenkov telescopes are limited by their total mirror collection area. Larger mirrors yield lower energy thresholds.

It has been recognized for some time that existing fields of solar mirrors (heliostats) represent an extremely promising resource for achieving lower energy γ-ray thresholds. Starting in early 1994, our group developed this idea and made the first detection of atmospheric Cherenkov radiation at a heliostat facility [6]. We also constructed a full-scale prototype detector which we operated at the heliostat field of the NSTTF [7]. Based on the success of our development work, we formulated the design of STACEE (Solar Tower Atmospheric Cherenkov Effect Experiment). The experimental concept is relatively simple. Atmospheric Cherenkov light created in a γ-ray air shower is reflected by heliostat mirrors to a secondary mirror on a central tower. In turn, this secondary mirror focuses the light onto cameras employing photomultiplier tubes (PMTs). Fast timing circuitry and coincidence multiplicity logic are used to trigger the experiment.

STACEE-32 Optics

The heliostat layout for the initial STACEE-32 configuration is shown in Figure 1. We used two clumps of heliostats with 16 heliostats in each clump. The Cherenkov photons from each clump were mapped onto a separate 2 m diameter secondary mirror on the central tower and subsequently onto a separate PMT camera. The secondary mirrors consist of seven hexagonal facets. The facets are made of ground and polished float glass that is aluminized on the front surface. The facets are held in position by an aluminum framework so that the overall secondary mirror has a focal length of 2 m. A metal support structure supports the mirror assembly and provides rails for the movable camera.

The camera holds an array of PMT cans. The cans are positioned so that the image from a single heliostat is centered on the entrance aperture of a can and so each can points to the center of the secondary mirror. The can holds a light concentrator coupled to the PMT via a pliable silicone cookie. Each concentrator, called a DTIRC (Dielectric Total Internal Reflection Concentrator), consists of a precisely machined single piece of UV transmitting acrylic. The DTIRCs also define the field of view of each PMT. The PMTs (Photonis XP2282) are fast, bi-alkali tubes that are run at moderate gain ($\sim 1.2 \times 10^5$).

Electronics and Data Acquisition

The PMT signals are AC-coupled and amplified (x100) near the cameras. The amplified signals are transmitted via high quality coaxial cable (RG213) to a dedicated electronics room located inside the tower. Copies of the PMT signals are sent to discriminators and to conventional gated ADCs after long ($1\mu s$) analog delays. The outputs from the discriminator are used to start conventional multi-hit TDCs and to form the trigger. The trigger requires a time coincidence (~ 20 ns wide) of the discriminated signals that have been delayed to account for the timing differences existing between the PMTs. The delays are updated every few seconds as a particular source is tracked across the sky. The trigger serves to interrupt the DAQ computer, stop the TDCs, and gate the ADCs. To minimize deadtime, the trigger decision consists of two separate levels: Level 1, which requires a minimum number of channels within a cluster of eight heliostats to fire in time, and Level 2, which requires a minimum number of clusters to fire.

The data acquisition (DAQ) and online control systems reside on a Silicon Graphics workstation connected to the electronics crates via Ethernet. The online system controls the data-taking and monitors the detector performance. For STACEE-32, the digitizing electronics and the DAQ system were largely built from off-the-shelf components of an older vintage. For STACEE-64, state-of-the-art electronics will be used for both the trigger and digitization functions.

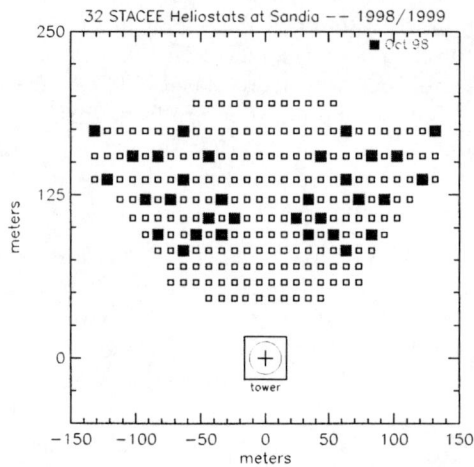

FIGURE 1. Plan view of STACEE-32, an initial version of STACEE that operated during 1998-99. STACEE-32 used 32 heliostats, two secondary mirrors, and 32 photomultiplier tubes. Each heliostat has a mirror area of 37 m^2.

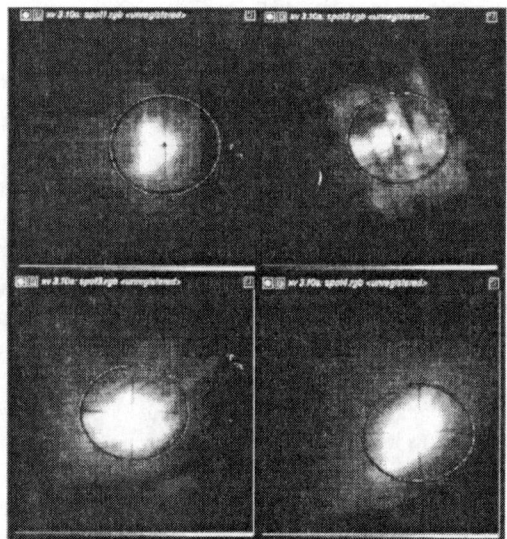

FIGURE 2. Images of the Moon reflected by four individual heliostats into the focal plane of one of the STACEE-32 cameras. The circles indicates the size of the entrance apertures of the PMT cans.

STACEE-32 OPERATIONS AND PERFORMANCE

Timeline

The hardware for STACEE-32 was installed during September, 1998, and the experiment was fully operational soon after. Observations with STACEE-32 took place between October, 1998 and April, 1999. In general, the weather was excellent; a large amount of both test and observational data was logged to disk.

Heliostats and Alignment

The heliostats at the NSTTF performed well during the seven months of data taking. Light from the Moon was used to determine the absolute headings for each of the 32 individual heliostats. Moonlight was reflected onto a projection screen on the tower, and the Moon images were recorded by a CCD camera system. Software was developed to allow us to reduce the data from the CCD image immediately and to correct the heliostat alignment. These corrections were made to a precision of one bit on the shaft angle encoders ($\sim 0.05°$).

Before the observing campaign started, the positions and angles of all the critical STACEE optical components, including mirrors, cameras, and PMTs, were carefully surveyed using a theodolite. The alignment was cross-checked using point sources of light in the field. Moonlight was used to cross-check the optical alignment of the secondary mirrors and the PMT can positions. Figure 2 shows four CCD images of the moonlight projected onto the PMT camera plane. The PMTs were positioned so as to be within 5% of the optimal light collection efficiency.

Detector Performance

For the most part, STACEE-32 operated smoothly during the entire observing period. Significant effort was made to characterize the detector performance. These studies were largely made by studying rates at different PMT threshold values and by taking Cherenkov test data at fixed positions in the sky.

Studies on timing properties of each channel were done to verify the correct performance of the trigger and to permit accurate reconstruction of the shower wavefront. Timing offsets and PMT gains for each channel were determined by two different methods: 1) a portable LED flasher which illuminated a single PMT at a time, and 2) a laser calibration system which transmitted a short optical pulse to fibers mounted at the centers of the secondary mirrors.

Typical operating conditions for STACEE-32 are summarized in Table 1. The accidental trigger rate (due to non-Cherenkov events) was calculated from the typical Level 1 rates. Accidentals were negligible at the operating threshold.

TABLE 1. Typical operating conditions for STACEE-32. The various items are described in the text.

Item	Typical Values
PMT threshold	6 p.e.
PMT rate	1-10 MHz
Trigger condition (L1)	5/8
L1 rate	1-20 kHz
Trigger condition (L2)	3/4
L2 rate	3 Hz
Accidental rate	$< 10^{-4}$ Hz

TABLE 2. Raw data (before cuts) taken by STACEE-32, in terms of the numbers of on-source hours.

Source	Data Collected (hours)
Crab	55.3
Markarian 501	10.8
Markarian 421	6.5
AGN 1219+285	7.2

Data Sample

STACEE-32 observed the Crab Nebula and three AGN. A summary of the amount of data taken on each source is shown in Table 2. The data were taken in hourly on/off pair intervals (i.e. 28 minutes on-source followed by 28 minutes off-source, or vice-versa). The off-source regions were at the same declination as the on-source regions, but ±7.5° away in right ascension.

CRAB ANALYSIS

Pair Cuts

The total Crab data sample consisted of 133 on/off pairs. We make cuts to remove those pairs that were affected by poor weather, poor or variable atmospheric clarity, or detector malfunctions. We also apply cuts requiring both halves of each pair to be closely matched with regards to their individual tube rates, cluster rates, and other quantities. These cuts ensure that any differences in the numbers of events recorded on-source and those recorded off-source are the result of changes in the Cherenkov rate due to the presence of γ-rays, as opposed to being due to trigger threshold variations. Overall, the cuts remove 27 pairs, leaving us with 106 pairs for the final data sample.

TABLE 3. On-source and off-source totals, the on-off differences, and the significance values for the STACEE-32 Crab data. The various categories are described in the text.

	Raw Data	Trigger	Shower Fit
ON	435,738	360,695	185,047
OFF	430,301	355,346	180,530
ON-OFF	5,437	5,349	4,517
Signif.	$+5.84\sigma$	$+6.32\sigma$	$+7.47\sigma$

Event Reconstruction

The first step of the event reconstruction is to re-impose the trigger condition in software using a 12 ns coincidence window. The software trigger condition is tighter than that in hardware, but it does not restrict the STACEE field of view. The measured arrival times of the shower at the heliostats are fit to a spherical wave front with a fixed core location (taken as the center of the STACEE-32 array). From the wavefront fit, the shower arrival direction in equatorial coordinates (α,δ) and a fit quality (χ^2) are determined. The average reconstructed positions in right ascension and declination agree with the position of the Crab to better than 0.1° accuracy. From simulations, we expect that γ-ray showers will have significantly lower values of χ^2 than most cosmic ray showers.

DC signal

In Table 3, we show the total numbers of events for the on-source and off-source portions of each pair. We also show the difference between the on and off portions. The numbers are given for three categories: the raw data, the data after re-imposing the trigger in software, and the data after subsequently requiring the shower fit χ^2 to be less than 1.0. The χ^2 cut value was chosen from Monte Carlo simulation work. In all three categories there is a statistically significant excess, which improves as successive cuts are applied. Thus, we have confidence that the excess events are due to a signal of γ-rays from the Crab.

Using the triggered data sample, we have an excess of 5,349 events recorded in a total on-source time of 2,648 minutes. Interpreting the excess events as a steady (DC) γ-ray signal, STACEE-32 detected $2.02 \pm 0.32\,\gamma$/min from the Crab. Preliminary Monte Carlo simulations indicate that the peak energy of the detected γ-rays is near 100 GeV. Figure 3 shows that the γ-ray event rate and the γ-ray fraction (γ-rays/cosmic rays) remained stable over time.

For each on/off pair, we calculate the significance, $(\text{ON-OFF})/\sqrt{\text{ON}+\text{OFF}}$, in the excess number of events. The distribution of these 106 significance values is shown in Figure 4. The significance distribution is consistent with a unit-width Gaussian displaced from zero. This distribution gives us confidence that our DC

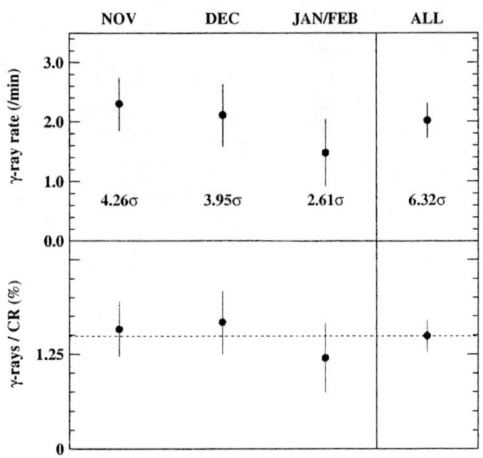

FIGURE 3. Top: STACEE-32 Crab γ-ray rate for three different observing periods in 1998-99. Bottom: γ-ray fraction (γ-ray rate / cosmic ray rate) for the different periods.

signal is not a statistical artifact or the result of a gross systematic error. After imposing the software trigger, 77 (29) of the pairs have positive (negative) excess.

Crab Pulsar Analysis

The mechanism that produces the pulsed γ-ray emission from the Crab is not fully understood. There are two broad classes of models for the pulsed emission: polar cap [8] and outer gap [9]. Determining the pulsed γ-ray spectrum above 10 GeV could provide a definitive test between the two models. Using the STACEE-32 data, we have made a search for the pulsar component. The arrival times of the on-source events are corrected to the reference frame of the solar barycenter and are folded with the radio pulsar ephemeris obtained by the Jodrell Bank group [10]. Figure 5 shows a histogram where the STACEE-32 events are binned in phase relative to the 33 ms pulsar period. The phase distribution is consistent with being uniformly distributed (H-test has chance probability for uniformity of 0.24). No strong enhancement is seen in the phase intervals of (-0.06,0.04) and (0.32,0.43) which are those portions of the phase where the γ-ray pulsed emission peaks [11]. Using these phase portions, we set an upper limit on the pulsed γ-ray fraction of the emission from the Crab of < 0.025 (90% C.L.) at a γ-ray energy of $\sim$100 GeV.

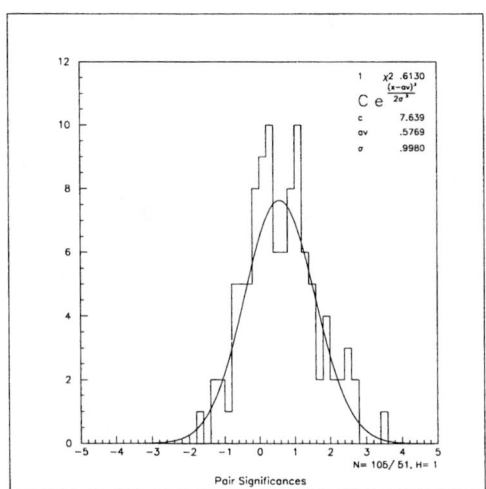

FIGURE 4. On/off pair significances for the STACEE-32 Crab data. The significance are calculated using $(\text{ON-OFF})/\sqrt{\text{ON}+\text{OFF}}$. The curve represents a Gaussian fit to the data.

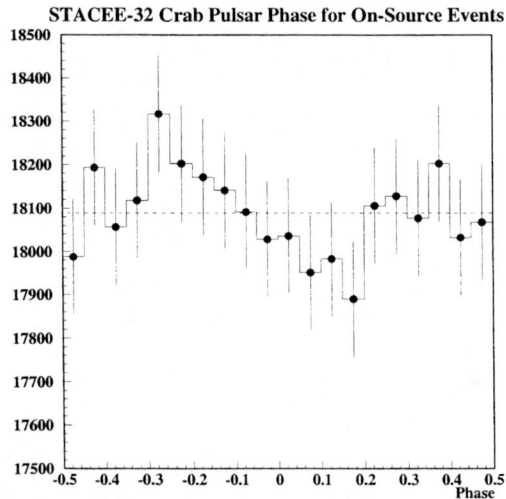

FIGURE 5. Phase diagram for STACEE-32 Crab data. The details of the analysis are given in the text. No strong evidence for pulsed emission from the Crab is seen.

CONCLUSIONS

STACEE is a new experiment designed to carry out γ-ray observations in the unopened window between 25 and 250 GeV. A portion of the experiment (STACEE-32) was constructed and operated during the 1998-99 season. From 44 hours of on-source observations (along with an equal amount off-source data), STACEE-32 detected a statistically significant excess ($+7\sigma$) from the Crab. The observed excess is consistent with being due to γ-rays at peak energy of $\sim 100\,\text{GeV}$. We are in the process of carrying out detailed simulation work in order to accurately estimate the energy response of the detector and in order to determine the γ-ray flux. No significant evidence for pulsation is seen in our data sample, and we accordingly set an upper limit on the pulsar fraction of the γ-ray signal. With the addition of more heliostats and new electronics, the full STACEE experiment will be completed in 2000. The full instrument will have significantly improved energy threshold and sensitivity relative to STACEE-32.

ACKNOWLEDGEMENTS

STACEE-32 is a collaboration involving the following scientists: D. Bhattacharya, L. Boone, M.C. Chantell, Z. Conner, C.E. Covault, M. Dragovan, D.T. Gregorich, D.S. Hanna, R. Mukherjee, R.A. Ong, S. Oser, K. Ragan, R.A. Scalzo, C.G. Theoret, T.O. Tumer, D.A. Williams, and J.A. Zweerink. We thank the staff of the NSTTF for their excellent support. This work was supported in part by the National Science Foundation, the Natural Sciences and Engineering Research Council, Fonds pour la Formation de Chercheurs et l'Aide a la Recherche, the Research Corporation, and the California Space Institute.

REFERENCES

1. D. Thompson et al., Astrophys. J. Supp. **101**, 259-286 (1995).
2. Rene A. Ong, Physics Reports **305**, 93-202 (1998).
3. F.W. Stecker, Astropart. Phys. **11**, 83-91 (1999).
4. J.R. Primack et al., Astropart. Phys. **11**, 93-102 (1999).
5. D.A. Smith, Proceedings of this Workshop.
6. R.A. Ong et al., Astropart. Phys. **5**, 353-365 (1996).
7. M.C. Chantell et al., Nucl. Inst. Meth. **A408**, 468-485 (1998).
8. See, for example: J.K. Daugherty and A.K. Harding, Astrophys. J. **458**, 278-292 (1996).
9. See, for example: J. Chiang and R. Romani, Astrophys. J. **400**, 629-637 (1994).
10. *Jodrell Bank Crab Pulsar Timing Results Monthly Ephemeris*, A.G. Lyne, R.S. Pritchard, M.E. Roberts, April 28, 1999.
11. See for example: P.V. Ramanamurthy et al., Astrophys. J. **450**, 791-804 (1995).

STACEE: Instrument Performance and Future Plans

C.E. Covault[1], D. Bhattacharya[4], L. Boone[3], M.C Chantell[1], Z. Conner[1], M. Dragovan[1], D. Gingrich[7] D. Gregorich[6] D.S. Hanna[2], R. Mukherjee[5], R.A. Ong[1], S. Oser[1], K. Ragan[2], R.A. Scalzo[1], C.G. Théoret[2], T.O. Tumer[4], D.A. Williams[3], J.A. Zweerink[4],

[1] *Enrico Fermi Institute, University of Chicago, Chicago, IL 60637, USA*
[2] *Department of Physics, McGill University, Montreal, Quebec H3A 2T8, Canada*
[3] *Santa Cruz Institute for Particle Physics, Univ. of California, Santa Cruz, CA 95064, USA*
[4] *Institute of Geophysics and Planetary Physics, Univ. of California, Riverside, CA 92521, USA*
[5] *Dept. of Physics & Astronomy, Barnard College & Columbia Univ., New York, 10027, USA*
[6] *Department of Physics and Astronomy, California State Univ., Los Angeles, CA 90032, USA*
[7] *Department of Physics, University of Alberta, Edmonton, Alberta, Canada*

Abstract. The Solar Tower Atmospheric Cherenkov Effect Experiment (STACEE) is a new instrument for observing astrophysical sources of gamma-rays in the energy range from 50 to 250 GeV. STACEE is currently under construction. The first phase of STACEE, using 32 large heliostat mirrors, was completed in the fall of 1998. We describe the performance of STACEE during the 1998-1999 winter observing season. The instrument is performing very well. We have detected the Crab Nebula with high significance ($\sim 7\sigma$). This result demonstrates that the STACEE concept is sound and that we can expect to make sensitive measurements of gamma-ray sources at energies below 100 GeV. The full STACEE instrument, with 64 heliostats, will be completed during the year 2000. The first three years of observations with the complete instrument will include a range of sources with an emphasis on AGN and supernova remnants.

INTRODUCTION

The energy range from 20 to 250 GeV corresponds to an "unopened window", inaccessible to previous ground and space-based experiments. STACEE is a ground-based experiment designed to operate in this energy region above the reach of the EGRET detector and below the energy threshold of current very high energy (VHE) atmospheric Cherenkov detectors. Astrophysical sources of interest in this energy range include galactic sources, such as gamma-ray plerion/pulsars, and extragalactic sources such as EGRET blazars and BL Lac type active galactic nuclei (AGN). Measurements of gamma-ray pulsars and the associated nebula provide critical information on their particle acceleration and emission processes. In the Crab Nebula, for example, the transition from synchrotron to inverse Compton emission occurs in the few GeV energy range. Likewise, measurements of the energy spectra of AGN in this energy region are crucial in understanding the mechanism for production of gamma-rays, which are believed to result from relativistic jets coming from the central black hole. Expected spectral cutoffs in this energy region could indicate absorption of gamma-rays by the intergalactic infrared radiation field [1,2].

THE STACEE INSTRUMENT

The STACEE experiment has been described in detail elsewhere [3,4]. STACEE collects Cherenkov light from gamma-ray air showers using an array of "heliostat" solar-collector mirrors at the National Solar Thermal Test Facility (NSTTF), Sandia National Laboratories, Albuquerque, NM. The facility is used during daylight hours for solar power energy research. We use this facility at night for gamma-ray astronomy. Currently STACEE is using 32 of the 212 heliostats, each with a collection area of 37 m^2, with plans to expand to 64 heliostats in the near future.

The heliostats focus Cherenkov light onto secondary mirrors located at the top of a central tower; these in turn reflect the light onto an array of photomultiplier tubes (PMT) positioned within a camera box. Each PMT collects light from a single heliostat. A trigger is formed from a narrow time coincidence of several

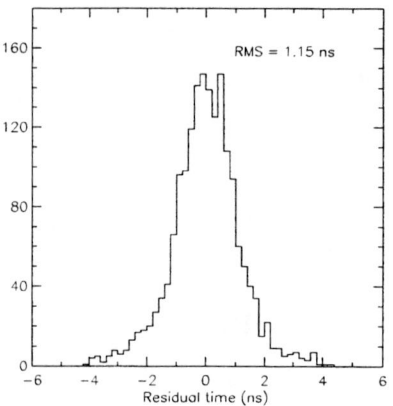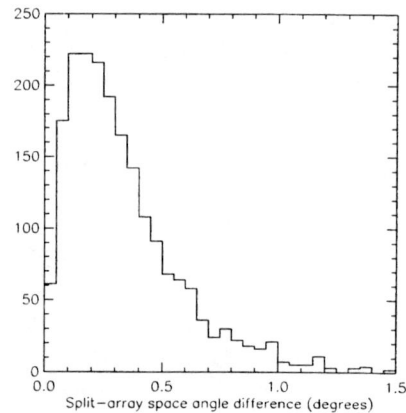

FIGURE 1. Left: Timing residuals for one STACEE channel relative to the best fit spherical wavefront. **Right:** Distribution of space-angle differences between overlapping sub-arrays for cosmic rays detected by STACEE. The median difference of about 0.2–0.3 degrees provides an independent estimate of STACEE's angular resolution.

discriminated phototube signals after appropriate time-of-flight delays are applied. Accurate pulse timing is used to determine the arrival direction of the primary, while pulse height measurements are used to estimate the primary's energy.

INSTRUMENT PERFORMANCE

The arrival direction of each shower is determined from the measured times at which the Cherenkov shower front passes each heliostat, so careful calibration (< 1 ns) of STACEE's timing apparatus is important. This has been accomplished using a combination of LED and laser calibration systems, and geometrical surveying of the optical components. Variations in timing with pulse height ("time slewing") have also been measured via laser calibration. RMS timing residuals of about 1 ns lead us to expect an angular resolution of 0.25 degrees (Figure 1). An independent estimate of this accuracy, in which the STACEE heliostats are divided into two sub-arrays and the angular difference between the reconstructed direction from the two sub-arrays is calculated, confirms this expectation. After all these corrections are applied, the residual pointing bias is less than $0.1°$, as shown in Figure 2.

An end-to-end verification of the STACEE experiment performance is demonstrated by our detection of gamma-rays from the Crab Nebula. The Crab was observed for approximately 50 hours during 1998/1999 with an signal detected at approximately 7 sigma. Details of this analysis are found elsewhere in these proceedings [5].

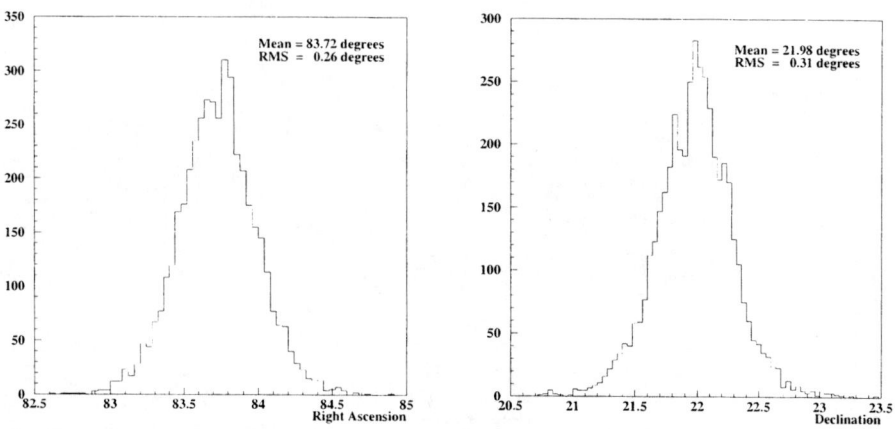

FIGURE 2. Reconstructed RA and DEC for a typical Crab on-source run. The displacement of the mean position is less than $0.09°$ from the expected Crab position of $83.64°, 22.01°$. Most of the events are background cosmic rays, and the widths of the distributions reflect both the angular resolution and the instrument field of view.

FUTURE PLANS

The STACEE instrument is still under construction. Data taken during 1998-1999 used a partially complete version of the experiment consisting of 32 heliostats in the field. During the year 2000, we will complete the construction of STACEE to the full experiment, upgrading to 64 heliostats (Figure 3) which will concentrate Cherenkov light onto five secondary mirrors.

We will also upgrade the electronics which will include a custom-designed trigger/delay and a new set of of high speed (1GHz) waveform digitizers. STACEE will use the Acqiris DC270 waveform digitizer, which is a 4-channel compact-PCI (cPCI) board (Figure 4). In comparison with STACEE-32, our simulations predict an increase in the trigger rate to $\sim 15\,\mathrm{Hz}$ for STACEE-64, and an improvement in gamma-ray sensitivity by a factor of five. We are also installing new IR monitors for detecting water vapor and an automated telescope/photometer for measuring sky clarity during observations.

By mid-2000, the construction of STACEE-64 will be completed and we will carry out several months of tests and calibrations to shake down the experiment. At this point, a three-year program of continuous astrophysical observation will

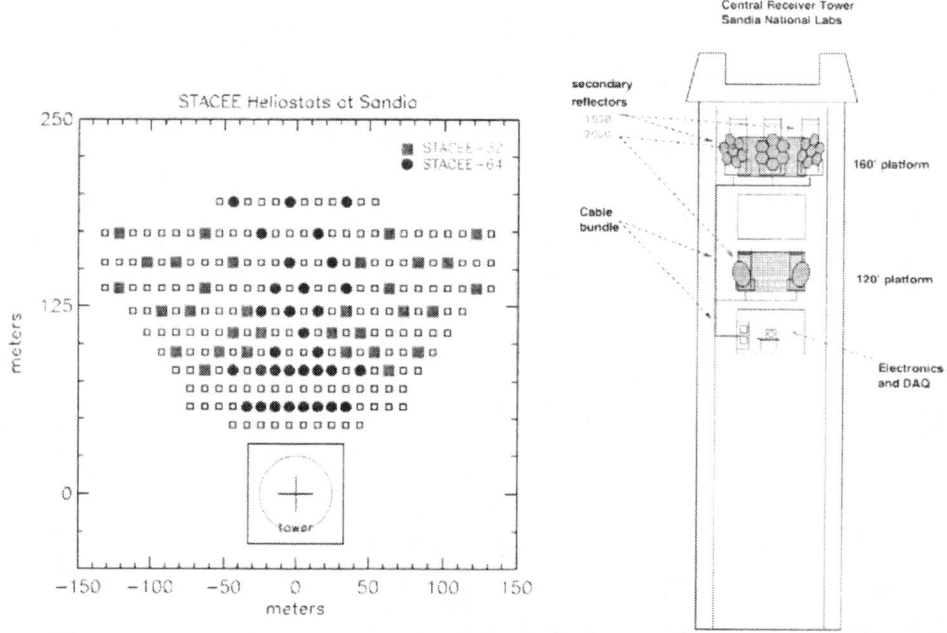

FIGURE 3. Left: Plan view of the Sandia heliostat field. The full experiment (STACEE-64) will use all shaded heliostats. (The coordinates are XY positions in meters.) **Right:** View of Sandia tower.

begin. The majority of targets planned during this interval will be AGN which will be observed over a range of redshift values. We will also observe selected pulsars, SNR, and EGRET unidentified objects that are easily visible at the STACEE site.

ACKNOWLEGEMENTS

We are grateful to the staff at the NSTTF for their excellent support. This work was supported in part by the National Science Foundation, the Natural Sciences and Engineering Research Council, FCAR, and the California Space Institute. CEC is a Cottrell Scholar of Research Corporation.

REFERENCES

1. For a summary of the scientific motivations for the field of gamma-ray astronomy, see for example, Ong, R.A., *Physics Reports*, **305**, 93-202 (1998).
2. Mukherjee. R. et al. *Proc. 26th ICRC (Salt Lake City, 1999)* OG 2.1.19 (1999).
3. Chantell, M.C. et al. *Nucl. Instr. Meth. A* **408**, 468 (1998).
4. The STACEE Website contains all published papers and conference proceedings: http://hep.uchicago.edu/~stacee
5. Ong, R.A. these proceedings (1999). See also Oser, S. et al. *Proc. 26th ICRC (Salt Lake City, 1999)* OG 2.2.07 (1999).

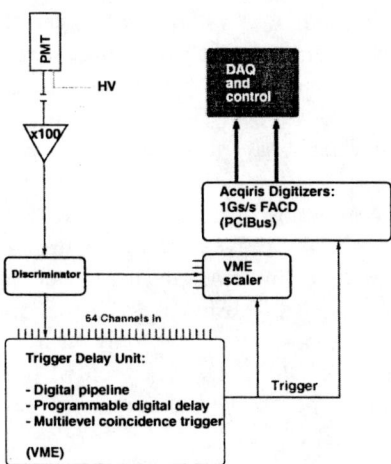

FIGURE 4. STACEE Version 2 Electronics: Analog signals from 64 PMT channels are discriminated and are fed into a trigger delay unit which will construct a trigger if the minimum number of tubes fire within 10 ns of each other (times reconstructed to the field). Eight-bit FADC's digitize triggered pulses to allow full reconstruction of time and pulse height for each channel.

First Observations with CELESTE

D.A. Smith[1] and M. de Naurois[2]

[1]*CEN de Bordeaux-Gradignan, 33175 France*
[2]*LPNHE, Ecole Polytechnique, Palaiseau 91128 France*

Abstract. The CELESTE solar farm gamma ray telescope detected the Crab Nebula near 80 GeV in early 1998, with 18 heliostats. CELESTE is now in its final configuration with 40 heliostats, with a trigger threshold below 30 GeV (7×10^{24} Hz) and a rate above 20 Hz. Overall, the detector is calibrated and aligned. Delays during construction combined with remarkably bad weather resulted in small data sets, and evidence for a Crab signal obtained when 25 heliostats were operational is weak. We describe the detector, present the current state of the data analysis and discuss our observations.

INTRODUCTION

Atmospheric Cherenkov astronomers come in two varieties: those who aim for a large improvement in flux sensitivity with a modest decrease in energy threshold, like HESS & VERITAS, and those who emphasize bridging the energy gap with EGRET. CELESTE, like her cousin STACEE, aims to move quickly into the 50 GeV regime by adapting the wavefront sampling technique to existing solar farms (Paré 1993; Smith 1997).

CELESTE announced a detection of the Crab nebula with data obtained early in 1998 (de Naurois 1998; Smith 1998). We then overhauled our acquisition electronics and software, while at the same time completing the detector to 40 heliostats. We essentially missed the recent Crab observing season. Since May 1999, 37 to 40 heliostats have been operational. However, only in July did we install the last 20 channels of 1 GHz Flash ADC's. Data presented in this paper was obtained by digitizing two phototubes per FADC channel, so that the night sky noise for analysis was $\sqrt{2}$ higher than it is now.

The first part of this paper reviews the main features of the CELESTE detector. We then present the key steps in event reconstruction. Showers observed simultaneously by both CELESTE and the CAT imager give insight into the data quality. We describe the data sample accumulated during the Spring of 1999, and conclude by discussing our observation strategy for the upcoming season.

EXPERIMENTAL APPARATUS

In broad terms, CELESTE in its final form is as described in our proposal (see reference), and test results with a full scale prototype are presented in (Giebels 1998). Here, we review the main detector elements.

Heliostats, Optics, and Phototubes

CELESTE uses 40 heliostats at the Thémis site in the eastern Pyrenees (N. 42.50°, E. 1.97°, 1650 m. a.s.l.), divided into 5 trigger groups. Each back-silvered heliostat mirror is 54 m^2 with an alt-azimuth mount guided by computers at the top of the 100 meter tall central tower.

Secondary optics replace the heat receiver in the five-by-five meter opening at the top of the tower. The spherical mirrors are divided into six segments (three levels) to optimize light collection: one views the farthest heliostats; two view the intermediate heliostats; and three view the heliostats at the foot of the tower.

At the secondary mirror focus is one two-inch Philips XP2282 photomultiplier for each heliostat. A solid Winston cone glued to each phototube defines a field-of-view of 10 milliradians. The small field-of-view requires us to aim the telescope not at the gamma ray source itself, but at the region in the atmosphere where the Cherenkov light is generated. In practice, we point at a distance of 11 km/$\cos\theta$ in the direction of the source, where θ is the source zenith angle.

The phototubes are equipped with active bases connected to control cards that provide various features: two stages of amplifiers for a combined gain of 100; anode current measurement upstream of the capacitive coupling; a trip circuit that cuts high voltage to the first dynode in the event of overillumination; a charge injection capacitor to allow gain and timing calibration. Heaters are integrated into this system to avoid problems of condensation, et cetera.

We aligned the heliostats by maximizing the phototube anode currents while scanning in one milliradian steps around a star. An alignment cross-check is provided by the gamma-ray induced air showers detected by the CAT imager and CELESTE simultaneously, discussed in (de Naurois 1999).

Single photoelectron peaks are measured two ways: via the differential rate versus threshold curve using discriminators with scalers; and by reconstructing peaks in the Flash ADC's. The charge seen in the FADC pulses, per unit time, agrees well with the measured anode currents, as can be seen in figure 3. For low illumination, that is, with the tower door closed, this latter technique also provides an absolute measurement of the photocathode illumination. For the data sample used in this paper the phototube gain was 10^5. Where the cables enter the control room we have 20 mV/p.e.

Data Acquisition

A counting room just below the mirrors houses electronics and acquisition computers. An HP 712/100 orchestrates the acquisition tasks, delegated to secondary computers. Figure 1 provides an overview. Heliostat tracking and electronics timing depend on the celestial rotation, while the electronics requires access to a calibrations database. This work is performed by the HP. All computers dialogue via TCP/IP. Each secondary task generates a data file, and an Event Builder combines the various files into one.

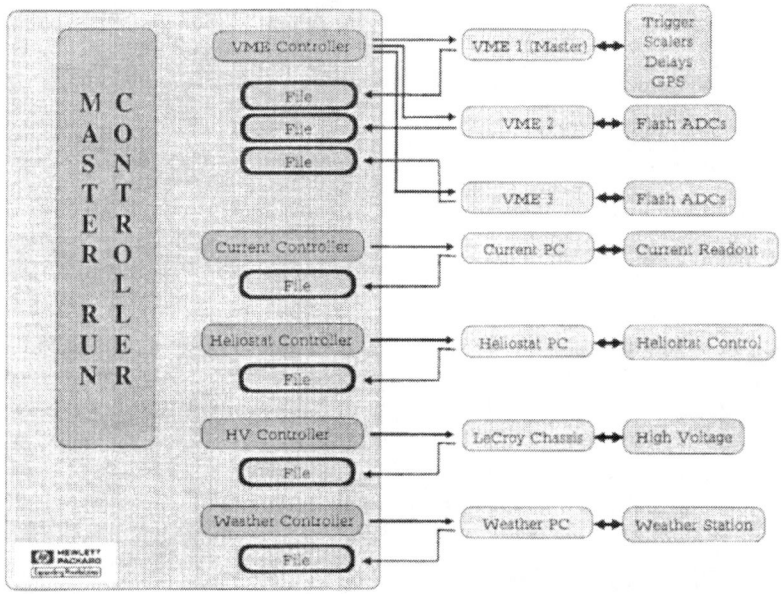

FIGURE 1. Overview of the data acquisition system.

8-bit Flash ADC's in two VME crates digitize the phototube signals at 930 MHz (Etep 301c, see reference). A third VME crate accomodates scalers, a GPS clock, and controls the trigger timing. The overall gain is 3 digital counts (dc) per photoelectron in the Flash ADC's, and a pedestal offset of 25 dc is used. Each VME crate is controlled by a Motorola 68040 running the Lynx OS [1].

All phototube anode currents are recorded twice per minute by a PC. Weather information (wind, pressure, humidity) from various places on the site is also recorded. A separate PC controls the heliostats, on the basis of instructions received from the HP, via a serial dialogue. The aging cables being a source of breakdowns, we are developing a radio control system to replace them.

---

[1] We gratefully acknowledge the programming work performed by Mr. N. Briand

At each trigger, FADC digitization is stopped, and a 100 ns window of FADC memory is read out for each phototube. The center of the window corresponds to the arrival time expected for a shower centered in the field-of-view. The dispersion due to lateral and angular shower-to-shower positions is of the order of ± 10 ns. The large window provides us with night sky noise information useful during data analysis. The disadvantage is a large readout time (0.3 ms/channel). We are implementing random triggers to preserve noise data, and will shrink the window width to decrease the dead time. For the data presented in this paper, all FADC channels were in a single VME crate and the total readout time was 13 ms, for a deadtime of 25%. In July, 20 additional channels were installed in a separate VME crate reducing the deadtime by half. Reduction of the window size and implementation of a DMA readout will ultimately lead to a deadtime of 8%.

A fast laser illuminates the heliostat field from just below the secondary optics. The laser was installed to calibrate the timing of the Themistocle electronics chain (Baillon *et al*). CELESTE uses it for the same purpose. The phototube anode currents and the FADC pedestal widths provide detailed information about the night sky light. Relative stellar photometry will give some information on the transparency of the entire atmosphere. A LIDAR, to measure transparency as a function of altitude, will become operational in 2000 (Snabre 1998).

Trigger

The trigger is designed to reach the lowest possible energy threshold. Programmable analog delays (switched cables) compensate for the varying path length differences as the source crosses the sky. When the longer cables are switched out, fixed attenuators are switched in to keep the attenuation variation below 10% independent of the delay value. The Cherenkov signals are then summed, in five groups of eight heliostats each. The resulting signal is discriminated, path length differences between the groups are compensated for using programmable logic delays (CAEN V486), and the final trigger is a threefold coincidence (CAEN V495). The V486 introduces of order 5% deadtime into the trigger, with typically less than a 2% difference in the ON and OFF deadtimes, which we correct for offline using measured parameters. (We are building a new programmable logic delay circuit with deadtime is less than 250 ns, to replace the V486 at the end of the current observing season.)

Figure 2 shows the final trigger rate as a function of threshold. The points were measured while tracking a source near the zenith. The histogram is a Monte Carlo of hadronic showers passing through a simulated detector, including the trigger, and incorporating the measured night sky background light. The agreement between the two is quite compelling and suggests that both the electronic calibration and our modeling of the detector response are good. We trigger routinely below three photoelectrons per heliostat. The trigger rate is 20 Hz.

Figure 2 also subtleshows a Monte Carlo prediction of the energy response for the

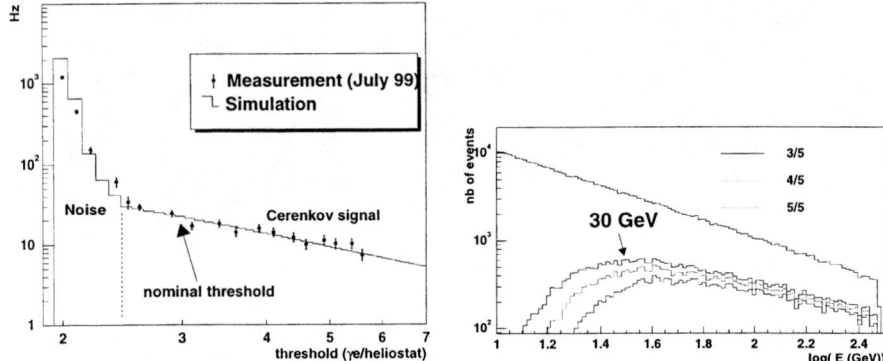

FIGURE 2. (Left) Trigger counting rate as a function of threshold. The solid dots show a typical measurement. The histogram is a full Monte Carlo simulation of hadronic showers, which dominate the right hand part of the curve, and includes night sky background light. (Right) Simulated detector response to a 3 photoelectron per heliostat threshold. The input hadron spectrum is shown (straight histogram), as well as three different trigger multiplicities.

three photoelectron per heliostat trigger threshold. The cosmic ray input spectrum used (protons, alphas, etc.) is shown by the power law histogram. The remaining curves are after the trigger simulation. We use a three-of-five multiplicity, which peaks at 30 GeV, although good sensitivity extends below 20 GeV. Our trigger studies indicate that the incremental hadron rejection obtained with a higher multiplicity trigger does not compensate for the decrease in gamma ray acceptance.

DATA ANALYSIS

Cherenkov data acquired by CELESTE from February through June 1999: Crab, 3 hours; Mrk 421, 6.5 hours; Mrk 501, 1.5 hours; 1ES1426+428, 2.5 hours; PSR1951, 3 hours. Comparison of the anode currents versus time in the ON and OFF data, as well as the rate of Cherenkov peaks and the overall trigger rate allow us to judge whether sky conditions were stable enough to justify data analysis.

Crab data are sparse because of technical delays followed by bad weather, the latter being particularly severe for Mrk 501. Mrk 421 illustrates a weakness in the solar-plant approach: coincident with this BL Lac object is a binary star with B-magnitudes of 6.2 and 7.8. For convergent viewing, only heliostats near the center of the field view the source directly. For these, the anode current is about 15% higher during the ON run as compared to the OFF run. The effect of increased background light is slightly larger background fluctuations, amounting to an effective shift in the relation between the number of observed photoelectrons and the shower primary energy. Thus the trigger and reconstruction rates change, albeit slightly, and even in the case of perfect atmospheric conditions data analysis requires mastery of this kind of bias. Work on adapting the padding concept (Cawley) to our FADC data is in progress, as are different approaches to handling this problem.

The solar farm geometry limits observations to ±2.5 hours before and after transit, and outside a declination range of $5° < \delta < 55°$ the threshold increases as compared to the Crab direction, due to decreased light collection efficiency. Fixed cable delays now in use further limit us to $\delta > 15°$. Thus even at this early stage in our experiment we began observations of new source candidates. 1ES1426+428 is typical of the observing strategy we aim for: this X-ray selected BL Lac was chosen using criteria inspired strongly by Whipple's discovery of 1ES2344+514 (Catanese 1998) and also by the work of (Ghisellini 1998) and his co-workers. Our lower energy range allows us to consider source candidates with larger redshift than for imagers, considering infrared absorption of gamma rays. ICRC papers also discuss AGNs in CELESTE (Münz 1999) and in STACEE (Mukherjee 1999). We plan to observe a relatively large sample of BL Lacs, and in parallel, to accumulate a large data sample for a small selection of SNRs, favoring Cas A. The pulsar is the subject of an ICRC paper (Musquère 1999).

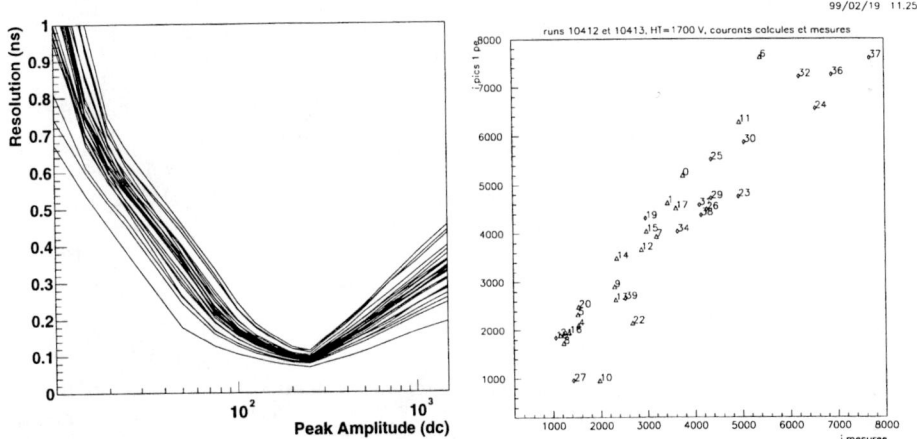

FIGURE 3. (Left) Timing resolution for peaks reconstructed from the Flash ADC data, for a range of phototube illuminations. Three photoelectrons is nine digital counts. (Right) Current reconstructed by summing the pulse charges in the Flash ADC data, versus the measured anode currents, in nanoamperes. The numbers are heliostat identifiers.

Flash ADC data

The first step in analysis is to find peaks in the FADC data, and thereby to estimate the Cherenkov wavefront amplitude and arrival time for each heliostat, on a shower-by-shower basis. The methods range from very simple and very fast (a few minutes to process a single 30 minute data run), to elaborate and slow (three hours for the same). For "normal" peaks, that is peaks well above the night sky fluctuations but not saturated, results of the different peak finders are comparable. The most sophisticated fitter allows good reconstruction beyond 500 dc (saturation occurs at 255 dc, or about 80 photoelectrons above pedestal), and also does a good job finding peaks below 5 p.e. Ultimately, we will use a hybrid scheme: a fast simple algorithm for most peaks, and fancier reconstruction for the largest and smallest peaks.

The Etep 301c 1 GHz Flash ADC was developed for CELESTE (see reference). Performance is excellent: some problems reported at the workshop have been resolved as this paper is going to press.

We evaluate the time and pulse height resolution in different ways. The basic method is that developed by (de Naurois 1998): graft software pulses onto FADC night sky data with illuminations covering the range of those encountered in the Cherenkov data, as determined from the anode current data. Reconstructing the pulses gives the curves shown in figure 3. In addition, we have evaluated the resolution using Themistocle laser pulses, and single muons emitting Cherenkov light in the solid Winston cones. The different methods agree to within the precision now available from these ongoing studies. For medium pulses, the timing resolution is adequate for wavefront reconstruction. Pulseheight resolution measured in the same manner is seen to be dominated by the phototube response, and not by the 8-bit sampling.

Wavefront Timing Reconstruction

Next, we use the pulse times to reconstruct a spherical wavefront of radius $(11km)/\cos\theta$, where θ is the zenith angle. (The fit results are insensitive to the choice of the radius, to within a few kilometers). A given heliostat is used in the χ^2 fit if the pulse height is above 5 photoelectrons. The residual distributions have widths consistent with the resolution described in the previous section but the mean values deviate from zero by as much as 1.5 ns. We attribute this to cumulated errors in the summed cable and electronic delays used in calculating the nominal pulse time, and we set these averages to zero in the subsequent analysis. A heliostat is then included in the χ^2 fit only if its pulse is within ± 10 ns of the nominal arrival time.

The night sky phototube illumination is typically 0.9 p.e. per ns [2]. For a 5 p.e. analysis cut, one naively expects 22 heliostats to have a cherenkov-dominated pulse, and 15 to have a night sky light pulse. Now that we have 40 FADC channels, this last number decreases to 7 heliostats.

We evaluate the reconstruction of the position of the shower maximum using the wavefront arrival times at the heliostat with the Monte Carlo, from the χ^2 of the fit, and by the "split detector" approach. All give comparable results. Figure 4 illustrates the split detector study. The 32 heliostats that were working in this

---

[2] More than the 0.7 in our proposal. The optical simulation in use was too simple.

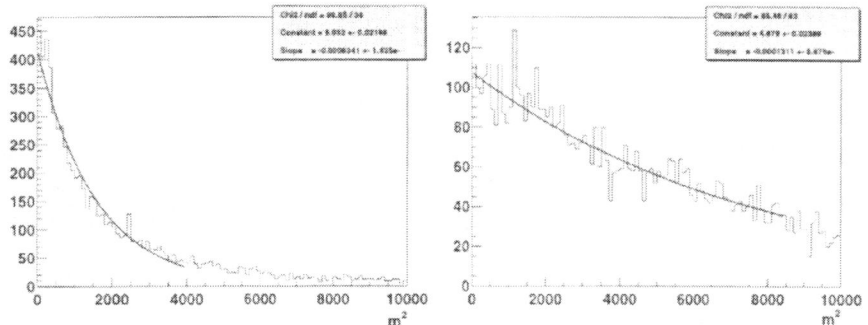

FIGURE 4. (Left) "Split detector" reconstruction of shower maximum: the arrival times for half the heliostats are used to reconstruct the wavefront, and compared with the result using the other half. The resolution for 20 heliostats is 15 meters. (Right) Reconstructed shower positions using timing measurements from the complete heliostat array. The width of the distribution is dominated by the 10 mr optical field-of-view.

run are divided into two groups, both distributed over the field. For each group the center $\vec{R}$ is calculated, as is the difference of the two results, $\Delta R = |\vec{R}_1 - \vec{R}_2|$. Figure 4 then shows the distribution of ΔR, as well as a fit to $exp(c + s\Delta R^2)$. The slope s of the fit, assuming a two-dimensional gaussian distribution for ΔR, is $-1/2\sigma^2$, hence $\sigma = 27$ meters for this convolution of the two detector halves, or 19 meters for a single detector half. Requiring 20 heliostats to reconstruct the wavefront, the expected spatial resolution is 14 meters.

Also in figure 4 the entire heliostat array is used to reconstruct the shower position $\vec{R}$, and the distribution of $|\vec{R}|^2$ is plotted. The exponential fit yields $\sigma = 1/\sqrt{2|s|} = 62$ meters. The heliostats aim at 11 km above the site, this width thus corresponds to 5.6 mr. This agrees nicely with the optical field-of-view of half-angle 5 mr.

Impact Parameter, and Direction Reconstruction

Since the wavefront is quasi-spherical, using the timing information to find the shower core does not give the shower direction. Our strategy has been to use the pulse height information to find the shower impact point in the plane of the heliostats. This second point then gives the shower direction, with a lever arm of about 11 km.

The lateral distribution of light on the ground is like a camembert cheese with rounded edges: finding the edge gives the center. Unfortunately, the optical transmission of each heliostat depends on its orientation with respect to the shower. This changes not only as the source tracks across the sky, but with the primary particle position and angle. At this time we have only a crude direction estimation.

CAT/CELESTE Joint Data

Whenever possible, the CAT imager and CELESTE track the same source at the same time. This is discussed in an ICRC paper (de Naurois 1999). A minimal statement is that the presence of CAT-identified gamma rays among the CELESTE trigger confirms that the telescope is pointed properly. One also finds that loosening the standard CAT analysis cuts and then imposing a CELESTE trigger reduces the OFF level in the α range away from the source direction, presumably because muons are rejected. The result is that a signal appears at $\alpha = 180°$, presumably due to low energy events poorly reconstructed by CAT, and the overall signal significance improves.

Search for 30 GeV gamma rays

Different workers apply different cuts to reject hadrons and extract a gamma ray signal. The results for the early 1999 data are sensitive to the choice of cuts and some pairs have negative excesses, that is, more events in the OFF than in the ON. Thus, we have no confidence in those pairs where ON exceeds OFF. We note that we take care to match the trajectories of the ON and OFF source across the sky, and we correct for the trigger and acquisition deadtime differences between the ON and OFF runs. The latter correction can be as much as 5%.

For the Mrk 421 data (13 ON/OFF pairs) and for 1ES1426+428 (5 pairs), the distribution of $\sigma = (ON - OFF)/\sqrt{ON + OFF}$, where ON and OFF are the numbers of events remaining after cuts, is inconclusive. We believe that the cause of the ON vs OFF rate differences is instrumental and not intrinsic to the solar farm approach. In addition to our own Crab result from last year, we are encouraged by the recent success of STACEE (Oser 1999). At the workshop we mentioned some problems in our programmable analog delays, which have since been fixed.

After the end of the 1997/1998 observing season CELESTE reported a 5.6 sigma excess in 3.5 hours of data taken while tracking the Crab (de Naurois 1998; Smith 1998). Three key differences with the 1999 detector bear mention. First, there were eighteen heliostats, in three trigger groups of 6 each. A threefold logic coincidence was required. Also, the Etep FADC's were not yet available, we thus summed three phototubes onto each of three Struck FADC's, since retired from the experiment, so only 9 of the heliostats were available for analysis, with increased night sky noise in each channel. Finally, the data acquisition architecture described above did not exist. The analog delays were controlled by independent PC's instead of receiving instructions from the HP via the Lynx's.

If we interpret the 3 sigma excess that we see in the early 1999 data as a gamma ray signal, its strength is consistent with our understanding of the detector's performance at that time. At press time we are analysing Crab data from the new season and the data quality is good.

CONCLUSIONS AND PROSPECTS

The CELESTE collaboration has worked hard through a particularly difficult year. The community will recall that we lost our leader, Eric Paré, to an automobile accident one year ago. We have attained our goal of triggering at 3 photoelectrons per heliostat, that is, at 30 GeV. We have identified and corrected an electronics bug that introduced a trigger bias at the 1% level. Our 1 GHz Flash ADC's are working well. Gamma ray showers seen jointly by CELESTE and CAT confirm that our telescope is well-aligned and hold great promise for a better understanding of both telescopes. Split-detector studies confirm that our wavefront reconstruction works. So far we have not succeeded in reconstructing the shower direction, and subtle changes in the ON and OFF conditions bias our rates. But given the wealth of instrumentation available at Thémis, such as the Themistocle laser and the upcoming LIDAR, it is only a question of time before these systematic effects are brought under control. The upcoming Crab season should mark the beginning of CELESTE's scientific career.

REFERENCES

1. Baillon, P. *et al* 1993, Astrop. Phy. **1**, 341
2. CELESTE proposal, http://wwwcenbg.in2p3.fr/Astroparticule, or from the authors.
3. de Naurois, M. 1998, 16th *European Cosmic Ray Symposium*, Madrid.
4. de Naurois, M. 1999, 26th ICRC, OG 4.3.06.
5. Catanese, M. *et al* 1998, ApJ **501**, 616
6. Cawley, M. 1993, in *Towards a Large Cherenkov Detector - II*, Calgary, 176
7. Etep 301c 2-channel VME 1 GHz Flash ADC, see http://www.etep.com, based on the SPT 7760 integrated circuit, see http://www.spt.com.
8. Ghisellini, G. *et al* 1998, MNRAS **301**, 451G
9. Giebels, B. *et al* 1998, Nucl. Instr. Meth. **412A** 329
10. Mukherjee, R. 1999, 26th ICRC, OG 2.1.19.
11. Münz, F. 1999, 26th ICRC, OG 2.1.20.
12. Musquère, A. 1999, 26th ICRC, OG 2.2.31.
13. Oser, S. 1999, 26th ICRC, OG 2.2.7.
14. Paré, E. 1993, in *Towards a Large Cherenkov Detector - II*, Calgary, 250
15. Smith, D.A. *et al* 1997, Nucl. Phys. **54B** (Proc. Suppl.) 362-367.
16. Smith, D.A., in proc. of the *19th Texas Symposium on Ultrarelativistic Astrophysics*, Paris, December 1998.
17. Snabre, P. *et al* 1998, Astropart. Phy. **8**, 159-177

THE SOLAR TWO GAMMA-RAY OBSERVATORY: Astronomy between 20-300 GeV

J.A. Zweerink*, D. Bhattacharya*, G. Mohanty*, U. Mohideen*, A. Radu[†], R. Rieben*, V. Souchkov*, H. Tom*, T.O. Tumer*

*IGPP, University of CA-Riverside, Riverside, California 92521
[†]Institute of Space Sciences, Bucharest, Romania

Abstract. The Solar Two Gamma-Ray Observatory is designed to close the energy gap between 20-300 GeV that is inaccessible by current instruments, such as the satellite-borne EGRET detector and the ground-based air Cherenkov telescopes. Utilizing the facilities of the Solar Two Power Plant in Barstow, CA, the observatory will detect the Cherenkov light generated as high-energy gamma rays and charged cosmic-ray particles interact with the atmosphere. With over 2000 heliostats available, Solar Two has the largest heliostat mirror area in the world and, thus, the potential to be the most sensitive gamma-ray detector at these energies.

Construction of a secondary mirror system capable of imaging 32 heliostats is nearing completion with plans for the first observations of the Crab Nebula in late November. We report on the design, status and testing of this secondary mirror system including the optics, electronics, and heliostat field.

INTRODUCTION

Many of the questions in gamma-ray astronomy can only be answered by opening the unobserved 20-300 GeV energy window. Three different approaches are being pursued to open this currently inaccessible energy range. On the space side, GLAST is an instrument in development intended to observe between 20 MeV-300 GeV and is planned for flight in 2005. The ground-based VERITAS experiment is an array of atmospheric Cherenkov telescopes (ACTs) with an estimated energy threshold of ~75 GeV and a planned completion in 2003. A third type of detector uses solar power plant facilities to observe 20-300 GeV gamma rays by collecting the light from heliostats with photomultiplier tubes (PMTs) (Figure 1). While all three types of detectors plan to observe the same energy region, the solar power plant detectors have two advantages: 1.) they will be ready for observations earlier [4,2] and 2.) they can be built for a fraction of the ACT array and satellite costs by using existing facilities.

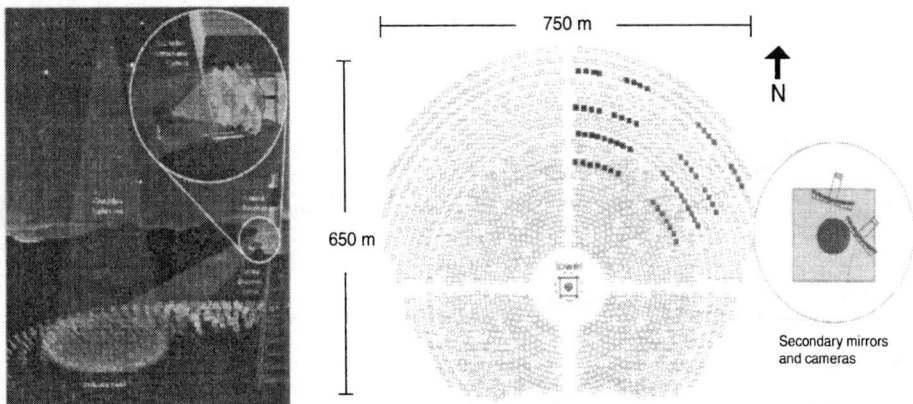

FIGURE 1. Left For a solar power plant gamma-ray telescope, the heliostat mirrors on the ground reflect the Cherenkov light to a secondary mirror located on the central receiver tower. The secondary mirror focuses the light from each heliostat onto a separate PMT. **Right** The filled squares show the 64 heliostats that will be viewed at the Solar Two Observatory–the first secondary will view the 32 heliostats on the left. Eventually, we hope to view more heliostats closer to the tower to cover the full 300 m diameter light pool of the Cherenkov shower.

Two gamma-ray detectors under construction have provided proof of the solar tower concept by detecting the Crab Nebula with thresholds below 100 GeV. STACEE [4,1] operates at the National Solar Thermal Test Facility, Sandia National Laboratories and has detected the Crab Nebula with an estimated energy threshold of ~75 GeV in 50 hours of observation using 32 heliostats. The full STACEE detector will use 64 heliostats. CELESTE [2] is located at the THEMIS site in France and uses 40 heliostats that are ~25% larger that those used by STACEE and Solar Two. Observations taken by CELESTE during 1997-98 using only 18 heliostats showed a signal after 3.5 hours of observation with an estimated energy threshold of ~80 GeV. Poor weather precluded any observations with the full 40 heliostat system. We are currently building a similar detector at the Solar Two Power Plant in Barstow, CA. By using 64 heliostats, Solar Two will have $> 2600 m^2$ of mirror area giving it the largest light collection power of any air Cherenkov telescope.

DETECTOR OVERVIEW

Initially, we plan to view 32 heliostats in the NE quadrant at Solar Two (dark filled squares in Fig 1) and expand to 64 heliostats within a year. Eventually, we hope to view ~250-300 heliostats that uniformly cover a 350 m diameter circle in the NE quadrant so that the complete Cherenkov light pool of the shower will be sampled. Tests described by Ong, et al. [3] demonstrate the feasibility of using the Solar Two site to build the gamma-ray detector described here.

Optics

The secondary mirrors (see Fig 2) used at Solar Two have a 6 m radius of curvature and are made from thirteen 1 m hexagonal facets. The total secondary size is 4.5 m wide by 3.0 m high. Although measurements by ourselves and Ong, et al. [3] confirm that the heliostat spot sizes at the secondary are between 3-5 m, space limitations on the central tower at Solar Two require the secondary mirror to have a height <3 m. The mirror dimensions coupled with the heliostat field geometry insure that each heliostat is viewed with an off-axis angle less than 7°. Since the secondary will be located inside the tower for protection against the elements, a door has been installed in the tower wall that allows the secondary to view the heliostat field.

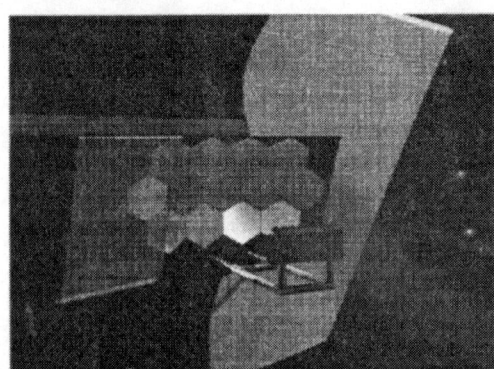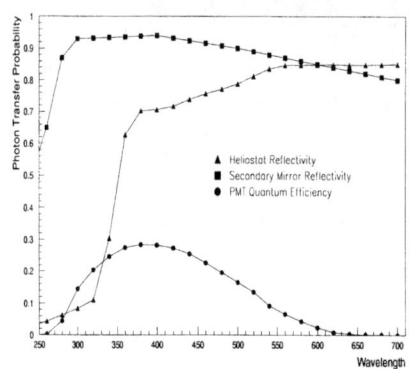

FIGURE 2. Illustration of the secondary mirror which has a radius of curvature R_c=6 meters and is 4.5 meters x 3 meters. The figure on the right shows the reflectivity of the heliostat and secondary mirrors as well as the quantum efficiency of the Lanco XP2280B photomultiplier tubes.

Light from each heliostat is reflected off the secondary and concentrated onto a PMT using a hollow Winston cone. We chose to use hollow instead of solid ones in the hope of eventually exploiting the UV content of the Cherenkov shower. Consequently, we are researching ways to modify the back-silvered heliostats since they are the only optical component that severely attenuates UV light (See Fig. 2).

Electronics

Unlike the imaging ACTs, solar power plant detectors rely on accurate reconstruction of the Cherenkov wavefront to discriminate hadrons and gammas. As such, it is vital to precisely correct for the various delays as the Cherenkov signals travel from the interaction region in the atmosphere, through the electronics and are combined to form a trigger. At Solar Two, the signals from each PMT are RC-filtered, amplified 100 times, individually discriminated and digitally delayed. Certain delays such as the time of flight from the heliostat to the PMT are fixed and can be corrected by using cables of the appropriate length. However, the delays caused by the gamma-ray source moving across the sky are variable and must be continually changed as the night progresses. The only commercial available, high-speed, programmable delay modules with suitably long delays were the LeCroy 4518 series.

Currently, each PMT signal passes through three of these modules, two of which have delay ranges from 8-128 ns (8 ns steps) and one with a 1-16 ns (1 ns steps) range. The modules have double pulse resolutions of 30 ns and 10 ns, respectively. While the delays for each setting on each module are steady, the delay for a given setting varies on the order of a few nanoseconds from channel to channel and module to module as shown in Fig. 3.

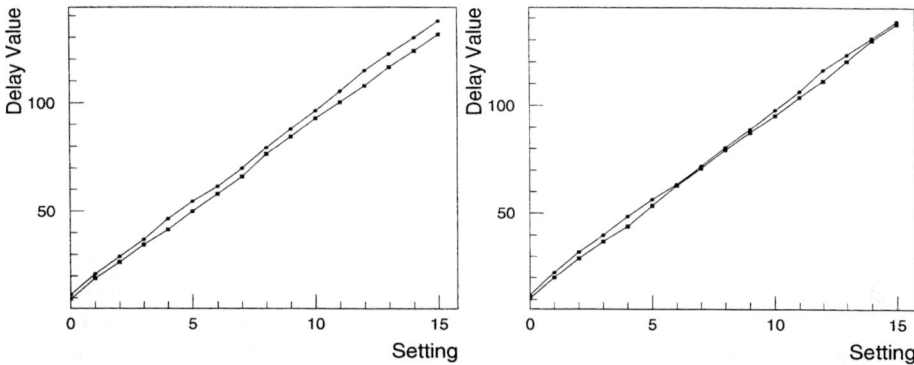

FIGURE 3. Comparisons of the delays for 16 settings for the modules with the longer delay ranges. On the left, 2 different channels on the same module are shown. On the right, the same channel on 2 different modules are compared. For the same setting, delay differences of 5 ns are common and differences as large as 8 ns are found as well.

The delays programmed into each channel account for the source motion on the sky as well as the time differences that arise as the signals travel through various CAMAC and NIM modules. Since the pulses that go into the coincidence module are between 5-10 ns wide, the signals from each PMT must be aligned in time to better than 5 ns or else a trigger will not be generated. Thus, a lookup table for the delay modules is used to provide the settings for a given delay to better than 1 ns.

After being delayed, the signals are again discriminated and combined in a two level process to form a detector trigger. For the first level, the digital signals from 8 heliostats in a subcluster are summed and fire a discriminator if a given number of pulses are coincident within a 10 ns window (currently, we require pulses from 5 heliostats in the subcluster). For the second level, which is the detector trigger, 3 out of the 4 subclusters must fire. For each detector trigger, precise timing information for each PMT pulse is recorded using a 32-channel TDC.

One major drawback of the current trigger system is the large deadtimes associated with talking to the CAMAC crate. Even using a list processor from Hytec, the deadtime/event to readout the TDCs and scalars and program the delays is $\sim$30 ms which limits the trigger rate to 3-4 Hz. However, we have purchased 32 channels of GHz digitizers (Acqiris model DC270) which operate on a compact PCI bus and have a readout time of $\sim$2.5 ms. Since the digitizers provide both precise timing and pulse height information for each PMT signal, we can use the CAMAC system to generate the trigger pulse, but do the readout over the compact PCI bus. With this scheme, trigger rates of $\sim$40 Hz give the same deadtime/event as the CAMAC system operating at 3-4 Hz.

To have uniform sensitivity across the detector the gains of the different PMTs are needed. We are currently determining the absolute gains by measuring the single photoelectron peak for each PMT by placing the PMT under a low current and digitizing the signals using an oscilloscope which are then stored on disk. The integrated pulse from each trace is binned to find the single photoelectron peak. A method developed by STACEE using a blue LED to measure the relative gains *in situ* is being implemented at Solar Two so that it is possible to measure how the PMT gains change from month to month over the course of the observing season.

CONCLUSIONS

Construction is nearly complete for the first secondary system capable of viewing 32 heliostats at the Solar Two site in Barstow, CA and the various subsystems of the detector are being aligned and calibrated. A CAMAC-based data acquisition system being tested with plans to move to compact PCI. We plan to begin observations of the Crab Nebula in December 1999 or January 2000.

ACKNOWLEDGEMENTS

We are thankful for the grant from the Keck Foundation and for the technical assistance and support provided by the STACEE collaboration. We acknowledge the support and assistance of Southern California Edison in coordinating the design and construction at the Solar Two Power Plant. We also thank the advisory committee of our Keck Foundation grant for their guidance and Prof. Steven Ahlen from Boston Univ. for allowing us to use the ASAP software.

REFERENCES

1. Chantell, M., et al., *Nucl. Inst. Meth. A*, **408**, 468 (1998).
2. de Naurois, M., et al., Proc. 26th ICRC. **5**, 211 (1999).
3. Ong, R.A., et al., *Towards a Major Atmospheric Cherenkov Detector-IV* (Padova), 295 (1995).
4. Oser A., et al., Proc. 26th ICRC. **3**, 464 (1999).

The Physics Potential of Ground-Based Gamma-Ray Astronomy below 50 GeV

Norbert Magnussen

Fachbereich Physik, Universität Wuppertal, Gaußstr. 20, 42097 Wuppertal, Germany

Abstract. Based on the expected performance of the MAGIC Telescope an overview of the scientific prospects of ground-based high energy Gamma-Ray Astronomy in terms of astrophysics, cosmology and particle physics questions is given.

INTRODUCTION

Technical developments have so far allowed to observe the universe from radio waves to γ-rays up to about 10 GeV and from about 200 GeV up to 100 TeV. A gap has remained between 10 GeV and 200 GeV which is going to be investigated with the MAGIC Telescope currently under construction [1]. This instrument will be an Imaging Air Cherenkov Telescope (IACT) employing advanced technology for all of its ingredients. Other ground-based projects aiming at the energy domain below 100 GeV are the solar array projects CELESTE and STACEE.

THE COSMOLOGICAL GAMMA-RAY HORIZON

In spite of an energy-flux sensitivity superior to the EGRET instrument onboard CGRO (for energy spectra extrapolated to higher energies), a much smaller number of sources has been discovered with the IACT technique above 200 GeV implying that most of the EGRET sources have spectra turning over between 10 GeV and 200 GeV. For extragalactic sources this might either be due to external absorption on the diffuse cosmological background or due to internal absorption in the sources.

That above some critical energy defining the γ-ray horizon the visible universe in high energy photons should be limited because of pair production on the cosmological low-energy diffuse background photons was first pointed out by Gould & Schréder [2]. With increasing γ-ray energy, the threshold condition is fulfilled for an increasing number of low-energy photons from the diffuse radiation background, resulting in a decreasing γ-ray horizon. Conversely, triggering at γ-ray energies lower than current IACTs can observe one will have access to a much larger fraction of the Hubble volume and thus to a much larger source population. According to the current best estimates of the diffuse background current IACTs should only be able to observe the universe out to redshifts of z ≈ 0.1.

With the MAGIC Telescope operating above 10 GeV a large number of AGN will be observable, and a population study of all results should then reveal a plot of the highest energy seen by the MAGIC Telescope, vs redshift. The slope of the locus of energy maxima should then be proportional to the density of intergalactic target photons. However, those maximum energies which are below the locus of points, should be due to intrinsic source absorption, and can be used to constrain γ-ray production models.

The flux of the isotropic radiation background from the far-infrared to the ultraviolet is only poorly known from direct measurements. The measurement of turnover energies in the spectra of extragalactic sources will allow to infer the low-energy background flux in a manner completely independent of conventional methods. The background radiation flux is an important observable for models of cosmic structure formation because it constitutes a convolution of the star formation history, the dust extinction history, and the evolution of the initial mass function. In addition it has some sensitivity to the existence of massive neutrinos or stable particles from supersymmetric extensions of the Standard Model acting as dark matter. In Fig. 1 the shape of the cosmological γ-ray horizon as calculated by Mannheim [3] is shown. Also shown as horizontal lines are the lowest energies measurable by the Whipple telescope and the planned HESS array of 16 IACTs and the MAGIC Telescope in phase 1 (i.e., equipped with classical PMTs). As AGN activity seems to be linked to galaxy merger activity and the star formation era the volume density of AGN shows a prominent peak or plateau at $z \geq 1$. The MAGIC Telescope will thus have access to the bulk of cosmological AGN and it will be the only IACT with access to γ-rays with energies below the (possibly) asymptotic γ-ray horizon. The exact shape of the γ-ray horizon will also depend on the cosmological parameters and constitutes an important cross-check of their values determined by other means. Should the UV background be accurately determined, e.g., by the proximity effect, the distance to the horizon could possibly be used to determine the Hubble parameter, H_0, based only on the size of the Thomson cross section.

THE DIFFUSE GAMMA-RAY BACKGROUND

An important scientific question closely connected to the asymptotic γ-ray horizon is the understanding of the diffuse γ-ray background in the GeV energy domain in terms of contributions from point and diffuse sources. The only known point sources today are AGN. An analysis of the AGN γ-ray luminosity function based on the EGRET results by Chiang & Mukherjee [4] indicates that a significant fraction of the diffuse γ-ray background may be due to diffuse sources.

The electromagnetic emissivity of the universe at energies above the energy of the asymptotic horizon on Hubble length and Hubble time scales will appear in the diffuse γ-ray background at energies *below* the asymptotic value due to the

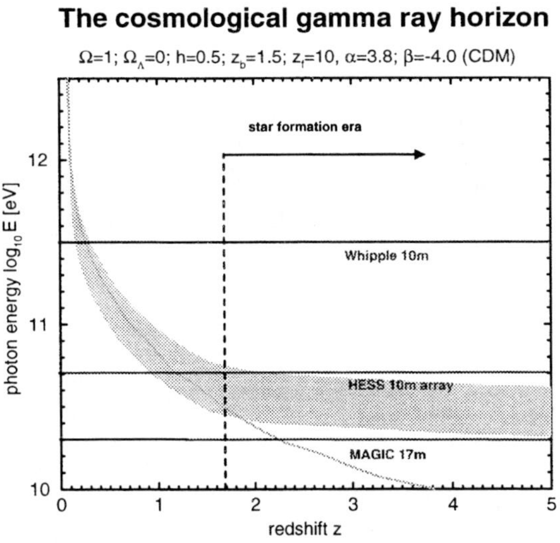

The cosmological gamma ray horizon

$\Omega=1; \Omega_\Lambda=0; h=0.5; z_b=1.5; z_f=10, \alpha=3.8; \beta=-4.0$ (CDM)

FIGURE 1. The shape of the γ-ray horizon resulting from a calculation by Mannheim (1999). The grey band indicates the uncertainty in the calculation due to uncertainty in the diffuse photon background level. The line is the prediction based on an extreme assumption of the star formation rate continuing at the maximum level beyond $z = 1.5$ and negligible dust absorption. For distant sources only γ-rays with energies less than the horizon can reach the observer. From [3].

pathlength for the cascading process (initiated at high energies due to interactions with the diffuse background) reaching a length scale of the order of the Hubble length. This also stresses the importance of the location and shape of the asymptotic γ-ray horizon. The determination of the AGN luminosity function from high sensitivity data as will be provided by the MAGIC Telescope thus is of fundamental importance for the understanding of the high energy emissivity.

GAMMA-RAY EMISSION FROM PULSARS

Of the more than 800 known radio pulsars EGRET has revealed 6 to emit pulsed γ-rays above 100 MeV. No steady pulsed emission from pulsars has yet been detected by ground-based IACTs above 200 GeV. To clarify the production mechanism, measurements in the 10 GeV to 100 GeV energy domain are crucial. The polar cap model for pulsed emission [5] predicts a sharp cutoff in the γ-ray spectra above a few GeV due to absorption in the strong magnetic field. As detailed phase resolved modelling showed, the bridging emission between the two pulses should exhibit harder spectra [6]. This prediction was recently confirmed by phase resolved spectroscopy of the Crab, Vela and Geminga pulsars [7] and provides a very low threshold IACT like the MAGIC Telescope with the unique opportunity to provide answers on the emission regions for the highest energy γ-rays from the neutron star magnetosphere. Phase resolved rates above 10 GeV of up to more than 100σ per hour per 0.1 phase interval (after image analysis), or more than $\sim 10\sigma$ if no background cuts are made, are estimated based on the MAGIC Telescope's sensitivity. Note that the locking onto the pulsed signal will be easily achieved and thus suppress any systematic background effects. In addition this high sensitivty will make the MAGIC Telescope a unique instrument for the discovery and study of further γ-ray pulsars [8].

GAMMA-RAY SUPERNOVA REMNANTS

Supernova remnants (SNRs), possible sites of cosmic ray acceleration favoured in most models of the cosmic ray origin, seem to be more complex than previously believed [9]. Although four SNRs have been observed above 200 GeV (Crab nebula, Vela, PSR1706-44, and SN1006), the question of the origin of cosmic rays is far from answered. More sensitive measurements at lower energies will be of great importance in identifying the spectral component showing up above 200 GeV in the four sources above and to discover γ-ray emission in more SNRs. With the low energy threshold of the MAGIC Telescope it may be possible to observe a *two component γ-ray spectrum*, which should then allow to decouple the predicted leptonic and hadronic components in SNR shells.

SEARCH FOR A COLD DARK MATTER CANDIDATE

Among the candidates for the dark matter in the universe, weakly interacting massive particles (WIMPs) such as the neutralinos arising in supersymmetric extensions of the standard model are plausible. Their interaction cross section and expected mass naturally match to produce the dominant contribution to the energy density of an expanding Friedmann-Robertson universe. With a lower limit on their mass from particle physics of about 20 GeV, the interesting mass range of ≈20 GeV to 300 GeV implies that γ-rays from decay or annihilation (neutralinos are Majorana fermions) could be discovered with the MAGIC Telescope, e.g., as a γ-ray line from the region of the Galactic Centre. Note that from the northern site (La Palma) the effective photon collection area will be of the order of 10^6 m^2 which will yield sufficient sensitivity to cover a fair fraction of the MSSM parameter space, see e.g., [10].

GAMMA-RAY BURST COUNTERPARTS

The low moment of inertia is one of the main features of the MAGIC Telescope and it will allow rapid positioning towards observation targets (typically within 30 s). The telescope is thus ideally suited to search for high-energy counterparts of

GRBs. The low-energy threshold will allow observations out to large cosmological distances. Extrapolation of the energy spectra of the GRBs detected by EGRET leads to the prediction that even medium-strength bursts will yield very high γ rates detectable by the MAGIC Telescope ($\sim$ kHz) due to the very large effective collection area. For γ-rates of this magnitude the MAGIC Telescope (in phase 2) will measure energy spectra from about 5 GeV up to the highest energies.

TESTING LORENTZ INVARIANCE

Because of the high rates expected for GRBs observed with the MAGIC Telescope the high energy lightcurve of GRBs can be obtained with good temporal resolution. A number of Quantum Gravity (QG) models, e.g., [11], yield non-Lorentz-invariant terms which lead to modified laws of propagation and interaction of neutral particles as a result of interactions with the quantum gravity medium. Measurable time delays can be expected if the particles have energies close to the QG scale (expected to be close to the Planck scale, i.e., 10^{19} GeV) or if they have traversed cosmological distances. This last requirement is fulfilled for at least a subclass of GRB which have been observed to have redshifts up to more than 3.4. For an assumed time resolution for the lightcurve of 1 sec and a pathlength of more than several Gpc, the sensitivity of MAGIC Telescope's measurements for the QG scale will be of the order of the Planck scale. For comparison, the current best limits are about 1% of the Planck scale. Recent work within the Liouville string formulation of QG [12] yields a refractive index which increases linearly with the photon energy, i.e., the high energy photons will arrive *later* compared to the low energy photons. This distinctive signature will help distinguishing the QG effect from classical dispersion effects which yield increasing time delays for decreasing energies.

REFERENCES

1. Lorenz E., *these proceedings*.
2. Gould R. J., and Schréder G., *Phys. Rev. Lett.* **16**, 252 (1966).
3. Mannheim K., *Rev. Mod. Astrophy.* **12**, 101 (1999).
4. Chiang J., and Mukherjee R., *Astrophy. J.* **496**, 752 (1998).
5. Harding A. K., *Astrophy. J.* **245**, 267 (1981).
6. Daugherty J. K., and Harding A. K., *Astrophy. J.* **458**, 278 (1996).
7. Fierro J. M. et al., *Astrophy. J.* **494**, 734 (1998).
8. de Jager O. C. *Proc. 26th European Cosmic Ray Conference*, Madrid, Spain, 1998.
9. Jones T. W. et al., *PASP*, (February issue, 1998).
10. Bergström L., Ullio P., and Buckley J. H., *Astropart. Phys.* **9**, 137 (1998).
11. Amelino-Camelia G., Ellis J., Mavromatos N. E., and Nanopoulos D. V., *Mod. Phys.* **A12**, 607 (1997).
12. Ellis J., Mavromates N.E., and Nanoloulos D.V., *electronic preprint* gr-qc/9904068.

The CLUE experiment running with 8 telescopes; observations of gamma sources and runs on Moon.

D. Bastieri,[2] B. Bartoli,[1] C. Bigongiari,[3] R. Biral,[3] M.A. Ciocci,[3]
D. Cosulich,[2] M. Cresti,[2] D. Kartashov,[2] F. Liello,[5] N. Malakov,[3]
M. Mariotti,[2] G. Marsella,[3] A. Menzione,[3] R. Paoletti,[4]
G. Parlavecchio,[3] L. Peruzzo,[2] A. Piccioli,[3] F. Rosso,[3] R. Sacco,[2]
A. Saggion,[2] G. Sartori,[2] P. Sartori,[2] C. Sbarra,[2] A. Scribano,[1]
A. Stameira,[3] N. Turini[4] and F. Zetti.[3]

[1] Università and I.N.F.N - Napoli,
[2] Università and I.N.F.N - Padova,
[3] Università and I.N.F.N - Pisa,
[4] Università di Siena and I.N.F.N di Pisa, and
[5] Università and I.N.F.N - Trieste

Abstract. The CLUE experiment is presently an array of 8 telescopes detecting UV Cherenkov light from atmospheric showers. Preliminary results from the last 2 years of data taking are shown. 1998 campaigns on *Markarian 421*, *Markarian 501* and the *Crab Nebula*. The 1998-1999 high-statistics campaign with the telescopes aiming directly at the Moon will be also presented. The last item is related with the measurement of the ratio $\bar{p}/p$ in cosmic rays, in its turn of prominent importance to validate early Universe models.

I DETECTOR OVERVIEW

The CLUE detector is an array of 8 units with a 45-m pitch hosted by the Instituto de Astrofísica de Canarias, at Roque de Los Muchachos — La Palma — Canary Islands, at an altitude of 2200 m in the same site of the experiment HEGRA. The ninth unit, needed to complete the 3×3 matrix, is on its way towards the island. The optimal geographic position at 28°43′ N makes possible quasi-zenithal observations of both VHE gamma-sources (*Mkn 421*, *Mkn 501* and the *Crab Nebula*) and the Moon, the latter remaining above 60° for as long as 4 hours during winter nights.

Each unit consists of a parabolic mirror with 1.8 m of diameter and focal length. The mirror collects the UV Čerenkov light emitted by secondary particles during the development of atmospheric showers. This light is then focused in a MWPC with 24 × 24 copper pads, where it can be detected via photoconversion on TMAE vapours that enrich the chamber gas mixture. At the end of the DAQ chain, single photons appear as charge clusters on the chamber pad structure and are easily tagged via clustering algorithms tailored from collider experiments. All the equipment is sheltered inside a standard 6 m container, that can open itself as a clamshell during data acquisition. The telescope, that is the system made up by the mirror and by the proportional chamber, is remote-controlled on both slewing and source tracking. Analog to digital converters, signal-shaping electronics and trigger logics complete the outfit, together with two I/O cards for VME bus that take care of chamber thermoregulation, high voltage setting and other remote tasks. Most of the items outlined above are detailed elsewhere (see e.g. [1–3]).

II THE RECONSTRUCTION ALGORITHM

The basis for the direction reconstruction algorithm of the primary particle, is the list of all charge clusters found in all chambers. Each of these clusters, through a proper backtracking onto the parabolic mirror, is associated with a direction of an UV Čerenkov photon emitted by an ultrarelativistic secondary. Mainly due to the strong atmospheric UV absorption, UV images are made up of few scattered photons, to be confronted with images from Čerenkov experiment working in the visible band, with plenty of photons making up *rich-of-information* "fish" diagrams. This forced the development of a new algorithm and the chosen approach was a three-step maximum likelihood method in the direction plane:

1. a proper likelihood function is centred on each direction in the list;

2. these functions are multiplied together obtaining a global likelihood function;

3. the maximum of this global function (actually the minimum of − log of the function) is identified with the desired direction.

The likelihood function adopted is the photoelectron angular distribution histogram filled by many simulated showers and then fitted with two gaussians:

$$\mathcal{B}(\theta) = P1 \times e^{-\frac{1}{2}\left(\frac{\theta - P2}{P3}\right)^2} + P4 \times e^{-\frac{1}{2}\left(\frac{\theta - P5}{P6}\right)^2}$$

Reconstruction algorithm performances are evaluated via Montecarlo. On-going studies are consistent, both for protons and gammas, with $0.5°$ angular resolution and with an effective area of $10^4 \, \text{m}^2$ at 3 TeV.

III DATA ANALYSIS

Data were collected with 8 units, in the period between February and April/May 1998 for the sources *Mkn 421*, *Mkn 501* and the *Crab Nebula*, and between February 1998 and April 1999 for the Moon (details in Table 1). *Ghost* runs, are off-source

TABLE 1. Data acquisition information. Besides *real* sources, data refer also to *ghost* source.

	DAQ period	Entries	Time
Moon	Feb 98 – Apr 99	124580	138 hr
Markarian 421	Apr/May – 98	26300	34 hr
Markarian 501	Apr/May – 98	59000	76 hr
Crab Nebula	Feb – 98	27600	39 hr
Mkn 421 – Ghost	Apr/May – 98	9500	15 hr
Mkn 501 – Ghost	Apr/May – 98	4500	5 hr
Crab – Ghost	Feb – 98	7500	15 hr

runs and were taken shifting backward by 2 to 4 hours the right ascension of the given object, in such a way that it is possible to *re-track* the same path that the *real source* followed. Besides regular data runs, special runs were taken for calibration purposes in order to evaluate gains and performances of the chamber in different zones. Trigger requirements were as follows:

- at least two *superpads*, or three for the Moon, above threshold (refer to fig. 1);
- at least three (out of eight) units with a signal.

Before employing the reconstruction strategy outlined in section II, the data underwent a major *clean-up*, to remove odd behaviours caused by electronic noise and faults during data collection. This can be summarised in three major steps:

1. infer a new pedestal from data;
2. tag *whimsical pads* while pursuing the pedestal drift during all the run;
3. subtract the *pursued* pedestal and recover whimsical pads by substituting their charge with the average of non-whimsical neighbouring pads.

A Source analyses

The long offline analysis is applied to all events that triggered on at least three mirrors with at least five clusters. Once determined, for these events, the direction of the primary particle expressed in chamber coordinates α and β, the data analysis proceeded subdividing the α, β plane into concentric bins of equal area indexed by $\theta = \sqrt{\alpha^2 + \beta^2}$. Each "off-source" run was normalised to its "on-source" companion at more than $2°$, that means that we counted the events in the on-source and in the off-source with θ greater than $2°$, and we scaled the off-source data to the analogous on-source value. In fact, we noticed that all backgrounds had the same shape, mainly reflecting the angular acceptance of the detector, so the normalisation at $2°$, far away from the signal, could not do any harm. The excesses of "on-line" versus "off-line" runs in the first three bins were reported at the *1998 ECRS* [4]. Updated results show that while *Mrk 501* had no excess at all, *Mrk 421* manifested a strong activity, 2.3 ± 1.1 times the Crab one (see table 2).

TABLE 2. Preliminary results from source campaigns. Excess calculated on the first three bins.

	Excess	Time	Rate (evt/hr)
Crab Nebula	320±100	39 hr	8.2±2.5
Markarian 421	570±108	34 hr	16.7±3.2
Markarian 501	...	76 hr	<1.3

B The Moon and the ratio $\bar{p}/p$

The analysis of data collected in the Moon campaign underwent the same preliminary treatment of source data. We obtained for both Moon and Moon-*ghost* data a list of primary directions properly rotated in such a way to have the proton shadow fixed along the $\alpha-$ axis. Given the asymmetry due to the presence of the Moon dip, we then proceeded subdividing (α, β) histogram into slices $1.5°$ wide around the direction semi-axes $\alpha+$ and $\alpha-$. In fact the Moon dip lies along the $\alpha-$ axis (the *West* of the sky) so it is evidenced in the difference between the corresponding *Western* and *Eastern* bins of the slices (the detection of antimatter exploiting the Moon and the Earth magnetic field is the subject of [5]).

The histogram of the events is reported in figure 2, where error bars are drawn according only with their statistical error. On-going evaluation of systematic errors and comparison with *ghost* data will be the subject of further analyses when more statistics should become available.

If the asymmetry between $\alpha+$ and $\alpha-$ is entirely due to the Moon shadow, then the energy threshold of the CLUE detector should be around few TeV.

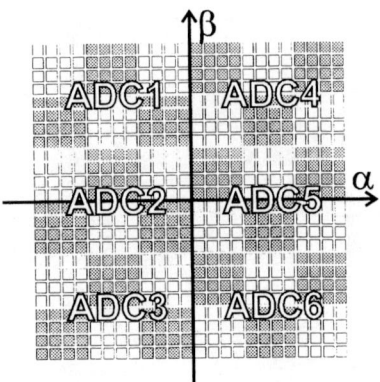

FIGURE 1. Chamber functional overview and coordinates α and β. Pad arrangement into 4×4 structure, *superpads*, and ADC layout (12×8) is evidenced.

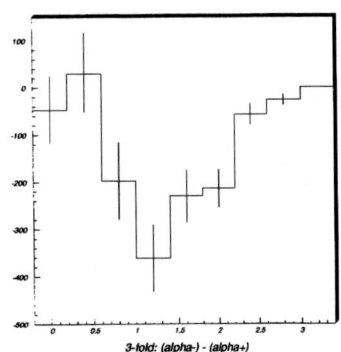

FIGURE 2. The difference between negative and positive α. The histogram comprehends all events with at least a triple coincidence of the CLUE units. Angular coordinate is expressed in degrees.

IV   CONCLUSION

Analysis algorithms, used in this preliminary analysis, are still in phase of tuning with a detailed Montecarlo simulation. Nevertheless, we were able to measure the fluxes from Mrk 421 and the Crab with the expected sensitivity of the apparatus at 2,200 metres of altitude. Preliminary analysis on Moon data, although with low statistics, shows a nice dip. We are confident that with the new data coming from the winter 1999/2000 campaign on both sources and the Moon and with a more refined analysis, we can inspect the energy threshold and the angular resolution of the apparatus. At the same time we should also set an upper limit on the relative abundance in cosmic rays of negatively charged particles in the TeV range.

REFERENCES

1. Alexandreas D. et al., *NIM* **A 409**, 488 491 (1998a).
2. Alexandreas D. et al., *NIM* **A 409**, 679–681 (1998b).
3. Cresti M. et al., *"The CLUE Mirrors"*, Proc. of *"Towards a major atmospheric Cherenkov detector - II"*, Calgary, R.C. Lamb ed., 169–175 (1993).
4. Bartoli B. et al., *Rayos Cósmicos 98:* Proc. of *"16th European Cosmic Rays Symposium"*, Alcalá de Henares, J. Medina ed., 405–408 (1999).
5. Urban M., Fleury P. et al., *Nuclear Physics B (Proc. Suppl.)* **14B**, 223 (1990).

Results from the Milagrito Experiment

A.J. Smith[9,3], R. Atkins[1], W. Benbow[2], D. Berley[3,10],
M.L. Chen[3,11], D.G. Coyne[2], B.L. Dingus[1], D.E. Dorfan[2],
R.W. Ellsworth[5], D. Evans[4], A. Falcone[6], L. Fleysher[7], R. Fleysher[7],
G. Gisler[8], J.A. Goodman[3], T.J. Haines[8], C.M. Hoffman[8],
S. Hugenberger[4], L.A. Kelley[2], I. Leonor[4], M. McConnell[6],
J.F. McCullough[2], J. E. McEnery[1], R.S. Miller[8,6], A.I. Mincer[7],
M.F. Morales[2], P. Nemethy[7], J.M. Ryan[6], B. Shen[9], A. Shoup[4],
G. Sinnis[8], G.W. Sullivan[3], T. Tumer[9], K. Wang[9], M.O. Wascko[9],
S. Westerhoff[2], D.A. Williams[2], T. Yang[2], G.B. Yodh[4]
(The Milagro Collaboration)

(1) University of Utah, Salt Lake City, UT 84112, USA
(2) University of California, Santa Cruz, CA 95064, USA
(3) University of Maryland, College Park, MD 20742, USA
(4) University of California, Irvine, CA 92697, USA
(5) George Mason University, Fairfax, VA 22030, USA
(6) University of New Hampshire, Durham, NH 03824, USA
(7) New York University, New York, NY 10003, USA
(8) Los Alamos National Laboratory, Los Alamos, NM 87545, USA
(9) University of California, Riverside, CA 92521, USA
(10) Permanent Address: National Science Foundation, Arlington, VA ,22230, USA
(11) Now at Brookhaven National Laboratory, Upton, NY 11973, USA

Abstract.
The Milagro water Cherenkov detector near Los Alamos, New Mexico is the first air shower detector capable of continuously monitoring the sky at energies between 500 GeV and 20 TeV. Preliminary results of the Milagro experiment are presented. A predecessor of the Milagro detector, Milagrito, was operational from February 1997 to May 1998. Milagrito consisted of 228 8" photomultiplier tubes (PMTs) arranged in a grid with a 2.8 meter spacing and submerged in 1-2 meters of water. During its operation, Milagrito collected in excess of 9 billion events with a median energy of about 3 TeV. The detector's sensitivity extends below 1 TeV for showers from near zenith. The results of an all sky search for the Milagrito data for both transient and DC sources will be presented, including the Crab Nebula and active galaxies Markarian 501 and 421, which are known sources of TeV gamma-rays. Also presented will be a study of the TeV emission from gamma ray bursts (GRBs) in Milagrito's field of view detected by the BATSE experiment on the Compton Gamma-Ray Observatory.

I INTRODUCTION

The burgeoning field of Very High Energy (VHE) γ-ray astronomy has been revolutionized in the past decade by the development of imaging air Cherenkov telescopes [1–3]. These instruments detect γ-ray induced air showers by observing the Cherenkov light produced by cascading electromagnetic particles in the air. They have opened up a new window on the universe, observing γ-rays with energies greater than 1 TeV from Active Galactic Nuclei (AGN), plerions, and supernova remnants. Despite the success of this technique, observations are limited because the instruments have a relatively narrow field and can only operate on clear moonless nights. A second type of VHE air shower detector, extensive air-shower (EAS) arrays, detect air-showers by observing the particles that reach the ground level. A conventional EAS array consists of many small particle detectors spread over a large area. Typically, only 0.5-1.0% of the total area is instrumented, resulting in very high energy thresholds of 30 TeV or higher. Unlike ACTs, these instruments can operate continuously and observe the entire overhead sky. The Milagro collaboration has constructed an array sensitive to showers with energies below 1 TeV. The low threshold is achieved by constructing the detector at an altitude higher than most air shower detectors and using a large area water Cherenkov detector to sense particles from the shower. From February 1997 to May 1998, the Milagro collaboration operated a prototype, Milagrito.

II THE MILAGRITO DETECTOR

The Milagrito detector [4] exploited a 21-million-liter man-made pond located in the Jemez Mountains near Los Alamos, New Mexico. The site is located at altitude of 2650m above sea level (750 g/cm^2 atmospheric overburden). The detector consisted of 228 8" PMTs submerged in a covered pond. The PMTs were secured to a grid with a 2.8 meter spacing. The pond was filled approximately $\frac{1}{3}$ full with purified water covering the sensors. Data were collected at depths of 0.9, 1.5 and 2 meters above the PMTs.

Milagrito was operated with a trigger that required 100 PMTs to sense 1 or more photo-electrons (PEs) within a 300ns time window. The nominal trigger rate was 300 Hz, but varied from 180 Hz to 450 Hz depending upon operating conditions. Snow collecting on cover increased the overburden reducing the rate, and increasing the water depth increased the rate. The bulk of the triggers, roughly 90%, were due to electromagnetic showers induced by cosmic ray hadrons. About 7% of the triggers were produced by single muons with large zenith angles, and the remainder of the triggers, about 3%, were unaccompanied hadron events.

Milagrito detected EAS particles via the Cherenkov light emitted by charged shower particles as they pass through the detector. Using water as a detector medium allows for the direct detection of the charged component of the showers through the Cherenkov light they produce, and also converts the abundant γ-rays into electrons through pair production and Compton scattering. Roughly 50% of the particles reaching the array are detected, with an average of 2 PMTs hit per particle. This efficiency for particle detection is much larger than that of sparse scintilator arrays that typically detect <2% of the particles reaching the ground level.

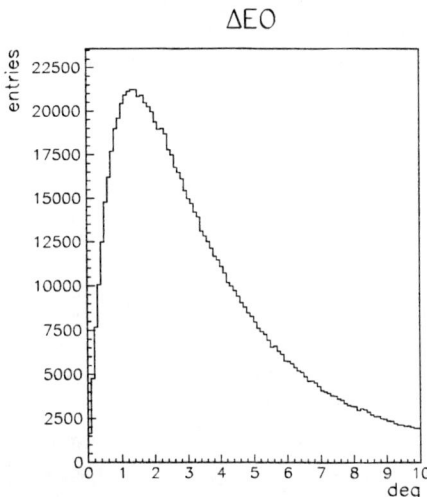

FIGURE 1. The angular resolution is estimated by dividing the array into 2 smaller arrays. Plotted is the space angle difference between the incident shower angle reconstructed by the two independent detectors. The angular resolution of the detector is approximately equal to half of ΔEO.

The PMT hit times were measured to an accuracy of about 1 ns. These times are used to reconstruct the shower front of the incoming extensive air shower. The statistical error in the reconstructed shower angle is estimated by dividing the sensors grid into two overlapping sub-arrays of equal size. The space angle difference between the reconstructed directions of the even and odd arrays (ΔEO) is twice as large as the point spread function of the entire detector. Figure 1 is a plot of ΔEO for the Milagrito detector. The distribution peaks at about $2°$, corresponding to a detector angular resolution of about $1°$. In practice, analyses are optimized by cutting on the number of hits include in the angle fit (n_{fit}) and choosing the bin-size that maximizes the sensitivity for the background level and the spectrum of the source.

Backgrounds are estimated [5] by measuring the event distribution in local detector coordinates (HA,δ) for a period of a few hours. This distribution is assumed to be constant over the entire period. The data for the period is then separated into intervals sufficiently short that the angular rotation of the earth during the interval is small compared to the detector point spread function. The background for each interval is then estimated by normalizing the (HA,δ) distribution to the number of events collected in the much shorter interval. The process is repeated summing the signals and backgrounds for the short intervals. The estimation of the angular resolution and the background subtraction method were verified by the detection of the shadow of the Moon [6].

Milagrito was sensitive to showers produced by primary γ-rays with energies as low as $\sim$100 GeV, but the mode shower energy was about 1 TeV for showers from zenith and typically 2-4 TeV for showers from a source that tansits near zenith. The small size of the pond compared to the lateral extent of a typical air shower along with the poor ability of this instrument to measure the amount of energy deposited in the pond make the estimation of shower energy on an event by event basis nearly impossible.

III SEARCH FOR GAMMA-RAY POINT SOURCES

The Crab Nebula is a well known standard candle of TeV gamma ray astronomy. Unfortunately, the Milagrito detector lacks sufficient sensitivity to detect this source. Detector simulations predict an excess of 1-2 σ which is consistent with our observation of 0.8σ.

During the Milagrito data-taking period active galaxy Markarian 501 (Mrk 501) was observed by a number of groups [7–10] to be in a prolonged flare state. During this period, the flux and energy spectrum of γ-rays from Mrk 501 was well studied by the Whipple and Hegra Groups. Thus, Mrk 501 is an ideal source to use to study the sensitivity of Milagrito [11].

Figure 2 shows the significance of the observed signal in the vicinity of Mrk501. An 3.7σ excess of events was observed at the source position. An excess of 3624 ± 990 events was detected for the entire observational period corresponding to an excess of $9.8\pm2.7\ \frac{events}{day}$. The rate measured February 1997 and October 1997 can be directly compared to the measurements made by atmospheric Cherenkov telescopes. During this period, we observe a rate of $13.1 \pm 4.0\ \frac{events}{day}$ which is in good agreement with the predicted γ-ray rate of $12.5 \pm 3.8\ \frac{events}{day}$. We interpret this result as a detection of this well studied source and as a confirmation of our detector's performance.

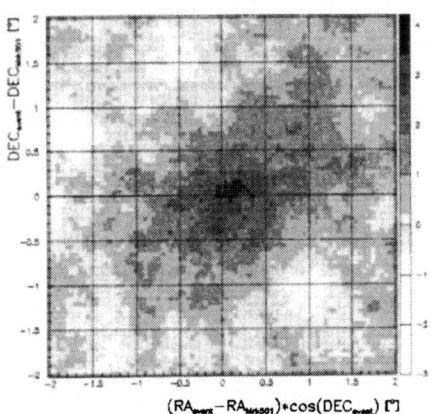

FIGURE 2. Sky map for the vicinity of Mrk 501.

A search for γ-ray point sources was conducted over the northern sky between $\delta = 10°$ and $\delta = 60°$ [12]. In this search the analysis was nearly identical to the Mrk 501 analysis with, for computational reasons, square search bins instead of round. As before, the size of the bin was optimized to maximize the sensitivity of the search. The sky was divided into a $0.1°\times0.1°$ grid in RA and δ defining a total of 18 million candidate source positions. Each position was searched for an excess as a signature for point source γ-ray emission. Figure 3 shows the distribution of significances for the entire sky and for a subset of 6197 independent (not overlapping) bins. In both cases, the distribution of significances has a mean consistent with 0 and a width consistent with 1. For the 6197 non-overlapping bins, the number of total "trials" is well defined. The bin of highest significance has an excess corresponding to 3.8σ. The probability of detecting a signal this large in 6197 trials 0.36. For the non-overlapping bins, the number of trials is less than the total number of entries, because neighboring bins are highly correlated. The number of effective trials for this search has not yet been calculated, but the data shows no evidence for the detection of a γ-ray point source. As simulations predict a 1.3σ excess from the position of the Crab nebula, a 6.5σ signal would correspond to a signal strength 5 times that of the Crab. No such signal is observed, so we conclude that no source with a Crab-like spectrum had an average flux greater than 5 times the flux of the Crab nebula during the observational epoch of Milagrito. A detailed study of the total number of effective trials taken in the search and a more stringent declination dependent upper limit will be calculated.

IV SEARCH FOR GAMMA-RAY BURSTS

During the lifetime of Milagrito, 54 gamma-ray bursts (GRBs) detected by the Burst and Transient Source Experiment (BATSE) [13] were within 45° of Milagrito's zenith [15,14]. A search was conducted for TeV emission coincident in time and space with the BATSE detections [14,15]. The search time used was T90 measured by BATSE and defined as the time over which from 5% to 95% of the detected γ-rays were detected. As the angular resolution of Milagro is much better than that of BATSE, a search was conducted for point source emission within the the BATSE 90% error radius. For this search, no n_{fit} cut was applied to the data and bins of radius 1.6 deg were used. Unlike the search for DC sources on large backgrounds, when the backgrounds are small, larger binsizes and smaller n_{fit} cuts are optimal. The background was estimated from 2 hours of data collected before and after the

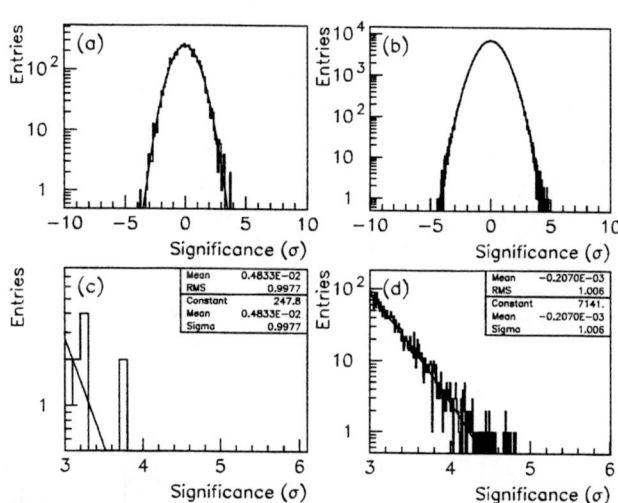

FIGURE 3. Plots (a) and (b) show the distribution of significances for 6197 independent bins and 18,000,000 overlapping bins respectively. Plots (c) and (d) show the same data as (a) and (b) with the region from 3σ to 6σ expanded and the fit statistics displayed

GRB. A grid search was conducted, and the point of highest Poisson probability was taken to be the candidate GRB position. In order to establish the significance of the candidate position, an ensemble of "fake" skymaps was generated by randomly selecting events from the background distribution. These maps were searched in the same manner as the data. The probability of a background fluctuation producing a candidate of the measured excess is simply the ratio of the number to fake skymaps that contain a candidate position of equal or greater significance to the total number of fake skymaps thrown. This search procedure was fixed prior to the analysis of the data.

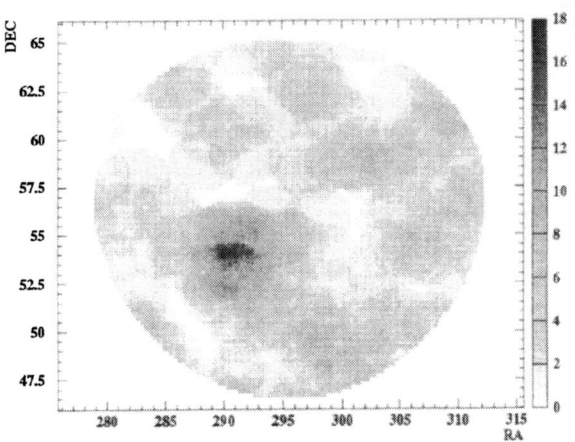

FIGURE 4. Skymap of GRB 970417a search region. The number of events within a the 1.6 deg radius search bin is plotted for each candidate position in RA and δ.

For one burst, GRB 970417a, a substantial excess above background was detected. Figure 4 is a map of the search region GRB 970417a. The search area corrected probability for this GRB is 2.4×10^{-5} As the search was conducted for 54 GRBs, the probability of obtaining such a low probability due to a background fluctuation in any one of the searches is 1.5×10^{-3}.

A search for 10 second duration bursts of γ-rays was conducted over the entire Milagrito data set [16]. The search time scale, 10 seconds was chosen to be sensitive to both GRBs and primordial black hole evaporation. The search was not confined to the BATSE defined search time and region. In this analysis, the computationally difficult search was made more efficient by only over-lapping the bins by 50% in RA,δ and time. In all, 1 745 105 940 bins were searched. No significant excess was observed within this data set.

V   THE MILAGRO DETECTOR

The heart of the Milagro detector [17] is of an array of 450 PMTs deployed 1.5 meters below the pond surface and a second layer of 273 PMTs located 7m below the surface. The top layer (shower layer) measures the arrival times of air-shower particles reaching the ground, while the bottom layer (muon/hadron layer) is used to measure the spatial distribution of the shower particles and is particularly sensitive for the detection of penetrating muons and hadrons. The shower layer is principally used to measure the direction of the incoming shower, while the muon/hadron layer will be used primarily for gamma-hadron separation. Conical "baffles" have been added to the PMTs to attenuate the late light tails that reduced the angular resolution of the prototype. Baffles also reduce the number of hits on the shower layer for single muon tracks. This will allow us to trigger the detector at a much lower threshold without being swamped by a background of single muon triggers.

In addition to the pond detector, an array of 5000 liter water tanks instrumented with a single PMT will be deployed. Because the pond is small compared with the lateral extent of typical air-showers, the additional detector array will allow for the identification of the core position dramatically improving the angular and energy resolution of the instrument.

Construction of the pond element of the Milagro detector was completed in the Fall of 1998. Data taking began in February 1999 at a reduced trigger rate to shake down the electronics and online software. Simulations indicate that Milagro will be able to measure an excess of events with significance greater than 5σ from the Crab Nebula without gamma-hadron separation. Studies of our ability to separate γ-ray initiated showers from hadron initiated showers predict quality factors of 1.5 for the whole data set and greater than 2.5 for events with many PMTs hit [18].

REFERENCES

1. Ong, R.A. 1998, *Phys. Rep.*, 305, 93.
2. Catanese,M., Weekes,T.C. to appear in Publications of the Astronomical Society of the Pacific.
3. Hoffman, C.M., Sinnis, C., Fleury, P., and Punch, M. 1999, to appear in Reviews of Modern Physics, 71.
4. Atkins, R., et al. 1999, submitted to Nucl. Instr. Meth. A.
5. Alexandreas, D.E., et al. 1992, Nucl. Instr. Meth. A, 311, 350.
6. Wascko M.O. et al., Proc.26th ICRC (Salt Lake City, 1999, OG2.1.11.
7. Samuelson, F.W., et al. 1998, *Astrophys. J.*, 501, L17.
8. N. Hayashida et al., *Ap. J. L.* **504** (1998) 71.
9. Aharonian, F., et al. 1999, *Astron. & Astrophys.*, 342, 69.
10. Djannati-Atai, A. et al., 19th Texas Simposium Proceedings
11. Westerhoff, S. et al., Proc.26th ICRC (Salt Lake City, 1999, OG2.6.06.
12. Wang, K. et al., Proc.26th ICRC, Salt Lake City, 1999, OG2.4.31.
13. Meegan, C. A. et al, Nature,355,143 (1992).
14. McEnery, J.E. et al., This Proceedings, 1999.
15. Leonor, I. et al., Proc.26th ICRC Salt Lake City, 1999, OG2.6.06.
16. Sinnis, G. et al., Proc.26th ICRC Salt Lake City, 1999, OG2.3.07.
17. McCullough,J.F. et al., Proc.26th ICRC Salt Lake City, 1999, HE6.1.02.
18. Yodh G.B. et al., These Proceedings.

Computer Animation of Extensive Air Showers Interacting with the Milagro Water Cherenkov Detector

Miguel F. Morales

*Santa Cruz Institute for Particle Physics and Department of Physics,
University of California, Santa Cruz, CA 95064, USA*

I employ advanced computer animation to visualize the interaction of Extensive Air Showers (EAS) with the Milagro water reservoir. The animations help conceptualize the evolution of the EAS particle front as it hits a large volume of water and converts to a front of Cherenkov photons. Expected effects such as refraction and curvature are easily seen, as well as a number of novel behaviors. A simple model that explains the observed dynamics of the shower front is presented.

Introduction

In recent years advanced computer visualization tools have become available to the scientific community. Visualization offers a very powerful way of analyzing and understanding scientific data, particularly when examining new or poorly understood situations. To create useful fits and histograms for automated data analysis, one needs a conceptual understanding of the experiment. It is in developing this conceptual understanding and in formulating models that data visualization can be an indispensable research tool. The models derived from the visualization can then be used for traditional data analysis.

The Milagro experiment is a case in point. Milagro is a water Cherenkov gamma ray telescope, and is the first air shower telescope of its kind (1). The particle front of an EAS enters the Milagro reservoir at relativistic speeds and is quickly absorbed by the water, being replaced by a front of Cherenkov photons. In principle it is a very simple design, but there are a number of subtleties involved. For example, the Cherenkov photons are not emitted parallel to the incoming EAS particles and are travelling at the speed of light in water, significantly slower than the particles in the EAS. In order to examine how these subtleties affect the formation of the Cherenkov light front, I developed a method for creating 3D animations of the EAS as it interacts with the Milagro pond.

These animations show a number of interesting and novel behaviors we had not anticipated, and have led to a new model of the EAS-pond interaction.

This paper will focus on explaining the model of the EAS-pond interaction that was developed using insights gained from the computer animations. Unfortunately, the computer animations themselves are nearly impossible to present in printed form. All of the animations are available on the web, and there is a separate web site for the animated figures referred to in this paper. Animations viewable on the web will referred to as Web Figures (e.g. Web Figure 2), and can be seen at http://scipp.ucsc.edu/milagro/Animations/Snowbird99.html. This site also includes a full description of how the animations were produced.

Bowl-Ring Model

Computer animation techniques were used to create a set of movies showing the interaction of an EAS and the Milagro Pond. After studying the animations in some depth, it became apparent that there were several unexpected phenomenon. This is most dramatically seen in Figure 1a, where an EAS initiated by a 2 TeV gamma ray from zenith is interacting with the Milagro pond. Since we are viewing the pond and shower from the side, we would expect the classic ring Cherenkov light ring to appear as a thin band. However, in addition to the Cherenkov ring a bowl of Cherenkov light can easily be seen. In Figure 1b the perspective is changed slightly to clearly show a thin bowl of Cherenkov light with a single bright band circling the bowl.

The bowl and ring structure is coincident with the core of the EAS shower, and can

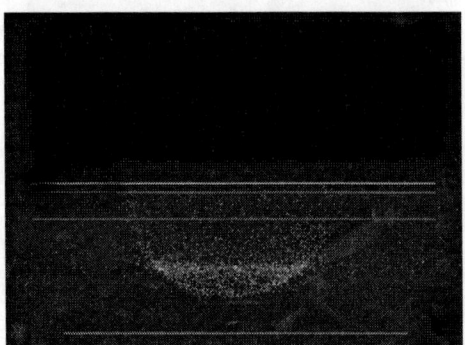 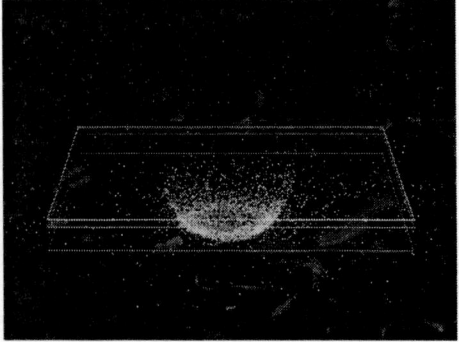

FIGURE 1. These stills show an EAS from a 2 TeV gamma primary as it interacts with the Milagro pond. **a)** The pond viewed from the side with the top horizontal line representing the surface of the pond, and the next two lines representing the layers of photo detectors. Almost all of the particles seen in this image are Cherenkov photons. Notice the bowl structure of Cherenkov light. **b)** The same shower from a slightly elevated perspective to clearly show the three dimensional structure of the Cherenkov light emission. Note the distinctive bowl and ring shape, not just the classic Cherenkov ring. The full color animation these stills are derived from can be seen in Web Figure 1 at http://scipp.ucsc.edu/milagro/Animations/Snowbird99.html.

be explained by multiple scattering. The particles of the EAS are moving relativistically as they enter the Milagro pond, each emitting light in the classic Cherenkov ring pattern (2). When the core of the shower first enters the pond, the particles' paths are coaxial and the rings of Cherenkov light from all of the particles add to form one bright ring as depicted in Figure 2. However, as the particles penetrate deeper into the detector, they suffer multiple scattering and can be deflected through large angles. The particles of the core are no longer collimated and are in fact travelling in all directions. Because of this scattering the Cherenkov rings from the particles in the core are no longer coincident, but are oriented in all directions. This scatter in the orientation of the Cherenkov rings leads to the formation of a bowl structure as depicted in Figure 3.

This model very naturally describes the formation of the bowl and ring structure of Cherenkov light seen in the computer animations. When the core first enters the pond all of the emitted Cherenkov light goes into the bright ring. However, as the core penetrates deeper, multiple scattering takes over, scattering the particles of the core so that the Cherenkov light emitted forms a thin bowl instead of a ring. We can confirm that multiple scattering forms the bowl structure by returning to the Monte Carlo and creating a movie with multiple scattering turned off. This animation can be seen in Web Figure 2, where the ring is still apparent, but the bowl structure is absent.

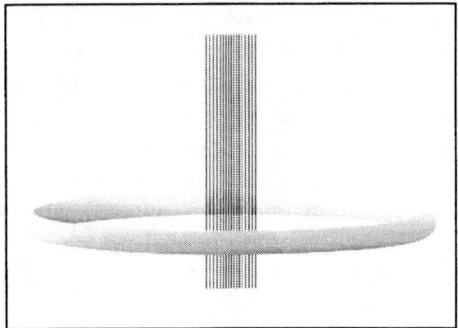

FIGURE 2. The particles of the EAS core are travelling colinearly as they enter the pond. The Cherenkov rings from each particle add to form one bright Cherenkov ring.

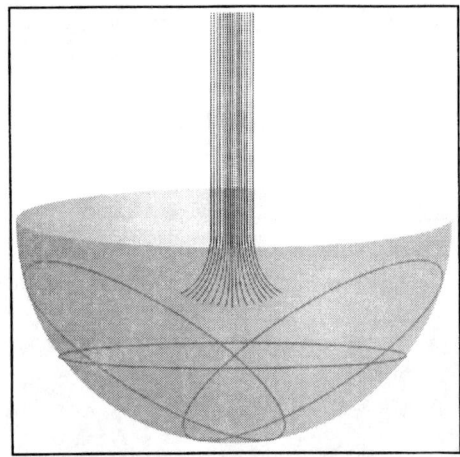

FIGURE 3. As the particles in the EAS core penetrate into the pond they suffer multiple scattering. This scatters the orientation of the individual Cherenkov rings. The individual rings no longer add to form one ring, but instead form the surface of a bowl. Combining the ring from when the particles first enter the pond and the bowl formed due to multiple scattering gives the bowl and ring structure observed in the computer animations.

The bowl-ring pattern applies to more than just the core. Because every particle emitting Cherenkov light will suffer multiple scattering, the probability density of Cherenkov light emission for a single particle will be the bowl-ring structure. This qualitative bowl-ring pattern is not confined to the core, but in fact describes the normal pattern of light production in the detector.

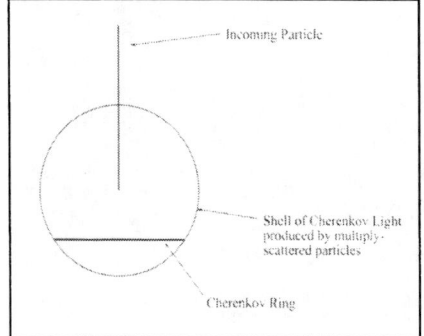 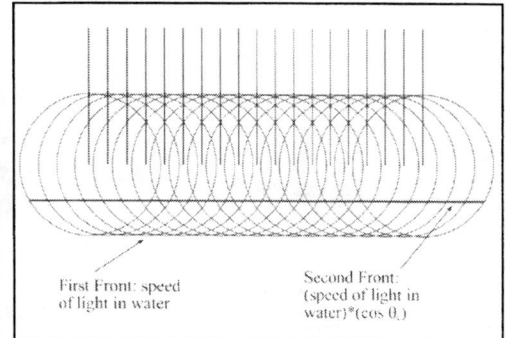

FIGURE 4. a) This diagram sketches the probability density of Cherenkov light emission from a single incoming particle. **b)** A plane particles from zenith with all emit photons in the same bowl-ring structure. The bowls add to form a leading photon front that travels at the speed of light in water. The rings add to form a secondary front that propagates more slowly. This diagram also shows that there is a lot of diffuse light that not associated with either front.

This new model for the Cherenkov emission pattern not only explains the structure seen in the first animation, but also explains a number of other novel behaviors observed in other animations. Before looking at more animations, let's study the behavior of a shower front assuming that the new pattern of Cherenkov emission is valid.

In order to study some of the subtler effects of a shower interacting with the Milagro pond, let me introduce a simplified model of an EAS. My toy model EAS will consist of a plane of equally spaced, identical, monoenergetic particles. By removing effects caused by variations in particle density and energy with distance from the shower core, some of the more subtle effects can be easily observed.

Figure 4a shows a simple cartoon for the probability of Cherenkov emission from a single incoming particle. If we consider a plane of identical particles coming from zenith, each will emit Cherenkov light in the same bowl-ring pattern. The combination of both a bowl and ring in the emission pattern leads to the formation of two separate fronts as shown in Figure 4b. The first front is formed by the bottom of the bowls, and travels at the speed of light in the medium. The second front is formed by the rings and travels at cosine of the Cherenkov angle in the medium times the speed of light in the medium – or about 30% slower for a water based detector. This leads to bifurcation of the EAS particle front into two Cherenkov photon fronts – one for the bowls and another for the rings.

Now let us consider another toy model EAS incident at an angle as depicted in Figure 5a. Again the bifurcation of the shower front and the late light are apparent, but the second front is much broader than in the shower from zenith. This is because the Cherenkov rings are coaxial with the incoming particles, not the refracted shower fronts. The rings stack like dominoes when the shower is incident at an angle, and only fully overlap for a vertical shower. This stacking leads to an angle dependent broadening of the second Cherenkov light front.

All of these effects can be seen through close examination of Figure 5b, showing a

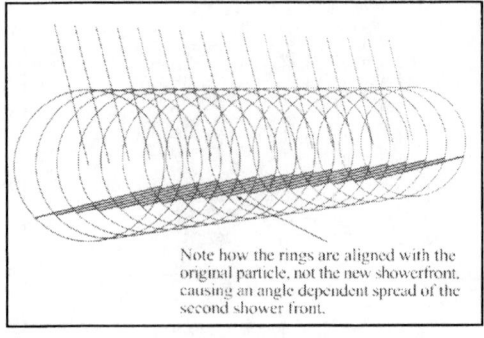 

FIGURE 5. a) Depicts a toy model EAS incident on the pond at an angle. Note again the double front, with the second front showing angle dependent smearing. **b)** A still from a full Monte Carlo animation of the toy model EAS. The angled line shows where the particles of the EAS would be if they had not encountered the pond. Close examination of this image shows the refraction, bifurcation, and angle dependent smearing predicted by the bowl-ring model of Cherenkov light emission. The full animation can be seen in Web Figure 3.

full Monte Carlo simulation of a toy model EAS incident at an angle. The refraction is clearly apparent, with close study revealing the second shower front. Web Figure 4 shows the same simulation without multiple scattering to isolate the second front and clearly show the angle dependent spreading.

The qualitative bowl-ring model of Cherenkov light emission explains both the formation of the bowl structure seen in the full EAS simulations, and the refraction, bifurcation and angle dependent spreading of the Cherenkov light fronts. This model gives us a much more robust understanding of the EAS-pond interaction, and will allow us to develop more effective automated data analysis routines.

Conclusion

Computer animation techniques are a powerful new way of analyzing experiments and the interplay of subtle physical effects. By applying animation techniques we were able to observe several unexpected phenomena, and to develop a new model that better explains the interaction of an EAS with the Milagro detector.

I would like to thank the UCSC Scientific Visualization Laboratory for enabling this research, and the Milagro collaboration for their suggestions and support.

References

1. McCullough, J. F., "Status of the Milagro Gamma Ray Observatory," presented at the 26th International Cosmic Ray Conference, Salt Lake City, UT, August 1999.
2. Jackson, J.D., *Classical Electrodynamics, Second Edition*, New York, John Wiley & Sons, 1962, pp. 638-641.

Gamma Hadron Separation in Milagro

Gaurang B. Yodh * for the Milagro Collaboration[1]

University of California Irvine, Irvine, CA, 92697-4575

Abstract.
All ground based high energy gamma ray detectors need to eliminate as many cosmic ray induced shower triggers as possible. This paper discusses techniques to achieve gamma-hadron separation and thereby improve the sensitivity of Milagro to gamma ray showers. These methods depend primarily on the two level structure of Milagro to recognize the penetrating power and clustering characteristic of hadron induced showers. Results from simulation studies are presented. The gamma-hadron cuts can improve the ratio, $\frac{signal}{\sqrt{background}}$, by more than a factor of two, which corresponds to rejecting more than 90 percent of cosmic ray induced showers.

INTRODUCTION

Air showers generated by nuclear cosmic rays dominate the trigger rates in ground based very high energy gamma ray telescopes. In order to enhance a gamma ray signal, techniques for rejection of hadronic generated showers need to be applied to the data. Air Cherenkov telescopes achieved rejection by using distinguishing characteristics of shower images from nuclear and gamma ray primaries, respectively [1] coming from the direction of the source being studied. These images reflect differences in longitudinal and lateral distribution of shower profiles from gamma ray and nuclear showers and ACT telescopes exploit these differences to reject unwanted cosmic ray background.

Milagro samples showers which survive to observation level. The lateral distributions of the shower particle densities for gamma ray initiated showers are expected to be relatively smooth over the detector, while nuclear showers can be clumpy due to the presence of muons, high energy gammas and hadrons in shower development. Milagro samples the Cherenkov light produced by shower particles in two layers: the air shower layer below about 1.5 meter of water and the 'muon' layer below 7

---

[1] This research was supported in part by the National Science Foundation, the U.S. Department of Energy Office of High Energy Physics, the U. S. Department of Energy Office of Nuclear Physics, the Los Alamos National Laboratory, the University of California, the Institute of Geophysics and Planetary Physics , the Research Corporation and the California Space Institute.

meters of water. The raggedness expected for nuclear showers is obscured by the general illumination by shower particles (e^+, e^- and MeV gamma rays) in the air shower layer(called the top layer) but their presence should show up in the muon layer(called the bottom layer). The presence of muons, high energy gammas and hadrons in the Milagro detector should lead to clusters of PMTs in the bottom layer having large pulse heights(pes). The bottom layer will also be illuminated by Cherenkov light from shower particles penetrating through 7 meters of water. We have searched for an algorithm which measures the presence of penetration over the general illumination by shower particles in the bottom layer.

CHARACTERIZATION OF MILAGRO TRIGGERS:

Showers which trigger Milagro, with its out-rigger water tank array, are characterized by number of shower tubes hit, timing and pulse heights of hit PMTs which depend on zenith and azimuth angles of the shower and the location of the core of the shower. We list some of the measured parameters below:

nhittop: Number of hit PMTS in the air shower(top) layer.
nhitbot: Number of hit tubes in the muon(bottom) layer.
petop: Total number of pes detected in the top layer.
pebot: Total number of pes detected in the bottom layer.
pemax: Maximum number of pes detected in a tube in the bottom layer.
theta: Zenith angle of the shower
rcore: distance of the core position from Milagro pond center.

There are other parameters which can be generated which quantify clustering, as well as, parameters which have to do with relative timings of hits. The algorithms developed here to reject hadronic showers are functions of rcore, theta and nhittop.

We have developed three different methods(not all independent) to reject nuclear showers: (A) Neural Net approach(Stefan Westerhoff), (B). A method using wavelet techniques(Richard Miller) and (C) A single cut parameter method(Gaurang Yodh and Robert Atkins)

In this paper, we will discuss method C, which uses the bold faced parameters (**nhitbot,pemax**) to generate a sensitive cut parameter for rejection of nuclear showers. It is based upon study of simulated Milagro triggers. A brief summary of method A will also be given.

SIMULATION STUDY OF MILAGRO TRIGGERS:

The Monte Carlo shower samples were generated for proton and gamma initiated showers using Corsika. Proton events were generated with a spectral slope of -2.75 and gamma events with spectral index -2.4. Spectra were generated with a minimum energy of 100 GeV and the zenith angle range was from 0-60 degrees. The cores of these showers were thrown over an area of 4×10^4 m^2 around the center of

Milagro pond and the response of Milagro was simulated using a Geant program. Triggers with more than 50 tubes in the top layer were used in this study.

In method C, two experimentally measured quantities are considered:(1) pe-max:the maximum pulse height recorded in the bottom layer PMTs (in equivalent photo-electron units - pes), abbreviated as **pm**; and (2) nhitbot:the total number of tubes hit in the bottom layer, abbreviated as **nb**. Figure 1 shows the distribution of nb for proton and gamma showers respectively.

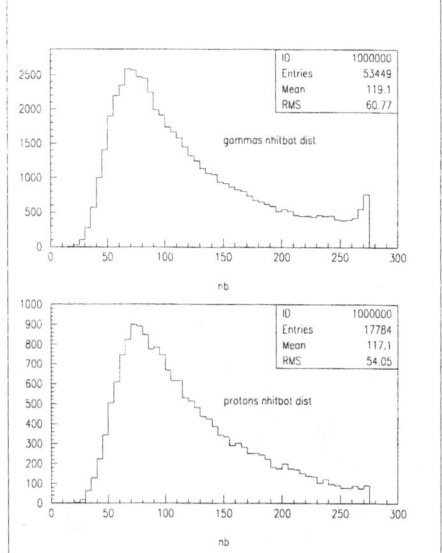 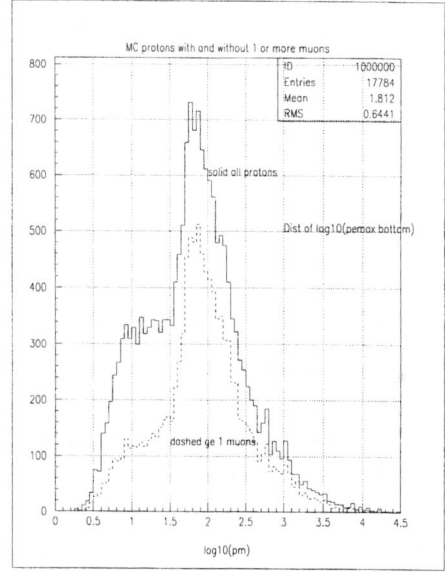

Figure 1: nhitbottom distributions   Figure 2:log(pm) with and without muons

The distribution of $\log_{10}(pm)$ for protons is shown in Figure 2. A large peak is seen at pm$\sim$ 100 and a shoulder at lower values. Also shown in the figure is the distribution when at least one muon penetrates to the bottom layer. Note that only about half of the peak is due to penetrating muons, the remaining events must have penetration of some other type of energetic component of the shower - either hadrons or high energy photons which shower in the 7meters.

In Figure 3, the $\log_{10}(pm)$ distributions of gamma initiated showers are compared with that of proton initiated showers. The gamma distribution does not show the peak seen in proton distribution. This difference is now used to develop the cut parameter.

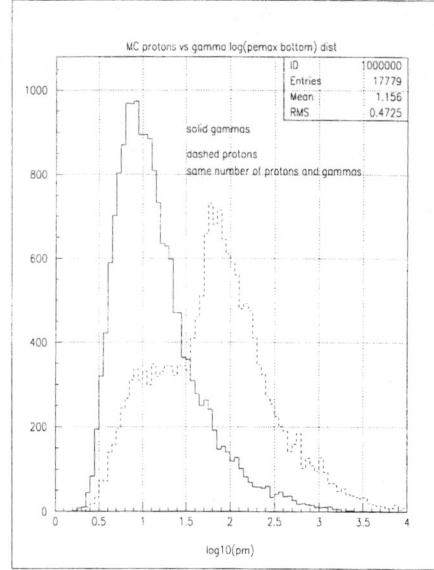 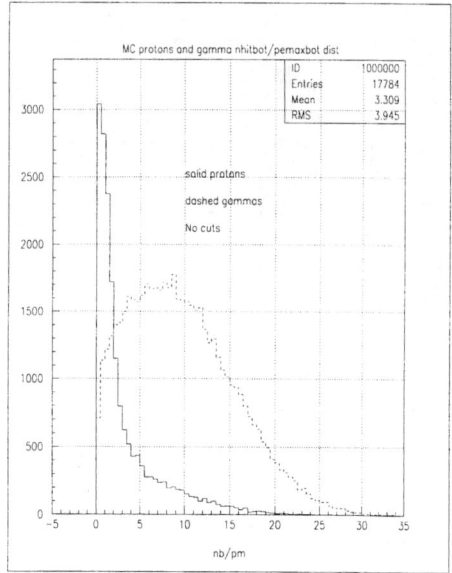

Figure 3: Graphs of log(pm) for p & g   Figure 4: x distribution for all p and g

The cut parameter, called x, is defined as $x = \frac{nhitbottom}{pemaxbottom} = \frac{nb}{pm}$. The x distributions for gamma and proton showers are superposed in Figure 4. This plot is for all triggers with nhittop > 50 and all angles. The figure shows that a simple cut on the variable x can result in rejection of a substantial fraction of proton induced showers.

As an illustration, place the cut at x=5 and reject all events with x<5, then one retains f_γ=75 % of the gamma showers and only f_p=20 % of the proton showers. The resulting Q factor, defined by $Q = \frac{f_\gamma}{\sqrt{(f_p)}}$, is close to 1.7. So a 4 sigma would be increased to 6.8 sigma in the same observation time. For every cut on the data sample - different ranges of nhittop, different core distances and different zenith angles- it is necessary to optimize the location of x -cut.

The Q factor depends on core distance and the number of hits in the air shower layer, nhittop. Higher energy particles in a shower are generally closer to the core, for either a gamma or a hadronic primary. One expects, therefore, that the gamma-hadron separation will decrease for showers whose core is on the pond, resulting in a smaller Q-factor. For showers whose cores are off the pond, simulations show that gamma-hadron separation improves because for a fixed nhittop, showers which trigger Milagro are of higher energy compared to those with cores on the pond. Higher energy proton showers contain a larger number of energetic particles away from the core which could produce a large pulse heights in the bottom layer. Gamma showers at large distances from the core are relatively smooth and contain smaller number of energetic particles than proton showers. Simulations show that:

1. Summed over all core distances, Q factor increases from 1.5 for nhittop between 50-100 hits to 4.7 for nhittop between 250 to 300 hits.

2. For showers whose cores fall outside the pond the Q factor reaches 6-7 for nhittop between 250 to 300 hits! For this case more than half of all gamma showers are retained.

Method C does not utilize all the measured parameters which could contribute to determining gamma-hadron separation. Method A is a neural net approach to work in a d dimensional space spanned by d parameters that have potential gamma-hadron separation power. In this method, fifteen parameters were used as input parameters for a three layer feed-forward neural network. These 15 parameters included nhittop, nhitbot, petop, pebot, pemax listed earlier and in addition the number of PMTs and pes in five rings centered on the fitted core position. The network is trained to give the output "0" for γ and "1" for proton showers. Neural net was trained on 1500 gamma showers and 1500 proton showers whose cores were thrown over 9×10^4 m^2 area with energies from 100 GeV to 100 TeV and a spectral slope of -2.4 for γs and -2.7 for protons and zenith angle from 0 to 45 degrees. A minimum of nhittop of 50 tubes was required. Neural net output histograms are shown in Figure 5 for all showers, for showers with cores on the pond and for showers with cores off the pond. The optimal cut in the network output is a function of nhittop. Using optimal cut in the network output, Figure 6 shows the variation of Q factor as a function of nhittop for the three cases of Figure 5. The Q values obtained using neural net output are comparable to those obtained by Method C.

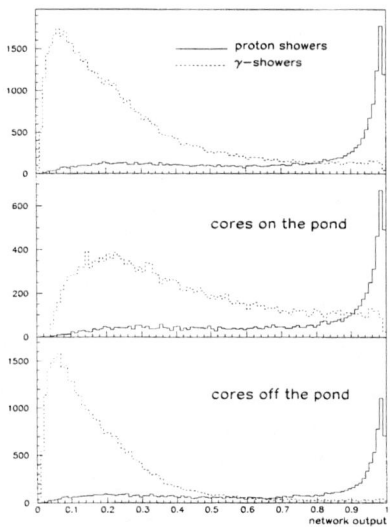

Figure 5: Neural network output

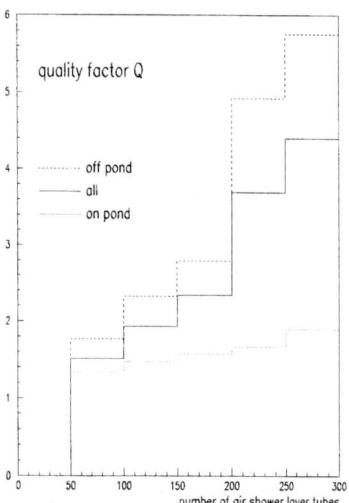

Figure 6: Q factor vs nhittop

CONCLUSIONS:

The performance of Milagro air shower detector can be enhanced by utilizing the ' muon ' layer of Milagro detector to reject a significant number of showers due to nuclear cosmic rays. A sensitive parameter which distinguishes proton showers from gamma showers, is x = $\frac{nb}{pm}$. Two methods are outlined in the paper, one which depends on rejecting hadron showers using this x parameter and another which uses a neural net technique. Applying optimal cuts, these methods can yield Q factors which can become as large as 6 for selected subset of triggers corresponding to rejection of greater than 99 percent of nuclear showers .

The work on method C was done in collaboration with Robert Atkins and method A was developed by Stefan Westerhoff [2,3]

REFERENCES

1. Weekes T. C., et al.,*ApJ* **342**, 379 (1989)
2. Atkins Robert and Gaurang B. Yodh, http://daneel.ps.uci.edu/yodh/MilagroMemos/ in files ghqfactor.ps, ghqfactor_add.ps and ghqfactor_data.ps. Printed copies of these memos can be obtained by contacting the author at e-mail address gyodh@uci.edu
3. Westerhoff Stefan, http://daneel.ps.uci.edu/yodh/MilagroMemos/ in file stefan_ghsep.ps.Printed copies of these memos can be obtained by contacting the author at e-mail address gyodh@uci.edu

Search for Multi-TeV Gamma-Ray Emission from Nearby SNRs with the Tibet Air Shower Array

The Tibet ASγ Collaboration

M. Amenomori[1], S. Ayabe[2], P.Y. Cao[3], Danzengluobu[4], L.K. Ding[5,9], Z.Y. Feng[6],
Y. Fu[3], H.W. Guo[4], M. He[3], K. Hibino[7], N. Hotta[8], Q. Huang[6], A.X. Huo[5],
K. Izu[9], H.Y. Jia[6], F. Kajino[10], K. Kasahara[11], Y. Katayose[9], Labaciren[4],
J.Y. Li[3], H. Lu[5], S.L. Lu[5], G.X. Luo[5], X.R. Meng[4], K. Mizutani[2], J. Mu[12],
H. Nanjo[1], M. Nishizawa[13], M. Ohnishi[9], I. Ohta[8], T. Ouchi[7], J.R. Ren[5],
T. Saito[14], M. Sakata[10], T. Sasaki[10], Z.Z. Shi[5], M. Shibata[15], A. Shiomi[9],
T. Shirai[7], H. Sugimoto[16], K. Taira[16], Y.H. Tan[5], N. Tateyama[7], S. Torii[7],
T.Utsugi[2], C.R. Wang[3], H. Wang[5], H.Y. Wang[5], P.X. Wang[12], X.W. Xu[5],
Y. Yamamoto[10], G.C. Yu[6], A.F. Yuan[4], T. Yuda[9], C.S. Zhang[5,9],
H.M. Zhang[5], J.L. Zhang[5], N.J. Zhang[3], X.Y. Zhang[3],
Zhaxisangzhu[4], Zhaxiciren[4] and W.D. Zhou[12]

[1] *Department of Physics, Hirosaki University, Hirosaki, Japan*
[2] *Department of Physics, Saitama University, Urawa, Japan*
[3] *Department of Physics, Shangdong University, Jinan, China*
[4] *Department of Mathematics and Physics, Tibet University, Lhasa, China*
[5] *Institute of High Energy Physics, Academia Sinica, Beijing, China*
[6] *Department of Physics, South West Jiaotong University, Chengdu, China*
[7] *Faculty of Engineering, Kanagawa University, Yokohama, Japan*
[8] *Faculty of Education, Utsunomiya University, Utsunomiya, Japan*
[9] *Institute for Cosmic Ray Research, University of Tokyo, Tanashi, Japan*
[10] *Department of Physics, Konan University, Kobe, Japan*
[11] *Faculty of Systems Engineering, Shibaura Institute of Technology, Omiya, Japan*
[12] *Department of Physics, Yunnan University, Kunming, China*
[13] *National Center for Science Information Systems, Tokyo, Japan*
[14] *Tokyo Metropolitan College of Aeronautical Engineering, Tokyo, Japan*
[15] *Faculty of Engineering, Yokohama National University, Yokohama, Japan*
[16] *Shonan Institute of Technology, Fujisawa, Japan*

Abstract. The Tibet air-shower array operating at Yangbajing (4300 m above sea level) is sensitive to gamma-ray air showers at energies as low as 3 TeV. The observation of the Moon's shadow has provided a direct check of the angular resolution, the energy estimation and the systematic pointing error of this air-shower array. Using these data, we have searched for multi-TeV gamma-ray emission from 27 SNRs located within 10 kpc distance in the declination band of $-10°$ to $+70°$. The signal from the Crab Nebula was detected at 5.5 σ level. The intensity observed is $(4.61 \pm 0.90) \times 10^{-12}$ cm^{-2} s^{-1} TeV^{-1} at 3 TeV. No significant DC excess was found from any of these SNRs except from the Crab Nebula.

INTRODUCTION

It is widely believed that expanding supernova remnants (SNRs) are responsible for accelerating particles up to energies of at least 100 TeV, and a fraction of the accelerated particles would interact within SNR and produce high energy gamma-rays[1,2]. Searches for such gamma-rays in the TeV energy region have been so far performed with atmospheric Čerenkov telescope techniques. Several groups have so far succeeded in detecting signals from the Crab Nebula and/or the other SNRs with high significance[3,4]. The systematic uncertainties in their intensity estimations, however, remain large because that their observations much depend on atmospheric conditions.

A high-density air-shower array (HD array) has been operating in Tibet since 1996. Highly unambiguous observations of multi-TeV gamma-rays has been realized by using this air-shower array. In this paper we present the results given by this HD array on the search for gamma-ray emission from 27 SNRs.

EXPERIMENT

The Tibet air shower array shown in Figure 1 is located at Yangbajing (4300 m above sea level, $90.52°E$, $30.11°N$) in Tibet, China[5]. The HD array consists of 109 scintillation counters of 0.5 m² each placed on a 7.5 m × 7.5 m grid, covering an area of 5,175 m². The recording system is triggered by any fourfold coincidence of these counters, which results in a trigger rate of about 110 Hz in average.

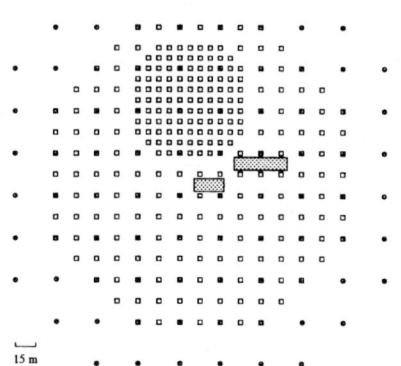

FIGURE 1. Schematical view of the Tibet air shower array. Squares and circles are scintillation counters. The dense part in arrangement indicates the HD array.

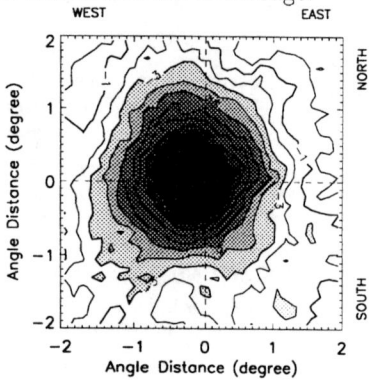

FIGURE 2. Moon's shadow.

ANALYSIS

The data used in the present search were collected during the period from November 1996 through May 1999. The effective running time is about 502.1 days. The event selection was performed by imposing the following four conditions to the recorded data: 1) Each of any four detectors should record a signal more than 1.25

particles; 2) the sum of the number of particles per m² detected in each detector $\Sigma \rho$ should be greater than 15; 3) among the four detectors recording the highest particles, two or more should be inside the innermost 9×9 detectors; and 4) the zenith angle θ of the incident direction should be less than 50°.

The air-shower events detected are mainly initiated by cosmic-ray protons and nuclei rather than gamma-ray. Those background cosmic-ray particles are isotropic and the gamma-rays from a source are apparently centered on the source direction. The reduction of background is accomplished via the good angular resolution. We checked the angular resolution by observing the shadow that the Moon casted in the cosmic-ray flux. The Moon's shadow was observed with a significance of 15 σ at the maximum deficit position for the events with $\theta < 45°$, as shown in Figure 2. The angular resolution is estimated to be better than 0.9° for those events.

The trajectories of cosmic ray protons and nuclei with positive charge are distorted by the geomagnetic field. The Moon's shadow should be, therefore, found in the direction somewhat away from the Moon to the west. The center of the Moon's shadow observed was found at a position shifted to the west by about 0.3°, as shown in Figure 2. According to the simulation calculation, the mean energy of cosmic-ray protons generating such events is estimated to be about 5.0 TeV. This value is compatible with the one expected from the observed shift of the shadow of about 0.3°. Thus, the observation of the Moon's shadow can provide a direct check of the energy determination method.

The displacement of the Moon's shadow in the north-south direction is almost free from the influence of the geomagnetic field. By examining a deviation of the shadow, the systematic pointing error is estimated to be smaller than 0.1°.

We searched for gamma-ray signals by counting the number of events coming from the on-source window. The background number of events was obtained by averaging over the events falling in the 10 off-source windows adjacent to the on-source one at the same zenith angle θ. We set a circular window. The radius of a window is set to optimize the significance of signals.

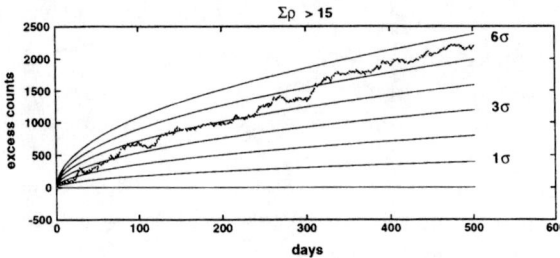

FIGURE 3. Cumulative excess of the events coming from the Crab direction.

RESULTS AND DISCUSSIONS

We used the events with $\theta < 30°$ to search for signals from the Crab Nebula. The cumulative excess counts of the events coming from the direction in terms of sidereal day are presented in Figure 3. We used there the search window with the radius of 0.8°. The statistical significance of the excess counts reached to 5.5 σ. This is the first clear observation of gamma-ray signals using a conventional air-shower array.

The statistical significances for the events with $\Sigma\rho > 40$, 70 and 130 are 3.9, 3.3 and 3.1 σ using search windows with the radii of 0.7°, 0.5° and 0.5°, respectively. The energy spectrum obtained from these results is shown in Figure 4. We fitted the results in the energy range between 3 and 15 TeV by a simple power law, yielding $dJ/dE = (4.61 \pm 0.90) \times 10^{-12}(E/3\text{TeV})^{-2.62\pm 0.17}\text{cm}^{-2}\text{ s}^{-1}\text{ TeV}^{-1}$.

We also searched for signals in the energy region above 10 TeV with another overlapping air-shower array consisting of 185 scintillation counters each placed on a 15 m × 15 m grid, covering an area of 36,900 m^2, as shown in Figure 1. Upper limit fluxes at the 90% confidence level are shown in Figure 4, together with other experiments[3,4,6]. Recent measurements with imaging Čerenkov telescopes give flux values lower than the present results in the multi-TeV energy region.

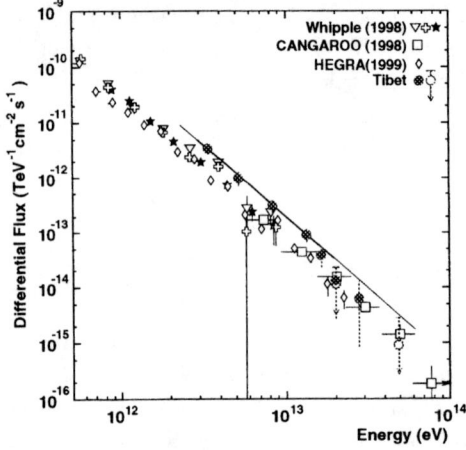
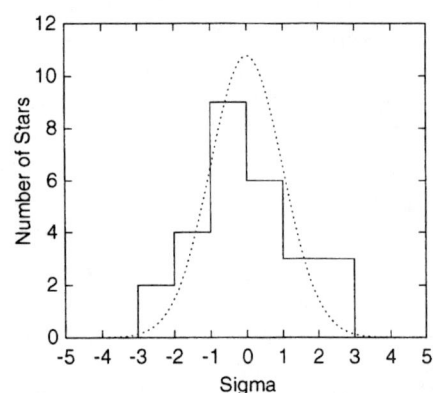

FIGURE 4. Differential energy spectrum of gamma-rays from the Crab Nebula.

FIGURE 5. Distribution of significance of 27 nearby SNRs except the Crab.

We also searched for continuous emission from the other 27 SNRs located within 10 kpc distance in the declination band of $-10°$ to $+70°$ in referring to the Green's Catalog[7], using the data during the period from Nobember 1996 through May 1998. In order to observe spread SNRs, we used search windows with the radii of 1°. A statistical significance was evaluated for each SNR, as shown in Figure 5, and the mean value was estimated to be -0.10 ± 0.26. No excellent excess was found for steady emission of 3 TeV gamma-rays from any object of these SNRs. There is a possibility of misdirecting in the analysis for widely spread sources. Also, it is hard to detect gamma-rays from distant sources. Only nearby small-sized SNRs are listed in Table 1. The distances of the SNRs in the table are less than 5 kpc and their apertures are smaller than 1°. In the table are given the background number of events, the statistical significance and the uperlimit flux at the 90% confidence level in the energy larger than 3 TeV. The flux values are estimated by a simulation calculation assuming a differential power-law spectrum of the form of $E^{-2.5}$[3]. It seems in the table that there is no correlation between significance and distance.

Table 1.

R.A.	Decl.		Type	kpc	N_b	σ	Flux*
6.33	64.14	Tycho	S	2.3	109896	−1.15	< 3.68
22.09	63.18	R5	S	1.2	115992	−0.48	< 4.98
31.42	64.82	3C58	F	3.2	104709	0.45	< 7.03
81.65	42.91		S	4.5	229535	−1.32	< 1.37
94.26	22.58	IC443	S	0.7	224208	2.01	< 2.70
284.01	1.37	W44	S	3.0	108450	0.05	< 4.18
290.95	14.10	W51		4.1	187191	−0.71	< 1.92
296.99	27.73		F	3.8	213569	1.13	< 2.31
350.86	58.81	Cas A	S	2.8	146778	−0.61	< 2.88
359.80	62.44	CTB 1	S	2.7	121731	−0.12	< 4.46

*Flux limits at 90 % confidence level in the unit of 10^{-12} cm^{-2} s^{-1}.

According to a simple model of shock acceleration in typical SNRs[1,2], the intensity of gamma-rays is expected to be a flux level of 10^{-11} cm^{-2} s^{-1} in the energy region greater than 3 TeV from a SNR of 3 kpc distant. The observation with the Tibet HD array seems to reach such level. The Whipple group reported the upper limits on the flux of gamma-rays of about 300 GeV given from the observations with an atmospheric Čerenkov telescope[8]. Under the assumption that the EGRET data give evidence for acceleration of cosmic rays, the present results on the flux upper limits lie below the corresponding expected values, same as the Whipple's ones.

The area of the present HD array will be extended by a factor of about five in 1999. Then, the Tibet experiment will result in significantly better statistics.

Acknowledgements : This work is supported in part by Grants-in-Aid for Scientific Research and also for International Scientific Research from the Ministry of Education, Science and Culture, in Japan and the Committee of Natural Science Foundation in China.

REFERENCES
1. T. Naito and F. Takahara, J. Phys. G: Nucl. Part. Phys., **20**, 477 (1994).
2. L.O'C. Drury et al., A&A, **287**, 959 (1994).
3. A.M. Hillas et al., ApJ, **503**, 744 (1998).
4. T. Tanimori et al., ApJ, **492**, L33 (1998).
5. M. Amenomori et al., 25th ICRC, **5**, 245 (1997).
6. A.K. Konopelko et al., preprint, astro-ph/9901094 (1998).
7. D.A. Green, 1998 September version, http://www.mrao.cam.ac.uk/surveys/snrs/
8. J.H. Buckley et al., A&A, 329, 639 (1998).

FUTURE OBSERVATORIES

The AGILE gamma-ray astronomy mission

S. Mereghetti[1], G. Barbiellini[2], G. Budini[2], P. Caraveo[1], E. Costa[3],
V. Cocco[4], G. Di Cocco[5], M. Feroci[3], C. Labanti[5], F. Longo[2],
E. Morelli[5], A. Morselli[4], A. Pellizzoni[6], F. Perotti[1], P. Picozza[4],
M. Prest[2], P. Soffitta[3], L. Soli[1], M. Tavani[1], E. Vallazza[2],
S. Vercellone[1]

[1] *Istituto di Fisica Cosmica G.Occhialini – CNR, Milano, Italy*
[2] *Università di Trieste and INFN, Trieste, Italy*
[3] *Istituto di Astrofisica Spaziale – CNR, Roma, Italy*
[4] *Università "Tor Vergata" and INFN, Roma, Italy*
[5] *Istituto TESRE – CNR, Bologna, Italy*
[6] *Agenzia Spaziale Italiana*

Abstract. We describe the AGILE gamma-ray astronomy satellite which has recently been selected as the first Small Scientific Mission of the Italian Space Agency. With a launch in 2002, AGILE will provide a unique tool for high-energy astrophysics in the 30 MeV - 50 GeV range before GLAST. Despite the much smaller weight and dimensions, the scientific performances of AGILE are comparable to those of EGRET.

INTRODUCTION

The AGILE satellite was proposed in June 1997 to the Program for Small Scientific Missions of the Italian Space Agency (ASI). AGILE (*Astro-rivelatore Gamma a Immagini LEggero*) is a mission devoted to gamma-ray (30 MeV–50 GeV) astrophysics during the years 2002-2005 After the initial ASI selection in December 1997, a Phase A study was carried out during 1998. AGILE was finally selected by ASI in June 1999 as the first Small Scientific Mission to be launched and is currently in the Phase B. The launch is foreseen in early 2002.

The AGILE scientific payload is based on the state-of-the-art and reliably developed technology of solid state silicon detectors [1–3]. The instrument is very light (~ 60 kg) and effective in detecting and monitoring gamma-ray sources (30 MeV–50 GeV) within a large field of view. The instrument is designed to achieve an optimal angular resolution (source location accuracy $\sim 5' - 20'$ for intense sources), a very large field of view ($\gtrsim 2$ sr), and a sensitivity comparable to that of EGRET for on-axis (and substantially better for off-axis) point sources. AGILE will also carry an imaging hard X-ray detector to simultaneously monitor in the 10-40 keV range the sources observed in the central part of the gamma-ray field of view.

Despite its simplicity and moderate cost, AGILE is ideal to perform a large number of tasks [4]: monitoring active galactic nuclei, detecting gamma-ray bursts with high efficiency, mapping the diffuse Galactic and extragalactic emission, studying pulsed gamma-ray emission from radiopulsars, monitoring the many unidentified sources and contributing to their unveiling, detecting energetic solar flares . Today, it is clear that successful investigations of gamma-ray sources rely on coordinated space and ground-based observations. The AGILE scientific program will be focussed on a prompt response to gamma-ray transients and alert for follow-up multiwavelength observations.

INSTRUMENT OVERVIEW

The AGILE scientific payload is made of three main detectors: a silicon/tungsten Tracker, a cesium iodide Mini-calorimeter, and a coded mask hard X-ray Imaging Detector. These three elements are integrated in a single structure, covered on five sides (top and lateral sides) with an Anticoincidence for charged particles rejection.

The Tracker, consisting of 14 planes of silicon strip detectors, will provide the unambiguous identification of incident gamma-rays by recording the characteristic track signature of the e^--e^+ that result from pair creation from the incident photons

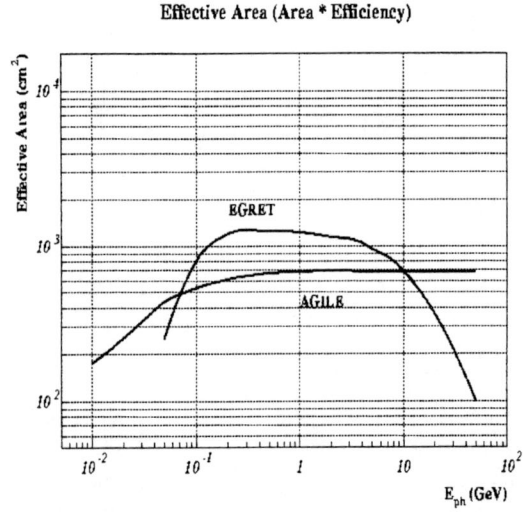

FIGURE 1. Effective area of the AGILE gamma-ray detector.

in thin layers converting material. Each plane of the Tracker (with the exception of the two at the bottom) is made of three layers: a 245 μm thick photon pair converter in tungsten (0.07 X_0) is followed by two planes of silicon strip detectors (thickness 410 μm) with the strips arranged in orthogonal directions to provide the plane coordinates of the particle tracks. The tungsten layer is absent in the last two planes, since the readout trigger requires a signal in at least three consecutive silicon planes to be activated. The Tracker has an on-axis total radiation length larger than 0.84 X_0. The resulting on-axis effective area is shown in Figure 1. The distance between planes has been fixed to 1.6 cm on the basis of extensive optimizations by Montecarlo simulations.

The fundamental unit for the silicon planes is a module of area 9.5×9.5 cm^2, and pitch (distance between strips) equal to 121 μm. Each silicon plane consists of 4x4 modules. The strips will be read out with a pitch of 242 μm, i.e. with a floating strip every two strips, yielding a total of ~43,000 channels for the whole Tracker. If a particle crosses a floating strip, image charges are induced on the two adjacent read-out strips through capacitive coupling. In this way, more than one strip has signal, thus enabling to use an interpolation algorithm to improve the spatial resolution. Special algorithms applied off-line to telemetered data will allow an optimal reconstruction of the photon incidence angle.

The Mini-calorimeter, consisting of 1.5 X_0 of Cesium Iodide (CsI), will allow to determine the energy of the incident photons imaged by the Tracker. In addition, the Mini-calorimeter will also be used to study gamma-ray bursts and other tran-

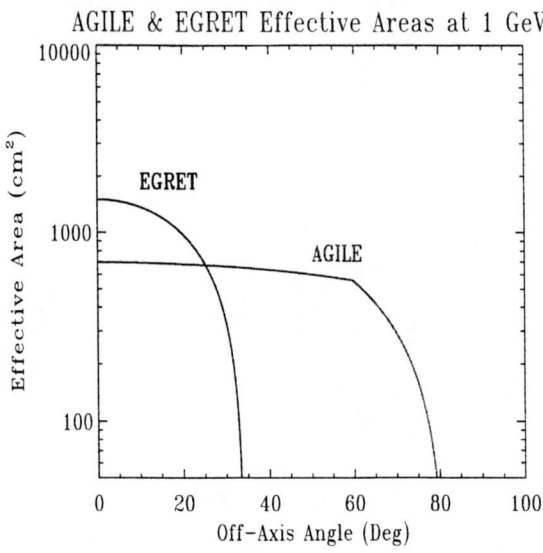

FIGURE 2. Comparison of the AGILE and EGRET effective areas for off-axis sources.

sient phenomena in the ~1-500 MeV energy range. The Mini-calorimeter is located below the Tracker and consists of two planes of CsI(Tl) bars. Each bar has dimensions of $2.3 \times 1.5 \times 40$ cm^3. The two planes have the bars arranged in orthogonal directions to provide the X and Y location of the showers. The scintillation light from each bar is collected by two photodiodes placed at both ends. We note that the problem of particle backscattering for this configuration is much less severe than in the case of EGRET, thus allowing a relatively efficient detection of photons up to 10 GeV.

The hard X-ray detector (Super-AGILE) is based on the 9.5x9.5 cm^2 silicon tiles that are used for the Tracker planes. These will be placed on the top of the Tracker (above the first tungsten layer) to form an additional detection plane sensitive in the 10-40 keV range and used in conjunction with a coded mask at a distance of about 10 cm (below the top Anticoincidence). Since the silicon microstrips provide pixels only along one-dimension, Super-AGILE will consist of four equal modules arranged in two pairs, giving monodimensional sky images along two orthogonal directions. The mask will be supported by an ultra-light structure in carbon fiber that will also serve as a collimator to reduce the Super-AGILE field of view. This is needed in order to limit the Super-AGILE background, which is dominated by the cosmic diffuse radiation.

The Anticoincidence system, aimed at both charged particle background rejection and preliminary direction reconstruction for triggered photon events, completely surrounds the top and lateral sides of the Super-AGILE, Tracker and Mini-calorimeter. The top panel is a single slab of plastic scintillator with thickness 0.5 cm. The scintillation light is collected by optical fibers glued on the four sides and directed to four photomultipier tubes at the corners. Each lateral face is segmented

TABLE 1. AGILE Scientific Performances

	Gamma-ray Detector	
Energy Range	30 MeV – 50 GeV	
Field of view	2 sr	
Sensitivity at 100 MeV	6×10^{-9} ph cm^{-2} s^{-1} MeV^{-1}	(5σ in 10^6 s)
Sensitivity at 1 GeV	4×10^{-11} ph cm^{-2} s^{-1} MeV^{-1}	(5σ in 10^6 s)
Angular Resolution at 1 GeV	36 arcmin	(68% containment radius)
Source Location Accuracy	~30 arcmin	for a source with S/N~10
Energy Resolution	$\Delta E/E \sim 1$	at 300 MeV
Timing Accuracy	25 μs	
	Hard X–ray Detector	
Energy Range	10-40 keV	
Field of view	60°×60°	Full Width at Zero Sensitivity
Sensitivity	~10 milliCrabs	(5σ in 1 day)
Angular Resolution	10 arcmin	
Source Location Accuracy	~2-3 arcmin	for a source with S/N~10
Energy Resolution	$\Delta E < 4$ keV	
Timing Accuracy	25 μs	

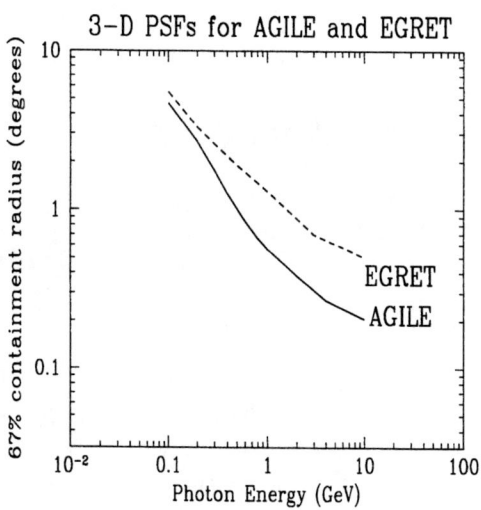

FIGURE 3. AGILE angular resolution.

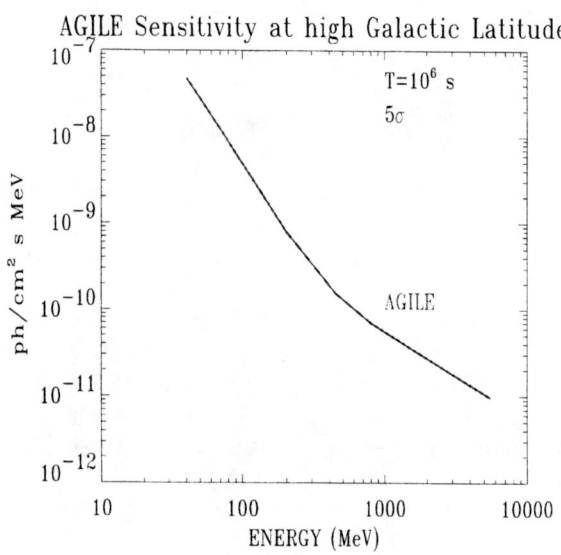

FIGURE 4. Expected sensitivity for the AGILE gamma-ray instrument.

with three partially overlapping plastic scintillator layers (0.5 cm thick) connected with photomultipliers placed at the bottom.

PERFORMANCES

The expected scientific performances of AGILE are summarized in Table 1. One of the main characteristics of the AGILE gamma-ray detector is the very large field of view. As shown in Figure 2 the AGILE effective area remains almost constant for large off-axis angles. This will allow to simultaneously monitor a large number of sources in a single pointing and will also result, at the end of the mission, in a large exposure factor for each region of the sky [5].

The great sky exposure factor, coupled with the good angular resolution (see Figure 3), will allow a detailed study of the diffuse Galactic and Extra-Galactic emission and to better locate the unidentified EGRET sources.

Another important characteristic of the AGILE Tracker is the small dead time. This will be crucial in the study of the high energy emission of gamma-ray bursts, which are expected in the field of view at a rate of ∼5-10 per year, based on the EGRET results.

The extension in the hard X-ray range of a gamma-ray mission, made possible by the Super-AGILE detector, is an innovative concept that will allow the study of correlated variability for sources of different classes, ranging from active galactic nuclei and blazars to unidentified galactic transients.

THE MISSION

The AGILE instrument will be carried in an equatorial (inclination $<5°$), circular orbit (altitude ∼550 km), by a spacecraft of the MITA class, which is currently being developed by Gavazzi Space as prime contractor. The total mass will be of the order of 180-200 kg, including ∼60 kg of scientific payload.

The satellite will point with a 3-axis attitude stabilization with an accuracy of the order of 1°. An attitude reconstruction at the arcmin level will be obtained a posteriori, by means of star sensors. A typical AGILE pointing will last 2-3 weeks. To maximise the observing efficiency, we are investigating the possibility of slewing to secondary pointing directions during the fraction of the orbit in which the primary target direction is occulted by the Earth. AGILE will also have the possibility to quickly repoint (within 1 day) in order to react to interesting targets of opportunity.

The operations center will have the primary responsibility of satellite operations and communications using the ASI ground base at Malindi (Kenya). The equatorial orbit will allow a single contact per orbit at a downlink rate of 500 kbit s^{-1}.

An AGILE data center will be devoted to the monitoring of the instrument scientific performance, to the quick look analysis, and to the processing, distribution and archival of scientific data products and the associated calibration data.

CONCLUSIONS

AGILE will provide crucial information complementary to the many lower energy detectors that will be operational during the first decade of the new Millenium (INTEGRAL, XMM, AXAF, ASTRO-E, SPECTRUM-X, and others). No other gamma-ray mission dedicated to the energy band above $\sim$ 30 MeV is planned before GLAST (which will most likely be operational after 2005-2006). AGILE might then be an ideal 'bridge' between old and new generation gamma-ray missions with an innovative design and efficient scientific management.

REFERENCES

1. Barbiellini G. et al., *Nuclear Physics B* **43**, 253 (1995).
2. Barbiellini G. et al., *Nucl. Instrum. & Methods* **354**, 547 (1995).
3. Barbiellini G. et al., *SPIE* **2478**, 239 (1995).
4. Tavani M., et al., *Mem.S.A.It.* **70**, 201 (1999)
5. Mereghetti S., et al. *Mem.S.A.It.* **70**, 229 (1999)

The Capabilities of the Alpha Magnetic Spectrometer as GeV γ-Rays Detector

R. Battiston

Sezione INFN and Dipartimento di Fisica dell' Università, Perugia, Italia 06100

Abstract. The modeled performance of the Alpha Magnetic Spectrometer (AMS) as a high-energy (0.3 to 100 GeV) gamma-ray detector is described, and its gamma-ray astrophysics objectives are discussed.

INTRODUCTION

Our knowledge of the γ-ray sky has increased dramatically during this last decade, due principally to the γ-ray instruments on board the Compton Gamma Ray Observatory (CGRO) [1]: EGRET, a spark chamber plus calorimeter instrument with γ-ray sensitivity in the energy interval 30 MeV to 30 GeV; COMPTEL,

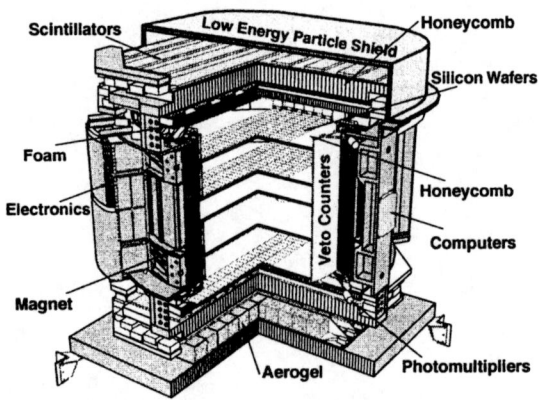

FIGURE 1. A cross section of the baseline AMS instrument.

a Compton telescope in the interval 0.1 to 30 MeV; BATSE, an omnidirectional x-ray and soft γ-ray "burst" detector consisting of large NaI scintillators sensitive to 30 keV to 2 MeV photons (with smaller spectroscopic NaI crystals for measurements up to 110 MeV); and OSSE, consisting of Nai-CsI phoswiches detecting photons of

0.1 to 10 MeV. Their observations have revolutionized our understanding of such extragalactic phenomena as blazars and gamma ray bursts (GRBs), as well those within our own Galaxy, such as pulsars.

Until recently, observations with CGRO and other space-based and ground-based telescopes have provided coverage of these and other sources up to the limiting sensitive energy of EGRET, approximately 30 GeV. From there, a gap in our knowledge of the γ-ray sky spectrum has existed up to 200-300 GeV, where ground-based γ-ray shower detectors presently have their energy thresholds. It is possible that within this gap there are novel features in the γ-ray sky, such as a gamma-ray line or continuum emission from postulated neutrino annihilation at the center of the Galaxy [22], [23], [24]. Future instruments with sensivity in this unexplored region, such as AGILE [3], recently approved by the Italian Space Agency (ASI), or GLAST [2] being proposed to NASA, may uncover exciting new phenomena.

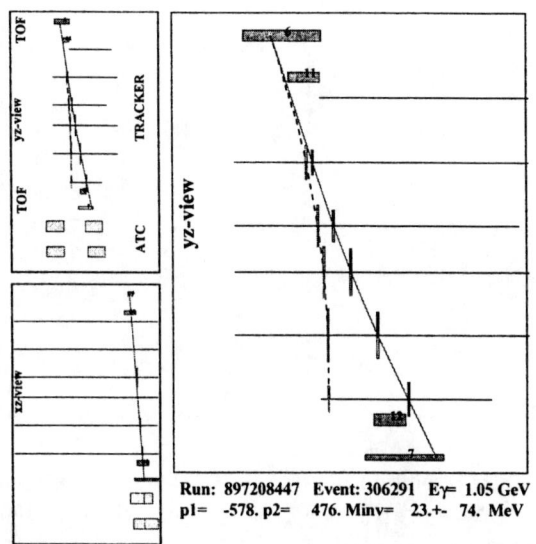

FIGURE 2. A $1 GeV$ γ-ray conversion detected during the STS91 AMS flight.

At the present time, however, our view of the γ-ray sky has diminished. With the effective turnoff of the EGRET γ-ray instrument on the CGRO due to the nearly complete consumption of its spark chamber gas, there is no operating instrument capable of observing the γ-ray sky in the energy interval $\sim 10^{-1}$ to $\sim 10^2$ GeV. Ground-based γ-ray detectors, based on the atmospheric Cerenkov technique (ACT) [4], turn on at current energy thresholds of $\sim$200-500 GeV. Within the next several years energy thresholds for some ACT observatories are expected to go to as low as 20 to 50 GeV, but lower thresholds than these are unlikely to be achieved due to the sizable effect of Earth's magnetic field on the γ-ray-induced air showers, and the lower Cerenkov photon yield which must be detected against the night sky background. For γ-ray energies much lower than 10^2 GeV, then, spaced-based detectors are required. This observational gap for the energy window $\sim 10^{-1}$ to 10^2 GeV will continue to exist for the next several years. Eventually, this gap will be eliminated with the launch of a next-generation γ-ray satellite, such as GLAST [2], but such a mission is unlikely to occur before the year 2005. Also the AGILE satellite [3], to be launched in 2002, will have a limited sensitivity above 50 GeV.

In this paper we describe how the Alpha Magnetic Spectrometer (AMS) can largely fill this gap by acting as a γ-ray detector with sensitivity in the energy interval of 0.3 to 100 GeV during its three-year mission on board the International Space Station Alpha (ISSA) from 2003 to 2006. AMS, described in detail elsewhere [8], has as primary mission the search for cosmic ray antinuclei as well as the search for dark matter studying anomalies in CR spectra and composition (e.g. $e^+, \bar{p}$).

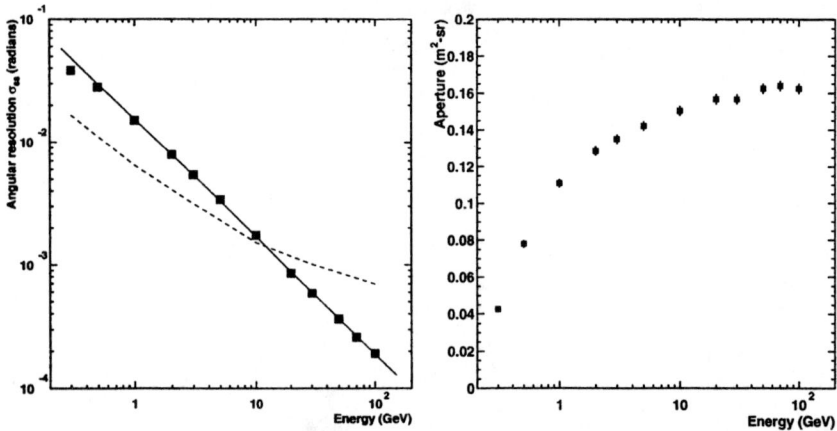

FIGURE 3. (a) Angular resolution of AMS/γ (filled squares) and of GLAST (dashed line) as a function of primary γ-ray energy in the interval 0.3 to 100 GeV; (b) AMS/γ aperture as a function of γ-ray energy.

There are at least two ways this modification could be done without significantly affecting the experiment sensitivity to antimatter: (a) adding a light (e.g. 0.3 X_0) converter at the entrance of the magnetic spectrometer, either passive (e.g., a high-Z thin plate) or active (e.g. a multilayer tracking detector), and/or (b) implementing an high granularity imaging shower detector at the bottom of the experiment. In this paper we present a study which has been perfomed on option (a). We will show that in this option, AMS can also detect γ-rays with performance characteristics similar to EGRET in the energy region of 0.3 to 20 GeV, and with significantly enhanced capabilities between 20 and ~ 100 GeV, a region which is not well explored. We refer to this modified instrument as "AMS/γ". We show that AMS/γ

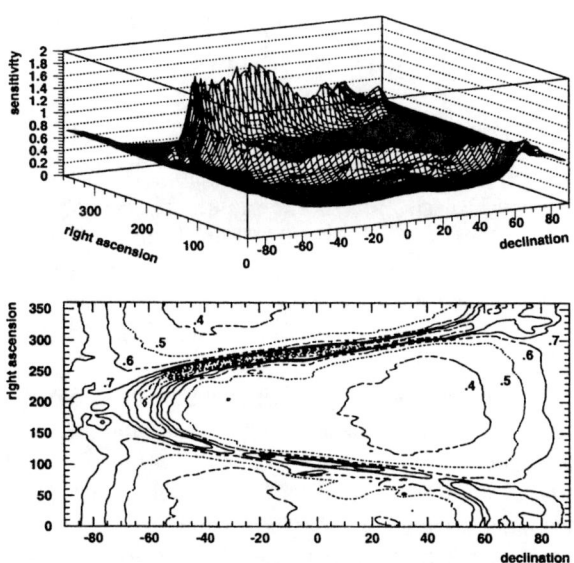

FIGURE 4. A 3-D plot and 2-D contour plot of AMS/γ's point source sensitivity n_0 versus celestial coordinates. Recall that n_0 is the minimum amplitude of the source's differential flux at 1 GeV required for a 5σ significance detection. The units for n_0 in the 3-D plot are 10^{-8} cm$^{-2} \cdot$s$^{-1} \cdot$GeV^{-1}. Large photon fluxes from the diffuse galactic background are responsible for the deterioration of sensitivity near the Galactic plane.

can continue the valuable work of EGRET by providing continued monitoring of extragalactic and galactic γ-ray sources and by participating in multiwavelength observational campaigns. In addition, AMS/γ will have unprecedented sensitivity to the γ-ray sky between the energies of 20 to $\sim$ 100 GeV (albeit at a level somewhat lower than required for detection of known sources with power law spectra), so that AMS/γ might provide us with unexpected discoveries in this region. The next section gives a brief description of the baseline AMS instrument and its experience during the first precursor flight. Then we describe the performance characteristics of AMS/γ in option (a) as determined from Monte Carlo analyses and we perform a comparison to EGRET. For a more detailed description of the results of these studies we refer to [5].

THE ALPHA MAGNETIC SPECTROMETER ON THE STS91 FLIGHT

The Alpha Magnetic Spectrometer has been built by a large international collaboration of high energy physics institutions from the U.S., Italy, China, Finland, France, Germany, Taiwan, Russia, and Switzerland. It recently had a successful test flight on the Space Shuttle mission STS-91 in June 1998 [29], when it was carried in the cargo bay and observed for several days in both the zenith and nadir directions, the latter for measurement of albedo cosmic ray backgrounds. The design of the baseline AMS instrument flown during the STS-91 mission is shown in Figure 1.

The magnet spectrometer consists of a permanent ring dipole magnet made of very high grade Nd-B-Fe rare earth material whose magnetic energy product and residual induction are respectively $(BH)_{max} > 50 \times 10^6$ G-Oe and 14,500 G, yielding an analyzing power of $\langle BL^2 \rangle = 0.14$ T-m^2 with less than 2 tonnes of magnet mass. Four high precision silicon strip detector tracking planes are located within the magnetic volume, with a fifth and sixth plane located just above and below the magnet [6], [7]. At low rigidities (below 8 GV) the resolution of the spectrometer is dominated by multiple scattering ($\Delta p/p \sim 7\%$), while the maximum detectable rigidity ($\Delta p/p \sim 100\%$) is about 500 GV. In addition to measuring particle rigidity, the silicon planes will provide six independent measurements of dE/dx for charge determination. Four time-of-flight (ToF) scintillator planes (two above and two below the magnet volume) measure the particle velocity with a resolution of 120 ps over a distance of 1.4 m. The ToF scintillators also measure dE/dx, allowing a multiple determination of the absolute value of the particle charge.

A solid state Cerenkov detector below the magnet provides an independent velocity measurement, useful to separate electrons and positrons from the hadronic CR components (protons, helium...) and residual background (pions). In addition, a scintillator anticounter system is located within the inner magnet wall, extending to the ToF scintillators.

The performance of this instrument as a charged cosmic ray detector, and its sensitivity to an antinucleus cosmic ray component, is discussed elsewhere [8].

During the STS-91 AMS has collected about 50 million of single CR events above $\sim 100 MeV/n$ kinetic energy, improving the existing cosmic antimatter limits [9] and measuring in details the structure of the CR flux at 400 km of height and over most of the surface of the earth [10].

A few tens of clean two tracks events compatible with GeV-range γ-ray converting in to an e^+e^- pair on the top layers ($\sim 5\%$ Xo) of the spectrometer have also been observed during the precursor flight (see Figure 2). This is, to our knowledge, the first time high energy ($E_\gamma > 1~GeV$)) γ-rays are observed in a magnetic spectrometer in space: thanks to the dipolar magnetic field the e^+e^- pair opens up in the bending plane view but it remains very collimated in the non-bending view as it is clearly shown in the Figure. In addition to the particle identification capabilites of the AMS detector, these distinctive topological features of a γ-ray conversion will be important in rejecting the background induced by the $O(10^5)$ times larger flux of charged CR.

AMS is scheduled to be secured to an external payload attachment point on the International Space Station Alpha in may 2003, where it will remain as a zenith-pointing instrument for 3 years of measurement time. We should also note that by the time AMS is attached to the International Space Station it may have undergone significant changes from the baseline design considered here. In particular, the permanent magnet may be replaced by a superconducting magnet, which would considerably improve AMS's γ-ray detection performance at high energy. Since the detector is still undergoing significant design changes from the baseline instrument flown on the Space Shuttle, we have chosen to fix on that design which currently exists as integrated hardware. The addition of other components discussed in the proposal and currently under developement, such as a transition radiation detector located at the entrance of the magnet and a solid state Čerenkov radiator followed by a segmented calorimeter located at the bottom of the instrument, will improve the particle identification capability of AMS as well as its performance as γ-ray detector.

THE PERFORMANCE OF AMS/γ

In option (a), the conversion of AMS to AMS/γ requires the addition of two hardware components: (1) A passive or active converter medium, located at the entrance of the spectrometer which converts γ-rays into electron-positron pairs. (2) A stellar attitude sensor gives the angular orientation of AMS with respect to the celestial sphere to an accuracy of better than 1.5×10^{-4} radians.

The determination of the converter thickness is based on an optimization between the probability of γ-ray convertion, the amount of bremstrahlung losses, which limit the energy resolution, and the amount of multiple scattering, which limits the angular resolution. In addition this modification should not degrade the sensitivity of AMS to the search of antimatter and its particle identification capability. Following the results of our study we choose the value $x = 0.3 X_0$, for which point source sensitivity is still optimal, the energy resolution is acceptable, and nuclear interaction losses are negligible. For the scope of our study we assumed a $x = 0.3 X_0$ tungsten plate converter located before the first ToF layer.

A full-instrument GEANT Monte Carlo code was run to determine the performance of AMS/γ. Gamma rays with fixed energies, ranging from 0.3 to 100 GeV, were thrown isotropically at the detector over an opening angle of 50°. All the physical processes for electrons and γ-rays were "on" in the GEANT code. Bremsstrahlung photons of energies < 20 MeV were not followed, although all bremsstrahlung energy losses were included. Cuts simulating the trigger conditions and the pattern recognition algorithm were applied on the converted events.

Primary γ-ray energy and incidence direction were determined by adding the fitted momenta vectors of all secondaries evaluated at the converter plate to obtain the primary momentum vector.

Particular care has been given to the study of the instrumental background generated by CR simulating a γ-ray conversion. As reference for this study we used the Extragalactic Gamma Ray Background (EGRB) at High Energies. Any measurement of the EGRB requires that instrumental background events be kept to a lower rate than the EGRB flux. To estimate the effects of background due to collisions of cosmic ray electrons, positrons, and protons with the AMS instrument, we divided the γ-ray spectrum into four bins per decade of energy (from 0.5 to 100 GeV) and required that we investigate backgrounds down to a level of 20% of the EGRB rate in *each* energy bin. For example, in the 25-40 GeV bin it will take AMS/γ 2.0×10^5 seconds to obtain 5 EGRB γ-rays. In our Monte Carlo analysis, therefore, we threw 2.0×10^5 seconds' worth of cosmic ray electron, positron, and proton flux at the instrument to generate a false γ-ray background. These cosmic rays were thrown isotropically over a zenith angle range of $0° < \theta < 110°$, where the largest zenith angle corresponds to the location of the Earth limb at an orbital altitude of 400 km. The energies were distributed according to known electron [16], [17], positron [18], and proton [19] energy spectra.

Using the same quality cuts to eliminate the electron-induced and proton-induced background events, we had a total of 2 electron-induced and 5 proton induced events; all events had reconstructed energies of < 1 GeV, leaving no background events in the interesting high energy region.

Comparison of AMS/γ with EGRET

Since the basic parameters of AMS/γ are close to those of EGRET in the following we present a comparison among the two detectors. The converter thickness largely dominates all multiple scattering effects, so that the angular and energy resolution of reconstructed primary photons will be completely dominated by multiple Coulomb scattering (MCS) and bremsstrahlung energy losses of the electrons within the converter plate. This is confirmed by the full MC simulation which gives an energy dependence on the angular resolution:

$$\sigma_{68}^{\text{AMS}}(E) = 0.88° \left(\frac{E}{1 \text{ GeV}}\right)^{-0.956}, \qquad (1)$$

which is to be compared to EGRET's angular resolution [12] of

$$\sigma_{68}^{\text{EGRET}}(E) = 1.71° \left(\frac{E}{1 \text{ GeV}}\right)^{-0.534}. \qquad (2)$$

In Figure 3a the expected AMS/γ angular resolution is compared to the corresponding figure for GLAST [30], where we note, as it should be expected, that at sufficiently high energy a precise pair spectrometer does eventually give a better angular resolution than a fine grain imaging calorimeter.

An integration of effective area $A(E,\theta)$ over solid angle gives instrument aperture, shown in Figure 3b. Below a γ-ray energy of 0.3 GeV, the converted electrons begin to have too small a radius of curvature to escape from the magnet volume, and detection efficiency plummets. Above 100 GeV the converted electron and positron often do not spatially diverge beyond the two-hit resolution distance of the silicon trackers, causing significant deterioration in γ-ray energy resolution. These considerations then define the limits of the energy window for AMS/γ.

One cannot directly compare the point source sensitivity of AMS/γ to that of EGRET, since AMS is not a pointable instrument. EGRET achieves a flux sensitivity of $I_{\min}(> 0.1 \text{ GeV}) \approx 10^{-7} \text{cm}^{-2}\text{s}^{-1}$ with a 2-week viewing period. Since it took EGRET one year to map the full sky, where each sky segment was viewed for roughly 2 weeks, we can compare EGRET's sensitivity with that achieved by AMS/γ after one year of operation. By assuming a source differential spectrum of E^{-2}, we can convert from EGRET's definition of sensitivity (in terms of integral flux above 0.1 GeV) to that of AMS/γ (a differential flux above 1 GeV). In our units the EGRET 5σ flux sensitivity is 1×10^{-8}. Over most of the sky, AMS/γ's mean sensitivity $\langle n_0 \rangle$ is estimated to be about a factor of 2 lower than that of EGRET (Figure 4).

The table below summarizes the performance characteristics of AMS as a γ-ray detector. In particular, by its comparison to EGRET, one sees that the two instruments perform similarly in many respects, one major difference being the energy windows: AMS's energy window is shifted up by roughly one order of magnitude from that of EGRET, thereby providing an improved view of the sky in the region $E_\gamma \sim 100$ GeV.

	AMS	EGRET
technique	magn.spectrometer	spark ch.+calorimeter
energy window (GeV)	0.3 to 100.	0.03 to 30.
peak effective area (cm^2)	1300	1500
angular resolution	$0.77°(E/1GeV)^{-0.96}$	$1.71°(E/1GeV)^{-0.534}$
half-area zenith angle	$\sim 30°$	$\sim 20°$
total viewing time (yr)	~ 3	~ 2
attitude capability	fixed	movable
flux sensitivity (ph/cm^2-s-GeV at 1 GeV)	$\sim 0.5 \times 10^{-8}$	$\sim 1.0 \times 10^{-8}$

Note that the point source sensitivity is nearly the same for both EGRET and AMS, with that of AMS being somewhat (factor ~ 2) lower. This is because EGRET and AMS have similar effective detection areas and angular resolutions. The fact that AMS's γ-ray energy threshold is an order of magnitude larger than EGRET's, thereby implying an integrated point source flux about an order of magnitude lower for AMS than for EGRET, is compensated for by the fact that AMS's angular aperture (half-angle area of 30°) is larger than EGRET's ($\sim 20°$), and by the

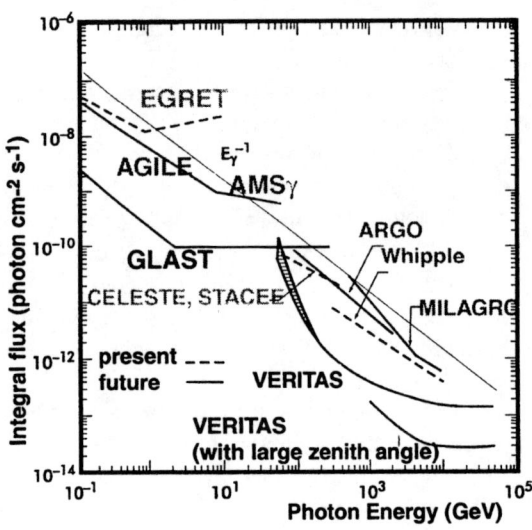

FIGURE 5. Sensitivity of present and future γ-ray detectors.

fact that AMS spends 100% of its time pointing to the sky (being attached to a gravity-gradient stabilized Space Station), while EGRET typically points one-third of the time to Earth. AMS lacks the low-energy end of EGRET's range due to the curvature of the electron-positron pair in AMS's magnetic field, which limits the detectable γ-ray energy to $\geq$ 300 MeV. On the other hand, EGRET lacks the high-energy end of AMS's range due to the effect of electromagnetic backsplash in EGRET's NaI calorimeter which vetos most events above 30 GeV.

Since AMS/γ and EGRET have very similar γ-ray detection capabilities, AMS/γ will be able to continue and extend the investigation of galactic and extragalactic γ-ray sources initiated by EGRET. One main difference between the two detectors, that of significant aperture above 30 GeV for AMS/γ, may lead to the observation of new phenomena in this relatively uncharted region.

Figure 5 shows a comparison between the sensitivity of present and future high energy γ-ray detectors.

CONCLUSIONS

We have shown that with minor modifications the Alpha Magnetic Spectrometer can become a powerful γ-ray detector as well, with overall performance characteristics being comparable, if not superior, to those of EGRET. With γ-ray energy resolution extending past 100 GeV, and with an aperture that is nearly flat above ~ 3 GeV, AMS/γ can address a number of outstanding issues in γ-ray astrophysics that relate to the relatively unexplored region of $E_\gamma = 20 - 200$ GeV. For one, AMS/γ will likely confirm or refute the hypothesis that unresolved blazars are responsible for the bulk of the extragalactic γ-ray background; AMS/γ will also extend the spectrum of the diffuse galactic background to above 100 GeV, helping to resolve current difficulties in interpreting the EGRET diffuse galactic background measurement [28].

AMS/γ should roughly double the total number of blazars detected in γ-rays, and will be enable multiwavelength observational campaigns to include the GeV region of blazar spectra during the flight years of 2003-2006. There is an additional possibility that an indirect detection of the cosmic UV and optical photon background can be made through the detection of extinctions in high-redshift blazars above ~ 20 GeV. AMS/γ will also likely observe GeV γ-ray emission from one or more gamma ray bursts during its operational lifetime.

AMS/γ will also search for both line and continuum emission of γ-rays from the region of the Galactic Center created by the annihilation of dark matter neutralinos. Although the sensitivity to line emission appears marginal, there is nevertheless a finite, though small, region of halo/MSSM phase space which allows a detection by AMS. However, a much larger region of dark-matter halo/MSSM parameter space can be constrained in a search for continuum γ-rays by AMS/γ.

Finally, we note that even higher sensitivities can be reached with the addition of a high granularity calorimeter below the magnet (option (b)); this will be the subject of future work.

REFERENCES

1. J.D. Kurfess, D.L. Bertsch, G.J. Fishman, and V. Schöfelder (1997) AIP Conf. Proc. 410, 509.
2. D. Engovatov et al. (1997) IEEE Nucl. Sci. Symp. (Albuquerque).
3. AGILE, Phase A Report, ASI Small Program Scientific Mission, October 1998, AR.DAS.98.501.
4. A.M. Hillas and J.R. Patterson (1990) J. Phys. G Nucl. Part. Phys. 16, 1271.
5. R. Battiston, M. Biasini, E. Fiandrini, J. Petrakis and M.H. Salamon, to be published on Astropart. Phys., astro-ph/9909432.
6. M. Acciarri et al. (1995) Nucl. Instr. Meth. A360, 103; R. Battiston (1995) Nucl. Phys. B Proc. Suppl. 44, 274.
7. W. Burger, in Proceedings of Vertex 1999.
8. S. Ahlen et al. (1994) Nucl. Instr. Meth. A350, 351.

9. AMS Collaboration, (1999) Phys. Lett. B461, 387.
10. U. Becker, invited talk at 26th ICRC (1999), (Salt Lake City); AMS Collaboration submitted to Phys. Lett. B.
11. R. Mukherjee, et al. (1997) Ap.J. 490, 116.
12. D.J. Thompson, et al. (1993) Ap.J.Suppl. 86, 629.
13. J.R. Mattox, et al. (1997) Ap.J. 481, 95.
14. R.C. Hartman, W. Collmar, C. von Montigny, and C.D. Dermer (1997) AIP Conf. Proc. 410, 307.
15. D.J. Thompson, et al. (1995) Ap.J.Suppl. 101, 259.
16. D. Müller and K.K. Tang (1987) Ap.J. 312, 183.
17. R.L. Golden, et al. (1994) Ap.J. 436, 769.
18. S.W. Barwick, et al. (1997) Ap.J. 482, L191.
19. W. Menn, et al. (1997) 25th ICRC 3, 409 (Dublin).
20. G. Alverson, et al. (1993) Phys. Rev. D48, 5.
21. M. Mori (1997) Ap.J. 478, 225.
22. G. Jungman, M. Kamionkowski, and K. Griest (1996) Phys. Rep. 267, 195.
23. P. Ullio and L. Bergström (1997) hep-ph/9707333.
24. L. Bergström and P. Ullio (1997) hep-ph/9706232.
25. L. Bergström, P. Ullio, and J.H. Buckley (1997) astro-ph/9712318.
26. T. Sjöstrand (1994) Comp. Phys. Comm. 82, 74.
27. M. Biasini (1999) in Proc. XIII Rencontres de Physique (La Thuile).
28. M. Pohl and J.A. Esposito (1998) Ap. J. 507, 327.
29. R. Battiston (1999) in Proc. XIII Rencontres de Physique (La Thuile); also as astro-ph/9907152.
30. E.D. Bloom (1996) Sp. Sci. Rev. 75, 109.

The CANGAROO-III Project

Masaki Mori[1*], S.A. Dazeley[2], P.G. Edwards[3], S. Gunji[4], S. Hara[5], T. Hara[6], J. Jinbo[7], A. Kawachi[1], T. Kifune[1], H. Kubo[5], J. Kushida[5], Y. Matsubara[8], Y. Mizumoto[9], M. Moriya[5], H. Muraishi[10], Y. Muraki[8], T. Naito[6], K. Nishijima[7], J.R. Patterson[2], M.D. Roberts[1], G.P. Rowell[1], T. Sako[8,11], K. Sakurazawa[5], Y. Sato[1], R. Susukita[12], T. Tamura[13], T. Tanimori[5], S. Yanagita[10], T. Yoshida[10], T. Yoshikoshi[1], and A. Yuki[8]

[1] *Institute for Cosmic Ray Research, University of Tokyo Tanashi, Tokyo 188-8502, Japan*
[2] *Department of Physics and Mathematical Physics, University of Adelaide, South Australia 5005, Australia*
[3] *Institute of Space and Astronautical Science, Sagamihara, Kanagawa 229-8510, Japan*
[4] *Department of Physics, Yamagata University, Yamagata 990-8560, Japan*
[5] *Department of Physics, Tokyo Institute of Technology, Meguro, Tokyo 152-8551, Japan*
[6] *Faculty of Management Information, Yamanashi Gakuin Univeristy, Kofu, Yamanashi 400-8575, Japan*
[7] *Department of Physics, Tokai University, Hiratsuka, Kanagawa 259-1292, Japan*
[8] *STE Laboratory, Nagoya University, Nagoya, Aichi 464-8602, Japan*
[9] *National Astronomical Observatory, Tokyo 181-8588, Japan*
[10] *Faculty of Science, Ibaraki University, Mito, Ibaraki 310-8521, Japan*
[11] *LPNHE, Ecole Polytechnique, Palaiseau CEDEX 91128, France*
[12] *Computational Science Laboratory, Institute of Physical and Chemical Research, Wako, Saitama 351-0198, Japan*
[13] *Faculty of Engineering, Kanagawa University, Yokohama, Kanagawa 221-8686, Japan*

Abstract. The CANGAROO-III project, which consists of an array of four 10 m imaging Cherenkov telescopes, has just started being constructed in Woomera, South Australia, in a collaboration between Australia and Japan. The first stereoscopic observation of celestial high-energy gamma-rays in the 100 GeV region with two telescopes will start in 2002, and the four telescope array will be completed in 2004. The concept of the project and the expected performance are discussed.

---

*) E-mail: `morim@icrr.u-tokyo.ac.jp`

INTRODUCTION

Following the CANGAROO-I (3.8 m) and CANGAROO-II (7 m) telescopes, CANGAROO-III is a project to study celestial gamma-rays in the 100 GeV region utilizing a stereoscopic observation of Cherenkov light flashes with an array of four 10-meter telescopes. The CANGAROO-II telescope (hereafter C-II), which has a 7-meter reflector and has been operational since 1999 May [1] [2] [3], is going to be expanded in early 2000 by adding more small mirrors and will be the first 10 m telescope of this array.

The CANGAROO-III project started in April 1999 and is planned as a five-year program. The schedule is shown in Figure 1. This year we will expand the 7 m telescope to 10 m, and the second year we will build the second telescope which will be installed in the third year. The other two telescopes will be installed in the fourth and fifth years. Each telescope will be set on a corner of a diamond of about 100 m side in order to have a maximum number of pairs of telescopes of the same baseline length. The first stereoscopic observation will be performed in 2002 and the full four telescope will be in operation in 2004.

EXPANSION OF CANGAROO-II

Expansion of the 7 m telescope to 10 m is simple. Since C-II is originally designed as a 10 m telescope, all we have to do is add 54 mirrors and tune their attitude (Figure 2). This work will be completed in early 2000.

Additional outer mirrors will worsen the point image at the focus but it is not a serious problem. Simulations show that the concentration of photons in one pixel will be reduced from 56% to 42% at the center, and from 50% to 36% at one-degree

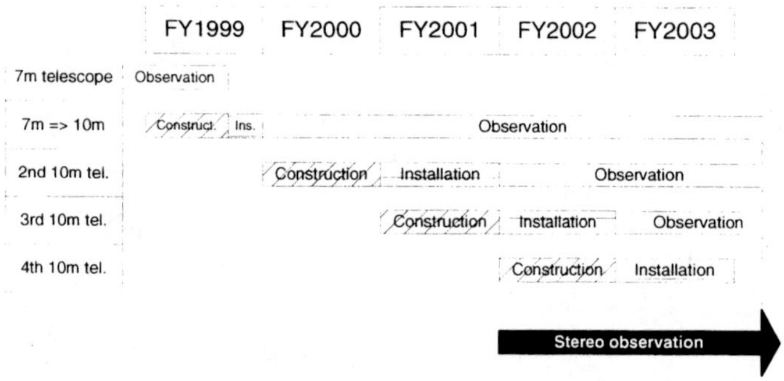

FIGURE 1. Schedule of the CANGAROO-III project. Note that the Japanese fiscal year is from April to March.

off-axis. In any case, the number of collected photons will be almost doubled, reducing the energy threshold by a factor of two.

CANGAROO-III TELESCOPE DESIGN

At this stage we will use basically the same design for the support structure and the driving mechanism as the C-II telescope, which is originally designed as a 10-meter telescope and has been proved to work well. The reflector will have a parabolic, composite mirror consisting of 114 small mirrors of 80 cm in diameter [4]. The focal length will be 8 m if we use the same mirrors.

Mirrors made of plastic laminates used for C-II are very light and pose little stress on the support structure. Observation of star images at various zenith angles showed the deformation of mirrors was negligible. But the image quality of these mirrors are not as good as glass-made mirrors since they are made by molding: thus we are still investigating other possibilities. The attitude of each mirror will be controlled by stepping motors as for the present 7 m telescope. Tuning this number of small mirrors to a common focus is not a simple task. For C-II we tuned the mirrors one by one using lids to cover the other mirrors, but this is not easy for larger numbers of mirror segments.

The prime focus camera will be similar to the present CANGAROO-II camera consisting of 512 half-inch photomultipliers and subtending about 3 degrees, but the optimization for stereoscopic observation is underway.

The electronics and data acquisition system will be improved to match higher data rates. In any case, we take timing information of each signal, in addition to pulse height, to utilize the isochronous nature of our parabolic reflector. For stereoscopic observation, we must introduce an inter-telescope trigger to compensate for geometrical delays using programmable delays between telescopes. The local trig-

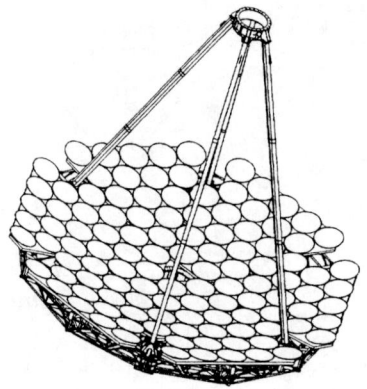

FIGURE 2. A sketch of the 10 m reflector.

gers will be as frequent as 1 kHz but the delayed coincidences at the main trigger will be reduced to about 100 Hz, we hope.

STEREO SIMULATION

Here we briefly show some results of simulations of stereo observations [6]. This work was done before the whole CANGAROO-III project was approved and takes only two telescopes into account, but the result is valid if we use a twofold coincidence in the inter-telescope trigger.

The detection efficiency as a function of baseline length between telescopes is given in Figure 3. If we cut some detected events using the core distance, the energy resolution will be better and an angular resolution less than 0.1° can be achieved if we use the baseline longer than 100 m. Thus we will adopt the baseline length of around 100 m, which agrees with other calculations. Figure 4 is a comparison of effective area and energy resolution between single and stereo observations.

EXPECTED PERFORMANCE

Expected sensitivity assuming one 10 m telescope shown in Figure 5 is around 10^{-12} cm^{-2}s^{-1} above threshold energy of $100 \sim 200$ GeV, and we may detect many EGRET sources in tens of hours of observation if their spectra extend to higher energies. Also shown are gamma-ray spectra of 22 X-ray selected BL Lacs which are predicted by Stecker et al. [5], however only 5 are in the southern hemisphere.

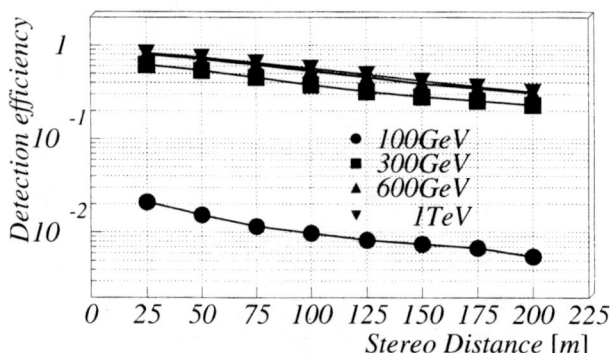

FIGURE 3. Detection efficiency as a function of the baseline separation of two telescopes. Here we define the efficiency as a fraction of triggered gamma-rays when we simulated gamma-rays going vertically and having cores in a circle of 180 m radius.

We note that observations at other wavelengths, especially ground-based ones, are rather biased to the northern hemisphere sky and there are undoubtedly more candidate XBLs in the southern sky.

OBSERVATION TARGETS

Table 1 shows the list of objects observed by the CANGAROO 3.8m telescope for its 6 years of operation. We had been given preference to Galactic sources because of the rather high threshold energy ($\sim$ 2 TeV) of the 3.8m telescope, but we may spend more time on extragalactic objects taking account of the lower threshold of new telescopes. One can see from the table we have needed more than 50 hours of observation at least as the necessary condition to conclude "positive detection" with sufficient statistics on the number of gamma-rays. In addition, the imaging Cherenkov technique still suffers from systematic errors which are not negligibly small when compared with gamma-ray signal strength from even "strong sources", and careful estimation on the experimental errors is indispensable by using the data spanning over a long period of observation. We performed survey observations of shorter duration on many sources, which possibly provide a chance of time varying activities of episodic flares, as well as the objects like X-ray binaries which might be "strong sources" if the claims in earlier days are true. The prime efforts of CANGAROO-III will be on those types of objects appearing as top-ranked sources in the table, extending a systematic survey on more sources. In the case of sources of soft spectra, better statistics in the 100 GeV energy region will enable us to detect them in 10 to 20 hours of observation. However, we still have shortage of total observation time available during a year. It is necessary to develop world-wide

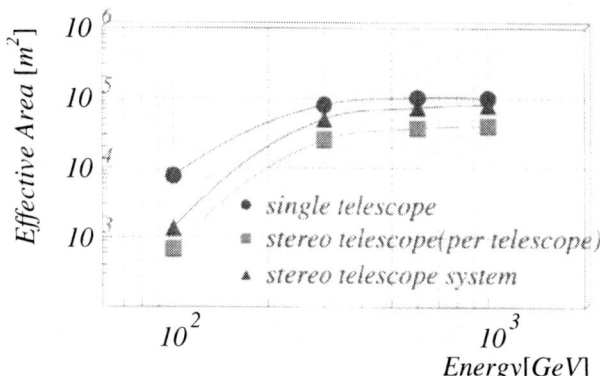

FIGURE 4. Detection area as a function of gamma-ray enegy for single and stereo observations.

efforts for more new types of high-energy gamma-ray sources in collaboration with other groups proposing next-generation telescopes.

SUMMARY

CANGAROO-III will start to explore the southern half of the 100 GeV gamma-ray sky in 2004, complementing projects located in the northern hemisphere to ensure the entire sky is covered at these energies.

REFERENCES

1. Tanimori, T. et al., *Proc. 26th ICRC* (Salt Lake City, Utah, USA), OG.4.3.04 (1999).
2. Mori, M. et al., *Proc. 26th ICRC* (Salt Lake City, Utah, USA), OG.4.3.31 (1999).
3. Kubo, H. et al., in these proceedings.
4. Kawachi, A. et al., in these proceedings; *Proc. 26th ICRC* (Salt Lake City, Utah, USA), OG.4.3.05 (1999).
5. Stecker, F.W., de Jager, O.C., and Salamon, M.H., *Astrophys. J.* **473**, L75 (1996).
6. Hara, S., Master thesis, Tokyo Inst. Tech. (1999); Hara, S., Kubo, H., and Tanimori, T., in preparation (1999).
7. Kifune, T. et al., *Astrophys. J. Lett.* **438**, L91 (1995).
8. Yoshikoshi, T. et al., *Astrophys. J. Lett.* **487**, L65 (1997).
9. Tanimori, T. et al., *Astrophys. J. Lett.* **429**, L61 (1994).

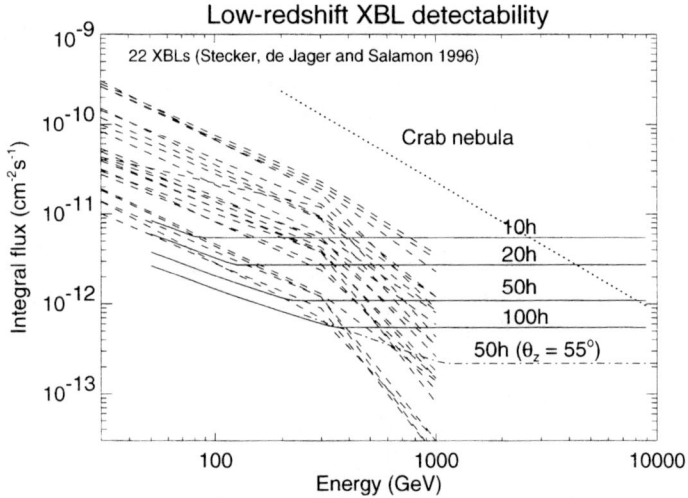

FIGURE 5. Sensitivity of the 10 m telescope and predicted fluxes from X-ray selected BL Lacs (dashed lines) [5]. Here "10h" means 10 hours of observation and so on.

10. Tanimori, T. et al., *Astrophys. J. Lett.* **492**, L33 (1998).
11. Sako, T. et al., *Proc. 25th ICRC* (Durban, South Africa), Vol. 3, 193 (1997).
12. Sako, T. et al., submitted for publication.
13. Mori, M. et al., *Proc. 25th ICRC* (Rome, Italy), Vol. 2, 487 (1995).
14. Rowell, G. et al., submitted for publication.
15. Roberts, M.D. et al., *Astron. Astrophys.* **337**, 25 (1998).
16. Roberts, M.D. et al., *Astron. Astrophys.* **343**, 691 (1999).
17. Susukita, R., Doctor thesis, Kyoto University (1997) (Preprint KUNS 1450/PN97-D09).
18. Rowell, G. et al., *Astropart. Phys.* **11**, 217 (1999).
19. Tanimori, T. et al., *Astrophys. J. Lett.* **497**, L25 (1998).
20. Muraishi, H. et al., *Proc. 26th ICRC* (Salt Lake City, Utah, USA), OG.2.2.20 (1999).

TABLE 1. A list of objects observed by the CANGAROO 3.8m telescope from July 1992 to September 1998 for more than 15 hours in the order of observation time, including bad weather runs. Total observation time is about 2,000 hours. Off-source runs are roughly the same duration but are not listed here.

Object	Observation time (hr)	Remark	Reference
PSR1706−44	308.5	Plerion	[7]
Vela	252.6	Plerion	[8]
Crab	193.8	Plerion	[9] [10]
PSR1259−63	167.4	Pulsar binary	[11]
PSR1509−58	161.5	Plerion	[11] [12]
W28	121.3	SNR	[13] [14]
PKS0521−322	104.0	AGN	[15]
PKS2005−489	94.5	AGN	[16]
Cen A	80.3	Radio galaxy	[17] [18]
PSR1055−52	78.6	Pulsar	[11] [17]
SN1006	63.7	SNR	[19]
RXJ1713.7−394	61.4	SNR	[20]
PKS0548−322	49.2	AGN	[16]
PKS2155−304	37.5	AGN	[16]
PKS2316−423	28.8	AGN	[15]
Sgr A*	28.2	Galactic center	
EXO0423.4−084	23.7	AGN	[15]
GROJ1317−44	22.7	Cen A?	
Vela X-1	20.8	X-ray binary	
GRB970402	19.6	Gamma-ray burst	
Cen X-3	17.7	X-ray binary	
2EGJ1746-2852	15.2	EGRET unID	

The Gamma-ray Large Area Space Telescope (GLAST)

D.A. Kniffen[1], D.L. Bertsch[2] & N. Gehrels[2]

1. Code S, NASA Headquarters, 300 E. St. SW, Washington, DC 20024
2. Code 661, NASA/Goddard Space Flight Center, Greenbelt, MD 20771 USA

Abstract. The Gamma-ray Large Area Space Telescope is planned as NASA's next major mission in high-energy gamma-ray astronomy. It is included in *the Space Science Enterprise Strategic Plan as a 2002* "New Start," with a planned 2005 launch. NASA commissioned a Facility Science Team (FST) of both advocates and non-advocates to establish the framework of the mission. The FST developed a Science Requirements Document that outlines the specifications of an instrument needed to make the next advance in this field. The adopted specifications will lead to an investigation with an energy response extending to 300 GeV, ten times higher than the EGRET instrument on the Compton Observatory, and with a source sensitivity 50 times greater. In addition there would be gains in spectral and spatial resolution. A heavy emphasis will be placed on multi-wavelength observations to maximize the science from the mission.

1. INTRODUCTION

The discoveries in high-energy gamma ray astronomy, largely attributable to the very successful Compton Gamma Ray Observatory (CGRO) mission, have led to new insights into a wide array of astrophysical processes. In the high-energy domain above 20 MeV, the EGRET instrument (Thompson et al. 1993) on this Observatory has continued the exploration of the earlier SAS-2 (Derdeyn et al. 1972) and COS-B (Bignami et al. 1975) missions and has greatly expanded our knowledge of both galactic and extra-galactic sources. The first sensitive survey of the full gamma-ray sky was performed by EGRET between April, 1991 and November, 1992. This was followed by a phase where telescope pointings were chosen from peer reviewed proposals. The instrument was operated at full duty cycle until 1996. After 1996 it is being operated at reduced duty cycle and reduced field of view to conserve spark chamber gas. A map of the point sources in the recently released 3rd EGRET catalog (Hartman et al. 1999) is shown in Figure 1.

The most important type of source seen by EGRET is the blazar sub-class of Active Galactic Nuclei (AGN). The prevailing view that blazars have jets aimed toward us indicates that the gamma radiation originates from particles accelerated in the jets. For pulsars, EGRET has tripled the number of spin-down pulsars (from 2 to 6) detected in high-energy gamma rays. In conjunction with ROSAT (Halpern et al. 1992), EGRET (Bertsch et al. 1992) has found that Geminga is a gamma-ray pulsar with little or no radio emission. Concerning transients, EGRET has made the important discovery of long lasting high-energy emission from both gamma-ray bursts

(Hurley et al. 1994) and solar flares (Kanbach et al. 1993). GLAST will build on these findings and open a large area of new discovery space.

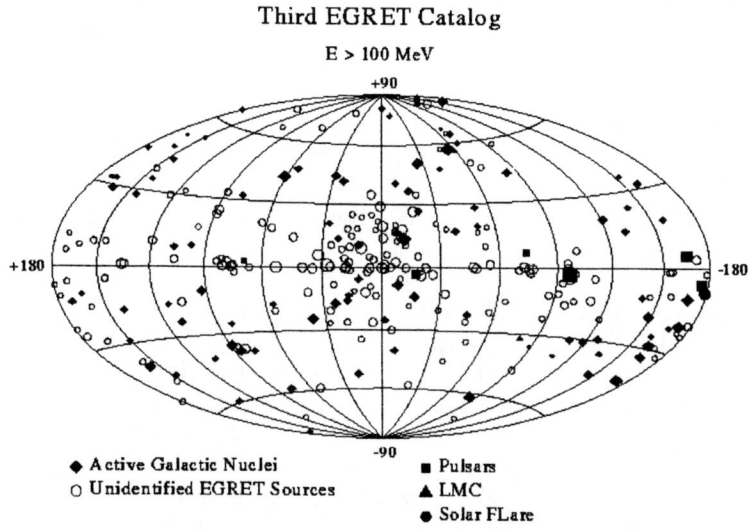

Fig. 1. Point sources detected by EGRET

Discoveries made with EGRET have at the same time generated scientific interest in a follow-on mission and have helped to define its emphasis and requirements. The Gamma-ray Large Area Space Telescope (GLAST) is planned as NASA's next major mission in high-energy gamma- ray astronomy. It is included in *the Space Science Enterprise Strategic Plan* as a 2002 New Start, with a planned 2005 launch. In the sections that follow, the science objectives of the GLAST mission are briefly summarized, and finally the top-level requirements of the GLAST instrument and spacecraft are presented.

More details of the GLAST program can be found at the Web site, [http://www.glast.gsfc.nasa.gov] that contains links to instrument design concepts, science discussions, and the *GLAST Science Requirements Document*.

2. SCIENCE OBJECTIVES

2.1. Blazars

The most interesting objects detected by EGRET are perhaps the blazars, a subclass of Active Galactic Nuclei identified as flat spectrum radio quasars (FSRQ's) or BL Lac sources. Certainly these are the largest set of identified sources, and they are powerful sources at great distances. The redshifts of the EGRET blazars in Fig. 1 range from z = 0.03 to 2.3 (Hartman et al. 1999), and their distance distribution is similar to the radio blazars. EGRET has shown that blazars can produce copious quantities of gamma rays, most likely from Compton up-scattering in their relativistic particle jets (Blanford & Rees 1978). To avoid $\gamma\gamma \rightarrow e^+e^-$ absorption, beaming factors of ~10 are required to allow the gamma rays to escape (Mattox et al. 1993; Dermer & Gehrels 1995). The peak in the $\nu F\nu$ spectral energy distribution is typically seen to be in the high-energy gamma-ray band.

Blazars are observed to vary on the shortest time-scales that can be observed by EGRET which for intense sources is about 1 day (Wehrle et al. 1998, and ground-based TeV detectors have seen even shorter time-scales (Buckley et al. 1996). The phasing of these variations is found to vary with the source and even with different outbursts for a given source. Consequently, it is particularly important to simultaneously study the time varying spectrum of blazars with multi-wavelength coverage from radio through TeV gamma rays (e.g., Wehrle et al. 1998). From these observations we have an indication that the BL Lac subclass of blazars has a different characteristic multi-wavelength spectrum than that of FSRQ's. With the small sample of flaring events that have had broad wavelength coverage, and the current poor statistics for very short time scales in the EGRET data and the ground-based TeV observatories, it is difficult to model the spectra in detail and to understand differences in subclasses. GLAST, combined with new-generation TeV instruments such as VERITAS, will tremendously improve blazar spectral studies, filling in the band from 20 MeV to 10 TeV with high significance data for hundreds of AGN.

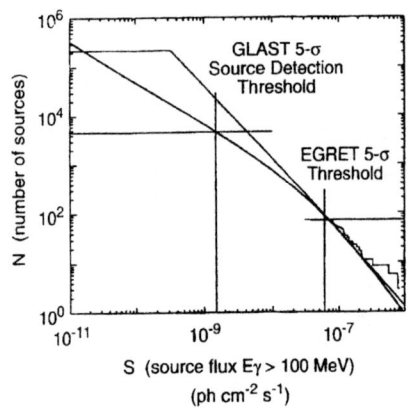

Fig. 2. Log N-log S distribution extension from the EGRET observations to the GLAST threshold. The curved line is calculated assuming that the luminosity function of gamma-rays is proportional to that of the radio.

EGRET has found that the gamma-ray emission from blazars is highly variable. For almost all blazars seen, the emission is detected in only one or two observations, and it is not seen in other good exposures. GLAST will have the capability to monitor most AGN in the sky at all times with its wide field of view. Also, its high sensitivity will allow the first observations of low-state emission from blazars. Based on the logN-logS plot shown in Figure 2, the number of blazars that GLAST will detect at 5σ is ~4000.

2.2. Unidentified Sources

More than half of the EGRET point sources are not correlated with known astrophysical objects (Hartman et al 1999), due in part to the large error circles of these detections. The unidentified EGRET sources in general do not exhibit variability on time-scales greater than a few days, and the statistics are too sparse to search for short periods that are characteristic of pulsars. Most likely, there are several classes of objects that have been detected among the unidentified sources. Included are perhaps molecular clouds (Houstan & Wolfendale 1983), supernova remnants (Sturner & Dermer 1995; Esposito et al. 1996), massive stars (Montmerle 1979), and radio-quiet pulsars (Thompson et al. 1994; Romani 1996; Kaaret & Cottam 1996). Both of these EGRET limitations will be greatly alleviated with GLAST. In particular, the combination of GLAST with the next generation X-ray telescopes should resolve a large part of this long-standing mystery.

Supernova Remnants (SNR) are an especially interesting galactic source since there is near consensus among scientists that they may be the sites of cosmic ray production. X-ray observations by ASCA give evidence for electron acceleration in SNR's such as SN1006 (Koyama 1995). Some EGRET sources appear to be associated with SNR's, but the moderate spatial resolution and sensitivity make the identifications uncertain (Esposito et al. 1996). GLAST could observe high-energy gamma rays from interactions producing pions characterized by a spectral feature at 70 MeV and thereby provide the crucial observations of cosmic-ray nucleon acceleration.

2.3. Pulsars

The energy output of the spin-powered pulsars that are detected by EGRET is predominantly in the gamma ray regime. Models of the production mechanism are based on particle acceleration in the high magnetic and electric fields in the magnetospheres of the spinning neutron star. The characteristics of the light curves in gamma ray and lower energy bands give specific information about physical processes in these extreme conditions. Models based on the EGRET pulsars such as polar cap (Daugherty & Harding 1996) and outer gap (Romani 1996) models make specific predictions that will be testable with the larger number of GLAST pulsars (Thompson et al. 1997). GLAST will detect approximately 50 radio pulsars and will greatly expand the search for more radio-quiet, Geminga-type pulsars. These studies will lead to understanding of acceleration mechanisms in the pulsar magnetosphere and multi-wavelength beaming.

2.4. Gamma Ray Bursts

EGRET discovered that some bursts can have high energy (> 50 MeV) emission, and that it can extend for up to an hour after the initial burst (Hurley et al. 1994; Dingus, Catelli, & Schneid 1997). The most remarkable event was GRB940217 during which the EGRET detector was saturated during the main peak, and the high-energy emission continued for about 6000 s. An 18 GeV photon was seen from the burst direction 4700 s after the BATSE detection. The presence of high energy gamma rays during and after bursts is of key importance in understanding the acceleration mechanisms during the burst and as the blast wave interacts with its surroundings. Although the EGRET findings are intriguing, they provide poor statistics with only 4 bursts having been imaged during the main burst event and with only 1 burst detection in long-lasting afterglow. GLAST will significantly expand this data set by detecting ~100 bursts/year.

2.5. Diffuse Gamma Radiation

EGRET observations of the residual emission after point sources have been subtracted provide all-sky maps of the predominantly diffuse emission. Within the plane, comparison of the observations with a model based on the matter distribution from radio surveys, and on an assumed correlation of cosmic ray matter furnishes insight into the cosmic ray distribution and on the Galaxy's structure (Hunter et al. 1997). GLAST will explore diffuse radiation on scales from molecular clouds to galactic arms with improved angular and energy resolution. Its increased sensitivity will allow much finer detail to be mapped in the diffuse emission along the galactic plane. Also, contributions from currently undetected and unresolved point sources can be subtracted.

At high latitude, GLAST will make high-sensitivity observations of the extra-galactic diffuse background. With better understanding of the galactic emission, this component can be accurately subtracted. The combination of improved diffuse component observations and improved measurements of AGN will allow GLAST to determine if the extra-galactic background is made up solely of unresolved AGN (Stecker & Salamon 1996) or if there are cosmological contributions.

2.6. Extra-galactic Background Light

Energetic photons propagating through space may interact with infrared, optical, and UV photons (the extra-galactic background light or EBL) in pair-production events. The origin of the lower-energy intergalactic photons is starlight, predominantly from starburst activity during the epoch of galaxy formation. The attenuation of gamma rays by this process, depends on the density of the EBL, the distance of the source, and on the energy of the photon. The effect of this attenuation may be discernible in the ground-based TeV observation of Mrk 421 (Stecker & DeJager 1997). The combination of GLAST and future ground-based instruments will be a powerful tool for systematically studying the attenuation of AGN spectra and thereby measuring the EBL to redshifts of z ~4.

2.7. Dark Matter

One of the leading candidates for the dark matter thought to dominate the Universe are stable, weakly-interacting massive particles (WIMPs). One candidate in super-symmetric extensions of the standard model in particle physics is the neutralino, which might annihilate into gamma rays in the 30-300 GeV range covered by GLAST (e.g., Jungman, Kamionkowski, & Greist 1996).

3.0. The GLAST Mission

The capabilities of GLAST compared to those of EGRET are listed in Table 1. The GLAST mission will be flown in low-Earth orbit and will operate in both a zenith pointing mode and a stare mode. The instrument will view >16 % of the sky at a time, and in zenith pointing mode, and scan ~75 % of the sky every orbit. Combining this field of view with the large effective area and good angular resolution, gives a sensitivity of 2×10^{-9} photons cm^{-2} s^{-1} > 100 MeV for a 2-year all-sky survey.

GLAST is currently in NASA's plans for "New Start" congressional approval in 2002 and launch in 2005. The GLAST mission is being managed at NASA's Goddard Space Flight Center. Scientific development of the mission has been led by a Facility Science Team that will be replaced by a Science Working Group after instrument selections are made. The total NASA mission cost (including spacecraft, launch, and 5 years of operations) is specified to be $325 M.

NASA has competitively selected three groups to develop instrumentation and perform studies for GLAST. The two main instrument concepts being developed are Silicon GLAST and Fiber GLAST. These are briefly described below. Both are pair-conversion telescopes with three principal components: 1) a "tracker" in which pair conversion occurs in foils producing an electron and positron whose trajectories are detected by particle tracking detectors; 2) a "calorimeter" in which the electron and positron and shower particles/photons from them are stopped to provide a measurement of energy; and 3) an "anticoincidence shield" that discriminates particle backgrounds from the gamma-ray signal.

Silicon GLAST (PI: P. Michelson) uses thin silicon solid state detectors for tracking. The detectors are ~400 μm thick and have strip contacts of 240 μm pitch. The tracker (detectors plus converter foils) is ~ 0.5 radiation lengths thick. This is followed by a segmented CsI calorimeter. The thickness of the calorimeter is 10 radiation lengths. The instrument is surrounded by a segmented anticoincidence detector made of plastic scintillator.

Table 1
GLAST Instrument Parameters Compared with Those of EGRET

Parameter	GLAST	EGRET
Energy Range	20 MeV to 300 GeV	20 MeV to 30 GeV
Energy Resolution	10%	10%
Effective Area	8000 cm^2	1500 cm^2
Field of View	>2 sr	0.5 sr
Angular Resolution	<3.5° at 100 MeV <0.15° at 10 GeV	5.8° at 100 MeV 0.54° at 10 GeV
Sensitivity (>100 MeV)	<4 × 10^{-9} cm^{-2} s^{-1}	~1 × 10^{-7} cm^{-2} s^{-1}
Source Location Accuracy	1 to 5 arcmin	5 to 30 arcmin
Event Dead Time	<100 μs	100 ms

Table 2.
GLAST Program Schedule

Item	Date
Instrument technology NASA Research Announcement	January 1998
Technology teams selected	March 1998
GLAST NASA Announcement of Opportunity release	June 1999
Instrument selection	February 2000
Spacecraft procurement	2002
Congressional "New Start"	2002
First Guest Observer Research Announcement	2004
Launch	2005

Fiber GLAST (PI: G. Pendleton) uses scintillating fiber technology for both the tracker and calorimeter. The square cross section fibers (made of polystyrene) are ~0.75 mm on a side and are read out at their ends by multi-anode photomultiplier tubes. The tracker and calorimeter are 2.2 and 5 radiation lengths thick, respectively.

The instrument is surrounded by an anticoincidence detector made of plastic scintillator.

In addition to the two instrument concepts, a third team (PI: A. Zych) investigated the benefits and feasibility of enhancing the low-energy (10 - 100 MeV) response of GLAST.

NASA released an Announcement of Opportunity (AO) in August, 1999 for selection of a single Large Area Detector, plus possibly one or two small secondary instruments, and for a few Inter-Disciplinary Scientists (IDS's). The proposal responses are currently being peer reviewed, and a selection is expected to announced in February 2000 (see Table 2).

REFERENCES

Bertsch, D.L. et al. 1992, Nature, 357, 306.
Bignami, G. et al. 1975, Space Sci. Instrum., 1, 245.
Blanford, R. & Rees, M.J. 1978, Pittsburgh Conference on BL Lac Objects, A.M. Wolfe, ed. (Pittsburgh University Press, Pittsburgh,), p. 328.
Buckley, J.H. 1996, ApJ, 472, L9.
Daugherty, J.K. & Harding, A.K. 1996, ApJ, 458, 278.
Derdeyn, S. M. et al. 1972, NIM, 1998, 557.
Dermer, C. & Gehrels, N. 1995, ApJ, 447, 103.
Dingus, B.L. Catelli, J.R., & Schneid, E.J. 1997, in: 25th ICRC Proceedings, Durban, S. Africa, Vol. 3, M.S. Potgieter, B.C. Raubenheiner, D.J. van der Walt eds. (ICRC Press), p.29.
Esposito, J.A. et al. 1996, ApJ, 461, 820.
Halpern, J.P. and Holt, S.S. 1992, Nature, 357, 222.
Hartman et al. 1999, ApJS, 123, 79.
Houstan, B.P. & Wolfendale, A.W. 1983, A&A, 126, 22.
Hunter, S.D. et al. 1997, ApJ, 481, 205.
Hurley, K. et al. 1994, Nature, 372, 652.
Jungman, G., Kamionkowski, M., & Griest, K. 1996, Phys Rev., 267, 195.
Kaaret, J. & Cottam, J. 1996, ApJ, 462, L35.
Kanbach, G. et al. 1993, A&A, 97, 349.
Koyama, K. et al. 1995, Nature, 378, 255.
Mattox, J. et al. 1993, ApJ, 410, 609.
Montmerle, T. 1979, ApJ, 231, 95.
Romani, R.W. 1996, ApJ, 470, 469.
Stecker, F.W. & De Jager, O.C. 1997, ApJ, 476,712.
Stecker, F.W. & Salamon, M.H. 1996, ApJ, 464, 600.
Sturner, S.J. & Dermer, C.D. 1995, ApJ, 293, L17.
Thompson, D. J. et al. 1993, APJS, 86, 629.
Thompson, D.J. et al. 1994, ApJ, 436,229.
Thompson, D.J. et al. 1997, Fourth Compton Symposium, C.D. Dermer, M.S. Strickman, & J.D. Kurfess, eds. (AIP, New York), p.39.
Wehrle, A. et al. 1998, ApJ 197, 178.

The High Energy Stereoscopic System (HESS) Project

Werner Hofmann, for the HESS collaboration

Max-Planck-Institut für Kernphysik, P.O. Box 103980, D-69029 Heidelberg, Germany

Abstract. The paper summarizes the status of the HESS project as one of the next-generation instruments for VHE gamma-ray astronomy. In its first phase, a system of four large Cherenkov telescopes with about 100 m² mirror area will be installed in the Khomas Highland of Namibia. A later expansion of the system to up 16 telescopes is foreseen.

In the following, I will briefly touch upon the physics goals of HESS, review the basic configuration choices governing the layout of the HESS telescope array, cover some details of the technical implementation, summarize the key performance parameters and close with the current status of the project. I will concentrate on the first phase of the project.

I PHYSICS GOALS OF HESS

The basic physics goal of HESS is to provide a comprehensive study of nonthermal phenomena in the universe, using TeV gamma-ray emission as a diagnostic tool, with emphasis on the precise spectral and spatial mapping of sources. In this context, "TeV gamma rays" stands for the $10^{12\pm 1}$ eV range. TeV gamma rays are (almost) always secondary products; one is primarily interested in the parent populations, including in particular galactic and extragalactic nonthermal electron populations, and the nucleonic component of the nonthermal universe. In addition, issues in observational cosmology and astroparticle physics can be addressed, and sky surveys will provide an unbiased view of the TeV gamma-ray sky.

Gamma rays from nonthermal electron populations are characterized by their double-humped spectra governed by synchrotron radiation at low (often keV) energies, and by the IC component at high energies. From the analysis of multiwavelength spectra, the electron spectrum and the local B fields can be determined. Interesting galactic sources include pulsar nebulae; a source of the strength of the Crab Nebula can be detected almost anywhere in the Galaxy. Among extragalactic sources, AGNs are of special interest. HESS will be able to detect a source like

Mrk 501 out to $z \approx 0.3$, limited mainly by the $1/r^2$ decrease in flux rather than by absorption of gamma rays in interactions with the IR/O background.

A central part of the HESS program will be the search for the elusive sources of the nucleonic cosmic rays. Here, gamma rays are used to probe the product of the local CR density times the local gas density. Targets include SNR as CR sources, but also Giant Molecular Clouds which allow to probe the distribution and spectrum of CR throughout the Galaxy. Probing the CR flux in external galaxies is also of great interest; here, starburst galaxies and clusters of galaxies are expected to generate a detectable gamma-ray flux.

Observational cosmology and astroparticle physics with HESS includes the measurement of the IR/O background density with its implications concerning the history of galaxy formation, the search for pair halos around AGNs, which allow to measure absolute distances, and the search for WIMP annihilation lines from the galactic center.

II CONFIGURATION CHOICES

Site. At the time of the Kruger Park meeting, the Calar Alto in Spain was considered the prefered site for HESS. Since then, the HESS collaboration has abandoned this option, both because of the climatic conditions, which would require expensive domes for the telescopes, and because of a preference for locations on the southern hemisphere, with direct view onto the Galactic center. A suitable site, with excellent and documented optical quality, mild climate, and a large and easily accessible area has been located in the Khomas Highland in Namibia, near the Gamsberg, about 1.5 driving hours from the capital, Windhoek. The Gamsberg, a table mountain at about 2300 m asl., has been discussed as a possible site for a European southern observatory, and the optical quality of the Gamsberg site has been monitored continuously from 1970 to 1975 and in 1994/1995. In terms of the number of spectroscopic and photometric nights, the Gamsberg area is equivalent to La Silla. Because of the difficult access, the top of the Gamsberg is not suitable to install large telescopes. However, apart from seeing, which is not an issue for Cherenkov telescopes, the highland around the Gamsberg, at about 1800 m asl., provides the same conditions. Therefore, the plan is to rent an area of about 10 km^2 of one of the large farms near the Gamsberg. The infrastructure required for HESS – a control building, a residential building, and power generators – is more or less identical to the infrastructure of a typical farm, and will be built by the MPG in a similar style. Fig. 1 shows a view of the candidate site, which is easily accessible from the main road.

Energy regime and telescope size. The layout of the telescope system, under certain funding constraints, represents a compromise between the wish to have a large number of telescopes for best reconstruction and large detection area, and the desire to reduce the energy threshold and maximize the intensity of the images, implying a large dish area. A critical parameter is the sharing of costs between

FIGURE 1. One of the candidate sites for HESS, on the Farm Goellschau in the Khomas Highland of Namibia. The Gamsberg is visible at the horizon (the flat-top mountain in the center).

the camera and the mount and dish; rather general arguments indicate that best overall performance is achieved when the camera costs are 1/2 to 1/3 of the total cost. Of course, physics arguments may enforce a different sharing of costs.

Unlike MAGIC, HESS never aimed for the lowest possible energy threshold. In particular, the identification of accelerators of nucleonic CRs will rely on the high-energy ends of spectra, where simultaneous electron acceleration is inefficient because of synchrotron losses. In order to cover extragalactic objects out to $z \approx 1$, a threshold for spectroscopy around 100 GeV seemed appropriate, which also implies detection capability down to 40 GeV, with reduced resolution. These thresholds are reached with a 100 m^2 dish, consistent with the cost sharing arguments given above.

Number of telescopes. In Phase I of the HESS project, four telescopes are foreseen, reflecting both the available funding and arguments concerning the sensitivity and the control of systematic effects.

Obviously, at least two telescopes are required for stereoscopy. Only with three telescopes, however, is the shower geometry overconstrained, allowing consistency checks of the shower reconstruction. With four telescopes, two independent measurements of the shower parameters are possible, a big help in understanding the characteristics of the instrument.

Concerning sensitivity, Monte Carlo simulations show a significant gain in background rejection for three-telescope and four-telescope events, as compared to events with only two views. The number of valuable three-telescope events is doubled for a four-telescope system, as compared to a three-telescope system. This is achieved at a cost increase of less than 30% (taking into account the fixed infrastructure cost).

The significant increase in sensitivity for three-telescope and four-telescope systems, predicted by the simulations, can be verified using the HEGRA IACT system, by turning off in software some of the telescopes. Using the Mrk 501 data set, one finds that the significance $S/\sqrt{B}$ for weak sources improves like 1 : 1.83 : 2.52 when using 2, 3, and 4 telescopes, respectively. Only in the limit of larger numbers N of telescopes is the significance expected to follow the asymptotic $\sqrt{N}$ behavior.

A four-telescope system has the additional advantage that two sources can be observed simultaneously in stereo mode. This is mainly relevant for time-variable sources; the sensitivities given above imply that for DC sources the best strategy is to observe each source for half of the time with the full system.

Arrangement of telescopes. The four telescopes will be arranged in a square with a base line of about 120 m. Simulations show that the best sensitivity is obtained for base lines between 100 m and 150 m; the exact value represents a tradeoff between energy threshold (which is slightly lower for the shorter base line) and performance at higher energies, but is quite uncritical. The 120 m spacing is large compared to intertelescope spacings of other existing or discussed systems. To verify the simulations, pairs of telescopes of the HEGRA system were used to study the significance $S/\sqrt{B}$ of the Mrk 501 signal as a function of telescope distances. The arrangement of telescopes in HEGRA provides pairs spaced at about 70 m, 90 m, 110 m and 140 m. In accordance with simulations, the measured performance depends only weakly on the spacing (Fig. 2).

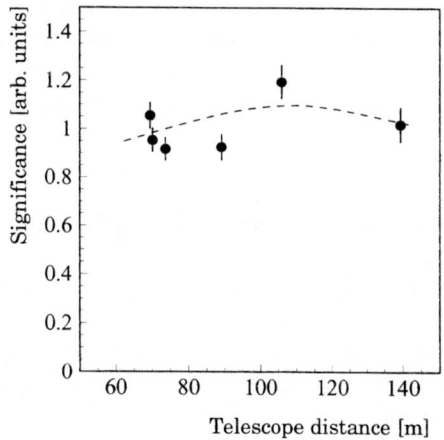

FIGURE 2. Significance $S/\sqrt{B}$ of the Mrk 501 signal, evaluated using pairs of HEGRA telescopes at different spacings.

Pixel size and field of view. Given the emphasis on the study of extended sources such as SNR, and on the survey capability, the HESS cameras need to provide a large field of view and a uniform pixel size throughout the field of view. As a compromise between cost and coverage, the cameras will initially have a 4.3°

field of view, with a later expansion to 5°. A relatively small pixel size of 0.16° will allow efficient triggering with a low threshold, and is particularly helpful for observations at larger zenith angles.

Trigger scheme. One configuration choice with significant implications on the readout electronics and its cost concerns the system trigger scheme. In stereo mode, at least two telescopes have to trigger in coincidence. Because of the propagation delays, a coincidence decision is only available after 1-2 μs. During this time, the PMT signals of the individual cameras need to be stored or delayed in some form. Cable delays are clearly excluded. One option is a deadtimeless storage, e.g. by digitizing signals in a Flash-ADC and storing the data in a ring buffer deep enough to hold a few μs. This technique is used in HEGRA; it has the advantage that also those telescopes which have not triggered, can be read out and included in the analysis.

An alternative is to use an analog sample-and-hold, or some other form of analog buffer, where following a local telescope trigger, the signals are stored for a few μs, after which the storage is reset unless a coincidence with other telescopes is detected. This approach is generally simpler and less expensive to realize, but has the disadvantage that only telescopes which trigger provide data, and that some deadtime is generated while a telescope waits for a intertelescope coincidence.

For HESS, this second option was adopted. Expected telescope trigger conditions – a coincidence of 4 pixels with at least 4 photoelectrons each – are so loose that telescopes which do not trigger will rarely contain useful images. At a predicted single-telescope trigger rate of a few kHz, the deadtime-induced losses are negligible. This choice for HESS also reflects the experience with HEGRA, where the Flash-ADC system provided little practical benefits, in particular once second-generation telescope trigger schemes allowed much lower trigger thresholds and lower noise rates than originally anticipated.

III TECHNICAL CHOICES

Mount and dish. The HESS telescopes will use a $f = 15$ m Davies-Cotton reflector composed of up to 380 individual mirror tiles (Fig. 3). The f/d ratio of 1.2 is large in order to provide good and uniform imaging over a large field of view. Mount and dish are realized as steel space frames. Both the azimuth and altitude drives are friction drives, acting on rails with a radius of about 7 m. This large lever arm reduces the forces and results in significant savings for the motors and gears.

The specifications require mirror support points to be stable to 0.14 mrad rms over the entire altitude range, such that the influence of gravity-induced or wind-induced deformations of the dish is at most comparable to the spot size of individual mirrors tiles (0.3 mrad rms), and small compared to the camera pixel size of 2.8 mrad. Bending of the camera masts is less than 2 cm, or 1/2 pixel size, and will be monitored and corrected using LEDs at the circumference of the camera, viewed

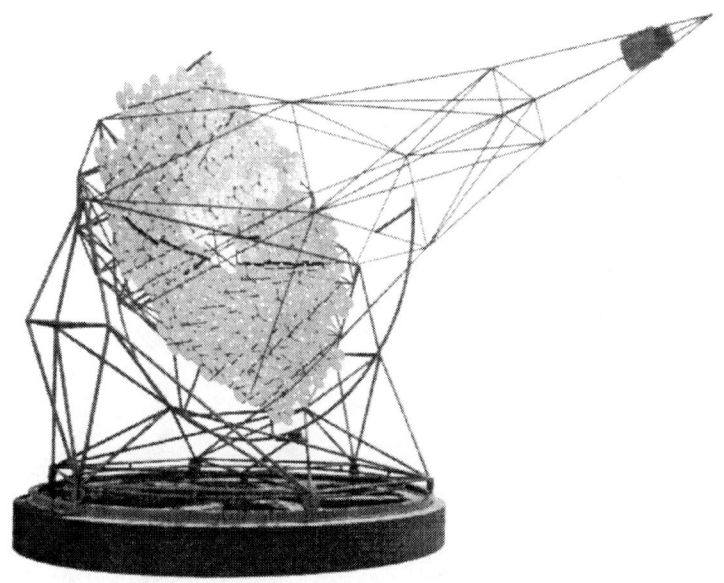

FIGURE 3. One of the HESS telescopes.

by a CCD camera on the dish.

Mirrors. The HESS Letter of Intent discussed diamond-machined aluminum mirror tiles, primarily for cost reasons. Since then, aluminum and glass mirrors from various manufacturers were acquired and tested. Two producers of glass mirrors, COMPAS and GALAKTICA, offered quartz-coated aluminized mirrors at a competitive cost. Given that the glass mirrors generally outperformed the aluminum mirrors available to us in terms of directed reflectivity and focus, and given the positive long-term experience with these mirrors in HEGRA and CAT, it was decided to use 60 cm diameter round glass mirrors. The production will be shared between COMPAS and GALAKTICA. The specifications require 80% reflectivity between 300 and 600 nm, and a focus spot of 1 mrad diameter containing 80% of the light. Mirrors for one telescope have been delivered to Heidelberg and tested, see Fig. 4. Whereas virtually all mirror surpass the specifications concerning the spot size, some of the mirrors show inferior reflectivity; they will be re-aluminized. On the dish, the alignment of each mirror can be adjusted by remote control, using two servo motors.

Photodetectors. The first four HESS telescope will use conventional 30 mm PMTs with bialkali photocathodes and 8 dynodes, operated at a gain of $2 \cdot 10^5$. PMT samples from various manufacturers were characterized with respect to gain and gain variations, cathode uniformity, afterpulsing rate, etc. On the basis of performance and cost, the Photonis PMT XP2960 was finally chosen.

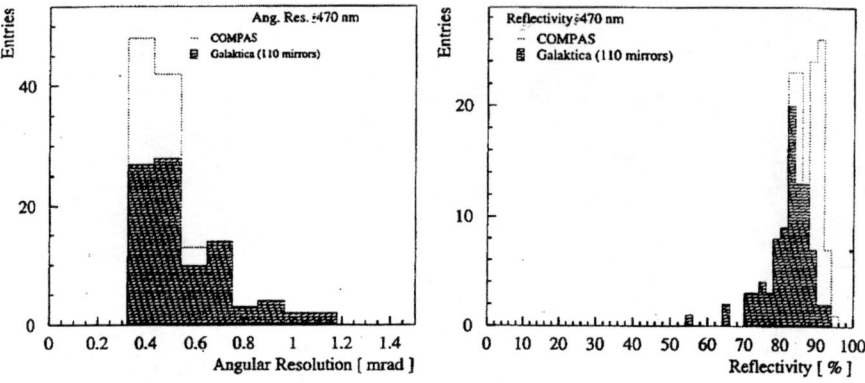

FIGURE 4. Measurements of first mirror samples. (a) Diameter of the spot containing 80% of the reflected light. (b) Directed reflectivity at 470 nm, averaged over the mirror area.

Camera electronics. Almost the entire trigger and readout electronics of the HESS telescopes will be contained in the cameras, eliminating the need for a wide-bandwidth signal transmission, and reducing the number of connectors and interfaces. Prototypes of two alternative electronics concepts have been developed: the "Smart Pixel" approach and a readout system based on the ARS analog sampling ASIC. Both concepts provide 13 to 14 bit dynamic range using dual-gain conversion, and a signal integration time of 15 to 20 ns. A modular construction allows easy maintenance and repair; pixels or pixel groups together with their electronics can be extracted from the front side of the camera. Both concepts are discussed in detail in other contributions to this conference.

The "Smart Pixel" (Fig. 5, see also the contribution by A. Kohnle) is a PMT pixel containing, behind the PMT, a DC-DC converter for the HV supply and a board with dual-gain gated charge integrators, a time-to-voltage converter, a PMT current monitor and rate meter, a discriminator, trigger and readout logic, as well as various monitoring functions. The individual pixels plug into a backplane, and are controlled and read out sequentially via an analog bus and a digital bus. To derive a trigger, each pixel receives the discriminator signals from its 6 neighbor pixels, and forms a coincidence of $N = 1...7$ pixels. These trigger decisions of all pixels are ORed, and are, via the backplane, distributed back to the pixels to provide a gate for the charge integrators. During the trigger decision time of about 50 ns, signals are delayed using a thin coax cable coiled up in each pixel.

The second readout concept, currently the favored option, uses the ARS (Analog Ring Sampling) ASIC to sample the PMT signal at a 1 GHz rate (Fig. 6, see also the contribution by M. Punch). The ARS stores 128 samples, or 128 ns, leaving

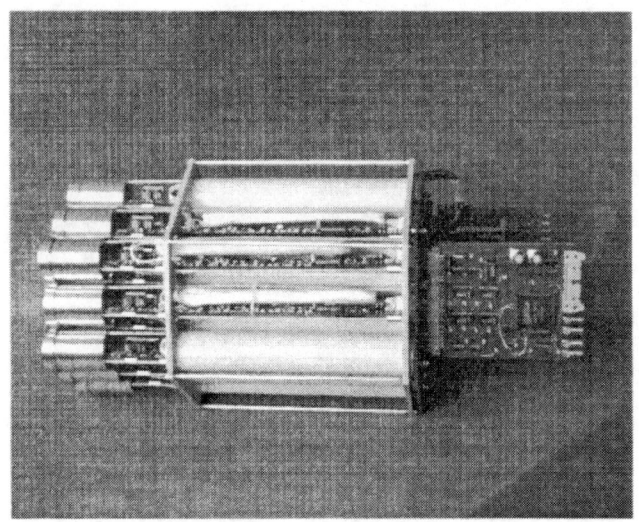

FIGURE 5. Prototype camera consisting of 19 "Smart Pixels" plugged into a backplane with boards to control the readout and triggering. The aluminum rings of each pixel serve to hold the PMTs, which are not yet mounted.

sufficient time for a trigger decision. The trigger circuitry is very similar to the approach used in the CAT telescope, and requires a coincidence of N (typically 3-5) pixels out of overlapping groups of 8x8 pixels. The length of the trigger signal provided by each PMT is determined by the time over threshold, resulting in very short effective coincidence windows around 1.5 ns, and low random-coincidence

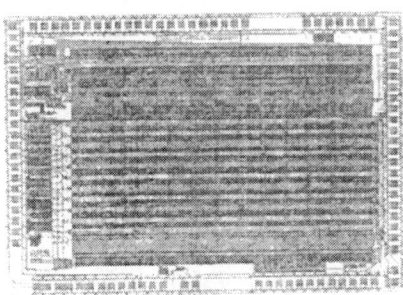 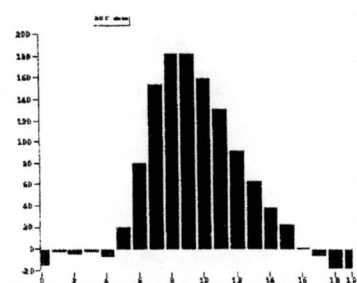

FIGURE 6. (left) The ARS analog ring sampling ASIC, providing four channels of analog storage of 128 samples each, at 1 GHz sampling frequency. (right) Typical sampled and digitized signal; each bar represents a 1 ns step.

rates. Following a camera trigger, the sampling is stopped and the samples are read out and digitized, with the option to either transmit all samples individually, or to form a digital sum over a certain signal integration window. In this concept, PMTs and electronics are packaged into modular 16-PMT units, again containing active HV supplies, current monitors, rate meters, temperature monitoring etc.

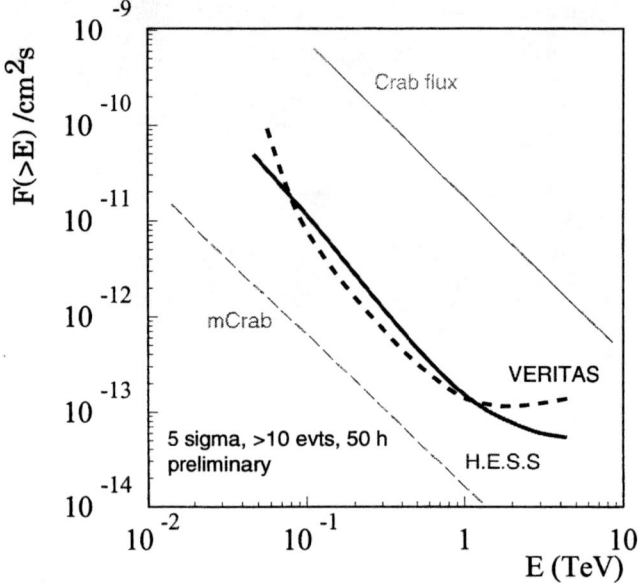

FIGURE 7. Minimal detectable flux as a function of energy threshold, for 50 h of observation time. Also shown is the sensitivity quoted in the VERITAS proposal, and for comparison the Crab flux levels. (Preliminary; telescope parameters assumed in the simulation differ slightly from the latest hardware parameters.)

IV PERFORMANCE

The initial four-telescope system will provide an energy threshold of about 40 GeV (defined by the maximal differential rate for typical sources), and full spectroscopic capability above 100 GeV, with an energy resolution of 20% or better, and an angular resolution of 0.1° or better. Fig. 7 shows (preliminary) estimates of the minimal detectable flux as a function of a (software) energy threshold, compared to the Crab flux. For 50 h of observations, a sensitivity around 10 mCrab is achieved. Also included is the VERITAS sensitivity, which is slightly better at medium energies, reflecting the larger number of telescopes. At high energies, the larger field of view of the HESS cameras results in an improved effective area. With the full 16-telescope HESS system and 100 h of observation time, limits of a 2-3 mCrab will be within reach. Illustrated in Fig. 8 is the angular resolution; HESS at 100 GeV has the identical characteristics as the HEGRA system at 1 TeV.

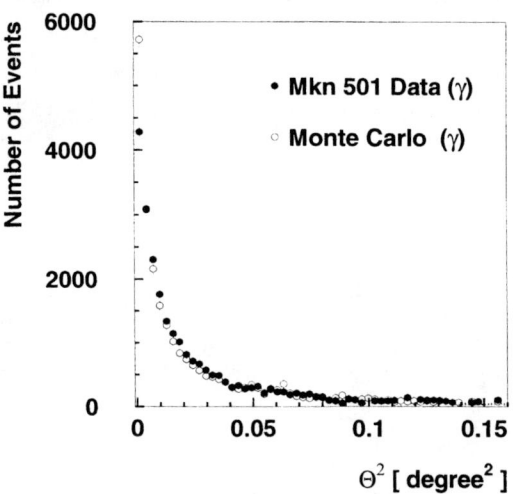

FIGURE 8. Angular distribution of reconstructed photon directions; HESS simulations at 100 GeV (open circles) compared to HEGRA data at 1 TeV (full points).

V STATUS

While decisions from some funding agencies are still pending at this time, a significant fraction of the funding required for the first four HESS telescopes is available, and the construction has started. The procurement of mounts and dishes is underway, the same holds for the PMTs. Mirrors are in production, with about 25% delivered. The electronics is in the stage of advanced prototyping. Concerning the installation in Namibia, an Exchange of Notes between the German and Namibian Governments is in preparation. Work on the infrastructure in Namibia should start early in the year 2000, with first light for the first telescope in 2001, and completion of the initial four-telescope system in 2002.

The MAGIC Telescope Project

E. Lorenz*, for the Magic collaboration

Max Planck Institute for Physics (Werner Heisenberg Institute)
Foehringer Ring 6, D 80805 Munich, Germany

Abstract. An overview of the design parameters and the status of the 17 m Ø MAGIC telescope project will be given. During phase I the telescope will reach a threshold of 30 GeV and a sensitivity of $6.10^{-11}/cm^2 sec$ (5 sigma, 50 hours on source). 'First light' is expected in mid 2001 and observations will begin in 2002. The telescope will be installed at the HEGRA site on La Palma.

INTRODUCTION

Gamma ray (in short γ) astronomy has successfully established the existence of a few galactic and extragalactic γ sources above 300 GeV. On the other hand satellite borne instruments, such as EGRET, with an about 5 orders of magnitude smaller collection area revealed at least 270 sources in the energy range between 20 MeV and 20 GeV. It is expected that future experiments in the up to now unexplored range between 20 and 300 GeV will reveal many new results of fundamental nature. A new satellite borne γ detector, GLAST (see contribution to this conference) will be launched in 2006. This instrument should explore the energy range up to 300 GeV with a nearly flat sensitivity above 20 GeV. On the other hand technical progress will allow one to increase the sensitivity and to lower the threshold of groundbased air Cherenkov telescopes (ACT) to below 50 GeV with classical photomultiplier (PM) cameras and eventually to below 15 GeV with high quantum (QE) efficiency photosensors. The sensitivity of these telescopes for point source studies will be significantly higher than that of GLAST, e.g.; these instruments are complementary.

Here we describe a new generation 17 m Ø ACT, dubbed MAGIC (Major Atmospheric Gamma Imaging Cherenkov) telescope with a threshold of 30 GeV (peak of the differential flux) in phase I and 10-15 GeV with high QE, red extended photosensors in phase II.

THE SCIENTIFIC OBJECTIVES

The main research targets for the MAGIC telescope will be:
 a) The study of active galactic nuclei (AGN) up to $z \sim 2.8$. Measurements of the spectra and fluxes between 20 GeV and 1 TeV will allow one to set stringent limits on the existence and size of the hitherto unquantified infrared (IR) background. The high sensitivity will allow one to measure flux variations down to a time scale of a few minutes. Conservative estimates predict that one will discover at least 50 – 100 new AGNs.
 b) Testing of gamma ray bursts (GRB) in the new energy window. A telescope like MAGIC, whose collection area is between 10^5 and 10^6 times larger than that of satellite borne γ-detectors, should be able to collect large data samples provided the γ-spectra extend to the GeV range, bursts last at least a minute and precise guidance from a satellite detector is available in due time.
 c) The systematic study of possible galactic γ emitters such as supernova remnants (SNR), plerions, X-ray binaries, unidentified galactic EGRET sources, etc., which will hopefully lead to the identification of the main sources of the galactic cosmic radiation up to about 10^{15} eV. A special trigger should permit to lower the threshold to about 10 GeV for time structure analysis of pulsars, i.e., to decide between the polar cap and outer gap model.
 d) The search for exotics such as for the lightest supersymmetric particle (SUSY), test of quantum gravity effects, etc.

THE TECHNICAL CONCEPT OF THE TELESCOPE

At groundlevel the photon density from air showers scales basically with energy. For imaging ACTs a minimum number of photoelectrons (PE) is required for efficient γ/hadron (γ/h) separation. Typically 60 - 100 PEs/image are needed, i.e., significantly above the night sky light background (NSB). A lower threshold can be achieved by increasing the mirror area, minimization of losses in the optical elements and by increasing the quantum efficiency (QE) of the PMs. For high γ/h separation the so-called stereo system is considered superior compared to a single dish. In a funding limited scenario a single large dish allows one to lower considerably the threshold compared to a same cost multi-telescope system. The MAGIC collaboration has chosen to concentrate at first on the single dish approach and to lower the threshold by exploring all the possible technical improvements. Later we intend to add more telescopes in the so-called DUO arrangement. It should be mentioned that below a few hundred GeV the determination of the impact parameter is less important compared to higher energies where MAGIC might be operated anyhow in the stereo mode in combination with some of the existing HEGRA ACTs. Below 100 GeV the energy resolution is dominated by intrinsic shower fluctuations

The main technical challenges for the construction of the MAGIC telescope are:
* Use of a very large mirror collector with high imaging quality
* A nearly 100 % active area camera of 4° diameter
* A fine granularity camera with 0.1° pixels in the central section and optimized PMs.
* A flexible trigger and a high rate DAQ of at least 1 kHz event rate
* Optimization of the optics and electronics for minimization of the NSB impact.

Additional goals set by physics are:
* Possible operation at large zenith angles (> 80°)
* Possible operation in the presence of moon light
* Fast positioning (< 30 sec) for GRB studies

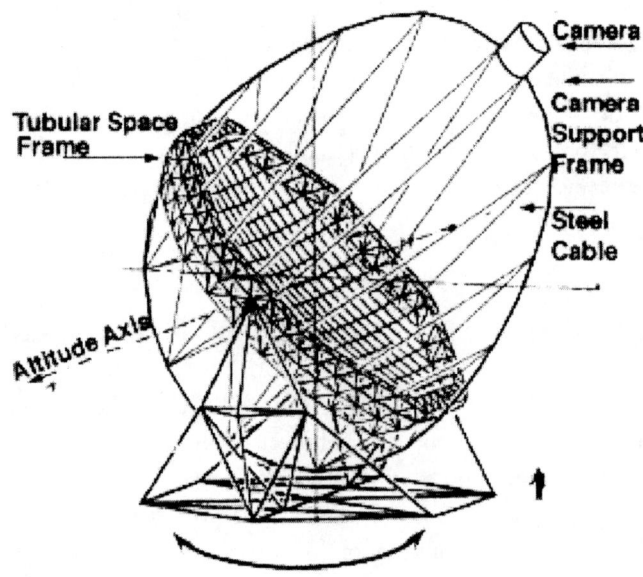

FIGURE 1. A model of the 17 m Ø MAGIC telescope.

Fig 1 shows an engineering drawing of the telescope. Technical details of the telescope can be found in the design report (1).

In the following some technical elements are discussed briefly:
- I) The telescope space frame is made from carbon fiber tubes for stiffness and lightweight for rapid positioning.
- ii) The 234 m^2 mirror (f/d = 1) is composed of 1000 small all-aluminum mirrors of 50x50 cm^2 area and with integrated heating. The gross mirror profile is parabolic, i.e., isochronous. Therefore individual mirrors differ in radius; this is achieved by diamond turning the surface of the blanks on a NC machine (2). Prototype mirrors are successfully operated since one year on the HEGRA CT1 telescope.
- iii) In order to counteract small dish deformations and to optimize focusing an active mirror control system has been developed and successfully tested (3). Similar arrangements are now considered for all new large ACTs.
- iv) For the camera (phase I with bialkali PMs) new, compact, hemispherical PMs with optimal photoelectron (PE) collection efficiency and minimal time dispersion are used. In order to minimize the impact of the DC like NSB and moon light the dynodes were reduced to six and operated with a total gain of only 15 000. Low noise, wideband, large dynamic range transimpedance amplifiers make up for the reduced gain. The light trajectory of most photons passes the semitransparent photocathode twice, thus increasing the QE by 15-20%. The camera of 4° Ø is composed of 600 PMs with coarser sampling in the outer section.
- v) The analog PM signals are transported to the 100 m away counting house by optical fibers. The motivation for this concept stems from the needs to minimize the camera weight, size as well as heat dissipation inside the camera by placing the main readout and trigger electronics and service critical electronics in the counting room. The optical analog transfer system has be designed in part in collaboration with members of the Whipple/VERITAS collaboration (4). Some PM channels of both the HEGRA CT1 and the Whipple telescope are running successfully with such a readout since some time.
- vi) The DAQ is based on a 300 MHz, 8 bit FADC system which will eventually be replaced by a $\geq$ 1 GHz system necessary to make full use of the fast Cherenkov signals and the full potential of the new PMs and the analog transfer system. The trigger has multiple stages and make in part use of the data buffering of the FADCs. Sustained event recording rate will be 1 kHz (zero deadtime) with a possible prolonged burst rate of 5 kHz for GRB or low threshold pulsar studies. Dual ranging of the analog signals at the FADc inputs will extend the dynamic range of the FADCs to about 70-80 dB

SOME PERFORMANCE DATA

The telescope will have a threshold of about 30 GeV (peak of the differential flux) for phase I (using classical PMs in the camera). The collection area will flatten to around 10^5 m^2 above 100 GeV. The γ/h separation will vary between 200 and 1000 as a function of energy. In the presence of moonlight the threshold has to be increased to 60-100 GeV (half-moon, > 30° away). Operation at moon light will increase significantly the potential for rare GRB observations and other transient studies. Close to threshold the energy resolution, dominated by intrinsic shower fluctuations will be around 50 %, and improves to around 10% at 1 TeV. MAGIC will be set up on the HEGRA site on La Palma (28.8° N, 17.8° W, 2200 m asl). Stereo observations with some of the already existing HEGRA telescopes will significantly enhance the prospects for precision measurements above 1 TeV. The upper energy limit of MAGIC is about 50 TeV.

SOME AUXILIARY INFORMATION

The costs of the telescope have been estimated to be around 3.5 M$. Prototypes have been built and successfully tested for all technical components. The construction of some long-lead items, such as the camera has already been started. We estimate that the telescope will see 'first light' in summer 2001 and observations will begin in early 2002. Up to now 12 institutions from 6 countries are participating in the construction.

ACKNOWLEDGEMENTS

Herewith I would like to thank my colleagues from the MAGIC collaboration for providing information and for many fruitful discussions. The German BMBF and the Spanish CICYT have supported part of the development work for components.

REFERENCES

(1) Barrio, J.A. et al.,: The MAGIC Telescope. Max Planck Inst. Report MPI-PHE/98-5
(2) Barrio J. A. et al.,: Development of All-Aluminum Mirrors for Imaging Cherenkov Telescopes. Procs. Workshop: Towards a Major Atmospheric Cherenkov Detector-V, 1997, ISBN 1-86822-295-0
(3) Wacker A. et al., : Test of an Active Mirror Control for Cherenkov Telescopes. . Procs. Workshop: Towards a Major Atmospheric Cherenkov Detector-V, 1997, ISBN 1-86822-295-0
(4) Rose J. et al.,: Fast analog signal transmission for an air Cherenkov photomultiplier camera using optical fibers. To be published in Procs. Workshop New Developments in Photondetection II, Beaune 1999.

VERITAS: Very Energetic Radiation Imaging Telescope Array System

F. Krennrich[1], S.M. Bradbury[2], I.H. Bond[2], A.C. Breslin[3], J.H. Buckley[4], D.A. Carter-Lewis[1], M. Catanese[5], B.L. Dingus[6], D.J. Fegan[3], J.P. Finley[7], J. Gaidos[7], J. Grindlay[5], A.M. Hillas[2], G. Hermann[8], P. Kaaret[5], D. Kieda[6], J. Knapp[2], S. LeBohec[1], R.W. Lessard[7], J. Lloyd-Evans[2], D. Müller[8], R. Ong[8], J. Quinn[3], H.J. Rose[2], M. Salamon[6], G.H. Sembroski[7], S. Swordy[8], V.V. Vassiliev[5], T.C. Weekes[5]

[1] Iowa State University, Ames, IA 50011, U.S.A.
[2] University of Leeds, Leeds, LS2 9JT, U.K.
[3] University College, Dublin, Ireland
[4] Washington University, St Louis, MO 63130, U.S.A.
[5] Harvard-Smithsonian CfA, Amado, AZ 85645-0097, U.S.A.
[6] University of Utah, Salt Lake City, UT 84112, U.S.A.
[7] Purdue University, West Lafayette, IN 47907, U.S.A.
[8] University of Chicago, Chicago, IL 60637, U.S.A.

Abstract. The Very Energetic Radiation Imaging Telescope Array System (VERITAS) is a wide energy range (50 GeV - 50 TeV) atmospheric Cherenkov detector and will start operation in 2004. The design is driven by a major scientific interest in jets of active galactic nuclei, probing the intergalactic IR fields with TeV γ-ray beams, measuring the high energy spectrum of γ-ray bursts and galactic sources of cosmic rays. Also γ-ray signatures of quantum gravity, neutralinos and primordial black holes constitute the exotic scientific motivations to built a highly versatile detector that can be operated in various modes. The technical concept and design of the seven-telescope array system is described.

INTRODUCTION

The objective of this paper is to give an overview of the VERITAS instrument and discuss its design considerations. The VERITAS proposal is to built an array of seven imaging atmospheric Cherenkov telescopes (IACT hereafter) of 10 m aperture. The individual telescope will be based on the proven design of its predecessor, the Whipple 10 m reflector. The Whipple observatory 10 m telescope has been instrumental in establishing the field of ground-based γ-ray astronomy (Weekes et al. 1989). With its energy threshold of 300 GeV it is currently the most sensitive IACT, whereas sensitivity is measured in significance per hour ($7\,\sigma\,\sqrt{t[\text{hours}]}$) for a source strength of 1 Crab.

The VERITAS detector is aimed at a better flux sensitivity, larger effective area, reduced energy threshold, increased energy resolution, better angular resolution and increased field-of-view (FOV) over existing instruments. The design has been optimized for maximum sensitivity in the energy range of 100 GeV - 10 TeV. The minimum detectable flux (5 sigma in 50 hours) will be 0.5% of the Crab Nebula at 200 GeV, a factor of 20 better than any previous IACT. VERITAS provides an unprecedented angular resolution of 0.05° (0.03°) at 300 GeV (1 TeV), substantially better than any existing γ-ray telescope on the ground or in space.

The power of stereoscopic imaging is the basis for the major improvements over the Whipple 10 m telescope, showing a better angular resolution (Daum et al. 1997; Krennrich et al. 1998), improved energy resolution (Hofmann et al. 1999) and a lower energy threshold. Furthermore, the individual 10 m telescopes will be substantially improved over the Whipple 10 m telescope, providing adequate optical resolution, wide FOV and fast electronics, in order to maximize the performance of VERITAS as a stereoscopic system. The Whipple team has been working during the last three years on improving many aspects of the Whipple observatory 10 m telescope through a program called GRANITE-III (Finley et al. 1999).

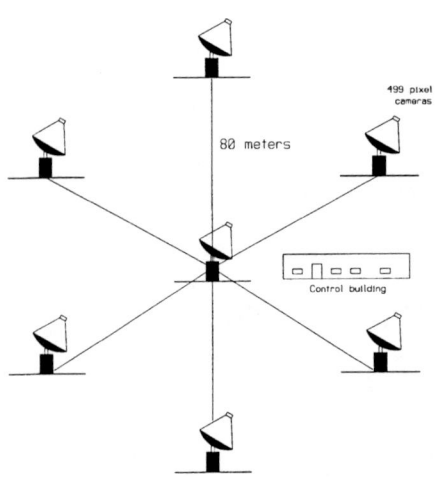

FIGURE 1. The Layout of the stereoscopic system of telescopes.

The VERITAS array will be located nearby the Whipple observatory, Arizona, at an elevation of $\approx$ 1390 m above sea level. Fig. 1 shows the geometrical layout of the VERITAS array. Six telescopes will be located at the corners of a hexagon of side 80 m and one will be at the center of the array. A control building will contain the central data acquisition system, a computer facility to perform the off-line analysis and the array trigger electronics. The estimated capital cost of VERITAS is $16.6M and it can be built in four years.

The detailed technical description of the elements of the design is given in the next three sections, followed by results from Monte Carlo simulations showing the performance of VERITAS.

MOUNT, OPTICAL SUPPORT STRUCTURE & OPTICS

The optical support structure (OSS) of each telescope will be made of trussed steel providing a dish with a 10 m aperture (Fig. 2a). The OSS will be rigid allowing the mirrors to be aligned with 0.01° accuracy. Effects from gravitational bending and slewing will be comparable or less than that. Slew speeds can be as high as 1°/second on both axes. The arms supporting the focus box (mass < 200 kg) will be made of steel providing a stiff structure to achieve a high pointing accuracy. The OSS will be designed to match a commercially available pedestal from radio antenna.

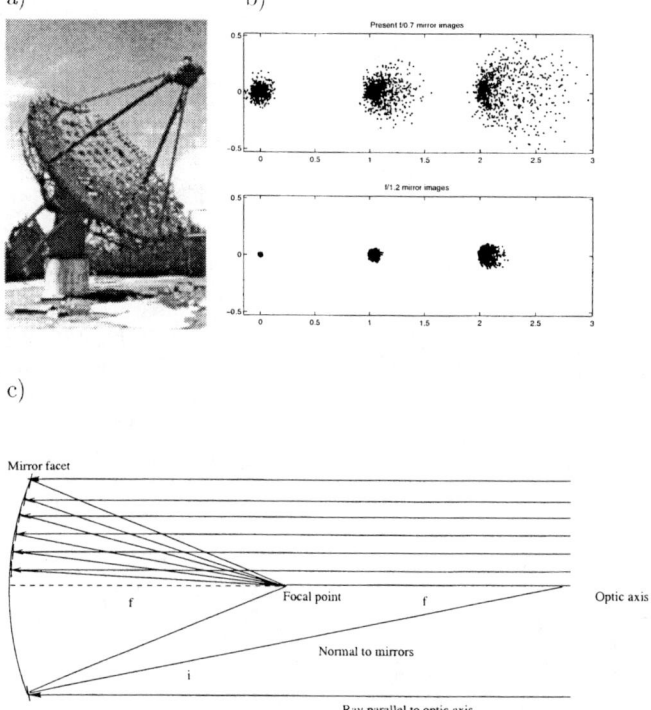

FIGURE 2. a) The Whipple 10 m telescope, b) the effect of increasing the f/number to 1.2 onto the image resolution. For the f/1.2 case, 100 % (75 %) of the light from a point source is contained in a 0.12° diameter circle out to 1° (2°) from the optical axis. In contrast, for f/0.7 optics, only 30 % of light from a point source is contained within 0.12° at 2°, c) the Davies-Cotton optical design.

The optical reflector will be a Davies-Cotton design (Davies & Cotton 1957) consisting of facet mirrors, each with a 24 m radius of curvature (see Fig. 1c). The mirrors will be positioned in a hexagonal pattern covering a spherical surface with 12 m radius of curvature. The mirror normals will intersect at the 2f (24 m) retro-reflection point on the optical axis. This design using identical mirrors facilitates fabrication and reduces cost. The Davies-Cotton design has off-axis aberrations smaller than a parabolic reflector, showing good image quality out to a few degrees from the optic axis (Lewis 1990).

In order to resolve the intrinsic characteristics of γ-ray Cherenkov images, a pixelsize of $\approx 0.03° - 0.1°$, across a 3.5° FOV would be desirable[1] and should be matched by the reflector. The off-axis resolution of the Whipple 10 m telescope is mostly limited by global aberrations inherent to the small f/number of 0.7, whereas the image quality near the center of the FOV is limited by the individual facets and the OSS.

Aberrations of the individual mirror facets, mostly from astigmatism scale as r/f^3 (r = radius of mirror facet), whereas the global aberrations go as $1/f^2$. Increasing the f/number from 0.7 to 1.2 substantially reduces the global aberrations, whereas facet aberrations become negligible for 0.6 m diameter mirrors. A comparison of simulated point source images using a 10 m reflector with f/numbers of 1.2 vs. 0.7 exhibits a dramatic improvement (see Fig. 2b). For the f/1.2 case, 100% of the light from a point source is concentrated in a 0.15° diameter circle out to 1.7° from the optical axis[2].

The Davies-Cotton reflector design is not isochronous and the time spread has a full width of $\approx D^2/2\,c\,f$ (D = aperture, c = speed of light). By means of increasing the f/number to 1.2, the time spread has been reduced to 3 - 4 ns, comparable to the intrinsic Cherenkov pulse width.

The f/1.2 design also improves the light collection efficiency in the focal plane. Because of the larger plate scale (linear displacement on the focal plane per angular displacement) of the f/1.2 design, photomultiplier can be more efficiently packed covering 55% of the surface area, instead of only 36% for a similar size tube in an f/0.7 design. In addition, the application of light concentrators is also more efficient due to a smaller average incidence angle of the photons impinging on the focal plane.

A large cost factor of the reflector is the cost of the mirror facets. After investigating a number of materials and fabrication techniques, a conventional approach has been chosen as the most cost effective: the mirrors will be made of glass, slumped and polished, aluminized and anodized.

CAMERA

Fig. 3 shows the focal plane detector of the VERITAS telescopes. It consists of 499 photomultipliers/camera with 0.15° spacing, corresponding to a FOV of 3.5°.

---

[1] Shower fluctuations limit the inherent accuracy of resolving the Cherenkov images, to $\approx 0.03°$ (300 GeV) as shown by Hillas (1989). However, for practical considerations, cost of phototubes and readout, the upper end of 0.1° seems reasonable.

[2] For triggering on small and narrow γ-ray images this measure is more appropriate than just the F.W.H.M., because the extended tails of the point spread function (PSF) contain a substantial amount of light. The F.W.H.M. of the PSF is $\approx 0.05°$ at 1° off-axis.

The pixelation and the FOV are two competing factors given a limited number of phototubes. The choice of pixel spacing is driven by the structure of γ-ray shower images.

In order to trigger efficiently on low energy γ-ray events (E $\approx$ 100 GeV), it is necessary that the image width is approximately matched by the pixel size - a pixelation larger than the image width would accept a large noise contamination from night sky background light, reducing the signal to noise ratio. In Fig. 3 the Cherenkov light image of a 100 GeV γ-ray shower is superimposed on a VERITAS camera. Images of sub-TeV showers have a RMS width and length (depending on energy) of $0.10° - 0.15°$ and $0.20° - 0.30°$, respectively.

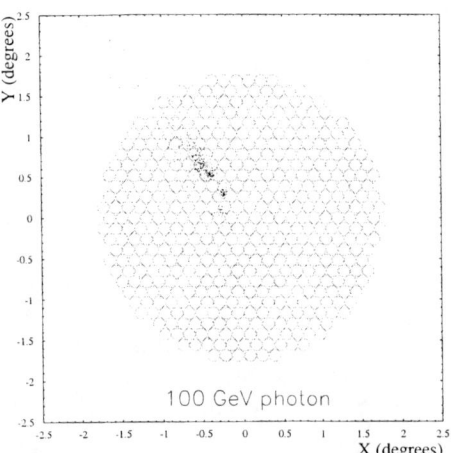

Figure 3. The 499-pixel photomultiplier matrix of the seven VERITAS cameras is shown. The pixel spacing is 0.15° and the FOV is 3.5°. A 100 GeV γ-ray shower image shown is projected into the focal plane.

The camera pixelation (0.15°) and matched optical quality of the reflector will provide sufficient resolution to trigger efficiently and resolve useful shower structure on this scale. The Cherenkov light from air showers is dominantly emitted from shower maximum (at 8-10 km atmospheric height) and the images are off-set from the arrival direction by $0.6° - 1.2°$, depending on the impact distance of the shower from the telescope. A 2.5° FOV is required to capture most of the Cherenkov image. However, for extended sources, stereoscopic operation and for the detection of high energy (E > 10 TeV) γ-ray events, a larger FOV is necessary.

We have investigated different options for the photosensitive detectors in the focal plane. Although, currently available photomultipliers have a relatively low quantum efficiency (avg. 20% @ 300 nm) they are at present the only viable detectors given cost and performance characteristics required for VERITAS. Several other detector technologies have been studied (hybrid photomultipliers with bialkali photocathode, avalanche photodiodes) and look very promising for future cameras (Lorenz 1994). Note, that avalanche photodiodes have a 3-4 times higher quantum efficiency than standard photomultipliers, which may provide a substantially lowered energy threshold for IACT arrays (20 GeV range). It should be emphasized that the VERITAS telescope optical design could accommodate future high quantum efficiency detectors with a pixelation of 0.05°.

ELECTRONICS

In order to produce a digital image of the Cherenkov light flash with minimal noise contamination, the electronics of the recording system has to preserve the short (6-10 ns) pulse until it is digitized. The electronics also has to form a trigger decision based on the characteristics of a typical γ-ray image in an array of IACTs. The atmospheric Cherenkov technique is ultimately limited by the fluctuations from the night sky background. To estimate the lowest possible energy threshold for the VERITAS array, a careful analysis of the expected trigger rates from accidentals due to the night sky is required[3]. The trigger threshold, given in photoelectrons (p.e., hereafter), determines the energy threshold of the IACT array. The trigger of VERITAS is formed in a sequence, based on the different levels in the electronic chain. Fig. 4 shows the expected rates for VERITAS at different trigger levels.

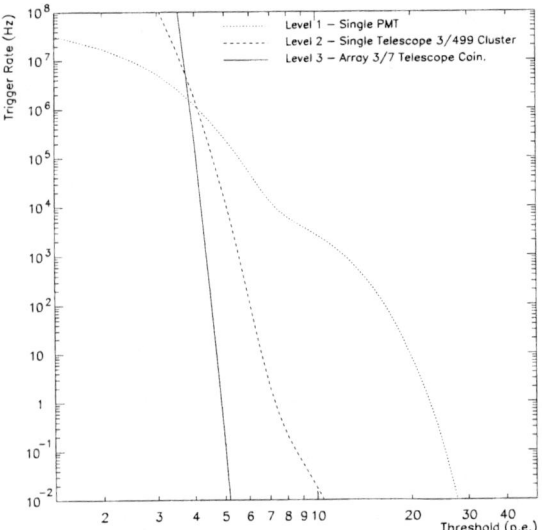

Figure 4. Accidental trigger rates for VERITAS due to night sky fluctuations.

<u>Level 1:</u> Each photomultiplier (Fig. 5) has a constant fraction discriminator with a programmable threshold (CFD, in Fig. 5), located at the base of each individual telescope. From measurements using the 10 m telescope, it has been deduced, that a threshold corresponding to 4.2 photoelectrons would result in a rate of $\approx$ 800 kHz.

<u>Level 2:</u> At each individual telescope a pattern trigger is used (Bradbury et al. 1999) to form a local intelligent trigger decision (Buckley 1994) based on the shape of γ-ray images.

To trigger efficiently on compact γ-ray images, the pattern trigger can be programmed to select patterns of $\geq$ N adjacent pixel (N = 2, 3 or 4) with a coincidence time window of 14 ns. This ensures that accidental trigger from random night sky fluctuations are at a minimum level. A trigger condition of $\geq$ 3 adjacent pixels reduces the rate to 100 kHz. The local trigger decisions (Level 2) from each telescope are transmitted by digital optical fiber cable to the central station.

<u>Level 3:</u> The individual telescope level 2 trigger can be combined in an array trigger decision requiring several telescopes. The level 2 trigger are sent through

---

[3] For practical consideration, the array must trigger at a rate < 1 kHz, a higher rate would introduce significant dead time for the data acquisition system.

individual digital delays to account for the orientation of the shower front (delay range: 0-500 ns). The solid line in Fig. 4 shows the background rate as a function of the photoelectron threshold if 3 out of 7 telescopes produce a level 2 trigger within 40 ns coincidence time. At a threshold of 5 p.e., this array trigger produces a negligible background rate, at 4.2 p.e., the rate would be 300 Hz. The array trigger is required to be flexible for various operation modes: single telescope trigger, using three and four telescopes independently, trigger requiring 3 out of seven telescopes. The central trigger station also contains a time to digital converter (TDC) to measure the arrival time for each telescope with 0.5 ns resolution with respect to the array trigger. This is important information as to which telescope contributed to the array trigger. It is also useful for the event reconstruction in estimating the arrival direction of the shower front independently from image analysis.

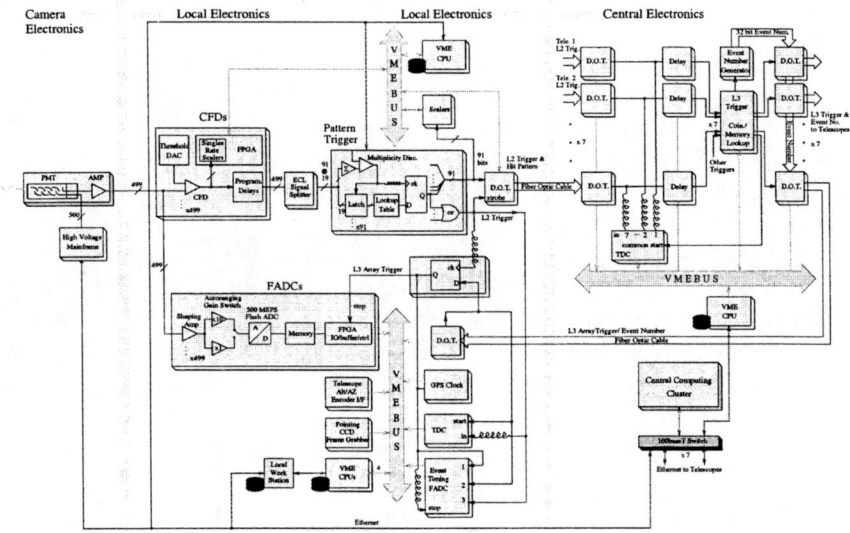

FIGURE 5. Detail of the VERITAS electronics.

Level 4: A level 4 trigger is currently under development, using the parallactic displacement expected for γ-ray images in different telescopes. Based on the location of the triggering pixels in the cameras of two telescopes, the position of an image in a third telescope can be predicted and tested in real time, e.g., using Field-Programmable-Gate-Arrays. The level 4 trigger should reduce accidentals by another factor of at least ten, making it possible to achieve stable array operation for trigger thresholds of 4-5 p.e.

The digitization of the signals can begin, once a trigger decision has been formed.

The trigger decision based on an array trigger of 3 telescopes can take up to $1.2\mu s$ assuming operation at large zenith angles. A Flash ADC system (Buckley et al. 1999) at each telescope allows the digitized output samples of each channel to be written into a circulating memory (depth $> 8\mu s$), while waiting for the trigger decision. If a trigger arrives, the writing stops and the memory contents are examined for a signal in the corresponding time bin. Besides an optimum signal-to-noise ratio (due to a minimum integration time width, real-time calibration of the photomultipliers and signal cable propagation times) a Flash ADC system provides also the possibility to search for short sub-GeV-range γ-ray burst phenomena in the 0.1 -8 μs regime (Krennrich et al. 2000).

SIMULATION OF PERFORMANCE

Monte Carlo simulations have been used to determine the optimum configuration and to characterize the performance of VERITAS (Vassiliev et al. 1999). The design optimization was performed by means of full Monte Carlo simulation of air showers and the telescopes. Subsequent event reconstruction is based on established methods developed for single telescope data (Fegan 1997) and extended through several new algorithms, allowing a lower reconstruction energy threshold[4]. The input parameters such as ADC integration time, pixel coincidence gate, the level 1 trigger threshold and level 3 (telescope array) trigger coincidence gate width are given in the second part of Table 1.

The VERITAS design is optimized for maximum point source sensitivity between 100 GeV - 10 TeV. To find the optimum configuration of an array of IACTs, the number of free parameters have to be limited in a sensible way: for cost reasons, the total number of pixels in all cameras was fixed to 3500. Several parameters were varied: number of telescopes, the spacing between the telescopes, the focal length, the aperture of the telescopes, and the FOV of the cameras[5]. The outcome of these studies has resulted in the so-called baseline design for VERITAS given in Table 1.

The stereoscopic technique was chosen because it provides a low energy threshold in addition to excellent angular resolution and background rejection. At minimum 3 telescopes are necessary to efficiently utilize the stereoscopic imaging technique. Two telescopes can be used to measure the arrival direction and core location, however, a third telescope is necessary to constrain the measurement by estimating its systematic uncertainties. Using only three telescopes results in a smaller effective area and higher energy threshold. The effective area increases with the number of telescopes steeply until an optimum is reached ($\approx$ 7 telescopes). Beyond

---

[4] It should be noted that the operation of an array of imaging telescopes allows to reconstruct faint, poorly defined images (Krennrich & Lamb 1995; Hillas 1998), effectively lowering the analysis threshold.
[5] Note that an optimization considering different telescope designs, has to allow the selection criteria in the subsequent event analysis to vary as well.

TABLE 1. The baseline design and MC input.

Number of telescopes	7 (hexagonal layout)
Telescope spacing	80 m
Mirror	Davies-Cotton
Reflector aperture/area	10 m / 78.6 m^2
Focal length	12 m
Facets	244, 61 cm hexagon
Field of View (FOV)	3.5°
Number of pixels	499
Pixel Spacing/Photocathode Size	0.148° / 0.119°
Array Trigger	3 telescopes out of 7
Pixel BS noise	1.1 pe pixel^{-1}
Optical efficiency	0.9× mirror reflectivity
Light-concentrator enhancement	1.35
Telescope Triggers	2, 3 pixels (adjacent)
ADC integration gate	8 ns
Trigger threshold	4.2 p.e.
Pixel coincidence gate	14 ns
Array Trigger	3 telescopes out of 7
Telescope coincidence gate	40 ns

seven telescopes, the increase in effective area/telescope begins to level off, and the sensitivity increases only as the square-root of the number of telescopes.

A separation between telescopes of 80 m provides the best combination of flux sensitivity and energy threshold. Increasing the spacing beyond 80 m does not improve flux sensitivity in the 200 GeV - 1 TeV energy range but it does increase the energy threshold. Decreasing the spacing below 80 m can reduce the energy threshold slightly, but the sensitivity is significantly reduced over the entire energy range: the stereoscopic view becomes less efficient because of the short baseline for triangulation of the images. The most relevant parameters determining the VERITAS sensitivity are energy threshold, effective area, angular resolution, and γ/hadron separation.

<u>Angular resolution:</u> The angular resolution is defined as the dispersion in the reconstructed arrival direction from photons arriving from a point source, fitted by a two-dimensional Gaussian. The angular resolution depends on the energy of the γ-ray primary and on the quality of γ-ray images accepted. Therefore, the following quoted detector properties (angular resolution, collection area, and energy threshold) refer to the same sample of events. At 100 GeV (300 GeV) VERITAS will achieve an angular resolution of 0.09° (0.05°), at the highest energies it will reach $\approx$ 0.02°. For comparison, using the Whipple telescope - the arrival direction perpendicular to the image axis can be deduced with 0.14° accuracy, the direction along the axis can be only roughly estimated (0.20°) due to uncertainty in location

of the shower impact point. The angular resolution of VERITAS 0.05° is achieved for individual events, allowing reduction of isotropic background from cosmic rays.

Energy threshold and collection area are closely related. The collection area for an IACT array increases as a function of photon energy (Fig. 7a)[6].

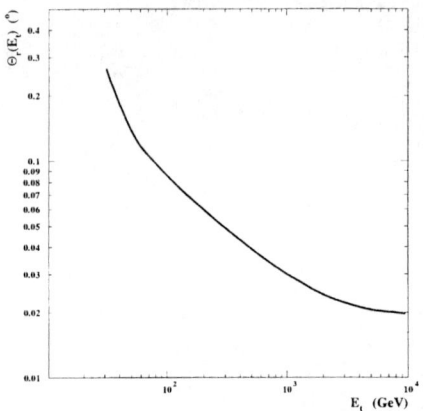

We define the energy threshold of VERITAS as the energy at which the differential rate of successfully reconstructed γ-rays from the Crab Nebula per unit interval of energy reaches its maximum (Fig. 7b), showing a value of ≈ 75 GeV. Above this energy reliable spectroscopy of a source can be achieved. Spectral measurements in the interval 50-75 GeV are possible if systematic effects are well understood.

FIGURE 6. Angular resolution of VERITAS as a function of threshold energy.

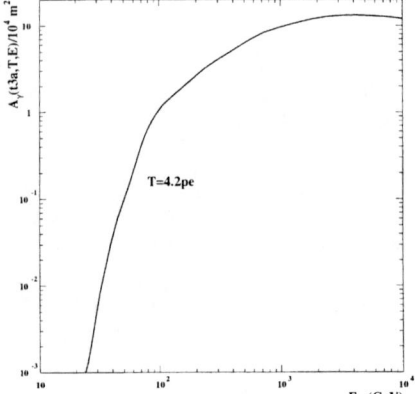

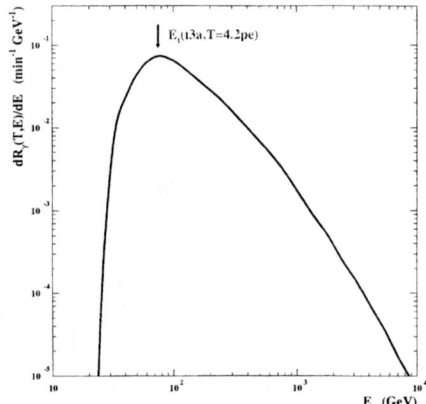

FIGURE 7.a The collection area of VERITAS as a function of energy using a 4.2 p.e. trigger threshold. Also the image cleanup and reconstruction algorithm has been applied.

FIGURE 7.b Differential detection rates of the Crab Nebula for VERITAS. The peak of the curve is the energy threshold of the array in stereoscopic observation mode.

---

[6] It is important for comparison with the other IACTs, that the collection area depends strongly on the reconstruction method used and subsequently on the sample of events: allowing for a reduced angular resolution, the collection area would be much larger. At the other extreme a sample of "gold plated" events for high angular resolution can be selected. The numbers quoted here provide maximum flux sensitivity to a point source.

Flux sensitivity: The ability to detect γ-ray sources can be described by the flux sensitivity of the detector. We define the minimum detectable flux of γ-rays requiring a 5σ excess above background (or at least 10 photons) for 50 hours of observation assuming a source spectrum given by $dN/dE \propto E^{-2.5}$. The γ-ray flux sensitivity of the VERITAS baseline design as a function of energy is shown in Fig. 8. The flux sensitivity is limited by different effects depending on energy. The region above ≈ 2 TeV is limited by the collection area and therefore by photon statistics. In this regime large zenith angle observations can substantially improve the sensitivity. Between 900 GeV - 2 TeV, a small fraction of cosmic ray proton showers, producing Cherenkov images similar to γ-ray showers constitute the major background. At lower energies (200 GeV - 900 GeV) the cosmic ray electrons (only rejectable by a high angular resolution) are the major source of background. The flux sensitivity below 200 GeV is again limited by night sky background and cosmic-ray protons[7].

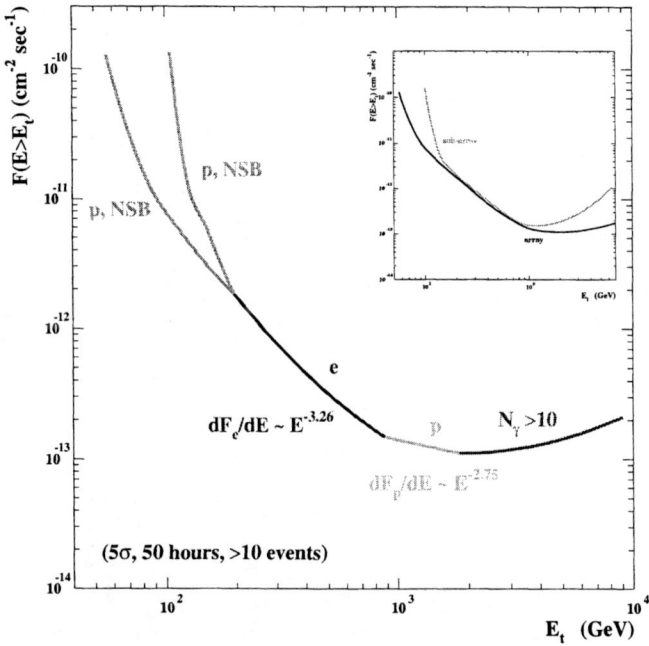

FIGURE 8. The sensitivity of VERITAS for a point-source in 50 hours. The inset shows a comparison of the sensitivity of a 3-reflector sub-arry with the full VERITAS array.

---

[7]) Note, that a cosmic-ray muon can trigger the VERITAS array, but is easily rejected by the presence of a hadronic shower halo in at least one of the cascade images or by parallax.

TABLE 2. Performance of VERITAS

Characteristic	E	Value
Energy threshold[a]		75 GeV
Flux sensitivity[b]	>100 GeV	$9.1 \times 10^{-12} \mathrm{cm}^{-2}\mathrm{s}^{-1}/15\,\mathrm{mCrab}$
	>300 GeV	$8.0 \times 10^{-13} \mathrm{cm}^{-2}\mathrm{s}^{-1}/5\,\mathrm{mCrab}$
	>1 TeV	$1.3 \times 10^{-13} \mathrm{cm}^{-2}\mathrm{s}^{-1}/7\,\mathrm{mCrab}$
Angular resolution	50 GeV	$0.14°$
	100 GeV	$0.09°$
	300 GeV	$0.05°$
	1 TeV	$0.03°$
Effective area	50 GeV (100 GeV)	$1.0 \times 10^3 \mathrm{m}^2$ ($1.0 \times 10^4 \mathrm{m}^2$)
	300 GeV (1 TeV)	$4.0 \times 10^4 \mathrm{m}^2$ ($1.0 \times 10^5 \mathrm{m}^2$)
Crab Nebula γ-ray rate	>100 GeV	50/minute
Energy resolution[c]		21% @ 100 GeV; 18% @ 300 GeV; 10% @ 10 TeV

[a] Energy at which the rate of photons per unit energy interval from the Crab Nebula is highest for a 4.2 photoelectron trigger threshold.
[b] Minimum integral flux for detecting a 5σ excess (or a minimum of 10 events) in 50 hours of observations of a source with a Crab-like spectrum.
[c] RMS $\Delta E/E$

In the lower curve, a low night sky background is assumed, the less sensitive curve is for a bright region in the sky, e.g., the galactic plane. The inset of Fig. 8 shows the performance of a 3-telescope subarray. Between 200 GeV -1 TeV, the sensitivity is very similar (within 40%) to the 7-telescope array, allowing to split the array for specific observations. However, the full array provides sensitivity down to 50 GeV and substantially increased sensitivity above 2 TeV. Further performance details are given in Table 2.

CONCLUSIONS

A next generation imaging atmospheric Cherenkov detector with a flux sensitivity of a few mCrab and an energy range covering three orders of magnitude (50 GeV -50 TeV) is on its way to be constructed in the vicinity of the Whipple Observatory, nearby Tucson, Arizona. A VERITAS type instrument will provide the first highly sensitive and efficient view of the high energy universe beyond GeV γ-ray energies. VERITAS will complement GLAST (Gehrels & Michelson 1999), the next generation high energy γ-ray instrument in space: GLAST with wider FOV but smaller effective area, VERITAS with smaller FOV but larger effective area will provide for the first time a continuous coverage for low energy and high energy γ-ray observations from 20 MeV to 50 TeV.

REFERENCES

1. Bradbury, S.B. et al., *these Proceedings* (1999).
2. Buckley, J.H., in *"Towards a Major Atmospheric Cherenkov Detector III"* , ed. T. Kifune, 221 (1994).
3. Buckley, J.H. et al., *26th ICRC (Salt Lake City)* **5**, 267 (1999).
4. Daum, A. et al., *Astropart. Phys.* **8**, 1 (1997).
5. Davies, J.M., & Cotton, E.S., *J. Solar Energy Sci. and Eng.* **1**, 16 (1957).
6. Fegan, D.J., *J. Phys. G* , **23**, 1013 (1997).
7. Finley, J. et al., *these Proceedings* (1999).
8. Gehrels, N., & Michelson, P., *Astropart. Phys.* **11**, 277 (1997).
9. Hillas A.M., VERITAS, "Letter of Intent", 58
10. Hillas A.M., in *"Very High Energy Gamma Ray Astronomy"* , eds. A.A. Stephanian, D.J. Fegan & M.F. Cawley, 134 (1989).
11. Hofmann, W. et al., *these Proceedings* (1999).
12. Krennrich F. & Lamb, R.C., *Exp. Astron.* **6**, 285 (1995).
13. Krennrich F. et al., *Astropart. Phys.* **8**, 213 (1998).
14. Krennrich F., Le Bohec, S., & Weekes, T.C., *ApJ* , **529**, in press (2000).
15. Lewis D.A., *Exp. Astron.* **1**, 213 (1990).
16. Lorenz E., in *"Towards a Major Atmospheric Cherenkov Detector III"* , ed. T. Kifune, 341 (1994).
17. Vassiliev, V.V. et al., *26th ICRC (Salt Lake City)* **5**, 299 (1999).
18. Weekes T.C. et al., *ApJ* **342**, 379 (1989).

LIST OF ATTENDEES

Artur Alaverdian	P.N. Lebedev Physics Institute
Robert Atkins	University of Utah
Hussein M. Badran	Whipple Observatory
Matthew Baring	NASA Goddard Space Flight Center
Denis Bastieri	University di Padova
Roberto Battiston	INFN
Peter L. Biermann	MPI fur Radioastronomie
Rudolf K. Bock	CERN
Markus Böttcher	Rice University
Heidrun Bojahr	University of Wuppertal
Lowell Boone	University of California, Sants Cruz
Philippe Bruel	LPNHE
James H. Buckley	Washington University
Andrew Burdett	Whipple Observatory
Michael Catanese	Iowa State University
Juan Cortina	MPI For Physics, Munich
Corbin Covault	University of Chicago
Marcello Cresti	University and INFN Padova
Ocker C. de Jager	Potchefstroom University for CHE
Mattieu de Naurois	LPNHE Ecole Polytechnique
B. Degrange	IN2P3 / CNRS
Charles Dermer	NRL
Maria Jose Diaz Trigo	Max Planck Institut
Brenda Dingus	University of Utah
Michelle D'Vali	University of Leeds
Ernst A. Dorfi	Institute fuer Astronomie
Lev Dorman	Israel Cosmic Ray Center
Luke O'C. Drury	Dublin Institute for Advanced Studies
Charles Duke	Grinnell College
Daniele Fargion	Rome University
Stephen Fegan	Whipple Observatory
David Fegan	University College Dublin
John Finley	Purdue University
Patrick Fleury	Ecole Polytechnique - IN2P3
Esso Flyckt	PHOTONIS
Victoria Fonesca	U. Complutense Madrid
Gerard Rene Fontaine	IN2P3
Piron Frederic	LPNHE Ecole Polytechnique
James Gaidos	Purdue University
Kenneth Gibbs	University of Chicago
Niels Götting	University of Hamburg
Juan Carlos Gonzalez	MPI For Physics, Munich
Javier Bussons Gordo	University of Maryland
Philippe Goret	DAPNIA / SAp
Peter W. Gorham	JPL / NASA
Isabelle Grenier	University of Paris VII & CEA Saclay
Tony Hall	Purdue University
David Hanna	McGill University

Goetz Heinzelmann	II. Inst. fuer Experimentalphysik
Earl Hergert	Hamamatsu Corp.
German Hermann	University of Chicago
Jim Hinton	University of Leeds
Werner Hofmann	MPI for Nuclear Physics
Yasuyuki Horiuchi	Hamamatsu Photonics KK
Dieter Horns	University of Hamburg
Ivan Ivanov	P.N. Lebedev Physics Institute
Mead Jordan	Washington University
Fumiyohi Kajino	Konan University
Ron Kasulones	Hamamatsu Corp.
Akiko Kawachi	University of Tokyo
Yoshiya Kawasaki	Osaka City University
Martin Kestel	MPI For Physics, Munich
David Kieda	University of Utah
Tadashi Kifune	Institute for Cosmic Ray Research
Johannes Knapp	University of Leeds
Donald A. Kniffen	NASA Headquarters
Antje Kohnle	MPI fuer Kernphysik
Alexander Konopelko	MPI fuer Kernphysik
Daniel Kranich	MPI For Physics, Munich
Frank Krennrich	Iowa State University
Hidetoshi Kubo	Tokyo Institute of Technology
Richard C. Lamb	California Institute of Technology
Stephan L. LeBohec	ISU Physics Dept.
Rod Lessard	Purdue University
E. Lorenz	MPI-Physics, Munich
Karl Lyons	Univesity of Durham
Norbert Magnussen	University of Wuppertal
Oliver Mang	Universitat Kiel
Manel Martinez	IFAE
Apostolos Mastichiadis	University of Athens
Yutaka Matsubara	Solar-Terrestrial Environment Lab.
T.J.L. McComb	University of Durham
Julie McEnery	University of Utah
Hinrich Meyer	University of Wuppertal
Yoshihiko Mizumoto	National Astronomical Observatory of Japan
Kohei Mizutani	Saitama University
Gora Mohanty	LPNHE, Ecole Polytechnique
Masaki Mori	University of Tokyo
Patrick Moriarty	Galway-Mayo Institute of Technology
Anita Muecke	University of Adelaide
Reshmi Mukherjee	Columbia University
Filip Munz	PCC, College de France
A. Musquere	Ctr d'Etude Spat. des Rayonnements
Sergey Nikolsky	P.N. Lebedev Physics Institute
Rene Ong	University of Chicago
J.L. Osborne	University of Durham
Scott Oser	University of Chicago

Laura Pagano	Hamamatsu Corp.
Michael Panter	MPI Fuer Kernphysik
Eric Perlman	STScI
Luigi Peruzzo	INFN
Rainer Plaga	MPI fuer Physik
Martin Pohl	Ruhr-Universitat Bochum
Michael Punch	College de France
Jorg P. Rachen	Utrecht University
Kenneth Ragan	McGill University
Olaf Reimer	MPE Garching
Gavin Rowell	University of Tokyo
Michael Salamon	University of Utah
Rita Sambruna	Penn State University
Manfred Samorski	Universitat Kiel
Mereghetti Sandro	Istituto Di Fisca Cosmica G Occhialini
Gerd Schatz	
Martin Schilling	Universitat Kiel
Frank Schroeder	University of Wuppertal
Glenn Sembroski	Purdue University
Simon Shaw	University of Durham
Ramin Sina	University of Maryland
Ryan Sincic	University of Utah
Matt Sinnott	University of Adelaide
David A. Smith	Centre diEtudes Nucleaires
Andrew Smith	UC Riverside
Floyd W. Stecker	NASA Goddard Space Flight Center
Simon Swordy	University of Chicago
Vera G. Synitsyna	P.N. Lebedev Physics Institute
Vera Y. Synitsyna	P.N. Lebedev Physics Institute
Toru Tanimori	Tokyo Institute of Technology
Jean-Paul Tavernet	LPNHE
Tumay Tümer	UCR, IGPP
K. E. Turver	University of Durham
Vladimir Vassiliev	FLWO
Heinrich J. Völk	Max-Planck-Institute
Scott P. Wakely	University of Minnesota
Trevor C. Weekes	Smithsonian Astrophysical Obs.
David A. Williams	University of California, Santa Cruz
Yoshiaki Yamamoto	Konan University
Tokomatsu Yamamoto	Tokyo University
Shohei Yanagita	Ibaraki University
Guarang B. Yodh	UC Irvine
Takanori Yoshikoshi	Institute for Cosmic Ray Research
Yuji Yoshizawa	Hamamatsu Photonics KK
Jeffrey Zweerink	University of CA-Riverside

Author Index

A

Amenomori, M., 139, 459
Atkins, R., 243, 441
Ayabe, S., 139, 459

B

Barbiellini, G., 467
Baring, M. G., 173, 238
Barrio, J. A., 348, 368
Bartoli, B., 436
Bastieri, D., 436
Battiston, R., 474
Benbow, W., 243, 441
Berley, D., 243, 441
Bertsch, D. L., 492
Bhattacharya, D., 411, 426
Bigongiari, C., 436
Biral, R., 436
Bojahr, H., 81
Bond, I. H., 129, 301, 515
Boone, L. M., 100, 411
Böttcher, M., 31, 66, 119
Bradbury, S. M., 129, 301, 515
Breslin, A. C., 129, 301, 515
Buckley, J. H., 129, 301, 515
Budini, G., 467
Bulian, N., 378
Bullock, J. S., 100
Burdett, A. M., 129, 301

C

Cao, P. Y., 139, 459
Caraveo, P., 467
Carson, M., 129, 301
Carter-Lewis, D. A., 129, 301, 515
Catanese, M., 129, 301, 515
Cawley, M. F., 129, 301
Chadwick, P. M., 86, 91, 96, 210, 271, 276, 353, 393
Chamoto, N., 134
Chantell, M. C., 411
Chen, M. L., 243, 441

Chiang, J., 225
Chikawa, M., 134
Chounet, L.-M., 373
Ciocci, M. A., 436
Cocco, V., 467
Connaughton, V., 119
Conner, Z., 411
Contreras, J. L., 388
Cortina, J., 348, 363, 368
Costa, E., 467
Cosulich, D., 436
Covault, C. E., 411
Coyne, D. G., 243, 441
Cresti, M., 436

D

Danzengluobu, 139, 459
Dazeley, S. A., 187, 313, 485
Degrange, B., 373
de la Calle, I., 388
Denance, J.-P., 373
de Naurois, M., 416
Dermer, C. D., 225
Di Cocco, G., 467
Ding, L. K., 139, 459
Dingus, B. L., 243, 441, 515
Dorfan, D. E., 243, 441
Dragovan, M., 411
Drury, L. O'C., 183
Duffy, P., 183
Dunlea, S., 129, 301
D'Vali, M., 129, 301

E

Edwards, P. G., 187, 313, 485
Ellsworth, R. W., 243, 441
Espigat, P., 373
Evans, D., 243, 441

F

Falcone, A., 243, 441
Fegan, D. J., 129, 301, 515
Fegan, S. J., 129, 301

Fen, Z. Y., 139
Feng, Z. Y., 459
Feroci, M., 467
Finley, J. P., 129, 301, 515
Fleury, P., 373
Fleysher, L., 243, 441
Fleysher, R., 243, 441
Fonseca, V., 348, 388
Fu, Y., 139, 459

G

Gaidos, J. A., 129, 301, 515
Gehrels, N., 492
Gingrich, D., 411
Gisler, G., 243, 441
González, J. C., 124, 348
Goodman, J. A., 243, 441
Goret, P., 373
Gorham, P. W., 165
Gregorich, D., 411
Grenier, I. A., 261
Grindlay, J., 515
Gunji, S., 187, 313, 485
Guo, H. W., 139, 459

H

Haines, T. J., 243, 441
Hall, T. A., 129, 301
Hanna, D. S., 411
Hara, S., 313, 485
Hara, T., 187, 313, 485
Hayashi, S., 134
Hayashi, Y., 134
Hayashida, N., 134
He, M., 139, 459
Heck, D., 343
Hermann, G., 515
Hibino, K., 134, 139, 459
Hillas, A. M., 129, 301, 515
Hirasawa, H., 134
Hirsch, T., 378
Hoffman, C. M., 243, 441
Hofmann, W., 318, 328, 333, 378, 500
Holder, J., 187
Honda, K., 134
Horan, D., 129, 301

Hotta, N., 134, 139, 459
Huang, Q., 139, 459
Hugenberger, S., 243, 441
Huo, A. X., 139, 459

I

Ibarra, A., 348
Inoue, N., 134
Ishikawa, F., 134
Ito, N., 134
Izu, K., 139, 459

J

Jacobs, C. S., 165
Jia, H. Y., 139, 459
Jinbo, J., 313, 485

K

Kaaret, P., 515
Kabe, S., 134
Kajino, F., 134, 139, 459
Kakizawa, S., 134
Kartashov, D., 436
Kasahara, K., 139, 459
Kashiwagi, T., 134
Katayose, Y., 139, 459
Kawachi, A., 187, 313, 383, 485
Kawakami, S., 134
Kawasaki, Y., 134
Kawasumi, N., 134
Kelley, L. A., 243, 441
Kieda, D., 515
Kifune, T., 187, 313, 485
Kihm, T., 378
Kirk, J., 183
Kitamura, H., 134
Knapp, J., 129, 301, 515
Kniffen, D. A., 492
Kohnle, A., 378
Konopelko, A. K., 159, 215, 323
Kranich, D., 124
Krawczynski, H., 154
Krennrich, F., 129, 253, 301, 515
Kubo, H., 313, 485

Kuramochi, K., 134
Kusano, E., 134
Kushida, J., 313, 485

L

Labaciren, 139, 459
Labanti, C., 467
Lafoux, H., 134
Lamb, R. C., 288
Lampeitl, H., 328
Le Bohec, S., 129, 253, 301, 515
Leonor, I., 243, 441
Lessard, R. W., 129, 301, 515
Li, J. Y., 139, 459
Liello, F., 436
Lloyd-Evans, J., 515
Loh, E. C., 134
Longo, F., 467
Lorenz, E., 358, 363, 510
Lu, H., 139, 459
Lu, S. L., 139, 459
Luo, G. X., 139, 459
Lyons, K., 86, 91, 96, 210, 271, 276, 353, 393

M

Macomb, D. J., 288
Magnussen, N., 431
Malakov, N., 436
Mang, O., 71, 76
Mariotti, M., 436
Marsella, G., 436
Mase, K., 134
Masterson, C., 129, 301
Matsubara, Y., 187, 313, 485
Matsuyama, T., 134
McComb, T. J. L., 86, 91, 96, 210, 271, 276, 353, 393
McConnell, M., 243, 441
McCullough, J. F., 243, 441
McEnery, J. E., 243, 441
McKernan, B., 129, 301
Meng, X. R., 139, 459
Menzione, A., 436
Mereghetti, S., 467
Miller, R. S., 243, 441

Mincer, A. I., 243, 441
Mirzoyan, R., 358, 363
Mizumoto, Y., 187, 313, 485
Mizutani, K., 134, 139, 459
Mohanty, G., 426
Mohideen, U., 426
Morales, M. F., 243, 441, 448
Morelli, E., 467
Mori, M., 187, 313, 485
Moriarty, P., 338
Moriya, M., 313, 485
Morizane, Y., 134
Morselli, A., 467
Mu, J., 139, 459
Mücke, A., 149
Mukherjee, R., 66, 411
Müller, D., 515
Muraishi, H., 187, 313, 485
Muraki, Y., 187, 313, 485

N

Nagano, M., 134
Naito, T., 187, 313, 485
Nanjo, H., 139, 459
Nayman, P., 373
Nemethy, P., 243, 441
Nishijima, K., 313, 485
Nishijimi, K., 187
Nishikawa, D., 134
Nishimura, J., 134
Nishiyama, T., 134
Nishizawa, M., 134, 139, 459

O

Ogio, S., 187
Ohnishi, M., 134, 139, 459
Ohoka, H., 134
Ohta, I., 139, 459
Ong, R. A., 401, 411, 515
Orford, K. J., 86, 91, 96, 210, 271, 276, 353, 393
Osborne, J. L., 86, 91, 96, 210, 271, 276, 353, 393
Oser, S., 411
Osone, S., 134
Ouchi, T., 134, 139, 459

P

Panter, M., 154, 378
Paoletti, R., 436
Parlavecchio, G., 436
Patterson, J. R., 187, 313, 485
Pellizzoni, A., 467
Perlman, E. S., 53
Perotti, F., 467
Peruzzo, L., 436
Petry, D., 119
Piccioli, A., 436
Picozza, P., 467
Piron, F., 113, 308
Pohl, M., 249
Prest, M., 467
Primack, J. R., 100
Protheroe, R. J., 149
Pühlhofer, G., 215
Punch, M., 373

Q

Quinn, J., 129, 301, 515

R

Rachen, J. P., 41
Radu, A., 426
Ragan, K., 411
Rauterberg, G., 71, 76, 368
Rayner, S. M., 86, 91, 96, 210, 271, 276, 353, 393
Ren, J. R., 139, 459
Renault, C., 373
Rieben, R., 426
Rivoal, M., 373
Roberts, M. D., 187, 313, 485
Rose, H. J., 129, 301, 358, 515
Rosso, F., 436
Rowell, G. P., 187, 313, 485
Ryan, J. M., 243, 441

S

Sacco, R., 436
Saggion, A., 436
Saito, T., 134, 139, 459
Sakaki, N., 134
Sakata, M., 134, 139, 459
Sako, T., 187, 313, 485
Sakurazawa, K., 187, 313, 485
Salamon, M., 515
Sambruna, R. M., 19
Samuelson, F. W., 129, 301, 338
Sartori, G., 436
Sartori, P., 436
Sasaki, T., 139, 459
Sasano, M., 134
Sato, Y., 313, 485
Sbarra, C., 436
Scalzo, R. A., 411
Schilling, M., 71, 76
Schlickeiser, R., 249
Schröder, F., 343
Scribano, A., 436
Sembroski, G. H., 129, 301, 515
Shaw, S. E., 86, 91, 96, 210, 271, 276, 353, 393
Shen, B., 243, 441
Shi, Z. Z., 139, 459
Shibata, M., 139, 459
Shimodaira, H., 134
Shiomi, A., 134, 139, 459
Shirai, T., 139, 459
Shoup, A., 243, 441
Sinitsyna, V. G., 205, 293
Sinnis, C., 243
Sinnis, G., 441
Smith, A. J., 243, 441
Smith, D. A., 416
Soffitta, P., 467
Sokolsky, P., 134
Soli, L., 467
Souchkov, V., 426
Stamerra, A., 436
Stein, M., 378
Sugimoto, H., 139, 459
Sullivan, G. W., 243, 441
Susukita, R., 187, 313, 485
Swordy, S., 515

T

Taira, K., 139, 459
Takahashi, T., 134

Takeda, M., 134
Tamura, T., 187, 313, 485
Tan, Y. H., 139, 459
Tanimori, T., 187, 313, 485
Tateyama, N., 139, 459
Tavani, M., 467
Tavernet, J.-P., 373
Taylor, S. F., 134
Teshima, M., 134
Théoret, C. G., 411
Thornton, G. J., 187
Tom, H., 426
Torii, R., 134
Torii, S., 139, 459
Toussenel, F., 373
Tsukiji, M., 134
Tumer, T. O., 243, 411, 426, 441
Turini, N., 436
Turver, K. E., 86, 91, 96, 210, 271, 276, 353, 393

U

Uchihori, Y., 134
Unwin, S. C., 165
Utsugi, T., 139, 459

V

Vallazza, E., 467
van Zee, L., 165
Vassiliev, V. V., 105, 129, 301, 515
Vercellone, S., 467
Vincent, P., 373
Völk, H. J., 197, 281

W

Wang, C. R., 139, 459
Wang, H., 139, 459
Wang, H. Y., 459
Wang, K., 243, 441
Wang, P. X., 459
Wascko, M. O., 243, 441
Weekes, T. C., 3, 129, 253, 301, 515
Westerhoff, S., 243, 441
Williams, D. A., 100, 144, 243, 411, 441

X

Xu, X. W., 139, 459

Y

Yamamoto, T., 134
Yamamoto, Y., 134, 139, 459
Yanagita, S., 187, 313, 485
Yang, T., 243, 441
Yasui, K., 134
Yodh, G. B., 243, 441, 453
Yoshida, S., 134
Yoshida, T., 187, 313, 485
Yoshii, H., 134
Yoshikoshi, T., 187, 192, 313, 485
Yu, G. C., 139, 459
Yuan, A. F., 139, 459
Yuda, T., 134, 139, 459
Yuki, A., 313, 485

Z

Zetti, F., 436
Zhang, C. S., 139, 459
Zhang, H. M., 139, 459
Zhang, J. L., 139, 459
Zhang, N. J., 139, 459
Zhang, X. Y., 139, 459
Zhaxiciren, 139, 459
Zhaxisangzhu, 139, 459
Zhou, W. D., 139, 459
Zweerink, J. A., 411, 426